Field Manual
No. 3-21.10 (7-10)

Department of the Army
Washington, DC, 27 July 2006

The Infantry Rifle Company

Contents

	Page
PREFACE	**xxi**

Chapter 1	**INTRODUCTION**	**1-1**
	Section I. OPERATIONAL ENVIRONMENT	1-1
	DEFINITION	1-1
	PHASES OF CONFLICT	1-2
	SYSTEMS-BASED WARFARE	1-4
	GENERATION AND FOCUS OF EFFECTS OF COMBAT POWER	1-5
	STRATEGIC AND OPERATIONAL PRINCIPLES OF FIGHTING	1-5
	CRITICAL, STRATEGIC, AND OPERATIONAL VARIABLES	1-6
	Section II. PREPARATION FOR WAR	1-7
	SOLDIER	1-7
	LEADER	1-7
	UNIT	1-8
	TRAINING PROGRAM	1-8
	Section III. MISSIONS, TYPES, CHARACTERISTICS, CAPABILITIES, LIMITATIONS, AND ORGANIZATION	1-8
	MISSIONS	1-8
	TYPES AND CHARACTERISTICS OF INFANTRY RIFLE COMPANIES	1-8
	CAPABILITIES AS COMPARED WITH OTHER INFANTRY	1-9
	ADDITIONAL CAPABILITIES AND LIMITATIONS	1-9
	ORGANIZATION	1-10
	Section IV. DUTIES AND RESPONSIBILITIES OF KEY PERSONNEL	1-13
	COMPANY COMMANDER	1-13
	EXECUTIVE OFFICER	1-13
	FIRST SERGEANT	1-14
	PLATOON LEADER	1-14
	PLATOON SERGEANT	1-15
	FIRE-SUPPORT OFFICER	1-15
	SENIOR RADIO OPERATOR	1-16
	RADIO OPERATOR	1-16
	SUPPLY SERGEANT	1-16

Distribution Restriction: Approved for public release; distribution is unlimited.

*This publication supersedes FM 7-10, 14 December 1990.

Contents

CHEMICAL, BIOLOGICAL, RADIOLOGICAL, OR NUCLEAR NCO 1-16
MORTAR SECTION LEADER 1-17
ARMORER 1-17
MEDIC 1-17

Section V. COMBAT POWER, LEADERSHIP, AND WARFIGHTING FUNCTIONS 1-18
LEADERSHIP 1-18
INTELLIGENCE 1-19
MOVEMENT AND MANEUVER 1-19
FIRE SUPPORT 1-19
PROTECTION 1-19
SUSTAINMENT 1-19
COMMAND AND CONTROL 1-20

Chapter 2 TROOP-LEADING PROCEDURES 2-1

Section I. OVERVIEW 2-1
BATTLE COMMAND 2-1
ARMY PLANNING PROCESS 2-2
COMMON PITFALLS 2-8

Section II. STEPS 1 AND 2--RECEIVE MISSION, ISSUE WARNING ORDER 2-8
STEP 1--RECEIVE MISSION 2-8
STEP 2--ISSUE WARNING ORDER 2-9

Section III. STEP 3--MAKE A TENTATIVE PLAN 2-10
MISSION ANALYSIS 2-10
METT-TC 2-11
COURSE OF ACTION DEVELOPMENT 2-42
COURSE OF ACTION ANALYSIS 2-46
COURSE OF ACTION COMPARISON AND SELECTION 2-50

Section IV. STEPS 4 THRU 8--INITIATE MOVEMENT, RECONNOITER, COMPLETE PLAN, ISSUE OPORD, SUPERVISE 2-51
STEP 4--INITIATE MOVEMENT 2-51
STEP 5--RECONNOITER 2-51
STEP 6--COMPLETE PLAN 2-52
STEP 7--ISSUE OPORD 2-52
STEP 8--SUPERVISE 2-53
PRECOMBAT CHECKS AND INSPECTIONS 2-54

Chapter 3 MOVEMENT 3-1
TACTICAL MOVEMENT AND ENEMY CONTACT 3-1
MOVEMENT TECHNIQUES 3-2
MOVEMENT FORMATIONS 3-5
CONTROL TECHNIQUES 3-14
SECURITY DURING MOVEMENT 3-15
MOVEMENT AS PART OF A BATTALION 3-16

Chapter 4 OFFENSIVE OPERATIONS 4-1

Section I. OVERVIEW 4-1
CHARACTERISTICS 4-1
TYPES 4-3

Contents

    FORMS OF MANEUVER 4-3
Section II. SEQUENCE 4-8
    ASSEMBLY AREA 4-8
    RECONNAISSANCE 4-8
    MOVEMENT TO LINE OF DEPARTURE 4-8
    MANEUVER 4-8
    DEPLOYMENT 4-9
    ASSAULT 4-9
    CONSOLIDATION AND REORGANIZATION 4-10
Section III. PLANNING CONSIDERATIONS 4-10
    INTELLIGENCE 4-10
    MOVEMENT AND MANEUVER 4-10
    FIRE SUPPORT 4-10
    PROTECTION 4-11
    SUSTAINMENT 4-11
    COMMAND AND CONTROL 4-11
Section IV. ACTIONS ON CONTACT 4-12
    FORMS 4-12
    CIRCUMSTANCES 4-12
    DEVELOPMENT 4-12
    TIME REQUIREMENTS 4-12
    STEPS 4-13
Section V. ATTACKS 4-15
    CHARACTERISTICS 4-15
    TYPES 4-16
    SPECIAL PURPOSE ATTACKS 4-18
    OTHER ATTACK TECHNIQUES 4-23
Section VI. MOVEMENT TO CONTACT 4-38
    DEFINITION 4-38
    PLANNING CONSIDERATIONS 4-39
    SEARCH AND ATTACK 4-39
    APPROACH-MARCH-TECHNIQUE 4-43
    LEAD COMPANY MOVEMENT 4-44
    OTHER COMPANIES 4-44
    FLANK GUARD AND REAR GUARD 4-44
    CONTACT 4-45
    TECHNIQUE CONSIDERATIONS 4-46
Section VII. COMMON ACTIVITIES 4-48
    INFILTRATION 4-48
    OVERWATCH 4-52
    FOLLOW AND SUPPORT 4-52
    BYPASS 4-53
    CLEARING OF AN OBJECTIVE 4-53
    COMPANY AS RESERVE 4-54

Contents

Chapter 5	DEFENSIVE OPERATIONS	5-1
	Section I. OVERVIEW	5-1
	TYPES	5-1
	PURPOSE	5-2
	CHARACTERISTICS	5-2
	Section II. SEQUENCE	5-3
	RECONNAISSANCE AND SECURITY OPERATIONS AND ENEMY PREPARATORY FIRES	5-3
	OCCUPATION AND PREPARATION	5-3
	APPROACH OF ENEMY MAIN ATTACK	5-4
	ENEMY ASSAULT	5-4
	COUNTERATTACK	5-4
	CONSOLIDATION AND REORGANIZATION	5-4
	Section III. PLANNING CONSIDERATIONS	5-5
	MOVEMENT AND MANEUVER	5-5
	FIRE SUPPORT	5-10
	PROTECTION	5-11
	SUSTAINMENT	5-11
	Section IV. PREPARATION AND INTEGRATION	5-11
	DEFENSIVE TECHNIQUES	5-11
	SECTOR DEFENSE	5-12
	BATTLE POSITION DEFENSE	5-14
	STRONGPOINT DEFENSE	5-18
	PERIMETER DEFENSE	5-20
	LINEAR DEFENSE	5-24
	NONLINEAR DEFENSE	5-25
	REVERSE SLOPE DEFENSE	5-27
	ENGAGEMENT AREA DEVELOPMENT	5-30
	PRIORITY OF WORK	5-39
	ADJACENT UNIT COORDINATION	5-43
	Section V. RETROGRADE OPERATIONS	5-43
	PURPOSE	5-43
	TYPES	5-44
Chapter 6	**STABILITY OPERATIONS**	6-1
	Section I. PLANNING CONSIDERATIONS	6-2
	INTELLIGENCE	6-2
	MOVEMENT AND MANEUVER	6-3
	FIRE SUPPORT	6-3
	PROTECTION	6-3
	SUSTAINMENT	6-4
	COMMAND AND CONTROL	6-4
	MEDIA	6-5
	OPERATIONS WITH OUTSIDE AGENCIES	6-5
	Section II. TYPES OF OPERATIONS	6-6
	PEACE OPERATIONS	6-6
	FOREIGN INTERNAL DEFENSE	6-6

Contents

SECURITY ASSISTANCE 6-7
HUMANITARIAN AND CIVIC ASSISTANCE 6-7
SUPPORT TO INSURGENCY 6-7
SUPPORT TO COUNTERDRUG OPERATIONS 6-7
COMBATTING OF TERRORISM 6-7
NONCOMBATANT EVACUATION 6-8
ARMS CONTROL 6-8
SHOW OF FORCE 6-8

Section III. COMPANY TASKS 6-8
ESTABLISH AND OCCUPY A LODGMENT AREA OR FORWARD OPERATING BASE 6-8
NEGOTIATE 6-11
MONITOR COMPLIANCE WITH AN AGREEMENT 6-13
SEARCH 6-19
PATROL 6-21
ESCORT A CONVOY 6-25
OPEN AND SECURE ROUTES 6-29
CONDUCT RESERVE OPERATIONS 6-29
CONTROL CROWDS 6-29

Chapter 7 CIVIL SUPPORT OPERATIONS 7-1
ROLES 7-1
DEFINITION 7-2
TYPES OF OPERATIONS 7-2
POSSIBLE TASKS 7-2
INTELLIGENCE 7-3
MOVEMENT AND MANEUVER 7-4
FIRE SUPPORT 7-5
PROTECTION 7-5
SUSTAINMENT 7-6
COMMAND AND CONTROL 7-6

Chapter 8 TACTICAL ENABLING OPERATIONS 8-1

Section I. RECONNAISSANCE 8-1
DEFINITION 8-1
CATEGORIES 8-1
TYPES 8-2
PLANNING CONSIDERATIONS 8-3
EXECUTION 8-3

Section II. SPECIAL PURPOSE OPERATIONS 8-5
LINKUP 8-5
RELIEF IN PLACE 8-8
PASSAGE OF LINES 8-13

Section III. SECURITY OPERATIONS 8-17
TYPES 8-17
PLANNING CONSIDERATIONS 8-18
SCREEN 8-21
GUARD 8-22
LOCAL SECURITY 8-26

Contents

Section IV. BREACHING 8-27
    DEFINITIONS 8-27
    TENETS 8-28

Section V. PATROLS 8-33
    DEFINITION 8-33
    TYPES 8-33
    COMPANY COMMANDER INVOLVEMENT 8-36
    ORGANIZATION 8-36
    RAID 8-39
    AMBUSH 8-44
    PATROL BASE 8-49

Chapter 9 DIRECT FIRE CONTROL 9-1

Section I. FIRE-CONTROL PRINCIPLES 9-1
    MASS EFFECTS OF FIRE 9-1
    DESTROY GREATEST THREAT FIRST 9-1
    AVOID TARGET OVERKILL 9-2
    EMPLOY BEST WEAPON FOR TARGET 9-2
    MINIMIZE FRIENDLY EXPOSURE 9-2
    PLAN AND IMPLEMENT FRATRICIDE AVOIDANCE MEASURES 9-2
    PLAN FOR EXTREME LIMITED VISIBILITY CONDITIONS 9-2
    PLAN FOR DIMINISHED CAPABILITIES 9-2

Section II. FIRE-CONTROL PROCESS 9-3
    IDENTIFY PROBABLE ENEMY LOCATIONS AND DETERMINE ENEMY SCHEME OF MANEUVER 9-3
    DETERMINE WHERE AND HOW TO MASS FIRES 9-4
    ORIENT FORCES TO SPEED TARGET ACQUISITION 9-5
    SHIFT FIRES TO REFOCUS AND REDISTRIBUTE 9-6

Section III. PLANNING CONSIDERATIONS 9-7
    OVERVIEW 9-7
    STANDING OPERATING PROCEDURES 9-8

Section IV. CONTROL 9-9
    MEASURES 9-9
    COMMANDS 9-20

Chapter 10 MANEUVER SUPPORT 10-1

Section I. COMMAND AND SUPPORT RELATIONSHIPS 10-1
    COMMAND RELATIONSHIPS 10-1
    SUPPORT RELATIONSHIPS 10-2

Section II. FIRE SUPPORT 10-3
    INDIRECT FIRE CAPABILITIES 10-4
    FIRE-SUPPORT TEAM 10-4
    FIRE-SUPPORT PLANS AND COORDINATION 10-5
    MANEUVER COMMANDER'S INTENT 10-10
    PLANNING PROCESS 10-11
    FIRE-SUPPORT EXECUTION MATRIX 10-15
    FINAL PROTECTIVE FIRES 10-16
    SPECIAL MUNITIONS 10-17

	SMOKE SUPPORT	10-17
	OBSERVER POSITIONS	10-18
	REHEARSALS AND EXECUTION	10-18
	COMMUNICATIONS	10-18
	QUICKFIRE CHANNEL	10-20
	INDIRECT FIRES IN CLOSE SUPPORT	10-20
	ECHELONMENT OF FIRES	10-21
	EXECUTION CONSIDERATIONS	10-24
	MORTARS	10-28
	MORTAR POSITIONS	10-31
	MORTAR EMPLOYMENT	10-31
	MORTAR DISPLACEMENT	10-32
	MORTAR ENGAGEMENTS	10-33
Section III. ENGINEERS		**10-39**
	ORGANIZATION	10-39
	MISSIONS	10-41
Section IV. AIR DEFENSE ARTILLERY		**10-46**
	SYSTEMS, ORGANIZATION, AND CAPABILITIES	10-46
	EMPLOYMENT	10-47
	WEAPONS CONTROL STATUS	10-47
Chapter 11	**SUSTAINMENT OPERATIONS**	**11-1**
Section I. PLANNING CONSIDERATIONS		**11-1**
	OVERVIEW	11-1
	COMPANY RESPONSIBILITIES	11-2
Section II. SOLDIER'S LOAD		**11-4**
	PLANS	11-5
	CALCULATION	11-7
	MANAGEMENT	11-7
Section III. TRAINS		**11-9**
	OVERVIEW	11-9
	SECURITY	11-9
Section IV. SUPPLY AND TRANSPORTATION OPERATIONS		**11-9**
	CLASSES	11-9
	ROUTINE RESUPPLY	11-12
	EMERGENCY RESUPPLY	11-16
	PRESTOCKAGE OPERATIONS	11-16
	SUPPLY CONSIDERATIONS	11-17
	TRANSPORTATION	11-18
Section V. MAINTENANCE OPERATIONS		**11-18**
	MAINTENANCE REQUIREMENTS	11-18
	DESTRUCTION	11-19
Section VI. HEALTH SERVICE SUPPORT		**11-19**
	HEALTH AND HYGIENE	11-19
	FIRST RESPONSE	11-20
	CASUALTY EVACUATION	11-21
	SOLDIERS KILLED IN ACTION	11-23

Contents

Section VII. REORGANIZATION AND WEAPONS REPLACEMENT 11-23
REPLACEMENTS AND CROSS-LEVELING OF PERSONNEL 11-23
ENEMY PRISONERS OF WAR, DETAINEES, AND OTHER RETAINED PERSONS 11-24

Chapter 12 URBAN OPERATIONS 12-1

Section I. INTRODUCTION 12-1
DEFINITIONS 12-1
CONDITIONS 12-2

Section II. URBAN BATTLESPACE 12-3
TYPES 12-3
ZONES 12-5
BUILDING ANALYSIS 12-6

Section III. CHARACTERISTICS 12-8
CHANGING CONDITIONS 12-8
SMALL-UNIT BATTLES 12-8
COMMUNICATIONS 12-9
NONCOMBATANTS 12-9
AMMUNITION 12-9
CASUALTIES 12-9
MANEUVER SPACE 12-10
THREE-DIMENSIONAL TERRAIN 12-10
COLLATERAL DAMAGE 12-10
HUMAN INTELLIGENCE 12-10
COMBINED ARMS 12-10
CRITICAL POINTS 12-10
SNIPERS 12-10
SUPPORT-BY-FIRE POSITIONS 12-11

Section IV. WEAPONS AND DEMOLITIONS 12-11
SURFACES 12-11
ENGAGEMENT RANGES 12-11
ENGAGEMENT TIMES 12-11
DEPRESSION AND ELEVATION 12-11
REDUCED VISIBILITY AND INCREASED NOISE 12-11
FRIENDLY FIRE 12-11
CLOSE COMBAT 12-12
MAN-MADE STRUCTURES 12-12
MODERN BUILDINGS 12-12

Section V. FUNDAMENTALS 12-12
PERFORM FOCUSED INFORMATION OPERATIONS AND AGGRESSIVE INTELLIGENCE, SURVEILLANCE, AND RECONNAISSANCE 12-12
CONDUCT CLOSE COMBAT 12-13
AVOID ATTRITION APPROACH 12-13
CONTROL ESSENTIALS 12-13
MINIMIZE COLLATERAL DAMAGE 12-13
SEPARATE COMBATANTS FROM NONCOMBATANTS 12-13
RESTORE ESSENTIAL SERVICES 12-13

PRESERVE CRITICAL INFRASTRUCTURE 12-14
UNDERSTAND HUMAN DIMENSION 12-14
CONTROL TRANSITION 12-14

Section VI. ARMOR **12-14**
EMPLOYMENT CONSIDERATIONS FOR COMPANY-SIZE
  COMBINED-ARMS TEAMS 12-14
STRENGTHS AND LIMITATIONS OF INFANTRY AND ARMORED
  VEHICLES 12-15
EMPLOYMENT OF INFANTRY AND ARMORED VEHICLES 12-16
TASK ORGANIZATION WITH TANKS AT COMPANY TEAM LEVEL 12-17
ARMORED VEHICLE POSITIONS 12-22
TRANSPORTATION OF INFANTRY 12-25
CONSIDERATIONS FOR ARMORED VEHICLES, WEAPONS, AND
  MUNITIONS 12-29
TASK ORGANIZATION WITH BRADLEYS AT COMPANY TEAM LEVEL ... 12-29

Section VII. OFFENSE **12-30**
PLANNING CONSIDERATIONS 12-30
TROOP REQUIREMENTS 12-30
MANEUVER 12-31
LIMITATIONS 12-31
METT-TC FACTORS 12-31
COMMAND AND CONTROL 12-36
TASK ORGANIZATION INTO THREE ELEMENTS 12-38
MOVEMENT 12-39
DELIBERATE ATTACK 12-41
ISOLATION OF URBAN OBJECTIVE 12-43
ASSAULT OF A BUILDING 12-45
ATTACK OF BLOCK OR GROUP OF BUILDINGS 12-48
CONSOLIDATION AND REORGANIZATION 12-49

Section VIII. DEFENSE **12-50**
METT-TC FACTORS 12-50
COMMAND AND CONTROL 12-60
HASTY DEFENSE 12-61
COMPANY DEFENSE OF A VILLAGE 12-62
DEFENSE OF A BLOCK OR GROUP OF BUILDINGS 12-63
DEFENSE OF KEY TERRAIN 12-64
DEFENSE OF AN URBAN STRONGPOINT 12-65
DELAY 12-66

Appendix A   **RISK MANAGEMENT, FRATRICIDE AVOIDANCE, AND EFFECTS OF
CONTINUOUS OPERATIONS** **A-1**
Section I. RISK MANAGEMENT **A-1**
TYPES OF RISK A-1
STEPS A-2
IMPLEMENTATION A-5
CHALLENGES A-5
COMMAND CLIMATE A-5

Contents

Section II. FRATRICIDE AVOIDANCE A-6
- EFFECTS A-6
- CAUSES A-7
- PREVENTION A-8
- GUIDELINES AND CONSIDERATIONS A-8

Section III. EFFECTS OF CONTINUOUS OPERATIONS A-9
- COMBAT STRESS CONTROL A-10
- RESPONSIBILITIES A-11
- SLEEP DEPRIVATION A-12

Appendix B   TOW AND JAVELIN EMPLOYMENT B-1

Section I. OVERVIEW B-1
- INFANTRY BATTALION WEAPONS COMPANY B-1
- ORGANIZATION AND EQUIPMENT B-1
- PRINCIPLES B-3
- CAPABILITIES AND LIMITATIONS B-7

Section II. JAVELIN CLOSE COMBAT MISSILE SYSTEM B-8
- COMMAND LAUNCH UNIT B-9
- MISSILE B-9
- LETHALITY B-10
- SURVIVABILITY B-10
- AGILITY AND FLEXIBILITY B-10
- LIMITATIONS B-11
- EMPLOYMENT CONSIDERATIONS B-11
- JAVELIN FIRING POSITIONS B-14
- DETECTION, RECOGNITION, AND CLASSIFICATION OF TARGETS ... B-15
- SELF-DEFENSE AGAINST HELICOPTERS B-16

Appendix C   HEAVY AND STRYKER EMPLOYMENT C-1
- VEHICLES C-1
- TANKS C-1
- INFANTRY FIGHTING VEHICLE C-2
- STRYKER INFANTRY CARRIER VEHICLE C-3
- SAFETY C-4
- PLANNING CONSIDERATIONS C-7
- COMBINED OPERATIONS WITH ARMORED VEHICLES C-8
- MOVEMENT TO CONTACT C-8
- ATTACKS C-9
- DEFENSE C-12
- RETROGRADE OPERATIONS C-14
- LOGISTICAL SUPPORT C-15
- INFANTRY ON TANKS C-15
- COMMUNICATION WITH TANKS C-15

Appendix D   AVIATION SUPPORT D-1
- EMPLOYMENT D-1
- HELICOPTER TYPES D-1
- GROUND TACTICAL PLAN D-2
- LANDING PLAN D-2

	AIR MOVEMENT PLAN	D-6
	LOADING PLAN	D-6
	STAGING PLAN	D-12
	DUTIES OF KEY PERSONNEL	D-12
	AIR MISSION BRIEFING	D-13
	ATTACK AVIATION CONSIDERATIONS	D-15
	SAFETY	D-21
Appendix E	**SNIPER EMPLOYMENT**	**E-1**
	SNIPER TEAM	E-1
	SQUAD DESIGNATED MARKSMAN	E-1
	OFFENSIVE EMPLOYMENT	E-3
	ACTIONS AGAINST FORTIFIED AREAS	E-4
	DEFENSIVE EMPLOYMENT	E-5
	RETROGRADE EMPLOYMENT	E-6
	URBAN OPERATIONS	E-7
	STABILITY AND RECONSTRUCTION OPERATIONS	E-8
	PEACE OPERATIONS	E-9
	RIVER CROSSINGS	E-10
	PATROLS	E-10
Appendix F	**OPERATIONS WITH ARMY SPECIAL OPERATIONS FORCES**	**F-1**
	UNITED STATES SPECIAL OPERATIONS COMMAND	F-1
	UNITED STATES ARMY SPECIAL OPERATIONS COMMAND	F-1
	SPECIAL FORCES	F-2
	75TH RANGER REGIMENT	F-3
	SPECIAL OPERATIONS AVIATION	F-4
	CIVIL AFFAIRS	F-5
	PSYCHOLOGICAL OPERATIONS	F-7
	PLANNING CONSIDERATIONS	F-7
	COORDINATION	F-8
	SPECIAL OPERATIONS COMMAND AND CONTROL ELEMENT	F-8
	CIVIL AFFAIRS PLANNING TEAM A	F-8
	RANGER DEPLOYABLE PLANNING TEAMS AND CROSS-FUNCTIONAL TEAMS	F-8
	REQUEST FOR SUPPORT	F-8
Appendix G	**IMPROVISED EXPLOSIVE DEVICES, SUICIDE BOMBERS, UNEXPLODED ORDNANCE, AND MINES)**	**G-1**
	Section I. IMPROVISED EXPLOSIVE DEVICES	**G-1**
	TYPES	G-1
	CHARACTERISTICS	G-2
	INGREDIENTS	G-2
	CAMOUFLAGE	G-2
	VEHICLE-BORNE DEVICES (CAR BOMBS)	G-3
	EMPLOYMENT	G-4
	COUNTERMEASURES	G-8
	FIVES C'S TECHNIQUE	G-9
	Section II. SUICIDE BOMBERS	**G-10**
	DEFINITION	G-10

	DELIVERY METHODS	G-10
	INDICATORS	G-11
	SPECIAL CONSIDERATIONS	G-11
	COMPLICATIONS	G-12
	Section III. UNEXPLODED ORDNANCE	**G-12**
	RECOGNITION	G-12
	IMMEDIATE ACTION	G-13
	BOOBY TRAPS	G-14
	Section IV. MINEFIELDS	**G-15**
	TYPES	G-15
	STANDARD MINEFIELDS	G-15
	MINEFIELD PATTERNS AND MARKINGS	G-16
	UNCHARTED MINEFIELDS	G-16
	MINE INDICATORS	G-16
	REPORTS	G-17
	EXTRACTION	G-17
Appendix H	**OPERATIONS IN A CHEMICAL, BIOLOGICAL, RADIOLOGICAL, OR NUCLEAR ENVIRONMENT**	**H-1**
	DEFENSE	H-1
	CHEMICAL AGENTS	H-4
	TREATMENT OF CHEMICAL CASUALTIES	H-4
	BIOLOGICAL AGENTS	H-8
	RADIOLOGICAL WEAPONS	H-9
	NUCLEAR WEAPONS	H-10
Appendix I	**MEDIA CONSIDERATIONS**	**I-1**
	OBJECTIVE	I-1
	REALITY	I-1
	OBJECTIVES AND INTERESTS	I-1
	CAPABILITIES	I-2
	COMMAND CONSIDERATIONS	I-2
	GUIDELINES	I-2
	INTERVIEWS	I-3
	TRAINING FOR MEDIA AWARENESS	I-4
	MEDIA CARDS	I-4
Appendix J	**PATTERN ANALYSIS AND SITUATIONAL UNDERSTANDING**	**J-1**
	GATHERING OF INFORMATION	J-1
	SOURCES	J-2
	ASSESSMENT	J-2
	OPERATIONAL VARIABLES	J-2
	COLLECTION	J-2
	RECORDS	J-4
	POPULATION CONSIDERATIONS	J-6
	RESPONSIBILITIES	J-8
	PLATOON LEADER	J-9
	PLATOON MEMBERS	J-9

Appendix K	MOTORIZED OPERATIONS	K-1
	Section I. WHEELED VEHICLE PLANNING CONSIDERATIONS	K-1
	MAINTENANCE	K-2
	VEHICLE WEIGHT, SURVIVABILITY, AND ARMOR	K-2
	FIREPOWER AND OBSERVATION CONSIDERATIONS	K-5
	COMMUNICATION CONSIDERATIONS	K-5
	CREW AND PASSENGER DESIGNATED VEHICLE POSITIONS	K-5
	VEHICLE EQUIPMENT LOAD PLANS	K-8
	Section II. PATROLLING CONSIDERATIONS IN URBAN OPERATIONS	K-9
	URBAN PATROLLING CONSIDERATIONS	K-9
	TACTICAL VEHICLE EMPLOYMENT AND URBAN PATROLS	K-9
	BRADLEY AND STRYKER CONSIDERATIONS	K-10
	MOUNTED HASTY CHECKPOINT OPERATIONS	K-10
	Section III. LONG-RANGE OPERATIONS CONSIDERATIONS	K-11
	LONG-RANGE OPERATIONS	K-12
	ROUGH TERRAIN DRIVING	K-12
	VEHICLE RECOVERY	K-13
	Section IV. OEF AND OIF VEHICLE MODIFICATIONS	K-13
	UNIT-INSTALLED WIRE CUTTERS AND WIRE GUARDS	K-13
	URBAN PATROL	K-14
	UNIT COMMAND VEHICLE	K-15
	UNIT-PRODUCED TRIPOD AND MODIFIED ARMOR	K-15
	UNIT-INSTALLED STEEL ARMOR	K-17
GLOSSARY		Glossary-1
REFERENCES		References-1
INDEX		Index-1

Figures

Figure 1-1. Enemy operations in OE. ... 1-2
Figure 1-2. Components of combat system. ... 1-5
Figure 1-3. Variables in operational environment. ... 1-7
Figure 1-4. Infantry. ... 1-11
Figure 1-5. Heavy. ... 1-11
Figure 1-6. Stryker. ... 1-12
Figure 1-7. Ranger. ... 1-12
Figure 1-8. Elements of combat power. ... 1-18
Figure 2-1. Army planning process within TLP. ... 2-3
Figure 2-2. Key planning concepts. ... 2-4
Figure 2-3. Parallel planning. ... 2-5
Figure 2-4 Nesting of concepts ... 2-7
Figure 2-5. Analysis of mission using METT-TC. ... 2-11

Figure 2-6. Areas of operation and interest. ... 2-15
Figure 2-7. Military aspects of terrain. ... 2-16
Figure 2-8. Analysis of obstacles and restricted terrain. ... 2-19
Figure 2-9. Analysis of mobility corridors and avenues of approach. ... 2-22
Figure 2-10. Analysis of key terrain. ... 2-24
Figure 2-11. Analysis of "IV" line. ... 2-25
Figure 2-12. Light matrix. ... 2-28
Figure 2-13. Example enemy composition. ... 2-31
Figure 2-14. Example enemy disposition. ... 2-32
Figure 2-15. Example enemy strength. ... 2-32
Figure 2-16. Example enemy situation template. ... 2-36
Figure 2-17. Example company timeline. ... 2-38
Figure 2-18. Risk assessment. ... 2-41
Figure 2-19. Example COA sketch. ... 2-46
Figure 2-20. Analysis of course(s) of action. ... 2-47
Figure 2-21. Box war-gaming technique. ... 2-48
Figure 2-22. Belt war-gaming technique. ... 2-49
Figure 2-23. Avenue-in-depth war-gaming technique. ... 2-50
Figure 3-1. Transition from movement techniques to maneuver. ... 3-2
Figure 3-2. Legend of company symbols. ... 3-3
Figure 3-3. Traveling technique. ... 3-3
Figure 3-4. Traveling overwatch technique. ... 3-4
Figure 3-5. Bounding overwatch technique. ... 3-5
Figure 3-6. Example company column formation. ... 3-7
Figure 3-7. Example company line formation. ... 3-8
Figure 3-8. Example company wedge formation. ... 3-8
Figure 3-9. Example company vee formation. ... 3-9
Figure 3-10. Example company file formation. ... 3-10
Figure 3-11. Example echelon right formation. ... 3-11
Figure 3-12. All-round security. ... 3-15
Figure 3-13. Strip map. ... 3-20
Figure 4-1. Envelopment. ... 4-4
Figure 4-2. Turning movement. ... 4-5
Figure 4-3. Infiltration. ... 4-6
Figure 4-4. Penetration. ... 4-7
Figure 4-5. Frontal attack. ... 4-7
Figure 4-6. Spectrum of attacks. ... 4-16
Figure 4-7. Movement to objective. ... 4-24
Figure 4-8. Isolation of objective. ... 4-26
Figure 4-9. Breaching and securing of a foothold. ... 4-27
Figure 4-10. Exploitation of penetration. ... 4-29

Figure 4-11. Linear assault. 4-35
Figure 4-12. Linear assault with support element. 4-36
Figure 4-13. Linear assault with follow and support. 4-37
Figure 4-14. Combination technique. 4-48
Figure 4-15. Company moving on single infiltration lane. 4-50
Figure 4-16. Company moving on multiple infiltration lanes. 4-50
Figure 5-1. Protective wire obstacles. 5-9
Figure 5-2. Company defense in sector, with platoon in a battle position. 5-13
Figure 5-3. Primary and alternate positions. 5-15
Figure 5-4. Supplementary position. 5-16
Figure 5-5. Defense from mutually supporting platoon battle positions. 5-17
Figure 5-6. Multiple engagement areas. 5-18
Figure 5-7. Company strongpoint. 5-19
Figure 5-8. Company perimeter defense. 5-20
Figure 5-9. Y-shape perimeter defense. 5-22
Figure 5-10. Modified Y-shape perimeter defense. 5-23
Figure 5-11. Linear defense. 5-24
Figure 5-12. Nonlinear defense. 5-26
Figure 5-13. Company defense on a reverse slope. 5-28
Figure 5-14. Likely enemy avenues of approach. 5-31
Figure 5-15. Example enemy scheme of maneuver. 5-32
Figure 5-16. Locations to kill enemy. 5-33
Figure 5-17. Emplacement of weapons systems. 5-35
Figure 5-18. Plans for and integration of obstacles. 5-36
Figure 5-19. Integration of direct and indirect fires. 5-38
Figure 5-20. Company defensive sector sketch. 5-41
Figure 5-21 Example company dismounted delay from subsequent positions. 5-46
Figure 5-22. Example company delay from alternating positions. 5-47
Figure 5-23. Example unassisted withdrawal. 5-49
Figure 6-1. Types of stability operations. 6-6
Figure 6-2. Example Infantry company lodgment area using existing facilities. 6-10
Figure 6-3. Example deliberate observation post. 6-14
Figure 6-4. Deliberate checkpoint layout. 6-17
Figure 6-5. Vehicular traffic stop. 6-19
Figure 6-6. Employment of checkpoints, OPs, and patrols to enforce a zone of separation. 6-22
Figure 8-1. Identification of intelligence requirements and use of patrols to reconnoiter. 8-4
Figure 8-2. Infantry company linkup. 8-7
Figure 8-3. Relief in place in sequence. 8-10
Figure 8-4. Relief in place (company graphics). 8-13
Figure 8-5. Infantry company conducting a forward passage of lines. 8-16
Figure 8-6. Infantry company conducting a rearward passage of lines. 8-17

Figure 8-7. Stationary guard with OPs forward. ... 8-25
Figure 8-8. Infantry company guarding flank during movement to contact. ... 8-26
Figure 8-9. Area reconnaissance patrol. ... 8-37
Figure 8-10. Zone reconnaissance patrol. ... 8-38
Figure 8-11. Combat patrol. ... 8-38
Figure 8-12. Organization of elements. ... 8-39
Figure 8-13. Security elements move into position. ... 8-43
Figure 8-14. Support and assault elements move into position. ... 8-44
Figure 9-1. Identification of probable enemy locations and determination of enemy scheme of maneuver. ... 9-4
Figure 9-2. Determination of where and how to mass (focus and distribute) fire effects to kill enemy. ... 9-5
Figure 9-3. Orientation of forces to speed target acquisition. ... 9-6
Figure 9-4. Shifting of fires to refocus and redistribute them. ... 9-7
Figure 9-5. Terrain-based quadrants. ... 9-11
Figure 9-6. Friendly based quadrants. ... 9-12
Figure 9-7. Frontal fire. ... 9-13
Figure 10-1. Command and support relationships. ... 10-3
Figure 10-2. Fire planning process. ... 10-12
Figure 10-3. Example fire support execution matrix. ... 10-16
Figure 10-4. Dimensions of final protective fires. ... 10-17
Figure 10-5. Company FIST communications. ... 10-19
Figure 10-6. QuickFire channel *illus*tration of sensor-to-shooter link. ... 10-20
Figure 10-7. Beginning of close air support. ... 10-24
Figure 10-8. 155-mm shaping fires, close air support shifts. ... 10-25
Figure 10-9. 155-mm shift, 81-mm, and supporting fires. ... 10-26
Figure 10-10. 81-mm shift, 60-mm mortars. ... 10-27
Figure 10-11. 60-mm cease fire, shift of supporting fires. ... 10-28
Figure 10-12. 60-mm mortar. ... 10-30
Figure 10-13. 81-mm mortar. ... 10-30
Figure 10-14. 120-mm mortar. ... 10-30
Figure 10-15. TTFACOR technique. ... 10-36
Figure 10-16. Example format for a nine-line close air support briefing. ... 10-38
Figure 10-17. IBCT engineer company. ... 10-40
Figure 10-18. Example sapper squad. ... 10-41
Figure 10-19. Stinger, man-portable and mounted (as "*Avenger*") on a HMMWV. ... 10-46
Figure 10-20. Machine-gun aim points against helicopters and high-performance aircraft. ... 10-49
Figure 11-1. Load echelon diagram. ... 11-6
Figure 11-2. Classes of supply. ... 11-11
Figure 11-3. Service station resupply method. ... 11-13
Figure 11-4. Tailgate resupply method. ... 11-14
Figure 11-5. In-position method. ... 11-14

Figure 11-6. Enemy prisoner of war detainee tag. 11-27
Figure 11-7. Unit record copy. 11-27
Figure 11-8. Enemy prisoner of war, document, and special equipment tag. 11-28
Figure 12-1. Tank in direct fire, supported by Infantry. 12-17
Figure 12-2. Graphic control measures for Infantry and heavy. 12-22
Figure 12-3. Hull-down position. 12-23
Figure 12-4. Hide position. 12-24
Figure 12-5. Building hide position. 12-25
Figure 12-6. Example positions for Infantry riding on a tank. 12-26
Figure 12-7. Danger areas around a tank firing a 120-mm main gun. 12-28
Figure 12-8. Artillery in direct-fire role. 12-35
Figure 12-9. Example numbering system. 12-37
Figure 12-10. Zones, boundaries, and phase lines. 12-38
Figure 12-11. Clearing of selected buildings within sector. 12-43
Figure 12-12. Isolation of an urban objective. 12-44
Figure 12-13. Direction-of-attack technique for direct-fire planning and control. 12-45
Figure 12-14. Assault of a building. 12-46
Figure 12-15. Example marking SOP. 12-48
Figure 12-16. Example of urban obstacles. 12-53
Figure 12-17. Platoon battle positions in a company sector. 12-55
Figure 12-18. Urban strongpoint. 12-66
Figure 12-19. Company delay in an urban area. 12-67
Figure A-1. Example completed risk management worksheet. A-3
Figure B-1. Infantry battalion weapons company. B-2
Figure B-2. Assault platoon of weapons company. B-2
Figure B-3. Overlapping sectors of fire. B-3
Figure B-4. Standoff ranges, TOW (top) and Javelin (bottom). B-5
Figure B-5. Dispersion between squads. B-7
Figure B-6. Command launch unit. B-9
Figure B-7. Launch tube assembly and missile. B-10
Figure B-8. Javelin flight profile in top-attack mode. B-12
Figure B-9. Javelin flight profile in direct-attack mode. B-13
Figure B-10. Minimum room enclosure for Javelin firing. B-14
Figure C-1. M1 tank danger zone. C-5
Figure C-2. BFV danger zone. C-6
Figure C-3. BFV TOW backblast danger zone. C-7
Figure C-4. Attacks along converging routes. C-10
Figure C-5. Attacking on same route. C-11
Figure C-6. Mounted forces support by fire. C-12
Figure C-7. Mounted forces integrated throughout position. C-13
Figure C-8. Mounted force held in reserve. C-14

Contents

Figure D-1. UH-60 unloading diagram. D-5
Figure D-2. Inverted "Y" marker. D-7
Figure D-3. Large, one-sided PZ. D-9
Figure D-4. Small, two-sided PZ. D-10
Figure D-5. UH-60 loading diagram. D-11
Figure D-6. Minimum planning requirements. D-16
Figure D-7. Bull's-eye technique: uses a known point or an easily recognizable terrain feature. D-16
Figure D-8. Grid technique: uses grid coordinates define point. D-16
Figure D-9. Sector and terrain technique: uses terrain and graphics, which are both available to air and ground units. D-17
Figure D-10. Phase line technique: uses graphics, which are available to both air and ground. D-17
Figure D-11. Example CCA nine-line briefing. D-20
Figure D-12. Example of CCA nine-line briefing. D-21
Figure F-1. ARSOF missions and collateral activities. F-2
Figure F-2. Special forces operational detachment A. F-3
Figure G-1. Example of IED detonation device with explosive. G-2
Figure G-2. Camouflaged UXO. G-3
Figure G-3. Vehicle IED capacities and danger zones. G-4
Figure G-4. Example of IED dropped into vehicles. G-5
Figure G-5. Typical IED combination ambush. G-6
Figure G-6. IED combination ambush in Iraq. G-7
Figure G-7. Deception or fake IED used to stop convoy in kill zone. G-7
Figure G-8. Suicide bomber vest. G-11
Figure G-9. Nine-line UXO incident report. G-14
Figure G-10. Example booby trap. G-15
Figure G-11. Example format for a mine incident report. G-17
Figure H-1. Nerve agent antidote Mark I and CANA. H-7
Figure H-2. Thigh injection site. H-8
Figure H-3. Buttocks injection site. H-8
Figure J-1. Example 1, pattern analysis. J-5
Figure J-2. Example 2, pattern analysis. J-6
Figure J-3. Example 1, population status overlay. J-7
Figure J-4. Example 2, population status overlay. J-8
Figure K-1A. Commercially produced and available gunner armor protection. K-3
Figure K-1B. Army-issue, roof-mounted, gun-ring armor. K-3
Figure K-2. Commercially produced HMMWV armored half-door and double-articulating, swing-arm mount for crew-served weapons. K-4
Figure K-3A. Possible platoon-seating technique. K-6
Figure K-3B. Variation of platoon-seating technique. K-6
Figure K-4. Modified M1025 (turtle-shell HMMWV). K-7
Figure K-5. Commercially produced version of an outboard-facing rear passenger seat and double-articulated swing arm for crew-served weapons. K-8

Figure K-6. Hasty checkpoints. K-11
Figure K-7. Unit-installed wire cutters and wire guards. K-14
Figure K-8. Urban patrol. K-14
Figure K-9. Unit command vehicle. K-15
Figure K-10. Unit-produced tripod and modified armor. K-16
Figure K-11. Closeup of sections of Bradley fighting vehicle spall liner armor applied to exterior of cargo bed. K-16
Figure K-12. Unit-installed steel armor. K-17
Figure K-13. Closeup, reverse-angle view of unit-fabricated and -installed steel plate armor. K-17

Tables

Table 1-1. Capabilities and limitations of Infantry rifle company. 1-10
Table 2-1. Factors to consider in analyzing obstacles and restricted terrain. 2-18
Table 2-2. Factors to consider in analyzing mobility corridors and avenues of approach. 2-21
Table 2-3. Factors to consider in analyzing key terrain. 2-23
Table 2-4. Factors to consider in analyzing observation and fields of fire. 2-26
Table 2-5. Considerations in cover and concealment. 2-27
Table 2-6. Example enemy capabilities. 2-33
Table 2-7. Example enemy recent activities. 2-34
Table 2-8. Recommended SITEMP items. 2-35
Table 2-9. Example PCC and PCI checklists. 2-55
Table 3-1. Comparison of movement formations. 3-11
Table 5-1. Obstacle effects. 5-8
Table 5-2. Selection of control measures. 5-12
Table 6-1. Example convoy briefing checklist. 6-26
Table 8-1. Relationship between breaching organization and fundamentals. 8-30
Table 9-1. Common fire-control measures. 9-9
Table 9-2. Weapons safety posture levels. 9-18
Table 10-1. Indirect fire capabilities. 10-4
Table 10-2. Artillery response times. 10-9
Table 10-3. Example battalion fire support execution matrix. 10-10
Table 10-4. Risk estimate distances for mortars and cannon artillery. 10-22
Table 10-5. Risk estimate distances for aircraft-delivered ordnance. 10-23
Table 10-6. Mortar ammunition characteristics. 10-29
Table 10-7. Advantages and disadvantages of direct lay. 10-33
Table 10-8. Advantages and disadvantages of direct alignment. 10-34
Table 10-9. Advantages and disadvantages of conventional indirect fire. 10-34
Table 10-10. Advantages and disadvantages of hip shoot. 10-35
Table 10-11. Close air support types for terminal attack attributes. 10-39
Table 10-12. Engineer missions. 10-42

Table 10-13. Emplacement authority.	10-44
Table 11-1. Five S's and T method of detainee field processing.	11-25
Table 12-1. Approximate frontages and depths in large built-up areas.	12-52
Table A-1. Examples of potential hazards.	A-2
Table A-2. Risk levels and impact on mission execution.	A-3
Table A-3. Combat stress behaviors.	A-10
Table A-4. Reduction of impact of continuous operations.	A-11
Table A-5. Effects of sleep loss.	A-13
Table A-6. Indicators of sleep deprivation and fatigue.	A-13
Table B-1. Javelin technical characteristics.	B-8
Table B-2. Range determination recognition method.	B-16
Table D-1. Air mission briefing format.	D-13
Table D-2. Techniques for marking of target or location.	D-18
Table F-1. Ranger force capabilities.	F-4
Table H-1. MOPP levels.	H-2
Table H-2. Comparison data for decontamination levels.	H-3
Table H-3. Characteristics of chemical agents.	H-4

Preface

Rather than providing rote solutions, this manual provides a doctrinal framework of principles; tactics, techniques, and procedures (TTP); terms; and symbols for the employment of the Infantry rifle company. This framework will help Infantry rifle company leaders effectively--

- Exploit capabilities unique to the Infantry.
- Reduce the vulnerability of the unit.
- Plan and conduct full-spectrum operations.
- Accomplish their missions in various tactical situations, from stability and civil support to high-intensity combat.
- Win on the battlefield.

The Infantry companies of the SBCT and HBCT mostly use the same doctrine, but cover more specific doctrine in their own manuals.

The main target audience for this manual includes Infantry rifle company commanders, executive officers, first sergeants, platoon sergeants and platoon leaders. Military instructors, evaluators, training and doctrine developers will also find it useful, as will other Infantry company commanders (HHC and weapons company), Infantry battalion staff officers, service school instructors, and commissioning source instructors.

This publication applies to the Active Army, the Army National Guard (ARNG), the National Guard of the United States (ARNGUS), and the US Army Reserve (USAR) unless otherwise stated.

Leaders must understand this manual before they can train their companies using ARTEP 7-10-MTP. They should use this manual as a set along with the publications listed in the References.

The *Summary of Change* lists major changes from the previous edition by chapter and appendix. Changes include lessons learned.

The proponent for this publication is the US Army Training and Doctrine Command. The preparing agency is the US Army Infantry School. You may send comments and recommendations by any means, US mail, e-mail, fax, or telephone, as long as you use or follow the format of DA Form 2028, *Recommended Changes to Publications and Blank Forms*. You may also phone for more information.

    E-mail.......... doctrine@benning.army.mil.
    Phone .... COM 706-545-7114 or DSN 835-7114
    Fax .... COM 706-545-7500 or DSN 835-7500
    US Mail........ Commandant, USAIS
                  ATTN: ATSH-ATD
                  6751 Constitution Loop
                  Fort Benning, GA 31905-5593

Unless this publication states otherwise, masculine nouns and pronouns may refer to either men or women.

Summary of Change

Chapter 1	ADDS	Full-spectrum operations in the COE. Warfighting functions. ES2.
	UPDATES	Organizational structure in the modular brigade design.
Chapter 2	EXPANDS	Battle command. TLP.
Chapter 4	ADDS	SUAS. Continuum of contact.
Chapter 6	ADDS	Stability operations.
Chapter 7	ADDS	Civil support operations.
Chapter 8	ADDS	Presence patrols for stability and civil support operations. Point reconnaissance. Tracking and contact patrols.
	EXPANDS	Reconnaissance and breaching as tactical enabling operations. Company commander's use of patrols to accomplish tactical tasks.
	DELETES	Friendly force use of non-command-detonated antipersonnel mines or booby traps.
Chapter 9	ADDS	Direct fire control and distribution.
Chapter 10	ADDS	Checklists such as the CAS 9-line briefing.
	EXPANDS	CAS, JTAC.
Chapter 11	ADDS	New terms.
	UPDATES	Sustainment. Unit trains. Resupply. Health service support. Weapons replacement operations.
Chapter 12	EXPANDS	Urban operations. Key planning issues.
Appendix B	ADDS	Risk management. Fratricide avoidance. Continuous operations. Safety and force protection.
Appendix C	COMBINES	TOW and Javelin employment.
Appendix D	ADDS	Employment with SBCT and HBCT elements.
Appendix E	ADDS	Sniper employment during tactical operations.
Appendix F	ADDS	Integration with Army SOF.
Appendix G	ADDS	IEDs, homicide bombers, UXO, and mines. Tactical-level countermeasures lessons learned.
Appendix H	UPDATES	CBRN defense operations.
	ADDS	Current CBRN concepts, terms, procedures, and equipment.
Appendix I	ADDS	Media considerations.
Appendix J	ADDS	Pattern analysis. Situational understanding for rapid planning.
Appendix K	ADDS	Motorized operations. Motorized patrolling.

Chapter 1
Introduction

"The unresting progress of mankind causes continual change in the weapons; and with that must come a continual change in the manner of fighting." --Alfred Mahan

The Infantry is an all-weather, all-terrain unit. Its mission is to close with the enemy by means of fire and maneuver to destroy or capture him, or to repel his assault by fire, close combat, and counterattack. Against this backdrop, the Infantry company must also be ready to adapt to various levels of conflict and peace in differing environments. This requires bold, aggressive, resourceful, and adaptive leaders who are willing to accept known risks to accomplish the mission. Infantry leaders must use their initiative and make rapid decisions to take advantage of unexpected opportunities. In order to succeed, Infantry companies must be aggressive, physically fit, disciplined, and well trained. The inherent strategic mobility of Infantry units dictates a need to be prepared for rapid deployment in response to the operational environment. This chapter discusses the operational environment (OE) and preparation of the Infantry company for war.

This chapter discusses the recent changes in the Infantry Company's organizational structure in the modular brigade design. It also adds a discussion of full spectrum operations in the contemporary operational environment (COE) and introduces the warfighting functions and the concept of *Every Soldier as a Sensor* (ES2).

Section I. OPERATIONAL ENVIRONMENT

This section defines the OE itself and discusses phases of the conflict; systems-based warfare; generation and focus of the effects of combat power; and the principles and variables of *full-spectrum operations*.

DEFINITION

1-1. In the OE, potential enemy state and nonstate actors see the United States as the world's dominant power. Potential enemies avoid US military strengths and focus on exploiting perceived US weaknesses. They hope this will enable them to achieve their own regional or international goals without US intervention or, failing this, without the US military defeating those goals. When potential enemies do not fight US forces the same as regional adversaries, asymmetry develops. Conditions that contribute to an asymmetric environment may include cultural and ideological differences, a technological or military imbalance, and a disparity in the application of combat power. In the context of military operations, an asymmetric threat means an adaptive enemy approach to avoid or counter US strengths without opposing them directly. It also seeks to identify, target, and exploit US weaknesses to achieve goals or objectives. Consequently, the Infantry company must be prepared to go into any region or operational environment and perform the full range of missions while dealing with a wide range of threats.

1-2. Such an OE changed the paradigm of the Soviet motorized rifle regiment to a new paradigm that encompasses the entire operational environment. Infantry Soldiers know that they will face adaptive and opportunistic enemies worldwide. Therefore, as part of their planning and execution of operations, Infantry company commanders must be acutely aware of the relationship between their tactical goals and the myriad of constantly changing factors that their units might encounter.

PHASES OF CONFLICT

1-3. The three general phases where the enemy will operate in the OE are regional, transitional, and adaptive (Figure 1-1). Knowing these phases helps commander anticipate how the enemy will fight. For example, enemy forces might operate differently, with or without cooperation, in different regions or even in commanders' areas of operations (AOs). The labeled phases in Figure 1-1 will help commanders analyze and better understand the OE. However, they should note that these phases are neither all inclusive nor mutually exclusive.

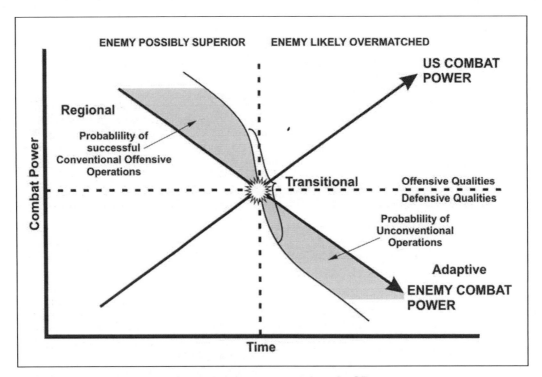

Figure 1-1. Enemy operations in OE.

1-4. Infantry companies can expect to operate against foreign conventional and unconventional forces striving to achieve regional superiority. Except for a few nation states, most notably the US, modern militaries and unconventional forces are designed to defeat regional threats and operate in and around the territory of their own nation. They train to achieve regional dominance. Their equipment is designed around the local environment. US military forces can expect to fight on foreign soil. When this happens, the enemy enjoys a "*home field advantage*," which offers--

- Better understanding of the terrain and weather.
- Better understanding of the population and language.
- knowledge of local religious and ethnic customs and courtesies.
- Combatants who can easily blend in with noncombatants.
- Potential access to cached arms and ammunition.
- Popular support.

REGIONAL

1-5. This phase is likely to occur in the beginning of a conflict. With limited or no US presence in a region, the regional militaries are likely to enjoy conventional military superiority. At the beginning of a conflict, US combat power is likely to be relatively lower than enemy combat power (Figure 1-1). During this phase, the enemy might be able to succeed with conventional offensive operations against regional or international threats in the immediate area. To negate his advantage, US forces begin building combat power in the region. This also helps them meet US goals and objectives. Enemy conventional offensive operations focus on consolidating gains or denying US entry to the region. The enemy state and allies work strategically and politically to prevent US interference.

TRANSITIONAL

1-6. On orders from the National Command Authority, US forces deploy to and build combat power in the region. As US forces enter the region, US combat power rises, which increases the ratio of US combat power to enemy combat power. At some point, the ability of US forces to generate combat power equals, then overmatches, that of enemy forces. (This is where the arrows cross in Figure 1-1.) For this reason, the enemy favors the use of unconventional, adaptive operations to achieve his goals.

ADAPTIVE

1-7. When US conventional forces gain clear superiority, and when the enemy realizes this, the conflict moves into the adaptive phase of the OE. In this phase, the enemy adopts unconventional, guerilla, insurgent, or terroristic tactics to counteract the US combat power advantages. This might be the most dangerous phase for US forces, because the enemy avoids conventional engagements and instead blends into and uses the local population. He also uses information and psychological operations and warfare to achieve his goals. These goals usually include removing US forces from the region. The enemy will likely adopt a systems-based approach to combat to negate US regional and technological superiority.

> ### EXAMPLE
>
> In early 1990, the US military was clearly more powerful overall. Iraqi forces dominated the Persian Gulf region, and so they easily, quickly, and successfully invaded Kuwait.
>
> In response, the US began deploying light forces to the region. However, even with US air and naval support, the Iraqi's multiple armor and mechanized divisions remained the superior combat power.
>
> However, the US presence grew, and by the time combat operations began in 1991, US and coalition forces easily defeated Iraqi conventional forces.
>
> The US has maintained a military presence in the Persian Gulf since then, which keeps the Iraq-US situation in the *transitional phase*.
>
> When the US massed forces and invaded Iraq in 2003, the US enjoyed a tremendous combat power advantage and destroyed all of the conventional Iraqi forces that tried to resist. Those who remained loyal to the old regime were forced into the adaptive phase, during which they have limited their attacks on US combat forces. They have instead concentrated on vulnerable subsystems. They do this with IEDs, ambushes with rocket-propelled grenades (RPGs), and other unconventional tactics. These adaptive tactics give them a temporary advantage in combat power.

SYSTEMS-BASED WARFARE

1-8. When employing the characteristics of systems-based warfare, the enemy will try to---

- Identify and target critical subsystems or components.
- Determine the best time and, or place of greatest vulnerability for maximum destruction.
- Conduct unexpected or random attacks to achieve shock and demoralize the population.

1-9. Enemy elements will attack vulnerable or unprepared segments of the US force to weaken the entire force structure. Commanders at all levels must know exactly how they and their forces look to the enemy, and identify and shield these critical or vulnerable systems before the enemy can attack them. Commanders must understand and analyze objectively every component in their combat system, looking for exploitable trends, characteristics, or vulnerabilities. The enemy focuses on splitting up the US combat system as a whole by rendering its parts (subsystems and components) ineffective. He identifies and attacks the friendly force's weakest or most critical links.

1-10. The enemy carefully analyzes US successes and modifies his operations accordingly. This requires Infantry company commanders to adapt the status and tactics of their force protection plan continually. They must maintain a strong and offensive force protection status to guard against a continually adapting enemy. Force protection extends to all elements of the commanders' combat systems. In accordance with FM 3-0, they channel the collection and processing of this information by clearly expressing which of their commander's critical information requirements (CCIR) are most important to them.

1-11. Figure 1-2 shows an example of systems-based warfare. The combat power of an Armor or Infantry company clearly exceeds that of a small guerilla team. If a guerilla(s) attacks an Armor or Infantry company conventionally, he can be destroyed easily. However, the Armor or Infantry company relies on subsystems, such as Class I, Class III, Class V, manpower, and communications networks, to function properly and generate combat power. The guerilla can identify and destroy one subsystem, thus degrading the combat power of the friendly force without the cost of a conventional attack. The guerilla might also ambush an unprotected company logistics package (LOGPAC). He might target Class I supplies biologically, or he might directly or electronically attack the C2 nodes. By destroying or isolating any one of the supporting sub-systems, he can hurt the entire friendly combat system. Finally, he capitalizes on these isolated successes in his information operations (IO) campaign by highlighting his victory against the US Army "as a whole."

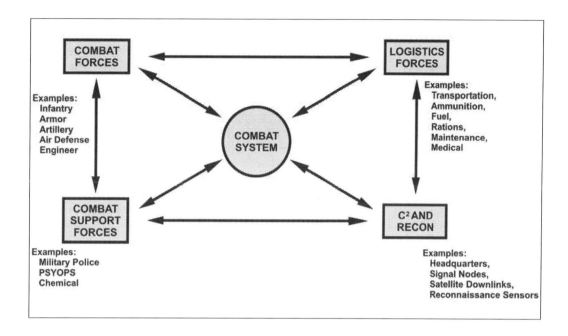

Figure 1-2. Components of combat system.

GENERATION AND FOCUS OF EFFECTS OF COMBAT POWER

1-12. Infantry company commanders must remember that the enemy will patiently study and evaluate their forces to identify their most vulnerable subsystems. During the adaptive phase of enemy operations, time favors the enemy because he knows the area. He will most likely attack only when, where, and how he can to avoid or negate US combat power advantages. This also lets the enemy create or generate his own combat power at that particular time or place. For example, the enemy might mass as an ambush force to quickly overwhelm a US convoy or patrol operating without Army aviation or armored vehicle support. The enemy uses the protection of the urban or restricted terrain to target soft-skin vehicles or a small numbers of personnel who are vulnerable to small arms and RPG fire. The ambush force likely disperses along preplanned routes before an effective US counterattack can occur. The ambush force reduces US advantages in protection, mobility, and firepower provided by armored vehicles, aviation, or heavy weapons. The enemy creates a momentary advantage in the ability to generate and focus the effects of combat power.

1-13. First, commanders at all levels must continually analyze both their essential elements of friendly information (EEFI) and PIR to anticipate where the enemy might try to gain greater combat power. Then, they (Infantry commanders) must act to deny him (the enemy) the chance. Second, commanders must continually analyze, plan, and implement procedures to eliminate enemy advantages.

STRATEGIC AND OPERATIONAL PRINCIPLES OF FIGHTING

1-14. Few combatants want to confront US combat forces directly. In fact, the enemy prefers to keep or get the US out of the conflict altogether. If confrontation is unavoidable, he will fight US forces differently than he fights regional peers or lesser forces. The US can expect the enemy to mix the following principles.

CONTROL AND LIMIT ACCESS INTO REGION

1-15. The enemy will likely target sea ports of debarkation (SPODs) and aerial ports of debarkation (APODs) in the region to prevent the initial or subsequent entry of US combat forces. These and other lines of communication (LOC) are likely to remain the focus of enemy action because they are difficult to secure.

CHANGE THE NATURE OF CONFLICT

1-16. The enemy will try to change the nature of the conflict. Naturally, his goal is to shape it to favor his strengths and focus on perceived US weaknesses. He will avoid open conventional engagements where massed US systems can work to their full potential. This means he will focus on highly populated urban and restricted terrain areas where he can generate more combat power.

CONTROL TEMPO

1-17. The enemy will also try to set and control the tempo of an operation. He does so in order to take the initiative, if he can, and to act within the decision cycle of the US force(s). For him to control the tempo, he must generate enough combat power for large-scale offensive operations or for many smaller, widespread, well-coordinated attacks. US forces aggressively counter the enemy's effort to control the tempo of an operation by assuming an offensive posture. This disrupts enemy operations and prevents their synchronization.

NEUTRALIZE TECHNOLOGICAL OVERMATCH

1-18. US enemies usually have a technological disadvantage. The greatest advantages US forces are likely to enjoy are in weapons ranges and destructive power, target acquisition, and information-sharing capabilities. The enemy might respond by moving the conflict where friendly weapons ranges are limited and commanders cannot use the most destructive weapons. The enemy will blend with local populations and confuse sophisticated friendly sensors.

CAUSE POLITICALLY UNACCEPTABLE CASUALTIES

1-19. A US center of gravity is popular support for military action. The enemy will do what he can to affect regional, US national, and world support for US military action adversely. Such measures might include targeting his own infrastructures and civilians, or it might include trying to cause collateral damage and noncombatant casualties by US forces. By exploiting the images resulting from these measures, they endeavor to manipulate the media to their advantage (see also Appendix I). As a result, the US population must be assured that casualties inflicted by the enemy are relative to the importance of military and political objectives.

ALLOW NO SANCTUARY

1-20. In hopes of lowering the morale of US troops, the enemy will try to prevent the establishment of an area where our Soldiers can rest and refit with a lower force-protection level. He will also target any perceived vulnerability at every echelon.

CRITICAL, STRATEGIC, AND OPERATIONAL VARIABLES

1-21. Eleven variables define the operational environments in which US military conflicts occur. These variables represent the exact conditions, circumstances, and influences in the OE and vary based on the situation, region, and politics. Whether or not these variables significantly affect the environment, commanders must nevertheless consider them when analyzing the mission or the changing situation. Figure 1-3 shows the eleven variables.

1. THE PHYSICAL ENVIRONMENT.
2. NATURE AND STABILITY OF THE ENEMY STATE OR SPONSOR.
3. MILITARY CAPABILITIES OF THE ENEMY STATE OR SPONSOR.
4. TECHNOLOGY AVAILABLE TO US AND ENEMY.
5. INFORMATION WARFARE.
6. EXTERNAL ORGANIZATIONS (GOVERNMENTAL AND NONGOVERNMENTAL).
7. SOCIOLOGICAL DEMOGRAPHICS.
8. REGIONAL, GLOBAL, POLITICAL, AND STRATEGIC RELATIONSHIPS.
9. US AND ENEMY NATIONAL WILL.
10. TIME.
11. ECONOMICS.

Figure 1-3. Variables in operational environment.

Section II. PREPARATION FOR WAR

Infantry companies are organized and equipped to close with and kill the enemy, to destroy his equipment, and to shatter his will to resist. This close personal fight requires combat-ready units with skilled Soldiers and leaders. These units are developed into agile combat forces by tough, thorough, and demanding training. This takes leaders who understand the effective employment of Infantry forces in a complex OE. All units receive extensive training in reconnaissance techniques. This ensures a thorough situational understanding, which allows the Infantry company commanders to employ overwhelming, precise force within the enemy's decision cycle. This precise application of combat power and agility helps reduce collateral damage to facilities and noncombatants.

SOLDIER

1-22. The successful resolution of ground combat depends on the Infantry. Individual Soldiers, molded into a disciplined and well-led team, create a combat-ready force. No Soldier must master a more diverse set of skills than the Infantry Soldier. He is an authority on the employment of weapons from the basic bayonet to high-tech mortars and multipurpose missiles. As needed, he can simultaneously function as an engineer, doctor, air defender, senior radio operator, diplomat, computer expert, mechanic, and construction expert. He is a survivor, because he can conduct operations and attain victory against steep odds in any conditions. Furthermore, *Every Soldier as a Sensor* (ES2) means that Soldiers are trained to actively observe for details related to CCIR while in an AO, and they are competent, concise, and accurate in their reporting. Their leaders understand how to optimize the collection, processing, and dissemination of information in their unit to enable generation of timely intelligence. The individual Soldier is the Infantry's most precious resource.

LEADER

1-23. Infantry leaders do everything they can to prepare their Soldiers--and their units--for the rigors of close combat. In combat, leaders have to channel the efforts of individuals and small units to achieve victory. Leadership in combat is a leader's most important endeavor. It is the culmination of all his training, counseling, coaching, and preparation. Infantry leaders seek responsibility and insist on tough, realistic training. They understand the consequences of inaction and complacency. They understand the importance of initiative and aggressiveness, the study of their profession, and the setting of the proper examples for his Soldiers. Leading in combat is the capstone of all previous experience. The Infantry leader is a resourceful, tenacious, decisive, and adaptive warrior.

UNIT

1-24. The strength of Infantry units comes from the skill, courage, and discipline of individual Soldiers. Their individual capabilities are enhanced by the teamwork and cohesion in the squads and platoons. This cohesion is essential to the survival and success of Infantry units in close combat. It gives the Infantryman the will and determination to persevere, to accept hardships, and to refuse defeat. In the close fight, these factors decide the victor. Cohesion and teamwork enhance combat effectiveness most at squad and team level.

TRAINING PROGRAM

1-25. Individual and collective training focus on critical wartime operations. Performance-oriented training is conducted to measurable standards IAW published Army and doctrinal references. Complexity increases with mastery of each level. Reinforcement training maintains proficiency. Proper training instills discipline and transforms the difficult to the routine. Training events require subordinate leaders to use their initiative and take independent action in order to prepare for decentralized operations. Training must be hard, realistic, physically demanding, and mentally stressful to prepare Soldiers for combat. Training programs assume personnel turnover and continue during all types of operations, including combat.

Section III. MISSIONS, TYPES, CHARACTERISTICS, CAPABILITIES, LIMITATIONS, AND ORGANIZATION

The fundamental considerations for employing Infantry companies result from the missions, types, equipment, capabilities, limitations, and organization of these units. Other capabilities result from a unit's training program, leadership, morale, personnel strengths, and many other factors. These other capabilities constantly change based on the current situation.

MISSIONS

1-26. The combat mission of the Infantry rifle company is to close with the enemy by means of fire and maneuver to destroy or capture him, repel his assault by fire, close combat, and counterattack. The inherent versatility of Infantry also makes it well suited for employment against asymmetrical threats across the full spectrum of operations. During joint campaigns overseas, Army forces execute a simultaneous and continuous combination of offensive, defensive, and stability and reconstruction operations. They do this as part of joint, interagency, and multinational teams. Concurrently with overseas campaigns, Army forces within the United States and its territories may combine offensive, defensive, and civil support operations to support homeland security. This combination, which defines full-spectrum operations, is also well served by the flexibility of the Infantry company.

TYPES AND CHARACTERISTICS OF INFANTRY RIFLE COMPANIES

1-27. Infantry, Heavy, Stryker, and Ranger comprise the four types of Infantry rifle companies. Some of these have specialized capabilities such as airborne and air assault. Though differences exist between them, they share some similarities in organization, tactics, and employment. The main differences lie in the means of transportation to and on the battlefield, and in the organic supporting assets available to them. Most of the combat power of the Infantry rifle company lies in its highly trained squads and platoons. The company maneuvers in all types of terrain and in climatic and visibility conditions and capitalizes on all forms of mobility. It also uses night vision devices and surveillance equipment.

INFANTRY

1-28. Infantry units can operate effectively in most terrain and weather conditions. They might be the dominant arm in fast-breaking operations because of their rapid strategic deployability. In such cases, they can wrest the initiative early, seize and hold ground, and mass fires to stop the enemy. They are particularly effective in urban terrain, where they can infiltrate and move rapidly to the rear of enemy positions. The commander can enhance their tactical mobility by using helicopters and tactical airlift.

RANGER

1-29. Ranger units are rapidly deployable, airborne-capable, and trained to conduct joint strike operations with (or in support of) special operations units of all services in any environment. They plan and conduct special military operations to support national policies and objectives. They also conduct direct-action missions to support the geographic combatant commanders and operate as conventional Infantry units when integrated with other combined arms elements. (FM 7-85 is the capstone manual for Ranger operations.)

CAPABILITIES AS COMPARED WITH OTHER INFANTRY

1-30. This paragraph compares the capabilities of the Infantry company with those of the Heavy and Stryker companies.

HEAVY

1-31. Heavy Infantry units are mounted on Bradley fighting vehicles. These units are task organized with M1 Abrams tanks in combined arms battalions of the Heavy brigade combat team (HBCT). These heavy units are highly mobile with tremendous combined arms firepower. They are best suited to less restrictive terrain and combat against an armored enemy.

STRYKER

1-32. The battalions of the Stryker brigade combat team (SBCT) serve as its primary maneuver force. The battalion is organized three-by-three: three rifle companies, with three rifle platoons each. Companies fight as combined arms teams with a section of organic 60-mm and strap-on 81-mm mortars, mobile gun system (MGS) platoon, and sniper team. The SBCT units are equipped with the Stryker Infantry carrier vehicle (ICV). The SBCT battalion retains most of the capabilities of the other Infantry plus the additional mobility of Stryker vehicles. Stryker companies operate across the full spectrum of modern combat operations. They are organized to maintain tactical flexibility within restricted and severely restricted terrain.

ADDITIONAL CAPABILITIES AND LIMITATIONS

1-33. Table 1-1 shows the capabilities and limitations of the Infantry rifle company.

Capabilities	Limitations
Conduct offensive and defensive operations in all types of environments, primarily at night.	Limited combat support (CS) and sustainment assets.
Seize, secure, occupy, and retain terrain.	Limited vehicle mobility.
Destroy, neutralize, suppress, interdict, disrupt, block, canalize, and fix enemy forces.	Vulnerable to enemy armor, artillery, and air assets when employed in open terrain.
Breach enemy obstacles.	Vulnerable to enemy chemical, biological, radiological, nuclear, and high yield explosive (CBRNE) attacks with limited decontamination capability.
Feint and demonstrate to deceive the enemy.	
Screen and guard friendly units.	
Reconnoiter, deny, bypass, clear, contain, and isolate. (These tasks might be oriented on both terrain and enemy.)	
Conduct small-unit operations.	
Participate in air assault operations.	
Participate in airborne operations (airborne and Ranger companies).	
Operate in conjunction with mounted or special operations forces.	
Participate in amphibious operations.	

Table 1-1. Capabilities and limitations of Infantry rifle company.

ORGANIZATION

1-34. With the exception of Ranger units, all Infantry rifle company organizations (Figure 1-4, Figure 1-5, Figure 1-6, and Figure 1-7) have the same TOE. Air assault and airborne-trained companies require some special equipment associated with unique capabilities. However, despite these few differences, the mission and employment considerations and tactics are nearly the same.

Introduction

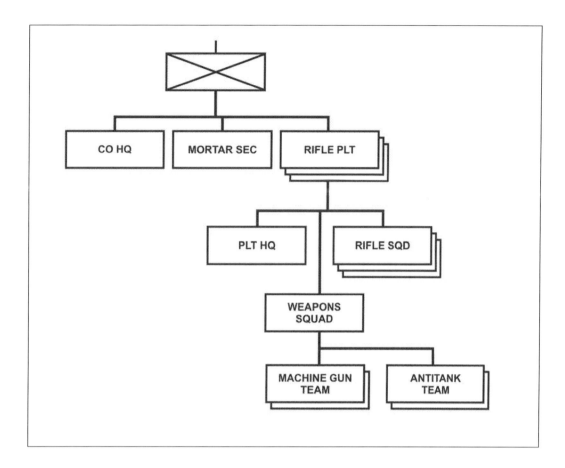

Figure 1-4. Infantry.

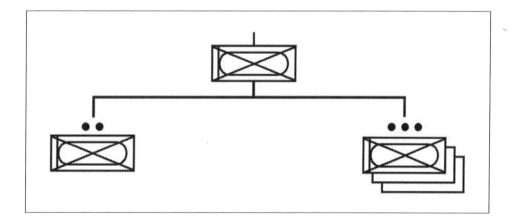

Figure 1-5. Heavy.

Chapter 1

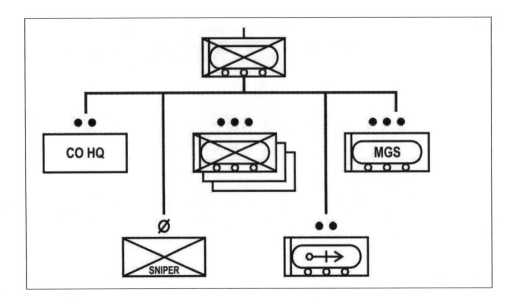

Figure 1-6. Stryker.

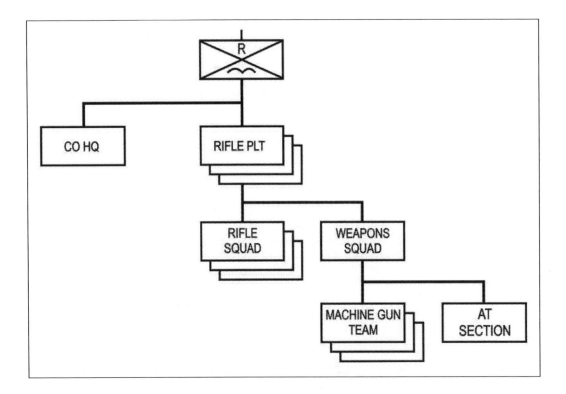

Figure 1-7. Ranger.

Section IV. DUTIES AND RESPONSIBILITIES OF KEY PERSONNEL

This section describes the duties and responsibilities of key personnel in the Infantry rifle company. Duties and responsibilities of the executive officer (XO) and first sergeant (1SG) may vary in the heavy and Stryker Infantry company.

COMPANY COMMANDER

1-35. The company commander leads by personal example and is responsible for everything the company does or fails to do. His principle duties include the key areas of tactical employment, training, administration, personnel management, maintenance, force protection, and sustainment of his company. Given the asymmetrical, noncontiguous environment, he must now integrate and synchronize a greater mix of forces for full spectrum operations including other combined arms and combat support elements, civil affairs (CA), psychological operations (PSYOP), interpreters, media, unmanned aerial system (UAS) and robotics teams. Among other things, he--

- Commands and controls through his subordinate leaders.
- Employs his company to accomplish its mission according to the battalion commander's intent and concept.
- Selects the best location to maneuver the platoons and other elements.
- Conducts mission analysis and troop-leading procedures (TLP) and issues operation orders for company tactical operations.
- Maintains and expresses situation awareness and understanding.
- Resources the platoons and other elements and requests battalion support when needed.
- Ensures that the company command post (CP) effectively battletracks the situation and status.
- Provides a timely and accurate tactical picture to the battalion commander and subordinate units.
- Implements effective measures for force protection, security, and accountability of forces and systems.
- Develops the leadership and tactical skill of his platoon leaders.

EXECUTIVE OFFICER

1-36. The XO is second in command. His primary role is to assist the commander in mission planning and accomplishment. He assumes command of the company as required and ensures that tactical reports from the platoons are forwarded to the battalion tactical operations center (TOC). The XO locates where he can maintain communications with the company commander and the battalion. He--

PLANS AND SUPERVISES

1-37. Plans and supervises, before the battle along with the 1SG, the company's sustainment operations; ensures that precombat inspections are complete. The XO plans and coordinates logistical support with agencies outside the company while the 1SG does the same internally. He prepares or aids in preparing paragraph 4 of the company operation order (OPORD). He may also help the company commander plan the mission.

COORDINATES

1-38. Coordinates with higher, adjacent, and supporting units. He may aid in control of critical events of the battle such as a passage of lines, bridging a gap, or breaching an obstacle; or, he may assume control of a platoon attached to the company during movement.

Performs as Landing Zone or Pickup Zone Control Officer

1-39. This may include straggler control, casualty evacuation, resupply operations, or air-ground liaison.

Leads Quartering Party or Detachment

1-40. The XO might lead a quartering party, an element consisting of representatives of various company elements. Their purpose is to precede the company and reconnoiter, secure, and mark an assembly area.

1-41. The XO might lead a detachment with other tactical tasks including shaping or sustaining force leader in a company raid or attack, control company machine guns, or mortar section. He may also--

- Lead the reserve.
- Lead the detachment left in contact during a withdrawal.
- Control attachments to the company.
- Serve as movement control officer.

FIRST SERGEANT

1-42. The 1SG is the senior noncommissioned officer (NCO) and normally the most experienced Soldier in the company. He is the commander's primary tactical advisor and the expert on individual and NCO skills. He helps the commander plan, coordinate, and supervise all activities that support the unit mission. He operates where the commander directs or where he can best influence a critical point or what is viewed as the unit's decisive point. The first sergeant--

- Supervises routine operations. This can include enforcing the tactical standing operating procedures (TSOP); planning and coordinating both training and full spectrum operations; and administering replacement operations, logistics, maintenance, communications, field hygiene, and casualty evacuation operations.
- Supervises, inspects, and influences matters designated by the commander as well as areas that depend on his expertise such as Soldier care, force protection, security, and accountability.
- Assists the XO and keeps himself prepared to assume the XO's duties, if needed.
- Leads task-organized elements or subunits for the company's shaping effort or other designated missions.

PLATOON LEADER

1-43. The platoon leader (PL) leads his soldiers by personal example. He is responsible for all the platoon does or fails to do and has complete authority over his subordinates. This centralized authority enables the PL to maintain unit discipline and unity and to act decisively. The demands of modern combat or full spectrum operations require the PL to exercise initiative without continuous guidance from higher commands. He must know his Soldiers as well as how to employ the platoon, its weapons, and its systems. He relies on the expertise of the Platoon Sergeant and regularly consults with him on all platoon matters. As part of his key tactical responsibilities, the PL--

- Leads the platoon in accomplishing its mission according to the company and battalion commanders' intent and concept.
- Performs TLP for missions assigned to the platoon.
- Locates where he can best maneuver the squads and the fighting elements, and then synchronizes their efforts.
- Anticipates the platoon's next tactical move.
- Requests and controls assets.
- Ensures force-protection measures are implemented.

- Maintains all-round, three-dimensional security.
- Controls emplacement of key weapon systems.
- Ensures security measures are implemented at the limit of advance (LOA).
- Provides a timely and accurate tactical picture to the commander.

PLATOON SERGEANT

1-44. The platoon sergeant (PSG) is the platoon's most experienced NCO and second in command. He is accountable to the platoon leader for the leadership, discipline, training, and welfare of the platoon's Soldiers. He sets the example in everything. His expertise includes tactical maneuver, employment of weapons and systems, logistics, administration, security, accountability, force protection, and Soldier care. As the second in command, the PSG assumes no formal duties except those prescribed by the PL. As part of his traditional tactical responsibilities, the PSG--

- Locates and acts where best to help control the fight or other platoon operations; may lead either the shaping or sustaining operation.
- Assures that the platoon is prepared to accomplish its mission by supervising precombat checks and inspections.
- Helps develop the squad leaders' tactical and leadership skills.
- Supervises platoon sustainment operations.
    - Receives the squad leaders' administrative, logistical, and maintenance reports and requests for rations, water, fuel, and ammunition.
    - Coordinates with the company 1SG or XO for resupply.
    - Runs the platoon casualty collection point (CCP); directs the medic and aid and litter teams; forwards casualty reports; manages personnel strength levels, receives orients replacements.

FIRE-SUPPORT OFFICER

1-45. The fire support officer (FSO) helps plan, coordinate, and execute the company's fire support. During planning, he develops a fire support plan based on the company commander's concept and guidance. He coordinates the fire support plan with the battalion fire support officer (FSO). During planning, the FSO--

- Advises the commander of the capabilities and statuses of all available fire support assets.
- Helps the commander develop the OPORD to ensure full integration of fires into the concept. Refines field artillery and mortar targets to support the maneuver plan.
- Designates targets and fire control measures and determines method of engagement and firing responsibility.
- Determines the specific tasks and instructions required to conduct and control the fire plan.
- Briefs the fire support plan as part of the company OPORD, and coordinates with PLs to ensure they understand their fire support responsibilities.
- Integrates platoon targets into the company target overlay and target worksheet, and sends the resulting products to the battalion fire support element (FSE).
- During the battle, normally locates near the commander. This allows greater flexibility in conducting or adjusting the fire support plan. At times, locates away from the commander to better control supporting fires. Informs the commander of key information on the radio net.
- Understands Infantry tactics in order to integrate fires effectively, and if the company commander becomes a casualty, may assume temporary control of the company until the XO can do so.
- Coordinates the employment of the joint air attack team (JAAT), close air support (CAS), attack helicopter, and UASs.
- Ensures the indirect fire plan is part of each company rehearsal.

SENIOR RADIO OPERATOR

1-46. The senior radio operator supervises operation, maintenance, and installation of organic wire, and FM communications. This includes sending and receiving routine traffic and making required communication checks. He--

- Supervises the company command post.
  - Relays information.
  - Monitors the tactical situation.
  - Establishes the CP security plan and radio watch schedule.
  - Informs the commander and subordinate units of significant events.
- Renders clear, accurate, and timely situation reports (SITREPS).
- Performs limited troubleshooting of organic communications equipment. Serves as the link between the company and the battalion for communications equipment maintenance.
- Supervises all aspects of communications security (COMSEC) equipment, to include requesting, receipting, maintaining, securing, employing, and training for COMSEC equipment and related materials.
- Advises the company commander in planning and employing the communications systems. Based on the commander's guidance, he prepares or helps prepare paragraph 5 of the OPORD.

RADIO OPERATOR

1-47. The radio operator uses and performs maintenance on his assigned radio, including preparation for special missions (cold weather, air assault, or waterborne) and construction of field-expedient antennas. He--

- Understands the company's mission. If the commander becomes a casualty, the radio operator might be the only Soldier on the radio for a time. If so, he must be prepared to call for and adjust artillery, or to request medical evacuation or resupply.
- Assists in OPORD preparation by copying overlays and building a sand table.
- Assists the senior radio operator and is prepared to assume his duties.

SUPPLY SERGEANT

1-48. The supply sergeant requests, receives, issues, stores, maintains, and turns in supplies and equipment for the company. He coordinates requirements with the 1SG and the battalion S-4. He--

- Controls the supply trucks that are organic to the company.
- Monitors tactical situation.
- Anticipates logistical requirements (Chapter 11, *Sustainment Operations*).
- If located in the Infantry battalion support area, the headquarters and headquarters company (HHC) commander may provide guidance and assistance to supply sergeants.
- Communicates using the battalion administrative/logistical (A/L) radio and digital network.

CHEMICAL, BIOLOGICAL, RADIOLOGICAL, OR NUCLEAR NCO

1-49. The CBRN NCO helps the company commander plan chemical, biological, radiological, or nuclear (CBRN) and high-yield (CBRNE) weapons operations. He conducts and supervises CBRN training within the company (decontamination, monitoring, survey, and equipment maintenance operations) and inspects detection and protective equipment for serviceability. He--

- Operates forward with the company CP and helps the senior radio operator with CP operations and security.

- Recommends mission-oriented protective posture (MOPP) levels to the commander (based on guidance from the battalion CBRN NCO and the current situation).
- Conducts continuous CBRN vulnerability analysis.
- Ensures connectivity with the joint warning and reporting network (JWARN).
- Acts as liaison with supporting chemical units.
- Reports, analyzes, and disseminates CBRN attack data manually or digitally using the NBC Warning and Reporting System (NBCWRS), and NBC1, NBC4, and spot reports from the FBCB2 system.
- Plans and supervises decontamination and monitoring and survey operations.
- Requisitions CBRN equipment and supplies.

MORTAR SECTION LEADER

1-50. The mortar section leader is responsible for employing the mortar section and ensures effective mortar support for the company. He--.

- Helps the company commander plan the employment of the mortar section.
- Coordinates with the company FSO and fire support team (FIST).
- Controls the section during tactical operations.
- Acts as the primary trainer for mortar systems.

ARMORER

1-51. The armorer is a supply specialist whose duties focus on organizational maintenance and repair of the company's small arms weapons. He assures accountability and security of weapons and ammunition under his control and evacuates weapons to the DS maintenance unit, if required. Normally, he helps the supply sergeant in the brigade support area (BSA), but he may operate forward with the company CP to support continuous CP operations.

MEDIC

1-52. The senior trauma specialist (senior company medic) is attached to the rifle company to provide emergency medical treatment (EMT) for sick, injured, or wounded company personnel. Emergency medical treatment procedures performed by the trauma specialist may include opening an airway, starting intravenous fluids, controlling hemorrhage, preventing or treating for shock, splinting fractures or suspected fractures, and providing relief for pain. The EMT performed by the trauma specialist is under the supervision of the battalion surgeon or physician's assistant (PA). The senior trauma specialist or company medic--

- Oversees and provides guidance to each platoon medic as required.
- Triages injured, wounded, or ill friendly and enemy personnel for priority of treatment as they arrive at the company casualty collection point (CCP).
- Oversees sick call screening for the company.
- Requests and coordinates the evacuation of sick, injured, or wounded personnel under the direction of the company 1SG.
- Helps train company personnel in first aid and combat lifesaver techniques, including in enhanced first-aid procedures.
- Requisitions Class VIII supplies from the BSA for the company according to the TSOP.
- Recommends locations for company CCPs.
- Provides guidance to the company's combat lifesavers as required.

- Monitors the tactical situation, and anticipates and coordinates health service support (HSS) requirement and Class VIII resupply as necessary.
- Advises the company commander and 1SG on mass casualty operations.
- Keeps the 1SG informed on the status of casualties, and coordinates with him for additional HSS requirements.

Section V. COMBAT POWER, LEADERSHIP, AND WARFIGHTING FUNCTIONS

Combat power is the ability to fight. It is the aggregate of a unit's disruptive or destructive force. It is made up of six warfighting functions (WFF) tied together by leadership (Figure 1-8, page 1-18).

LEADERSHIP

1-53. Leadership is the least tangible and most dynamic element of combat power. Confident, audacious, and competent leadership focuses the other elements of combat power. It serves as the catalyst that creates conditions for success. Leaders inspire Soldiers to succeed. They provide purpose, direction, and motivation in all operations. Leadership is crucial. It often makes the difference between success and failure, particularly in small units.

1-54. A warfighting function is a group of tasks and systems (people, organization, information, and processes) united by a common purpose that commanders use to accomplish missions and training objectives. The warfighting functions are intelligence, movement and maneuver, fire support, protection, sustainment, and command and control. These warfighting functions replace the battlefield operating systems.

1-55. Commanders visualize, describe, direct, and lead operations and training in terms of the warfighting functions. Decisive, shaping, and sustaining operations combine all the warfighting functions. No function is exclusively decisive, shaping, or sustaining.

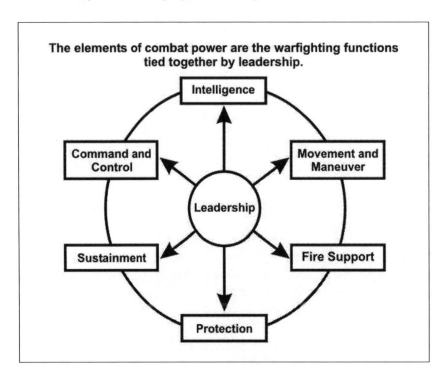

Figure 1-8. Elements of combat power.

INTELLIGENCE

1-56. The intelligence warfighting function is the related tasks and systems that facilitate understanding of the enemy, terrain, weather, and civil considerations. It includes those tasks associated with intelligence, surveillance, and reconnaissance. The intelligence warfighting function is a flexible and adjustable architecture of procedures, personnel, organizations, and equipment that provide relevant information and products relating to the threat, civil populace, and environment to commanders.

MOVEMENT AND MANEUVER

1-57. The movement and maneuver warfighting function is the related tasks and systems that move forces to achieve a position of advantage in relation to the enemy. It includes those tasks associated with employing forces in combination with direct fire or fire potential (maneuver), force projection (movement), and mobility and countermobility. Movement and maneuver are the means by which commanders concentrate combat power to achieve surprise, shock, momentum, and dominance.

FIRE SUPPORT

1-58. The fire support warfighting function is the related tasks and systems that provide collective and coordinated use of Army indirect fires, joint fires, and offensive information operations. It includes those tasks associated with integrating and synchronizing the effects of these types of fires with the other war-fighting functions to accomplish operational and tactical objectives.

PROTECTION

1-59. The protection warfighting function is the related tasks and systems that preserve the force so the commander can apply maximum combat power. Preserving the force includes protecting personnel (combatant and noncombatant), physical assets, and information of the United States and multinational partners. It includes the following task areas:

- Safety.
- Fratricide avoidance.
- Survivability.
- Air and missile defense.
- Antiterrorism.
- Counterproliferation and consequence management actions associated with chemical, biological, radiological, nuclear, and high-yield explosive weapons.
- Defensive information operations.
- Force health protection.

SUSTAINMENT

1-60. The sustainment warfighting function is the related tasks and systems that provide support and services to ensure freedom of action, extend operational reach, and prolong endurance. It includes those tasks associated with—.

- Maintenance.
- Transportation.
- Supply.
- Field services.
- Explosive ordnance disposal.

- Human resources support.
- Financial management.
- Health service support.
- Religious support.
- Band support.
- Related general engineering.

1-61. Sustainment allows uninterrupted operations through adequate and continuous logistical support such as supply systems, maintenance, and other services.

COMMAND AND CONTROL

1-62. The command and control warfighting function includes the related tasks and systems that support commanders in exercising authority and direction. It includes the tasks of acquiring friendly information, managing relevant information, and directing and leading subordinates.

1-63. Command and control has two parts: the commander and the command and control (C2) system. Information systems—including communications systems, intelligence-support systems, and computer networks—back the command and control systems. They let the commander lead from anywhere in their AO. Through command and control, the commander initiates and integrates all warfighting functions.

Chapter 2
Troop-Leading Procedures

The methods described in this chapter are guides for the commander to apply based on his situation, his experience, the experience of his subordinate leaders, and key planning concepts. The tasks involved in some steps, such as issuing the warning order, initiating movement, and reconnoitering, can recur several times. The last steps, supervising and refining the plan, occur throughout the ttroop-leading procedures.

The planning steps in the TLP reflect, but do not duplicate, those in the military decision-making process (MDMP) (FM 5-0). Some steps in the MDMP help the battalion or higher commander coordinate staff and commander responsibilities of units with staffs. However, leaders from company level down have no staff officers, so they must conduct their own planning. The TLP reflect this reality, yet they manage to incorporate the spirit, language, and general process of the MDMP. These help the commander prepare orders (WARNOs, OPORDs, and FRAGOs). The sections in this chapter discuss the steps of the TLP.

This edition expands the discussion of battle command and the TLP to address analysis of mission, enemy, terrain (and weather), troops (and support) available, time available, and civil considerations (METT-TC), key planning concepts, common pitfalls, and its integration into the military decision making process.

Section I. OVERVIEW

Battle command is the application of leadership to combat power. It is an art that relies on command skills. These skills are developed over time through study, practice, and judgment. The commander visualizes the operation, describing it in his intent and concept of the operation, and he directs the actions of subordinates *within* his intent. He directly influences operations by his personal presence and his command and control (C2) system. He uses the WFF to organize, prepare, coordinate, integrate, synchronize, and execute his plan. That is, he considers everything he has or knows for or about an operation for each WFF.

BATTLE COMMAND

2-1. Visualizing, describing, and directing are aspects of leadership common to every commander. Technology, the fluid nature of operations, the increased volume of information that commanders must process, and today's battlefield underline the importance of the commander's ability to visualize, describe, and direct operations. Assessment is also an integral part battle command. Commanders must continually assess the threat, friendly forces, and effects throughout all three aspects battle command.

VISUALIZE

2-2. On receipt of a mission, the Infantry company commander considers his battlespace and conducts a mission analysis. He uses the following to develop his initial vision, which he continually confirms or adapts.

- The factors of METT-TC.
- Elements of the operational framework.
- Staff estimates from higher headquarters.
- Input from other subordinates.
- His own experiences and judgment.

DESCRIBE

2-3. The commander uses the operational framework to relate decisive, shaping, and sustaining operations to time and space. For all operations, *purpose* and *time* determine the allocation of space. The commander clarifies his description as circumstances develop. He emphasizes how the combination of decisive, shaping, and sustaining operations relates to accomplishing the purpose of the overall operation. The commander describes his vision in the statement of his (commander's) intent and concept of the operation. He chooses terms and graphics that suit the nature of the mission and his experience.

DIRECT

2-4. The commander directs throughout the operations process. His directions take different forms during planning, preparation, and execution. Throughout the operation, he makes decisions and directs actions based on his understanding of the situation. He stays current on the situation by continuously assessing it.

ARMY PLANNING PROCESS

2-5. The standard Army planning process has five interrelated subprocesses: mission analysis, course of action (COA) development, COA analysis, COA comparison, and COA selection. Figure 2-1 shows this process within the TLPs. The goal of Army planning is to develop unique solutions to unique tactical problems. By tailoring solutions, leaders avoid setting patterns. This prevents the enemy from predicting friendly actions.

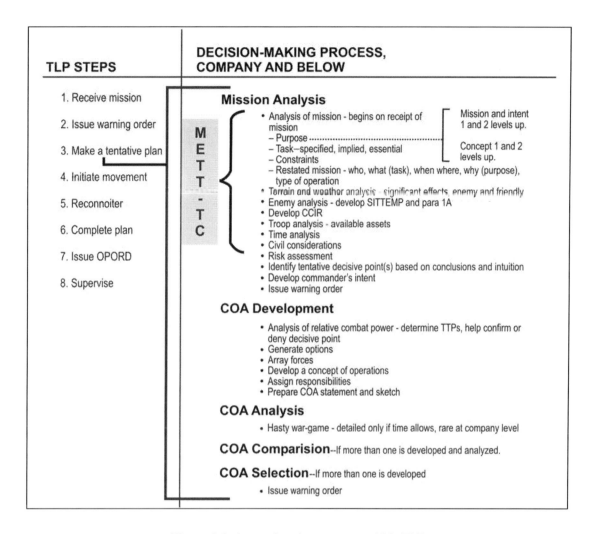

Figure 2-1. Army planning process within TLP.

KEY PLANNING CONCEPTS

2-6. Planning at the small Infantry unit level is seldom very detailed. Small-unit leaders rely more on thoroughly rehearsing unit SOPs, techniques, procedures, and drills. However, based on the mission and on the next higher commanders' concepts of the operations, company commanders *might* plan in detail. They use the acrostic "*PLANNING*" to remember these key planning concepts (Figure 2-2, page 2-4).

```
            PLANNING
  Parallel plan
  Limit (assess and manage) risk
  Approach–sequencing operations
  Nested concepts
  Necessary control measures
  ISR
  Never violate the 1/3-2/3 rule
  Go to the end state first–reverse plan
```

Figure 2-2. Key planning concepts.

Parallel Plans

2-7.   Parallel planning occurs when two or more echelons plan for the same operation at about the same time. Parallel planning is easiest when higher headquarters continuously shares information on future operations with subordinate units (FM 5-0 and Figure 2-3). Rather than waiting until higher finishes planning, effective leaders start developing their units' missions as they receive information. They flesh out their missions as they learn more. They start by identifying their units' missions, stating their intents, and ensuring that they reflect the operational concepts of their higher and second higher headquarters. They choose the tasks most likely to be assigned to their units, and then they develop mission statements based on the information they have received. At all levels, developing and describing the vision of Infantry leaders requires time, explanation, and ongoing clarification. All leaders understand that their next higher commander's concept of operations (CONOP) will continue to mature, and that they must continue parallel planning as it does so, up until execution.

Troop-Leading Procedures

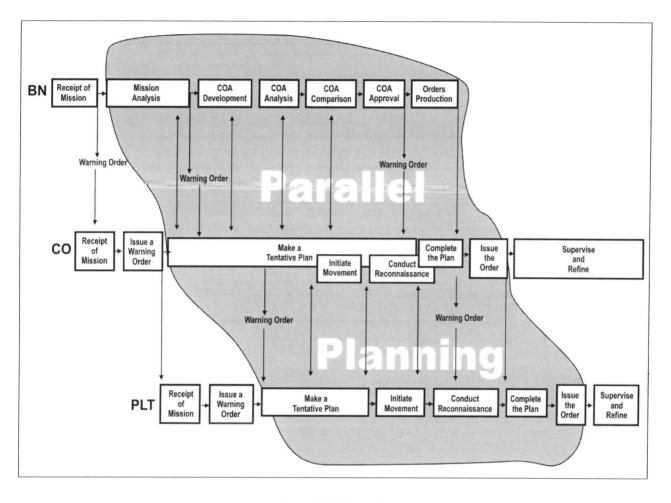

Figure 2-3. Parallel planning.

Risk Assessment

2-8. This starts during planning and is an inherent aspect of the operation process. On the battlefield, risk assessment (Appendix A, *Risk Management, Fratricide Avoidance, and the Effects of Continuous Operations*) is the leaders' best tool for identifying tactical hazards, and for reducing the risks to both their units and to the mission itself. While this tool helps control the accidental hazards present in all operations, its real value lies in helping the leaders identify and control *tactical* risks.

Approach–Sequencing Operations

2-9. Part of the art of planning is to determine the sequence of activities that will accomplish the mission most efficiently (FM 5-0).

Sequential Approach

2-10. This approach accomplishes tasks step-by-step. Conventional, sequential operations are easy to predict because they establish patterns.

Chapter 2

Simultaneous Approach

2-11. This approach accomplishes many things at once. Simultaneous operations are difficult to predict due to their suddenness. For the same reason, they are difficult to control.

Battlefield Organization

2-12. As part of the troop-leading procedures, commanders visualize their battlespace and determine how to arrange their forces. The battlefield organization is the allocation of forces in the AO by purpose. It consists of three all-encompassing categories of operations: decisive, shaping, and sustaining. Purpose unifies all elements of the battlefield organization by providing the common focus for all actions.

Decisive Operations

2-13. Decisive operations are those that directly accomplish the mission assigned by the higher headquarters. Decisive operations conclusively determine the outcome of major operations, battles, and engagements. There is only one decisive operation for any major operation, battle, or engagement for any given echelon. The main effort and the decisive operation are not always the same. Commanders anticipate shifts of main efforts throughout an operation and include those associated shifts in priorities, assets, and resources to them (subordinates) in the plan. In contrast, changing the decisive operation requires execution of a branch, sequel, or new plan. A shaping operation may be the main effort before execution of the decisive operation. However, the decisive operation becomes the main effort upon execution.

Shaping Operations

2-14. Shaping operations at any echelon create and preserve conditions for the success of the decisive operation. Shaping operations include lethal and nonlethal activities conducted throughout the AO. They support the decisive operation by affecting the enemy capabilities and forces, or by influencing enemy decisions.

Sustaining Operations

2-15. Sustaining operations are operations at any echelon that enable shaping and decisive operations by providing sustainment, rear area and base security, movement control, terrain management, and infrastructure development.

Nested Concepts

2-16. A nested concept is a planning technique to achieve unity of purpose. Each subsequent echelon's concept of operations is embedded in the other (Figure 2-4). When developing their concepts of the operation, leaders ensure that their concepts are nested within those of their higher headquarters. They also ensure that subordinate unit missions are unified by task and purpose to accomplish the mission.

Troop-Leading Procedures

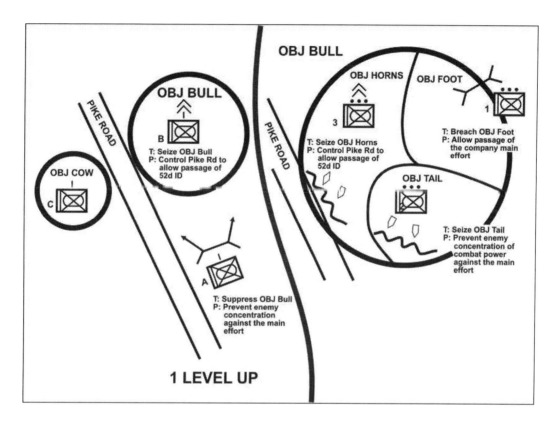

Figure 2-4 Nesting of concepts

Control Measures

2-17. Control measures are directives communicated graphically or orally by leaders to their subordinates. Commanders use them to assign responsibilities, coordinate fires and maneuver, and control operations. Each control measure can be shown graphically. In general, all control measures should be easily identifiable on the ground.

Intelligence, Surveillance, and Reconnaissance

2-18. Intelligence, surveillance, and reconnaissance (ISR) is an enabling operation that integrates and synchronizes all warfighting functions to collect and produce relevant information to help the commander make decisions (FM 1-02). Unit staffs use ISR operations to find the enemy. Although small Infantry units certainly conduct this critical function, conducting an ISR operation is beyond their scope and abilities. However, larger units can assign surveillance and reconnaissance missions to small Infantry units in support of the larger ISR operations.

One-Third/Two-Thirds Rule

2-19. Leaders follow the "*one-third/two-thirds*" rule to allocate time available for planning and preparation. This means leaders use no more than one-third of usable, available planning time. They leave the remaining two-thirds for their subordinates. This rule requires efficiency and discipline by Infantry leaders. However, they use the remaining two-thirds available time to further elaborate, refine, and strengthen their vision of the upcoming operation.

Chapter 2

End State

2-20. End state is what the commander wants the situation to be when operations conclude—both military operations, as well as those where the military is in support of other instruments of national power.

Reverse Planning

2-21. Reverse planning involves starting with the operation's end state and working backward in time. Leaders begin by identifying the last step, the next-to-last step, and so on. They continue until they reach the step that begins the operation. It answers the question--

Where do we eventually want to be?

Forward Planning

2-22. However, there are times when forward planning is the best technique for the situation. Forward planning involves starting with the present conditions and laying out potential decisions and actions forward in time, identifying the next feasible step, the next after that, and so on. In forward planning, the envisioned end state serves as a distant and general aiming point rather than as a specific objective. Forward planning answers the question--

Where can we go next?

COMMON PITFALLS

2-23. Some common planning pitfalls for leaders to avoid are—.

- Attempting to forecast and dictate events too far into the future.
- Delaying planning to gain more detailed information.
- Planning in too much detail.
- Using planning as a scripting process.
- Applying planning techniques inflexibly.

Section II. STEPS 1 AND 2--RECEIVE MISSION, ISSUE WARNING ORDER

In Step 1 of the TLP, the leaders *determine their units' missions* and *assess the time available to accomplish them*. They can conduct an initial (light) analysis of the order using the factors of METT-TC. They conduct detailed METT-TC analyses only *after* they issue the first warning order (Step 2). Rarely will they receive their missions until after higher headquarters issues the third warning orders or the OPORDs themselves. However, in the course of parallel planning, small-unit leaders will have already deduced their tentative missions.

STEP 1--RECEIVE MISSION

2-24. Leaders can receive their missions in several ways. They can get them in the form of warning orders (WARNOs) or, if higher chooses to wait for more information, an actual operation order (OPORD). Sometimes higher chooses not to send WARNOs, opting instead to wait and send a full OPORD. Worst case, leaders receive new missions due to situational changes that occur during the execution of a prior mission. In addition to receiving (or deducing) their missions during this step, the leaders must also--

- Assess the time available to prepare for and execute the mission.
- Prepare an initial timeline for planning and executing the mission.
- Conduct an initial planning-time analysis.
- Determine the total amount of time to plan and prepare.

- As planning continues, use the initial planning-time analysis to conduct a detailed time analysis.
- Analyze the time his unit has available.
- Prepare an initial timeline.

2-25. The most important element of the leaders' warning orders is the initial timeline for planning. They may also convey any other instructions or information that they think will help their subordinates prepare for the upcoming mission.

STEP 2--ISSUE WARNING ORDER

2-26. A warning order is a preliminary notice of an order or action that is to follow (FM 1-02). Though less detailed than a complete OPORD, a warning order aids in parallel planning. After the leaders receive new missions and assess the time available for planning, preparing, and executing the mission, they immediately issue warning orders to their subordinates. By issuing the initial warning orders as quickly as possible, they enable subordinates to begin their own planning and preparation (parallel planning) while he begins to develop the OPORD. When he obtains more information, he issues updated warning orders, giving subordinates as much as he knows.

2-27. Leaders can issue warning orders to their subordinates right after they receive higher headquarters' initial warning orders. In their own initial warning orders, they include the same elements given in their higher headquarters' initial warning orders. If practical, leaders brief their subordinate leaders face-to-face, on the ground. Otherwise, they use a terrain model, sketch, or map.

FIRST WARNING ORDER

2-28. The first warning order follows the five-paragraph OPORD format and includes the following items, at a minimum.

- Type of operation.
- General location of operation.
- Initial operational timeline.
- Reconnaissance to initiate.
- Movement to initiate.
- Planning and preparation instructions (to include planning timeline).
- Information requirements.
- Commander's critical information requirements.

SECOND WARNING ORDER

2-29. Infantry company commanders second warning orders include essential information from their mission analyses and additional guidance from battalion. They must understand the information from their highers' second warning orders. They assess the situation as best they can, but can probably complete detailed mission analyses only after they receive the actual OPORDs. Their second warning orders contain the following information from their own mission analyses.

- Terrain analysis.
- Enemy forces (para 1a of higher's OPORD, including enemy SITEMP).
- Higher headquarters' restated mission.
- Higher commander's intent.
- Areas of operation and interest.
- CCIR and EEFI.
- Risk guidance.
- Surveillance and reconnaissance to initiate.

Chapter 2

- Security measures.
- Deception guidance.
- Mobility and countermobility.
- Specific priorities.
- Updated operational timeline.
- Guidance on rehearsals.

2-30. Higher headquarters can issue additional information in the second warning order. For example, they can add a friendly forces paragraph (1b). Company commanders might also determine that they need to issue a second warning order, either after they receive the battalions' second warning order, or after they receive other pertinent information. Normally, they have received neither their battalion commanders' concepts of the operation nor their companies' missions. Depending on their situation, they might wait to issue their second company warning orders until after they receive the higher headquarters' third one.

THIRD WARNING ORDER

2-31. The third warning order is normally issued after the COA is finalized. For battalion level and up, that is, units with staffs, this occurs after COA approval. For company level and below, this normally occurs earlier, after COA development or analysis. This warning order contains the--

- Mission.
- Commander's intent.
- Updated CCIR and EEFI.
- Concept of operation.
- Areas of operation and interest.
- Principle tasks assigned to subordinates.
- Preparation and rehearsal instructions not covered in SOPs.
- Finalized operational timeline.

Section III. STEP 3--MAKE A TENTATIVE PLAN

Making a tentative plan is TLP Step 3. In a time-constrained environment, an Infantry company commander typically develops only one course of action (COA). However, as time permits, he can develop as many COAs, for comparison purposes, as time allows. He begins TLP Step 3 after he issues his own warning order, and after he has received higher headquarters' third warning order, or until he has enough information to proceed. He need not wait for a complete OPORD before starting to develop his own tentative plan. TLP Step 3 mirrors the five steps of the Army planning process.

MISSION ANALYSIS

2-32. The Infantry company commander begins his mission analysis when he receives the mission. During his mission analysis, he—.

- Restates the mission.
- Conducts an initial risk assessment.
- Identifies a tentative decisive point.
- Defines his own (commander's) intent.

2-33. He conducts the mission analysis to help him start developing his vision, and to confirm what he must do to accomplish his mission. At the lower levels, leaders conduct their mission analyses by evaluating the factors of METT-TC. They make significant deductions about the terrain, enemy, and own forces that most affect tactical operations. These significant deductions drive the planning process and the execution of operations. Leaders must convey to their subordinates the importance of these deductions, and the effect they will have on the units operations. In the end, the usefulness of mission analysis lies in recognizing and capitalizing on opportunities. The answers to the following questions become inputs into developing a COA. Mission analysis has no time standard. Leaders may take as much time as needed, while still adhering to the one-third/two-thirds rule. Mission analysis answers the four questions of the leader's battlefield vision.

- What is my mission?
- What is the current situation?
- How do we accomplish the mission?
- What are the risks?

METT-TC

2-34. Analyzing the factors of METT-TC is a continuous process (Figure 2-5). Leaders constantly receive information, from the time that they begin planning through execution. During execution, their continuous analyses enable them to issue well-developed fragmentary orders. They must assess if the new information affects their missions and plans. If so, then they must decide how to adjust their plans to meet these new situations. They need not analyze the factors of METT-TC in a particular order. How and when they do so depends on when they receive information as well as on their experience and preferences. One technique is to parallel the TLP based on the products received from the higher headquarters' MDMP. Using this technique, he would, but need not, analyze mission first; followed by terrain and weather; enemy; troops and support available; time available; and finally civil considerations.

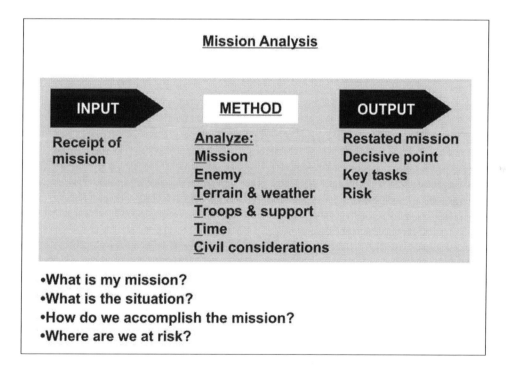

Figure 2-5. Analysis of mission using METT-TC.

Analysis of Mission

2-35. A mission is the task and purpose that clearly indicate the action to be taken and the reason for the action. In common usage, especially when applied to lower military units, a mission is a duty or task assigned to an individual or unit (FM 1-02). The mission is always the first factor leaders consider and the lens through which they view all the aspects of the operation. To analyze the mission, leaders answer the most basic question: What have I been told to do, and why?

2-36. Leaders at every echelon must understand the mission, intent, and operational concept one and two levels higher. This understanding makes it possible to exercise disciplined initiative. Leaders capture their understanding of what their units are to accomplish in their revised mission statements. They take five steps to fully analyze their assigned mission as directed from the higher headquarters.

- Higher headquarters' (two levels up) mission, intent, and concept.
- Immediate higher headquarters' (one level up) mission, intent, and concept.
- Unit's purpose.
- Constraints.
- Specified, implied, and the essential task(s).
- Restated mission.

Higher Headquarters' (Two Levels Up) Mission, Intent, and Concept

2-37. Leaders understand their second highers' concepts of the operation. They identify the tasks and purposes, and how their immediate highers are contributing to the fight. They must also understand the leaders' intent (two levels up).

Immediate Higher Headquarters' (One Level Up) Mission, Intent, and Concept

2-38. Leaders understand their immediate headquarters' concept of the operation. They identify their headquarters' tasks and purposes as well as their own contributions to this fight. They must clearly understand their immediate highers' intent from the OPORD. Also, they identify the tasks, purposes, and dispositions for all adjacent maneuver elements under that headquarters' control.

Unit's Purpose

2-39. Leaders find their units' purposes in the concepts of the operation in the immediate higher headquarters' OPORDs. The purpose of the decisive operation usually matches or achieves the purpose of the immediate higher headquarters. Similarly, shaping operation purposes must relate directly to those of the decisive operation. Sustaining operation purposes relate directly to those of the decisive and shaping operations. Leaders must understand how their units' purposes relate to higher's. They must understand why their leaders one level up assigned their units' particular purposes. Then, they determine how those fit into their commanders' concepts of the operation.

Constraints

2-40. Constraints either prohibit or require an action. Leaders identify all constraints the OPORD places on their units' ability to execute their missions. The two types of constraints are proscriptive (required; mandates action) and prohibitive (not allowed; limits action).

Tasks

2-41. Leaders must identify and understand the tasks required to accomplish a given mission. The three types of tasks are specified, implied, and essential.

Specified Tasks

2-42. Specified tasks are specifically assigned to a unit by a higher headquarters and are found throughout the OPORD. Specified tasks may also be found in annexes and overlays, for example--

- *"Seize OBJ FOX."*

- *"Reconnoiter route BLUE."*
- *"Assist the forward passage of B Company."*
- *"Send two Soldiers to assist in the loading of ammunition."*

Implied Tasks

2-43. Implied tasks are those that must be performed to accomplish a specified task, but that are not stated in a higher headquarters' order. Implied tasks derive from a detailed analysis of higher's order, from the enemy situation and COAs, from the terrain, and from a knowledge of doctrine and history. Analyzing the unit's current location in relation to future areas of operation as well as the doctrinal requirements for each specified task might reveal the implied tasks. Only those that require resources should be used. For example, if the specified task is "*Seize Objective Fox,*" and new intelligence has OBJ FOX surrounded by reinforcing obstacles, this intelligence would drive the implied task of "*Breach reinforcing obstacles vic Objecive Fox.*"

Essential Task

2-44. The essential task is the tactical mission task--it accomplishes the assigned purpose. It, along with the company's purpose, is usually assigned by the higher headquarters' OPORD in *Concept of Operations* or *Tasks to Maneuver Units*. For the decisive operation, since the purposes are the same (nested concept), the essential task also accomplishes the higher headquarters' purpose. For shaping operations, it accomplishes the assigned purpose, which shapes the decisive operation. For sustaining operations, it accomplishes the assigned purpose, which enables both the shaping and decisive operation (again, nested concept).

Restated Mission

2-45. Leaders conclude their mission analyses by restating their missions. To do this, they answer the five Ws:

- Who (the company).
- What (the unit's essential task and type of operation).
- When (this is the time given in the battalion OPORD).
- Where (the objective or location stated in higher's OPORD), and.
- Why (the company purpose, taken from higher's concept of the operation).

> **EXAMPLE**
>
> C Company/1-87 IN *(who)*
>
> Attacks to seize *(what)*
>
> OBJ FOX (NB123456) *(where)*
>
> NLT 010200 October ___ *(when)*
>
> To prevent enemy forces from counterattacking into the battalion's decisive operation *(why)*
>
> C Company/1-87 IN
>
> Essential task: seize, type of operation–attack
>
> OBJ Fox (NB123456)
>
> 010200 October ___ (attack time)
>
> To prevent enemy forces from counterattacking into the battalion's decisive operation

Analysis of Terrain and Weather

2-46. When they analyze terrain, leaders consider man-made features and their effects on natural terrain features and climate. Leaders also consider the effects of man-made and natural terrain in conjunction with the weather on friendly and enemy operations. In general, terrain and weather do not favor one side over the other unless one is better prepared to operate in the environment or is more familiar with it. The terrain, however, may favor defending or attacking. Analysis of terrain answers the question: *What is the terrain's effect on the operation?* Leaders analyze terrain using the categories of observation and fields of fire, avenues of approach, key terrain, observation, and cover and concealment (OAKOC).

2-47. From the modified combined obstacle overlay (MCOO) developed by higher headquarters, leaders already appreciate the general nature of the ground and the effects of weather. However, they must conduct their own detailed analyses to determine how terrain and weather uniquely affects their units' missions and the enemy. They must go beyond merely passing along the MCOO to their subordinate leaders and making general observations of the terrain such as "*This is high ground,*" or "*This is a stream.*" They must Figure out how the terrain and weather will affect the enemy and their units. Even more, they apply these conclusions when they develop COAs for both enemy forces and their units. At company level and below, leaders develop a graphic terrain analysis overlay (GTAO). This product is very similar to the MCOO in that it shows the critical military aspects of terrain. Not only does it facilitate planning, but it also aids in briefing subordinates.

Definition of Battlefield

2-48. For leaders to have *starting points* for terrain analysis, they must first define their battlefields. They must know their areas of operations and interest (AO and AI, Figure 2-6).

Troop-Leading Procedures

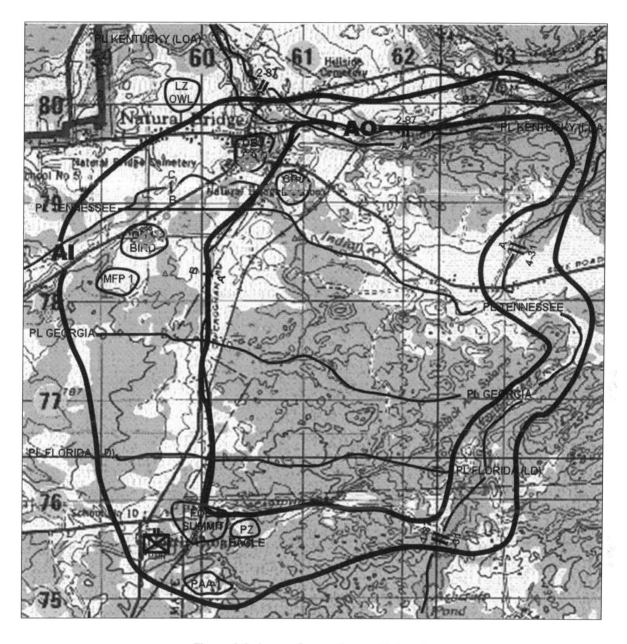

Figure 2-6. Areas of operation and interest.

Area of Operations

2-49. Higher commanders use boundaries to define their Infantry companies' AOs. Assigning AOs to subordinates lets the subordinates use their initiative and supports decentralized execution.

Area of Interest

2-50. An AI is a geographical area, usually larger than the leader's AO. The AI includes any threat forces or other elements that characterize the battlefield environment and that greatly influence the accomplishment of the mission.

Prioritization of Terrain Analysis

2-51. Limited planning time forces leaders to prioritize their terrain analyses. For example, in the conduct of attacks, leaders might prioritize the areas immediately around their objective for analysis, followed by the company's specific axis leading to the objective. Given more time, they might analyze the remainder of their companies' AOs and AIs.

Visual Aids

2-52. Leaders prepare a graphic depiction of terrain (GDOT) to help explain their findings about the effects of terrain and weather on the mission. The GDOT can be a photograph, overlay for a map sheet, or a terrain model. In it, leaders show terrain mobility classifications, key terrain, intervisibility lines, known obstacles, avenues of approach, and mobility corridors.

OAKOC

2-53. The military aspects of terrain (OAKOC) are used to analyze the ground. The sequence can vary (Figure 2-7). The leader determines the effects of each aspect of terrain on both friendly and enemy forces. These effects translate directly into conclusions that can apply to friendly or enemy COAs. Even if time is tight, the leader should allocate as much time as possible to factor, starting at the objective area, and then analyzing other aspects of key terrain. Terrain and weather are the most important aspects. Conclusions include at least the following.

- Effective templating of enemy forces and key weapon systems.
- Effective positioning of own assets.
- Understanding of time and space relationships of events, leading to thorough contingency plans.
- Effective echeloning and identifying of enemy observation and indirect fires.
- Effective selecting of movement techniques and formations, to include when to transition to tactical maneuver.

```
OAKOC

OBSTACLES
AVENUES OF APPROACH
KEY TERRAIN
OBSERVATION AND FIELDS OF FIRE
COVER AND CONCEALMENT
```

Figure 2-7. Military aspects of terrain.

Obstacles

2-54. The leader identifies existing (inherent to terrain and either natural or man-made) and reinforcing (tactical or protective) obstacles that limit mobility in his AO. Reinforcing obstacles are constructed, emplaced, or detonated by military force.

Existing Obstacles, Natural

2-55. Natural obstacles include--

- Rivers.
- Forests.
- Mountains.
- Ravines.
- Gaps and ditches over 3 meters wide.
- Tree stumps and large rocks over 18 inches high.
- Forests with trees 8 inches or more in diameter, with less than 4 meters between trees.

Existing Obstacles, Man-Made

2-56. The types of man-made obstacles include--

- Towns.
- Canals.
- Railroad embankments.
- Buildings.
- Power lines.
- Telephone lines.

Reinforcing Obstacles, Tactical

2-57. Tactical (reinforcing) obstacles inhibit the ability of the opposing force to move, mass, and reinforce. Examples include--

- Mine fields (conventional and situational).
- Antitank ditches.
- Wire obstacles.

Reinforcing Obstacles, Protective

2-58. Protective (reinforcing) obstacles offer close-in protection and are key to survivability.

Offensive and Defensive Considerations

2-59. Table 2-1 shows several offensive and defensive factors the leader can consider when analyzing obstacles and restricted terrain.

OFFENSIVE CONSIDERATIONS

- How is the enemy using obstacles and restricted terrain features?
- What is the composition of the enemy's reinforcing obstacles?
- How will obstacles and terrain affect the movement or maneuver of the Infantry company?
- If necessary, how can I avoid such features?
- How do I detect and, if desired, bypass the obstacles?
- Where has the enemy positioned weapons to cover the obstacles, and what type of weapons is he using?
- If I must support a breach, where is the expected breach site, and where will the enemy overwatch the obstacle?
- How will the terrain affect the employment of mortars, machine guns, and Javelin missiles?

DEFENSIVE CONSIDERATIONS

- Where does the enemy want to go? Where can I kill him? How do I get him to go there?
- How will existing obstacles and restricted terrain affect the enemy?
- How can I use these features to force the enemy into its engagement area, deny him an avenue, or disrupt his movement?
- How will the terrain affect the employment of mortars, machine guns, and Javelin missiles?

Table 2-1. Factors to consider in analyzing obstacles and restricted terrain.

Identified Obstacles

2-60. Figure 2-8 shows identified obstacles.

Troop-Leading Procedures

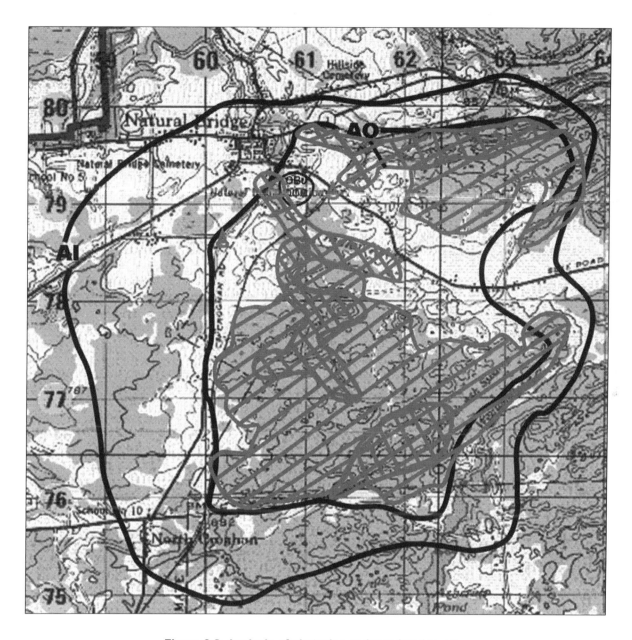

Figure 2-8. Analysis of obstacles and restricted terrain.

Categories of Terrain

2-61. Terrain is further classified in one of the following categories.

Unrestricted Terrain

2-62. This terrain is free of any restrictions to movement, so no actions are needed to enhance mobility. For armored and mechanized forces, unrestricted terrain is typically flat or moderately sloped, with scattered or widely spaced obstacles such as trees or rocks. This terrain generally allows wide maneuver and offers unlimited travel over well-developed road networks. It allows the Infantry company to move with little hindrance.

Restricted Terrain

2-63. This terrain hinders movement somewhat. Little effort is needed to enhance mobility, but units might have to zigzag or make frequent detours. They could have a hard time maintaining optimum speed, moving in some types of combat formations, or transitioning from one formation to another. For armor and mechanized forces, restricted terrain typically means moderate to steep slopes or moderate to dense spacing of obstacles such as trees, rocks, or buildings. Swamps and rugged ground are two examples of restricted terrain for Infantry forces. Poorly developed road systems may hamper logistical or rear area movement.

Severely Restricted Terrain

2-64. This terrain severely hinders or slows movement in combat formations unless some effort is made to enhance mobility. Engineer forces might be needed to improve mobility. Or, Infantry companies might have to deviate from doctrinal tactics. For example, they might have to move in columns rather than in lines. Or, they might have to move much more slowly than they would like. For armor and mechanized forces, steep slopes, densely spaced obstacles, and the absence of a developed road system characterize severely restricted terrain.

Avenues of Approach

2-65. An avenue of approach is an air or ground route of an attacking force leading to an objective or key terrain. Avenues of approach are classified by type (mounted, dismounted, air, or subterranean), formation, and speed of the largest unit that can travel on it. First, the leader must identify mobility corridors, if not provided by the higher headquarters. These are areas where a force can move in a doctrinal formation at a doctrinal rate of march (FM 34-130). Mobility corridors are classified by type and size of the force and formation employed, for example--

- A motorized rifle platoon (MRP) moving in column (MRP column).
- An enemy squad moving in a wedge (dismounted squad wedge).

2-66. The leader groups mutually supporting mobility corridors to form an avenue of approach. If he has no mutually supporting mobility corridors, then a single mobility corridor might become an avenue of approach. Avenues of approach are classified the same as mobility corridors. After identifying these avenues, the leader evaluates each and determines its importance. Table 2-2 shows several offensive and defensive considerations that the Infantry rifle leader can include in his evaluation of avenues of approach.

Offensive Considerations
• How can I use each avenue of approach to support my movement and maneuver?
• How will each avenue support movement techniques, formations and, once we make enemy contact, maneuver?
• Will variations in trafficability force changes in formations or movement techniques, or require clearance of restricted terrain?
• What are the advantages and disadvantages of each avenue?
• What are the enemy's likely counterattack routes?
• What lateral routes could we use to shift to other axes, and which could the enemy use to threaten our flanks?
• How will each avenue of approach affect the rate of movement of each type force?
Defensive Considerations
• What are all likely enemy avenues into my sector?
• How can the enemy use each avenue of approach?
• What lateral routes could the enemy use to threaten our flanks?
• What avenues would support a friendly counterattack or repositioning of forces?

Table 2-2. Factors to consider in analyzing mobility corridors and avenues of approach.

2-67. Figure 2-9 shows mobility corridors and a map analysis of an avenue of approach. (See Appendix K for more information.)

Chapter 2

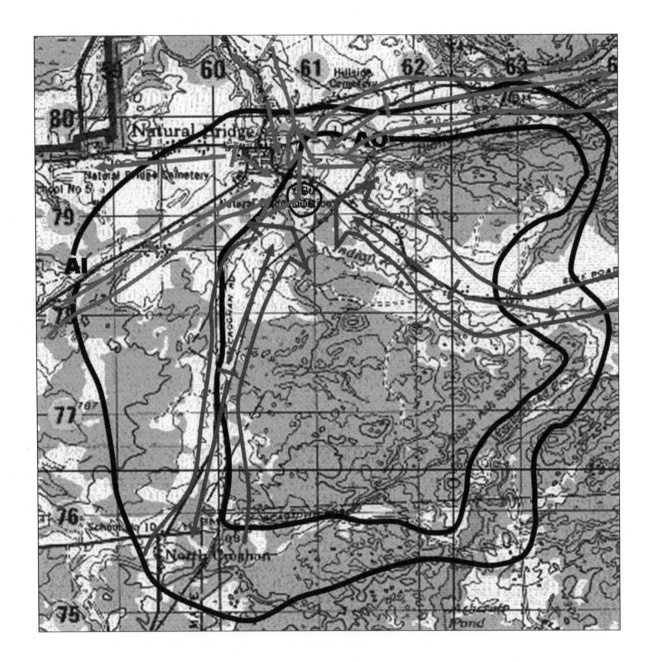

Figure 2-9. Analysis of mobility corridors and avenues of approach.

Key Terrain

2-68. Key terrain is any location or area whose seizure, retention, or control gives a marked advantage to either combatant. It is a conclusion, usually arrived at after enemy analysis and COA development, rather than an observation. For example, a prominent hilltop overlooking an avenue of approach might or might not be key terrain. Even if it offers clear observation and fields of fire, it offers nothing if the enemy can easily bypass it, or if the selected COA involves maneuver on a different avenue of approach. However, if it offers cover and concealment, observation, and good fields of fire on multiple avenues of approach, or on the *only* avenue of approach, then it offers a definite advantage to whomever controls it. The Infantry company commander must assess what terrain is *key* to mission accomplishment. Another example of key terrain for an Infantry rifle company in the attack is high ground overlooking the enemy's reverse slope defense. Controlling this area could prove critical in establishing a support-by-fire position to protect a breach force.

Decisive Terrain

2-69. The leader must also determine if any terrain is decisive. This is key terrain whose seizure, retention, or control is necessary for mission accomplishment. Some situations have no decisive terrain. If a leader identifies terrain as decisive, this means he recognizes that seizing or retaining it is necessary to accomplish the mission. Table 2-3 lists several factors the leader must consider in determining whether terrain is key.

Tactical Considerations

- What terrain is important for friendly observation, both for commanding and controlling and for calling for fire?
- What terrain is important to the enemy and why? Is it important to me?
- What terrain has higher headquarters named as key? Is this terrain also important to the enemy?
- Is the enemy controlling this key terrain?
- How do I gain or maintain control of key terrain?
- What terrain is key for communications nodes that could dictate the employment of digital communications equipment?

Table 2-3. Factors to consider in analyzing key terrain.

2-70. Figure 2-10 shows an analysis of key terrain using a map.

Chapter 2

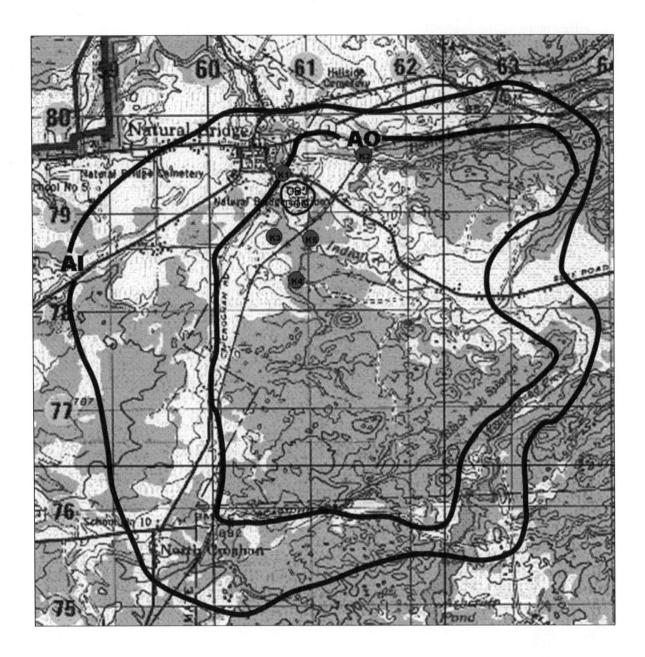

Figure 2-10. Analysis of key terrain.

Observation and Fields of Fire

2-71. The Infantry company commander identifies locations along each avenue of approach that provide clear observation and fields of fire for both the attacker and the defender. He analyzes the area surrounding key terrain, objectives, engagement areas (EAs), and obstacles. He locates intervisibility (IV) lines (ridges or horizons that can hide equipment or personnel from observation). He assesses the ability of the attacking force to overwatch or support movement (with direct fire). Intervisibility line analysis enables the leader to visualize the profile view of terrain when only a topographic product (map) is provided. Figure 2-11, page 2-25, shows intervisibility line analysis. In analyzing fields of fire, he considers the friendly and enemy potential to cover avenues of approach and key terrain, in particular, with direct fires. He also identifies positions where artillery observers can call for indirect fire. The observer must observe both the impact and effects of indirect fires. He analyzes whether or not vegetation will affect the employment or trajectory of the Javelin, or 60-mm mortars. It can do this by masking the target or by reducing overhead clearance. When possible, the observer conducts a ground reconnaissance from both enemy and friendly perspectives. He might do it personally, by map, or with his subordinate units, or he can use the assets and information provided by the Infantry battalion reconnaissance platoon.

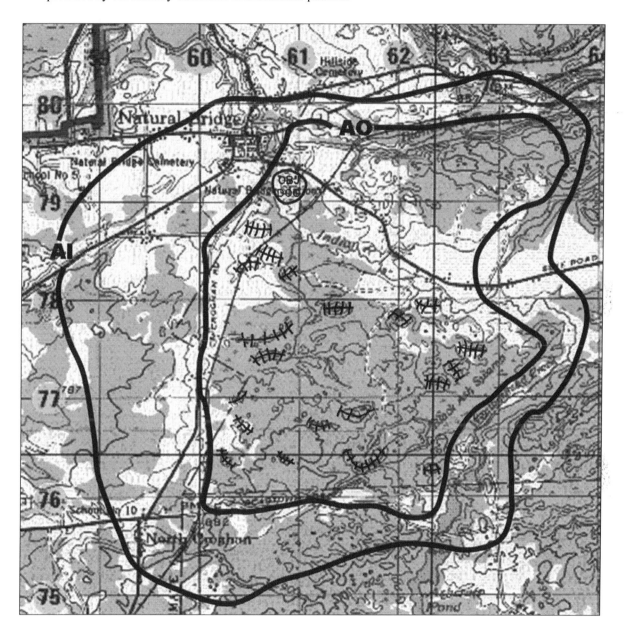

Figure 2-11. Analysis of "IV" line.

2-72. This reconnaissance helps him to see the ground objectively and to see how it will affect both forces (Table 2-4).

OFFENSIVE CONSIDERATIONS

- Are clear observation and fields of fire available on or near the objective for enemy observers and weapon systems?
- Where can the enemy concentrate fires?
- Where will the enemy be unable to concentrate fires?
- Where is the enemy vulnerable?
- Where can I support the movement of a friendly force with mortar, machine gun, or Javelin?
- Where can friendly forces conduct support by fire or attack by fire?
- Where are the natural TRPs?
- Where do I position indirect fire observers?

DEFENSIVE CONSIDERATIONS

- What locations have clear observation and fields of fire along enemy avenues of approach?
- Where will the enemy establish firing lines or support-by-fire positions?
- Where will I be unable to mass fires?
- Where is the dead space in my sector? Where am I vulnerable?
- Where are the natural TRPs?
- Where can I destroy the enemy? Can I observe and fire on that location with at least 2/3 of my combat power?
- How obvious are these positions to the enemy?
- Where do I position indirect fire observers?

Table 2-4. Factors to consider in analyzing observation and fields of fire.

Cover and Concealment

2-73. The leader looks at the terrain, foliage, structures, and other features along avenues of approach (and on objectives or key terrain) to identify sites that offer cover (protection from the effects of direct and indirect fire) and concealment (protection from observation). In the defense, weapon positions must be both lethal to the enemy and survivable for the Soldier. Effective cover and concealment is just as vital as clear fields of fire (Table 2-5, page 2-26). Cover and concealment can be either part of the environment or something brought in by the unit to create the desired effect.

Troop-Leading Procedures

OFFENSIVE CONSIDERATIONS • What axes afford both clear fields of fire and effective cover and concealment? • Which terrain provides bounding elements with cover and concealment while increasing lethality?
DEFENSIVE CONSIDERATIONS • What locations afford effective cover and concealment as well as good observation and fields of fire? • How can friendly and enemy forces use the available cover and concealment?

Table 2-5. Considerations in cover and concealment.

Conclusions from Terrain Analysis

2-74. A terrain analysis should produce several specific conclusions:

- Battle, support-by-fire, and attack-by-fire positions.
- Engagement areas and ambush sites.
- Immediate and intermediate objectives.
- Asset locations such as enemy command posts or ammunition caches.
- Assembly areas.
- Observation posts.
- Artillery firing positions.
- Air defense artillery system positions.
- Reconnaissance, surveillance, and target-acquisition positions.
- Forward area arming and refueling points.
- Landing and drop zones.
- Breach locations.
- Infiltration lanes.

Five Military Aspects of Weather

2-75. The five military aspects of weather are visibility; winds; precipitation; cloud cover; and temperature and humidity. Consideration of the weather's effects is an essential part of the leader's mission analysis. The leader goes past observing to application. That is, he determines how the weather will affect the visibility, mobility, and survivability of his unit and that of the enemy. He reviews his commander's conclusions and identifies his own. He applies the results to the friendly and enemy COAs he develops.

Visibility

2-76. The leader identifies critical conclusions about visibility factors such as light data, fog, and smog; and about battlefield obscurants such as smoke and dust. He considers light data and identifies critical conclusions about begin morning nautical twilight (BMNT), sunrise (SR), sunset (SS), end evening nautical twilight (EENT), moonrise (MR), moonset (MS), and percentage of illumination (Figure 2-12). Some additional visibility considerations include--

- *Will the sun rise behind my attack or in my eyes? Will I attack toward the sunrise?*
- *How can I take advantage of the limited illumination?*
- *How will this affect friendly and enemy target acquisition?*

- *Will the current weather favor the use of smoke to obscure during breaching?*
- *When are night vision devices effective?*

LIGHT	DATA	CONCLUSION
BMNT		
SR		
SS		
EENT		
MR		
MS		

Figure 2-12. Light matrix.

Winds

2-77. Winds of sufficient speed can reduce the combat effectiveness of a force downwind as the result of blowing dust, smoke, sand, or precipitation. The upwind force usually has better visibility. CBRN operations usually favor the upwind force (see more in Appendix H). Windblown sand, dust, rain, or snow can reduce the effectiveness of radar and other communication systems. Strong winds can also hamper the efficiency of directional antenna systems by inducing antenna wobble. Strong winds and wind turbulence limit airborne, air assault, and aviation operations. Evaluation of weather in support of these operations requires information on the wind at the surface as well as at varying altitudes. Near the ground, high winds increase turbulence and may inhibit maneuver. At greater altitudes, it can increase or reduce fuel consumption. Wind is always described as "*from...to*" as in "winds are from the east moving to the west." The leader must answer these questions:

- *Will wind speed cause smoke to dissipate quickly?*
- *Will wind speed and direction favor enemy use of smoke?*
- *Will wind speed and direction affect the employment of available mortars?*
- *What is the potential for CRBN contamination?*

Precipitation

2-78. Precipitation affects soil trafficability, visibility, and the functioning of many electro-optical systems. Heavy precipitation can reduce the quality of supplies in storage. Heavy snow cover can reduce the efficiency of many communication systems as well as degrade the effects of many munitions and air operations. The leader identifies critical factors such as type, amount, and duration of precipitation. Some precipitation questions to answer include--

- *How will precipitation (or lack of it) affect the mobility of the unit or of enemy forces?*
- *How can precipitation (or lack of it) add to the unit achieving surprise?*

Cloud Cover

2-79. Cloud cover affects ground operations by limiting illumination and the solar heating of targets. Heavy cloud cover can degrade many target acquisition systems, infrared-guided munitions, and general aviation operations. Heavy cloud cover often canalizes aircraft within air avenues of approach and on the final approach to the target. Partial cloud cover can cause glare, a condition that attacking aircraft might use to conceal their approach to the target. Some types of clouds reduce the effectiveness of radar systems. The leader identifies critical factors about cloud cover, including limits on illumination and solar heating of targets. Some cloud cover questions follow:

- *How will cloud cover affect unit operations at night? How will it affect the enemy?*
- *How will cloud cover affect the target acquisition of the command launch unit?*
- *How will cloud cover affect helicopter and close air support?*

Temperature and Humidity

2-80. Extremes of temperature and humidity reduce personnel and equipment capabilities and may require the use of special shelter or equipment. Air density decreases as temperature and humidity increase. This can require reduced aircraft payloads. Temperature *crossovers*, which occur when target and background temperatures are nearly equal, degrade thermal target acquisition systems. The length of crossover time depends on air temperature, soil and vegetation types, amount of cloud cover, and other factors. The leader identifies critical factors about temperature, including high and low temperatures, infrared crossover times, and the effects of smoke and chemicals. Some temperature considerations include—.

- *How will temperature and humidity affect the unit's rate of march?*
- *How will temperature and humidity affect the Soldiers and equipment?*
- *Will temperatures and humidity favor the use of nonpersistent chemicals?*

Analysis of Enemy

2-81. The second factor to consider is the enemy. Leaders analyze the enemy's dispositions, compositions, strengths, doctrine, equipment, capabilities, vulnerabilities, and probable courses of action. On the modern battlefield, the line between enemy combatants and civilian noncombatants is sometimes unclear. This requires the leader to understand the Laws of War, the ROE, and the local situation.

Questions

2-82. Analyzing the enemy answers the question, "*What is the enemy doing and why?*" Leaders also answer—.

- *What is the composition and strength of the enemy force?*
- *What are the capabilities of his weapons? Other systems?*
- *What is the location of current and probable enemy positions?*
- *What is the enemy's most probable COA? (DRAW-D [defend, reinforce, attack, withdraw, delay]).*

Phase

2-83. An important result of analyzing and templating the enemy is identification of his phase in the COE. Infantry company commanders must understand that the behavior of an enemy force operating in the regional phase of a conflict can differ greatly from that of an enemy in the adaptive phase. Commanders consider enemy trends, activities, and capabilities. They try to fully appreciate the thinking and will of the enemy. They must avoid ignoring enemy capabilities to save time or effort. An adaptive and capable enemy will fight to the best of his abilities. Commanders use (small-unit-level) situation templates

Chapter 2

(SITEMPs) and initial priority intelligence requirements (PIR) to add to their knowledge of the enemy. They need to know his doctrine, composition, disposition, strengths, and plan. Effective analysis helps commanders decide when, how, and where to apply overwhelming combat power. However, they must distinguish between what they know and what they template. Otherwise, they could base their plans on dangerous and unreliable assumptions.

Contemporary Operational Environment

2-84. Although COE doctrine is most closely associated with training doctrine, its constructs are useful in analyzing and understanding enemy forces in real world operations.

Assumptions

2-85. The leader must understand the assumptions the battalion intelligence staff officer (S-2) uses to portray the enemy's COA. Furthermore, his own assumptions about the enemy must be consistent with those of his higher commander. The leader must continually improve his situational understanding (SU) of the enemy and update his enemy templates as new information or trends become available. Any deviation or significant conclusions reached during his enemy analysis that could positively or negatively affect the battalion's plan should immediately be shared with the battalion commander and S-2.

2-86. In analyzing the enemy, the leader must understand the intelligence preparation of the battlefield (IPB). Although he does not usually prepare IPB products for his subordinates, he must be able to use the products of the higher headquarters' IPB effectively.

Doctrinal Analysis (How Enemy Will Fight)

2-87. The leader must know more than just the number and types of vehicles, Soldiers, and weapons the enemy has. The leader must thoroughly understand when, where, and how the enemy prefers or tends to use his assets. A doctrinal template is a visual *illus*tration of how the enemy force might look and act without the effects of weather and terrain. The leader looks at specific enemy actions during a given operation and uses the appropriate doctrinal template to gain insights into how the enemy may fight. Likewise, he must understand enemy doctrinal objectives. In doctrinal terms, he asks--

> *Is the enemy oriented on the terrain, for example, a reconnaissance force, his own force (assault force, terrorists, or insurgent forces), civilian forces or critical infrastructure (terrorist or insurgent forces, sabotage), or other supporting or adjacent friendly forces (as in a disruption zone)? What effect will this have on the way the enemy fights?*

2-88. However, as the global situation changes, the possibility of fighting adversaries who lack a structured doctrine increases. In such a situation, a leader must rely on information provided by battalion or higher echelon reconnaissance and surveillance assets and, most importantly, his and his higher headquarters' pattern analysis and deductions about the enemy in his AO (see Appendix J). He may also make sound assumptions about the enemy, human nature, and local culture.

Composition

2-89. The leaders analysis must determine the types of vehicles, Soldiers, and equipment the enemy could use against his unit (Figure 2-13, page 2-30). He should be familiar with the basic characteristics of the units and platforms identified.

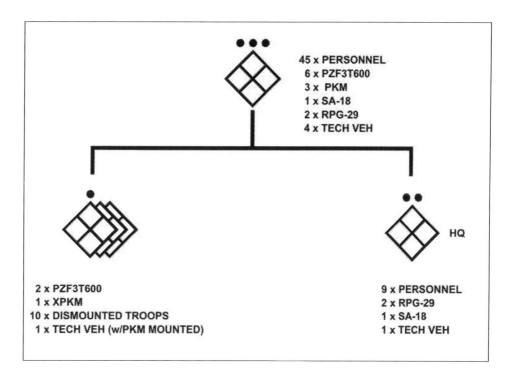

Figure 2-13. Example enemy composition.

Disposition

2-90.   From higher headquarters' information, he determines how the enemy is (or might be) arrayed. If the information is available, he determines the echelon force where the enemy originated. He determines the disposition for the next two higher enemy elements (Figure 2-14). From this analysis, he might be able to determine patterns in the enemy's employment or troops and equipment.

Chapter 2

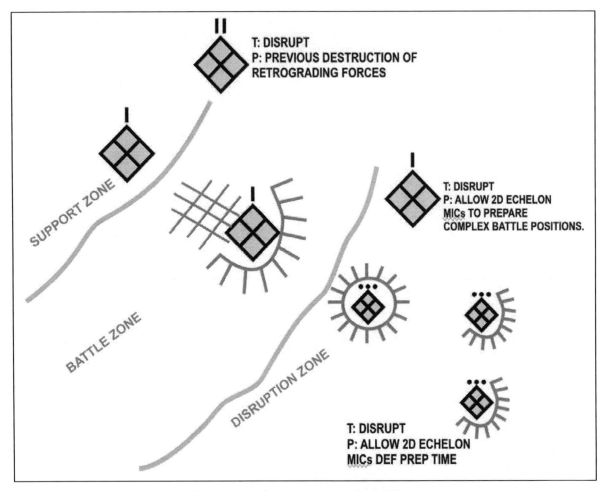

Figure 2-14. Example enemy disposition.

Strength

2-91. He identifies the enemy's strength by unit (Figure 2-15). He can obtain this information by translating percentages given from higher headquarters to the actual numbers in each enemy element or from information provided by the common operational picture (COP).

> Strength: The enemy platoon is estimated at 80 to 85 percent strength, which equates to about 30 personnel, 5 PZF3T600s, 3 PKMs, 1 SA-18, 2 RPG-29s, and 3 technical vehicles.

Figure 2-15. Example enemy strength.

Capabilities

2-92. Based on the S-2's assessment and the enemy's doctrine and current location, the leader must determine the enemy's capabilities. This includes studying the maximum effective range for each weapon

Troop-Leading Procedures

system, the doctrinal rates of march, and the timelines associated with the performance of certain tasks. One technique is to use the WFF as a checklist to address every significant element the enemy brings to the fight. The leader also determines the capabilities of the next higher enemy element. These capabilities should include reasonable assets the next higher element, or other higher enemy headquarters, may provide. This should include at least the employment of reserves, chemical weapons, artillery or mortar locations and ranges, and reconnaissance assets (Table 2-6).

Capabilities

INTELLIGENCE

Some commando and Kazarian militia are already in our AO and we can expect local, unconventional forces to execute disruption attacks using RPGs and IED ambushes against our C2 and log capabilities immediately.

MOVEMENT AND MANEUVER

The commander believes the MIB weapons platoon to be positioned near APOD Fullerton at WQ 045423. This unit can reinforce any MIP position within 25 to 45 minutes and Jetertown within 50 minutes. The trigger for their commitment is the loss of one MIP or two squads of commandos.

Expect rubbling effects within towns and villages as well as the commando's historical burning of large piles of tires to obscure our optics and deny routes within villas.

FIRE SUPPORT

The 1/54th 1st MIB has been reinforced with a battery of 2S3M in DS role. This artillery is believed to range as far South as our assault positions.

PROTECTION

Commandos have multiple RPG systems and small arms, which have shot down two ARFOR rotary aircraft within the past 72 hours.

SUSTAINMENT

Enemy's last LOGPAC believed to have occurred this morning. We do not expect the enemy to resupply until after his successful defense.

COMMAND AND CONTROL

The MIP leader is believed to be with his decisive operation on OBJ Fox. Most C2 is through FM and cell phones.

Table 2-6. Example enemy capabilities.

Recent Activities

2-93. Gaining complete understanding of the enemy's intentions can be difficult when his doctrinal templates, composition, and disposition are unclear. In all cases, the enemy's recent activities must be understood, because they can provide insight into his future activities and intentions (Table 2-7). If time permits, the leader might be able to conduct a pattern analysis of the enemy's actions to predict future actions. In the COE, this might be the most important analysis the leader conducts and is likely to yield the most useful information to the leader.

041800NOV05	Evidence of ethnic cleansing observed by discovery of mass grave vic grid WQ 058402
042200NOV05	Local arms dealer in town of Pitkin is molested He reports to the Kazarian Police Force the theft of 80 pounds of dynamite
052300NOV05	Arms cache suspected at location WQ 081385
060600NOV05	Local Kazarians report multiple murders, rapes, and thefts of POVs vic Jetertown and Huffton
161200Nov05	Two T-72Bs and multiple BMPs observed moving from Fullerton FLS to vic grid WQ 083408
161800NOV05	Enemy indirect fire splash vic grid WQ 099329
070600nov05	Unarmed protest involving 50 to 60 Kazarians vic TAA Bird

Table 2-7. Example enemy recent activities.

Enemy Situation Template

2-94. To identify how the enemy may potentially fight, the leader weighs the result of his analysis of terrain and weather against the higher headquarters' SITEMP. The refined product is a company SITEMP, a graphic showing how he believes the enemy will fight under specific battlefield conditions. This SITEMP is portrayed one echelon lower than that developed by the higher headquarters' S-2. For example, if a battalion SITEMP identifies a platoon-size enemy element on the company's objective, the leader, using his knowledge of both the enemy's doctrine and the terrain, develops a SITEMP that positions squad-size battle positions, crew-served weapons positions, or defensive trenches. He includes in this SITEMP the likely sectors of fire of the enemy's weapons and any tactical and protective obstacles, either identified or merely templated, which support the defense. Table 2-8, page 2-35, shows recommended SITEMP items. (For a more detailed list of items that must be situationally templated, see Appendix B, FM 34-13.) The leader must avoid developing his SITEMP independently of the higher commander's guidance and the S-2's product. The product must reflect the results of reconnaissance and shared information. Differences between the SITEMPs must be resolved before the leader can continue analyzing the enemy. Finally, given the scale with which the leader often develops his SITEMP, on a 1:50,000 map (Figure 2-16, page 2-36), the SITEMP should be transferred to a GDOT for briefing purposes, as the situation allows. This is not for analysis, but just to show subordinates the details of the anticipated enemy course of action (ECOA). Once he briefs the enemy analysis to his subordinates, he must ensure that they understand the differences between what he knows, what he suspects, and what he just templates (estimates). Unless given the benefit of reconnaissance or other intelligence, his SITEMP is only an estimate of how the enemy might be disposed. He must not take these as facts. This is why the leader must develop a tactically sound and flexible plan. It is also why he must clearly explain his intent to his subordinates. This allows them to exercise initiative and judgment to accomplish the unit's purpose. Reconnaissance is critical in developing the best possible enemy scenario.

Defense	Offense
• Primary, alternate, subsequent positions. • Engagement areas. • Individual vehicles. • Crew-served weapons. • Tactical and protective obstacles. • Trenches. • Planned indirect-fire targets. • Observation posts. • Command and control positions. • FPF and FPL. • Locations of reserves. • Routes for reserve commitment. • Travel time for reserve commitment. • Battle position, strongpoints, sectors. • Sectors of fire.	• Attack formations. • Axes of advance. • Firing lines. • Objectives. • Reserve force commitment. • Planned indirect-fire targets. • Situational obstacles. • Reconnaissance objectives. • Reconnaissance force routes. • Phase lines. • Planned point of penetration

Table 2-8. Recommended SITEMP items.

Chapter 2

Figure 2-16. Example enemy situation template.

Initial Priority Intelligence Requirements

2-95. A leader defines PIR as information about the enemy that leads to a critical decision, and he develops specific PIR for each situation. Answering the PIR questions lets him confirm or deny assumptions he made during planning. Although doing this helps him clarify the enemy situation, it also usually leads to answering the PIR of the next level higher.

Analysis of Troops and Support Available

2-96. Leaders study their task organization to determine the number, type, capabilities, and condition of available friendly troops and other support. Analysis of troops follows the same logic as that of analyzing the enemy by identifying capabilities, and vulnerabilities and strengths. Leaders should know the disposition, composition, strength, and capabilities of their forces one and two levels down. This information can be maintained in a checkbook-style matrix for use during COA development (specifically array forces). They maintain understanding of subordinates' readiness, including maintenance, training, strengths and weaknesses, leaders, and logistic status. Analysis of troops and support answers the question: What assets are available to accomplish the mission? Leaders also answer these questions:

- *What are the strengths and weaknesses of subordinate leaders?*
- *What is the supply status of ammunition, water, fuel (if required), and other necessary items?*
- *What is the present physical condition of Soldiers (morale, sleep)?*
- *What is the condition of equipment?*

- *What is the unit's training status and experience relative to the mission?*
- *What additional Soldiers or units will accompany?*
- *What additional assets are required to accomplish the mission?*

2-97. Perhaps the most critical aspect of mission analysis is determining the combat potential of one's own force. The leader must realistically and unemotionally determine all available resources and any new limitations based on level of training or recent fighting. This includes troops who are either attached to or in direct support of his unit. It also includes understanding the full array of assets that are in support of the unit. He must know, for example, how much indirect fire, by type, is available and when it will become available.

2-98. Because of the uncertainty always present in operations at the small unit level, leaders cannot be expected to think of everything during their analysis. This fact forces leaders to determine how to get assistance when the situation exceeds their capabilities. Therefore, a secondary product of analysis of troops and support available should be an answer to the question: How do I get help?

Analysis of Time Available

2-99. The fifth factor of METT-TC is time available. Time refers to many factors during the operations process (plan, prepare, execute, and assess). The four categories for the leader to consider include---

- Planning and preparation.
- Operations.
- Next higher echelon's timeline.
- Enemy timeline.

2-100. During all phases, leaders consider critical times, unusable time, the time it takes to accomplish activities, the time it takes to move, priorities of work, and the tempo of operations. Other critical conditions to consider include visibility and weather data, and events such as higher headquarters tasks and required rehearsals. Implied in the analysis of time is leader prioritization of events and sequencing of activities.

2-101. As addressed in the first step of the TLP, time analysis is a critical aspect to planning, preparation, and execution. Time analysis is often the first thing a leader does. The leader must not only appreciate how much time is available, but he must also be able to appreciate the time/space aspects of preparing, moving, fighting, and sustaining. He must be able to see his own tasks and enemy actions in relation to time. Most importantly, as events occur, he must adjust the time available to him and assess its impact on what he wants to accomplish. Finally, he must update previous timelines for his subordinates, listing all events that affect the company and its subordinate elements. Figure 2-17 shows an example company timeline.

Chapter 2

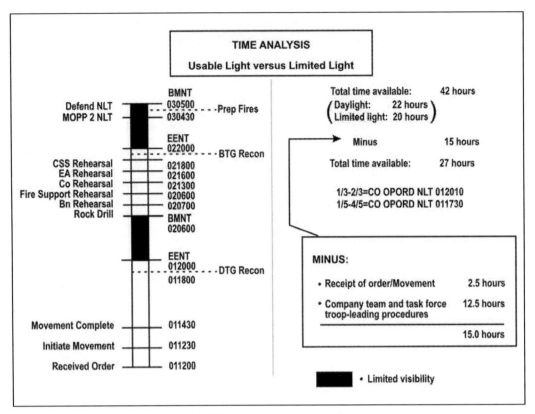

Figure 2-17. Example company timeline.

Analysis of Civil Considerations

2-102. Civil considerations include the influences of man-made infrastructure; civilian institutions; and the attitudes and activities of civilian leaders, populations, and organizations within an AO, with regard to the conduct of military operations. Civil considerations generally focus on the immediate impact of civilians on operations in progress. Civil considerations of the environment can either help or hinder friendly or enemy forces; the difference lies in which leader has taken time to learn the situation and its possible effects on the operation. Analysis of civil considerations answers three critical questions:

- How do civilian considerations affect the operation?
- How does the operation affect the civilians?
- How do our forces build national will in our AO?

2-103. The higher headquarters provides the leader with civil considerations that may affect the next echelon's mission. The memory aid the higher headquarters may use to analyze and describe these civil considerations is ASCOPE (Appendix B, FM 6-0).

Population Perceptions

2-104. The population within a prescribed AO comprises several different groups, both ethnically and politically. Leaders must understand each group's perceptions about the United States, the Army, and the specific unit operating within that area. Population status overlays can best describe groups and define what feelings that group has toward American forces. This is extremely important in understanding when and where to commit combat power, what relationships can be reinforced with certain groups versus what relationships need to start or cease, and ultimately what second and third order effects our actions will have in the AO. Information operations can also be properly focused with a healthy understanding of the perceptions of the civilian population.

2-105. This characteristic addresses terrain analysis from a civilian perspective. Analyze how key civilian areas affect the missions of respective forces and how military operations affect these areas. Factors to consider include political boundaries, locations of government centers, by-type enclaves, special regions such as mining or agricultural, trade routes, and possible settlement sites.

Structures

2-106. Structures include traditional high-payoff targets, protected cultural sites, and facilities with practical applications. The analysis is a comparison how a structure's location, functions, and capabilities can support operations as compared to costs and consequences of such use.

Capabilities

2-107. Assess capabilities in terms of those required to save, sustain, or enhance life, in that order. Capabilities can refer to the ability of local authorities to provide key functions and services. These can include areas needed after combat operations and contracted resources and services.

Organizations

2-108. Consider all nonmilitary groups or institutions in the AO. These may be indigenous, come from a third country or US agencies. They influence and interact with the populace, force, and each other. Current activities, capabilities, and limitations are some of the information necessary to build situational understanding. This becomes often a union of resources and specialized capabilities.

People

2-109. People is a general term describing all nonmilitary personnel that military forces encounter in the AO. This includes those personnel outside the AO whose actions, opinions, or political influence can affect the mission. Identify the key communicators and the formal and informal processes used to influence people. In addition, consider how historical, cultural, and social factors that shape public perceptions beliefs, goals, and expectations.

Events

2-110. Events are routine, cyclical, planned, or spontaneous activities that significantly affect organizations, people, and military operations, such as seasons, festivals, holidays, funerals, political rallies, and agricultural crop/livestock and market cycles and paydays. Other events, such as disasters and those precipitated by military forces, stress and affect the attitudes and activities of the populace and include a moral responsibility to protect displaced civilians. Template events and analyze them for their political, economic, psychological, environmental, and legal implications.

2-111. The leader must also identify any civil consideration that may affect only his mission. Civil considerations are important when conducting operations against terrorist or insurgent forces in urban areas. Most terrorists and insurgents depend on the support or neutrality of the civilian population to camouflage them. Leaders must understand the impact of their actions--as well as their subordinate's actions--on the civilian population, and the effect they will have on current and future operations. Considerations may include--

Ethnic Dynamics

2-112. Ethnic dynamics include religion, cultural mores, gender roles, customs, superstitions, and values that certain ethnic groups hold dear that differ from other groups. Leaders who analyze the ethnic dynamics of their AO can best apply combat power, shape maneuver with information operations, and ultimately find the common denominator that all ethnic varieties have in common and focus unit efforts at it. Gaining local support can best be accomplished by the commander who demonstrates dignity and respect to the civilian population he is charged to protect and train.

Organizations of Influence

2-113. Organizations of influence force the leader to look beyond preexisting civilian hierarchical arrangements. By defining organizations within the community, leaders can understand what groups have power and influence over their own smaller communities and what groups can assist our forces. After these groups have been defined, analyzing them and determining their contributions or resistance to friendly operations is easier. Many Eastern cultures rely upon religious organizations as their centers of power and influence, whereas Western cultures' power comes from political institutions by elected officials. Defining other influential organizations or groups of influence allows for effective information collection and intelligence gathering. For example, the educated persons in an Islamic culture include clerics and teachers. However, where clerics have power, teachers do not. However, teachers can be very useful in providing intelligence to US forces because they seek stability and value security for their people over the religious power arrangements within the area.

Patterns

2-114. Every culture, every group of people, has patterns of behavior. Whether it is set times for prayer, shopping or commuting, people follow patterns. Understanding these patterns helps leaders plan and execute combat and reconnaissance patrols and logistical resupply. Also, unit leaders who study the history of a people's culture can better understand and explain to others how--and why--the people have fought previous wars and conflicts. Starting with a baseline pattern and then keeping a running estimate on how the population is responding or have responded in the past under similar circumstances will assist leaders in using patterns to the unit's advantage.

Leaders and Influencers

2-115. Know who is in charge and who can influence and enable unit leaders to effectively exercise governance and monitor security within a prescribed area. Many times, the spiritual leader is not necessarily the decision maker for a community, but the spiritual leader must approve the decision-maker's actions. Commanders and staff officers who make link-diagrams of leadership that include religious, political, and criminal personnel allow focused planning and decentralized execution that bolsters legitimacy within the population. Using the targeting methodology of D3A (decide, detect, deliver, and assess) may prove useful in determining whether a leader or influencer would best facilitate an operation, when to engage them, and what to expect.

Economic Environment

2-116. Money and resources drive prosperity and stability. Leaders who identify the economic production base for their AOs can effectively execute civil-military campaigns within their AOs that bolster the economic welfare of the people. These campaigns include infrastructure rebuild projects, creation of labor opportunities, and education. By focusing on the motivations for civilian labor and creating essential services and prosperity where there was none, unit leaders/commanders can effectively win the support of the civilian who can now feed and clothe his family and now has clean running water. This aspect of civil considerations reinforces the security of the community against poverty and other enablers to instability.

RISK ASSESSMENT

2-117. Risk assessment is the identification and assessment of hazards that allows a leader to implement measures to control hazards (Appendix A, *Risk Management, Fratricide Avoidance, and the Effects of Continuous Operations*). Leaders assess risk to protect the force and aid in mission accomplishment. The leader must consider two kinds of risk: tactical and accident. Tactical risk is associated with hazards that exist due to the enemy's presence on the battlefield. The consequences of tactical risk take two major forms.

- Enemy action where the leader has accepted risk such as an enemy attack where the friendly leader has chosen to conduct an economy of force.
- Lost opportunity, such as movement across terrain that severely restricts the speed of traverse. This would then restrict the unit's ability to mass the effects of combat power.

2-118. Accident risk includes all operational risk other than tactical risk and can include hazards concerning friendly personnel, equipment readiness, and the environment. Fratricide is an example of an accident risk.

2-119. The leader must identify risks based on the results of his mission analysis. Once identified, risk must be reduced through controls. For example, fratricide is a hazard categorized as an accident risk; surface danger zones (SDZs; see also Appendix A) and risk estimate distances (REDs) are used to identify the controls, such as target reference points and phase lines, to reduce this accidental risk. When the leader decides what risks he is willing to accept, he must also decide in his COA how to reduce that risk to an acceptable level (Figure 2-18).

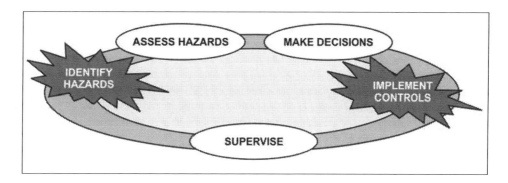

Figure 2-18. Risk assessment.

IDENTIFY TENTATIVE DECISIVE POINTS

2-120. Identifying a tentative decisive point and verifying it during COA development is the most important aspect of the TLP. Visualizing a valid decisive point is how the leader determines how to achieve success and accomplish his purpose. The leader develops his entire COA from the decisive point. Without determining a valid decisive point, the leader cannot begin to develop a valid or tactically sound COA. The leader, based on his initial analysis of METT-TC factors, his situational awareness, his vision of the battlefield, and insight into how such factors can affect the unit's mission, should visualize where, when, and how his unit's ability to generate combat power (firepower, protection, maneuver, leadership, and information) can overwhelm the enemy's abilities to generate combat power. The decisive point might orient on terrain, enemy, time, or a combination of these. The decisive point might be where or how, or from where, the unit will mass the effects of combat power against the enemy. The decisive point might be the event or action (with respect to terrain, enemy, or time, and the generation of combat power) that will ultimately and irreversibly lead to the unit achieving its purpose.

2-121. The decisive point does not simply restate the unit's essential task or purpose; it defines how, where, or when the unit will accomplish its purpose. The unit's decisive operation always focuses at the decisive point, and always accomplishes the unit's purpose. Designating a decisive point is critical to the leader's vision of how he will use combat power to achieve the purpose, how he will task organize his unit and how his shaping operations will support the decisive operation, and how the decisive operation will accomplish the unit's purpose. This tentative decisive point forms the basis of his planning and COA development; it also forms the basis of communicating the COA to his subordinates. The leader should clearly explain what the decisive point is to his subordinate leaders and why it is decisive; this objective, in conjunction with his commander's intent, facilitates subordinate initiative. A valid decisive point enables the leader to clearly and logically link how the application of combat power elements with respect to terrain, enemy, and time allows the unit to accomplish its purpose. If the leader determines that his tentative decisive point is not valid during COA development or analysis, then he must determine another decisive point and restart COA development.

COMMANDER'S INTENT

2-122. The commander's intent is a clear, concise statement of what the force must do to succeed with respect to the enemy, terrain and desired end state (a set of required conditions that, when achieved, attain the aims set for the operation). The commander's intent provides the link between the mission and the CONOP by stating key tasks or conditions that must exist to achieve the stated purpose of the operation. The commander's intent and mission statements form the bases for subordinates to exercise disciplined initiative and judgment in the face of new opportunities, or whenever the concept of operation ceases to apply. The commander's intent continuously evolves throughout the planning and preparation for the operation as the leader becomes more attuned to what he must do to accomplish his mission.

2-123. The key tasks and conditions specified in the commander's intent are not tied to a specific COA. They are not limited to tactical tasks. The operation's tempo, duration, and effect on the enemy, and the terrain that must be controlled are examples of key tasks or conditions.

2-124. The commander's intent does not include the method by which the force will get from its current state to the end state. The method is the CONOP. Nor does the intent contain acceptable risk. Risk is addressed in COAs. The example below shows key tasks and conditions related to enemy, terrain, and desired end state. The final commander's intent included in the OPORD is based on the sum of all of the analysis conducted during the TLP. This final intent can only be provided after the leader understands the end state of the mission. An example of commander's intent follows:

EXAMPLE

Commander's intent expressed as key tasks includes--

- All enemy forces on OBJ Atlanta destroyed.
- Bravo Company defeats enemy counterattacks.
- The company controls the west side of Bush Hill (OBJ BM312), in position and able to defeat counterattack in EA Red, TF 1-22 IN (M) passed through CP2.
- Commander's intent expressed as conditions includes--
- Enemy AT weapon systems on or around OBJ Hook should be unable to affect US vehicles on Route Bud between CPs 5 and 7.
- Alpha Company occupies terrain on or around OBJ Hook, successfully destroys enemy counterattacks, and prevents the disruption of US forces on Route Bud between CP 5 and 7.

COURSE OF ACTION DEVELOPMENT

2-125. From developing a strategy to analyzing, refining, and rehearsing the plan, a leader should be knowledgeable in the following areas detailed under this subheading to construct a solid COA (Figure 2-18). The purpose of COA development is to determine one or more ways to accomplish the mission that is consistent with the immediate higher commander's intent. A COA describes how the unit might generate the effects of overwhelming combat power against the enemy at the decisive point with the least friendly casualties. Each COA the leader develops must be detailed enough to clearly describe how he envisions using all of his assets and combat multipliers to achieve the unit's mission-essential task and purpose. To develop a COA, he focuses on the actions the unit must take at the decisive point and works backward to his start point (SP). The leader should focus his efforts to develop at least one well-synchronized COA; if time permits, he should develop several. The result of the COA development process is paragraph 3 of the OPORD. A COA should position the unit for future operations and provide flexibility to meet unforeseen events during execution. It should also give subordinates the maximum latitude for initiative.

SCREENING CRITERIA

2-126. According to Chapter 3, FM 5-0, a COA should be suitable, feasible, acceptable, distinguishable, and complete.

Suitable

2-127. If successfully executed, the COA accomplishes the mission consistent with the higher commander's concept and intent.

Feasible

2-128. The unit has the technical and tactical skills and resources to accomplish the COA successfully, with available time, space, and resources.

Acceptable

2-129. The military advantage gained by executing the COA must justify the cost in resources, especially casualties. This assessment is largely subjective.

Distinguishable

2-130. If more than one COA is developed, it must be sufficiently different from the others to justify full development and consideration.

Complete

2-131. The COA covers the operational factors of who, what, when, where, and how, and must show from start to finish how the unit will accomplish the mission. The COA must also address the doctrinal aspects of the mission. For example, in an attack against a defending enemy, the COA must address the movement to, deployment against, assault of, and consolidation upon the objective.

> *Note:* Leaders assess risk continuously throughout COA development.

ACTIONS

2-132. Next, the commander analyzes relative combat power, generates options, arrays his forces, develops a CONOP, assigns responsibility, and prepares a COA statement and sketch.

Analyze Relative Combat Power

2-133. During the first step of COA development, analyzing relative combat power, leaders compare and contrast friendly combat power with the enemy. There four goals include--

- Identify an enemy weakness to exploit.
- Identify friendly strengths to exploit the enemy weakness.

- Identify enemy strengths to mitigate.
- Identify friendly weaknesses to protect.

2-134. The purpose of this step is to compare the combat power of friendly and enemy forces. It is not merely a calculation and comparison of friendly and enemy weapons numbers or units with the aim of gaining a numerical advantage. Using the results of all previous analyses done during mission analysis, the leader compares his unit's combat power strengths and weaknesses with those of the enemy. He seeks to calculate the time and manner in which his force (and the enemy) can maximize the effects of maneuver, firepower, protection, leadership, and information in relation to the specific terrain, disposition, and composition of each force. The leader also determines how to avoid enemy strengths or advantages in combat power. In short, he strives to determine where, when, and how his unit's combat power (the effects of maneuver, firepower, protection, leadership, and information) can overwhelm the enemy's ability to generate combat power. An analysis of the ability to generate combat power will help the leader confirm or deny his tentative decisive point.

Generate Options

2-135. Most missions and tasks can be accomplished in more than one way. The goal of this step, generating options, is to determine one or more of those ways quickly. First, leaders consider TTP from doctrine, unit SOPs, history, or other resources to determine if a solution to a similar tactical problem already exists. If it does, the leader's job is to take the existing solution and modify it to his unique situation. If a solution does not exist, the leader must develop one. Second, leaders confirm the mission's decisive point. Then, using doctrinal requirements as a guide, the leader assigns purposes and tasks to decisive, and shaping, and sustaining operations.

2-136. This doctrinal requirement provides a framework for the leader to develop a COA. For example, a breach requires an assault element, support element, breach element, security element, and possibly a reserve. Beginning with the decisive point identified during mission analysis, the leader identifies the decisive operation's purpose and the purposes of his shaping and sustaining operations. The decisive operation's purpose is nested to his unit's overall purpose and is achieved at his decisive point. The shaping operations' purposes are nested to the decisive operation's purpose by setting the conditions for success of the decisive operation. The sustaining operations' purposes are nested to the decisive and shaping operation's purposes by providing sustainment, rear area and base security, movement control, terrain management, and infrastructure. The leader then determines the tactical mission tasks for the decisive, shaping, and sustaining operations. These tasks must be accomplished to achieve the subordinate units' given purposes.

Array Forces

2-137. Using the product from generating options, the leader then determines what combinations of Soldiers, weapons, and other systems are needed to accomplish each task. This is known as "*arraying forces*" or "*assigning troops to task*." This judgment call is unique to the specific METT-TC conditions the leader faces. He must then task organize his forces specific to the respective essential tactical tasks and purposes assigned to his subordinate elements. He determines the specific quantity of squads, weapons (by type), and fire support necessary to accomplish each task against the enemy array of forces. He allocates resources required for the decisive operation's success first and then determines the resources needed for shaping operations in descending order of importance.

Develop a Concept of Operations

2-138. The concept of operations (CONOP) describes how the leader envisions the operation unfolding, from its start to its conclusion or end state. Operations/actions are made up of numerous activities, events, and tasks. The CONOP describes the relationships between activities, events, and tasks, and explains how the tasks will lead to accomplishing the mission. The CONOP is a framework to assist leaders, not a script. The normal cycle for an offensive operation is tactical movement, actions on the objective, and consolidation and reorganization. The normal cycle for defensive operations is engagement area development and preparation of the battle positions, actions in the engagement area, counterattack, and consolidation and reorganization. In developing the CONOP, the leader clarifies in his mind the best ways to use the available terrain and to employ the unit's strengths against the enemy's weaknesses. He includes the requirements of indirect fire to support the maneuver. He then develops the maneuver control measures necessary to convey his intent, enhance the understanding of the schemes of maneuver, prevent fratricide, and clarify the tasks and purposes of the decisive shaping, and sustaining operations. He also determines the CS and sustainment aspects of the COA.

Assign Responsibilities

2-139. Infantry leaders assign responsibility for each task to a subordinate. Whenever possible and depending on the existing chain of command, they avoid fracturing unit integrity. They try to keep their span of control between two to five subordinate elements. The leader ensures that every unit in his command is employed, every asset is attached, and adequate command and control is provided for each element. The leader must avoid unnecessary complicated command and control structures and maintain unit integrity where feasible.

Prepare a COA Statement and Sketch

2-140. Leaders in small Infantry units primarily use the COA statement and COA sketch (Figure 2-19) to describe the CONOP. These two products are the basis for paragraph 3 of the operation order. The COA statement specifies how the unit will accomplish the mission. The first three steps of COA development provide the bulk of the COA statement. The COA statement details how the unit's operation supports the next higher leader's operation, the decisive point and why it is decisive, the form of maneuver or type of defensive operation, and the battlefield framework. The COA sketch is a drawing or series of drawings to assist the leader in describing how the operation will unfold. The sketch provides a picture of the maneuver aspects of the concept. Leaders use tactical mission task graphics and control measures (FM 1-02) to convey the operation in a doctrinal context. Both the COA statement and sketch focus at the decisive point. The COA statement should identify---

- Decisive point, and what makes it decisive.
- Form of maneuver or type of defensive operation.
- Tasks and purposes of the decisive, shaping, and sustaining operations.
- Reserve planning priorities.
- Purposes of critical WFF elements.
- The end state.

Chapter 2

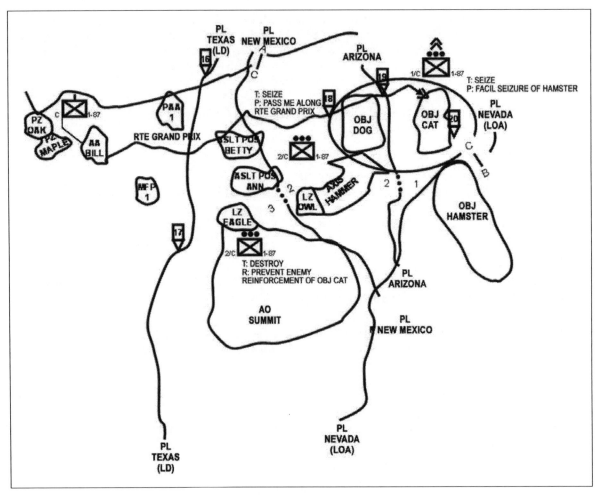

Figure 2-19. Example COA sketch.

2-141. The COA sketch should identify how the unit intends to focus the effects of overwhelming combat power at the decisive point. When integrated with terrain, the refined product becomes the unit's operations overlay.

COURSE OF ACTION ANALYSIS

2-142. COA analysis begins with both friendly and enemy COAs and, using a method of action-reaction-counteraction war game, results in a synchronized friendly plan, identified strengths and weaknesses, and updated risk assessment (Figure 2-20, page 2-47).

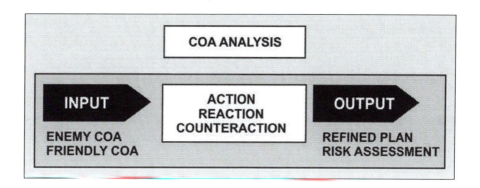

Figure 2-20. Analysis of course(s) of action.

2-143. After developing the COA, the leader analyzes it to determine its strengths and weaknesses, visualizes the flow of the battle, identifies the conditions or requirements necessary to enhance synchronization, and gains insights into actions at the decisive point of the mission. If he has developed more than one COA, he applies this same analysis to each COA developed. He does this analysis through war-gaming or "*fighting*" the COA against at least one enemy COA. For each COA, leaders think through the operation from start to finish. They compare their COA with the enemy's most probable COA. At small unit level, the enemy's most probable COA is what the enemy is most likely to do. During the war game, the leader visualizes a set of enemy and friendly actions and reactions. War-gaming is the process of determining "*what if?*" factors for the overall operations. The object is to determine what can go wrong and what decision the leader will likely have to make as a result. COA analysis allows the leader to synchronize his assets, identify potential hazards, and develop a better understanding of the upcoming operation. It enables him--

- To determine how to maximize the effects of combat power while protecting friendly forces and minimizing collateral damage.
- To anticipate battlefield events.
- To determine conditions and resources required for success.
- To identify additional control requirements.
- To identify friendly coordination requirements.

2-144. COA analysis (war-gaming) brings together friendly and enemy forces on the actual terrain to visualize how the operation will unfold. It is a continuous cycle of action, reaction, and counteraction. This process highlights critical tasks, stimulates ideas, and provides insights rarely gained through mission analysis and COA development alone. War-gaming is a critical step in the planning process and should be allocated more time than the other steps. War-gaming helps the leader fully synchronize friendly actions, while considering the likely reactions of the enemy. The product of this process is the synchronization matrix. War-gaming, depending on how much time is devoted to planning, provides---

- An appreciation for the time, space, and triggers needed to integrate fire support, smoke, engineers, air defense artillery (ADA), and CBRN with maneuver platoons (Infantry, antiarmor, or tank) to support unit tasks and purposes identified in the scheme of maneuver.
- Flexibility built into the plan by gaining insights into possible branches to the basic plan.
- The need for control measures, such as checkpoints, contact points, and target reference points (TRPs), that aid in control, flexibility, and synchronization.
- Coordinating instructions to enhance execution and unity of effort, and to ease confusion between subordinate elements.
- Information needed to complete paragraphs 3, 4, and 5 of the OPORD.

Chapter 2

- Assessments regarding on-order and be-prepared missions.
- Projected sustainment expenditures, friendly casualties, and resulting medical requirements.

2-145. The best way for small unit leaders to war-game is to start at their current location and go through the mission from start to finish, or start at a critical point such as the objective or engagement area. Using the action-reaction method, leaders can think through the engagement beforehand. As they proceed, they can either record their observations into a matrix or keep note in a notebook. The most important aspect of this process is not the method but the output, meaning a more in-depth understanding of the operation. Depending on the time available and his personal preference, the leader may use any of the following war-gaming techniques.

BOX TECHNIQUE

2-146. The box technique focuses the war game on a specific area of the battlefield (Figure 2-21). This might be the objective area, the EA, or some other critical location where decisive or critical actions will take place. It should include all of the units, friendly and enemy, that have a direct impact on those actions. This technique is used when time is limited and the enemy situation is relatively clear. However, a disadvantage is that when considering only the actions at the critical or decisive points, the leader may overlook other actions or events that could have a significant impact on the unit's mission.

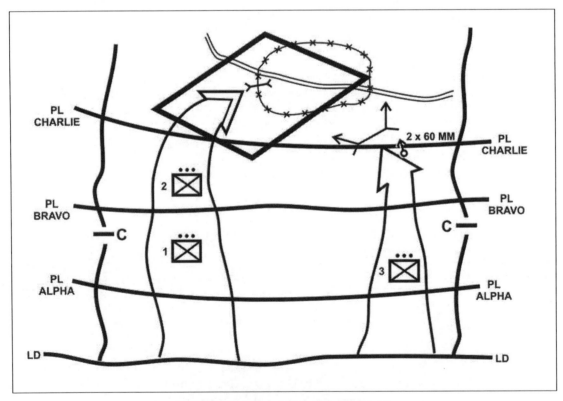

Figure 2-21. Box war-gaming technique.

BELT TECHNIQUE

2-147. The belt technique allows the leader to divide the COA into events or belts (Figure 2-22, page 2-49). He may do this in several ways, such as from phase line to phase line or by significant event. Each step then is war-gamed in sequence. This approach is most effective for offensive COAs. The leader can modify this technique by dividing the battlefield into belts that are not necessarily adjacent or overlapping but focus on the critical actions throughout the AOs.

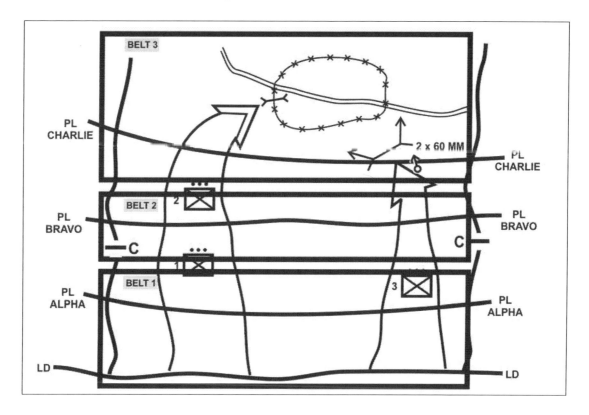

Figure 2-22. Belt war-gaming technique.

AVENUE-IN-DEPTH TECHNIQUE

2-148. This method is most effective for a defensive COA, especially when there are several avenues of approach to consider. Using the enemy's most probable COA, the leader analyzes friendly and enemy actions along one avenue of approach at a time (Figure 2-23).

2-149. To gain the benefits that result from war-gaming a COA, the leader must remain objective and record the results of the war game. He must remember the assumptions he made about the enemy, his unit, and the ground during the development of his tentative plan. He must avoid letting the enemy or his unit "win" to justify the COA. Also, he must avoid drawing premature conclusions about the war game or changing his COA until the war game is complete.

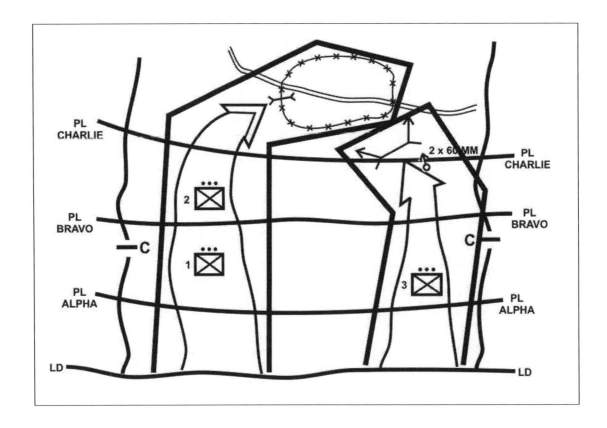

Figure 2-23. Avenue-in-depth war-gaming technique.

COURSE OF ACTION COMPARISON AND SELECTION

2-150. If the leader has developed more than one COA, he must compare them by weighing the specific advantages, disadvantages, strengths, and weaknesses of each COA as noted during the war game. These attributes may pertain to the accomplishment of the unit purpose, the use of terrain, the destruction of the enemy, or any other aspect of the operation that he believes is important. He uses these factors, gained from his relational combat power analysis (RCPA) matrix, as his frame of reference in tentatively selecting the best COA. He makes the final selection of a COA based on his own judgment, the start time of the operation, the AOs, the scheme of maneuver, and subordinate unit tasks and purposes.

2-151. The CCIR identify and filter information needed by leaders to support their vision and to make critical decisions, especially to determine or validate COAs. CCIR help commanders determine what is relevant to mission accomplishment. In one technique, they write the desired question, the quantified answer, and the reaction (critical decision to make). CCIR also help focus the efforts of subordinates and aid in the allocation of resources. Commanders should limit their CCIR to essential information.

PRIORITY INTELLIGENCE REQUIREMENTS

2-152. PIR are information that a leader needs to know about terrain or enemy to make a critical decision. PIR are best expressed in a question that can be answered yes or no.

Troop-Leading Procedures

> **PIR EXAMPLE**
> Can enemy wheeled vehicles cross the creek at NU12345678?
> - If yes, the company will reinforce the obstacle and establish an antiarmor ambush at this location.
> - If no, the company will emplace an OP, and then establish the antiarmor ambush along another route.

FRIENDLY FORCES INFORMATION REQUIREMENTS

2-153. Friendly force information requirements (FFIR) include information that leaders need to know about their units or about adjacent units to make critical decisions.

ESSENTIAL ELEMENTS OF FRIENDLY INFORMATION

2-154. Although EEFI are not part of the CCIR, they still become priorities when the leader states them IAW FM 3-0. EEFI are the critical aspects of a friendly operation that, if known by the enemy, would subsequently compromise or lead to failure of the operation. Consequently, this information must be protected from identification by the enemy.

Section IV. STEPS 4 THRU 8--INITIATE MOVEMENT, RECONNOITER, COMPLETE PLAN, ISSUE OPORD, SUPERVISE

This section discusses TLP Steps 4 thru 8.

STEP 4--INITIATE MOVEMENT

2-155. Leaders initiate any movement necessary to continue mission preparation or to posture the unit for the start of the mission. This step can be executed at any time throughout the sequence of the TLP. It can include movement to an assembly area, battle position (BP), or new AO, or the movement of guides or quartering parties.

STEP 5--RECONNOITER

2-156. To exploit the principles of speed and surprise, leaders should weigh the advantages of reconnoitering personally against the combat multiplier in the form of supplied information from the battalion's C2 information system. They realistically consider the dangers of reconnoitering personally, and the time required to conduct them. Leaders might be able to plan their operations using the unprecedented amount of combat information provided by the higher echelon reconnaissance and surveillance assets. However, if time permits, leaders should verify higher headquarters' intelligence by reconnoitering visually. They should seek to confirm the PIR that support their tentative plans. These PIR usually consist of assumptions or critical facts about the enemy. This can include strength and location, especially at templated positions. It can also include information about the terrain, for example, verification that a tentative support-by-fire position can suppress the enemy, or that an avenue of approach is useable.

2-157. If possible, leaders should include their subordinate leaders in their reconnaissance efforts. This allows the subordinates to see as much of the terrain and enemy as possible. The reconnaissance also helps subordinate leaders gain insight into the leaders' visions of the operation.

2-158. The leaders' recons might include moving to or beyond the line of departure (LD), reconnaissance of an AO, or walking from the forward edge of battle area (FEBA) back to and through the company AO or battle position (BP) along likely enemy avenues of approach. If possible, leaders should select vantage points with the best possible view of the decisive point.

2-159. In addition to the leaders' reconnaissance efforts, the units can conduct additional reconnaissance operations. Examples include surveillance of an area by subordinate elements, patrols to determine enemy locations, and establishment of observation posts (OPs) to gain additional information. Commanders can also incorporate Javelin command launch units (CLUs) as surveillance tools (day or night), based on an analysis of METT-TC factors.

2-160. The nature of the reconnaissance, including what it covers and how long it lasts, depends on the tactical situation and the time available. The leader should use the results of the COA development process to identify information and security requirements for the unit's reconnaissance operations.

2-161. The leader must include disseminating results and conclusions arrived from any reconnaissance into his time analysis. He must also consider how to communicate any changes in the COA to his subordinates and how these changes affect his plans, actions of the subordinates, and other supporting elements.

STEP 6--COMPLETE PLAN

2-162. During this step, leaders expand their selected (or refined) COAs into complete OPORDs. They prepare overlays, refine the indirect fire list, complete sustainment and C2 requirements and, of course, update the tentative plan based on the latest reconnaissance or information. They prepare briefing sites and other briefing materials they might need to present the OPORDs directly to their subordinates. They conduct final coordination with other units or staff members before issuing the order to their subordinates. Using the five-paragraph OPORD format helps them to explain all aspects of the operation: terrain, enemy, higher and adjacent friendly units, unit mission, execution, support, and command and control. The format also serves as a checklist to ensure that they cover all relevant details of the operation. It also gives subordinates a smooth flow of information from beginning to end.

STEP 7--ISSUE OPORD

2-163. The OPORD precisely and concisely explains both the leader's intent and concept of how he envisions the unit accomplishing the mission. The order does not contain unnecessary information. The OPORD is delivered quickly and in a manner that allows subordinates to concentrate on understanding the leader's vision and not just copying what he says verbatim. The leader must prepare adequately and deliver the OPORD confidently and quickly to build and sustain confidence in his subordinates.

2-164. When issuing the OPORD, the leader must ensure his subordinates understand and share his vision of what must be done and when and how it must be done. They must understand how all the company's elements work together to accomplish the mission. They must also understand how the company's mission supports the intentions of the immediate higher commander. When the leader has finished issuing the order, subordinate leaders should leave with a clear understanding of what the leader expects their elements to do. The leader is responsible for ensuring that his subordinates understand.

2-165. Also, and in many respects more importantly, the leader must issue the order in a manner that instills his subordinates with confidence in the plan and a commitment to do their best to achieve the plan. Whenever possible, he must issue the order in person. He looks into the eyes of his subordinate leaders to ensure each one understands the mission and what the element must achieve.

2-166. Complete the order with a confirmation brief. At a minimum, each subordinate leader should be able to backbrief the unit mission and intent, the immediate higher commander's intent, his own task(s) and purpose, and the time he will issue his unit's OPORD. Each subordinate should confirm that he understands the vision of the battlefield and how the mission is accomplished with respect to the decisive point. This confirmation brief also provides an opportunity to highlight any issues or concerns.

2-167. The five-paragraph OPORD format helps the leader paint a picture of all aspects of the operation, from the terrain to the enemy, and finally to the unit's own actions from higher to lower. The format helps him decide what relevant details he must include and in providing subordinates with a smooth flow of information from beginning to end. At the same time, the leader must ensure that the order is not only clear and complete but also as brief as possible. If he has already addressed an item adequately in a previous WARNO, he can simply state "*No change,*" or provide any necessary updates. The leader is free to brief the OPORD in the most effective manner to convey information to his subordinates.

STEP 8--SUPERVISE

2-168. This final step of the TLP is crucial. After issuing the OPORD, the commander and his subordinate leaders must ensure that the required activities and tasks are completed in a timely manner prior to mission execution. Supervision is the primary responsibility of all leadership. Both officers and NCOs must check everything that is important for successful mission accomplishment. This includes, but is not limited to--

- Conducting numerous back briefs on all aspects of the company and subordinate unit operations.
- Ensuring the second in command in each element is prepared to execute in his leaders' absence.
- Listening to subordinate's operation orders.
- Observing rehearsals of subordinate units.
- Checking load plans to ensure they are carrying only what is necessary for the mission or what the OPORD specified.
- Checking the status and serviceability of weapons.
- Checking on maintenance activities of subordinate units.
- Ensuring local security is maintained.

REHEARSALS

2-169. Rehearsals are practice sessions conducted to prepare units for an upcoming operation or event. They are essential in ensuring thorough preparation, coordination, and understanding of the commander's plan and intent. Commanders should never underestimate the value of rehearsals.

2-170. Effective rehearsals require leaders and, when time permits, other company Soldiers to perform required tasks, ideally under conditions that are as close as possible to those expected for the actual operation. At their best, rehearsals are interactive; participants maneuver their actual vehicles or use vehicle models or simulations while verbalizing their elements' actions. During every rehearsal, the focus is on the how element, allowing subordinates to practice the actions called for in their individual scheme of maneuver.

> *Note:* A rehearsal is different from a discussion of what is supposed to happen during the actual event. For example, in a rehearsal, platoon leaders send real spot reports (SPOTREPs) when reporting enemy contact, rather than just saying, "*I would send a SPOTREP now.*" The commander can test subordinate understanding of the plan by ensuring that they push the rehearsal forward rather than waiting on him to dictate each step of the operation.

2-171. The commander uses well-planned, efficiently run rehearsals to accomplish the following:

- Reinforce training and increase proficiency in critical tasks.
- Reveal weaknesses or problems in the plan, leading to further refinement of the plan or development of additional branch plans.
- Integrate the actions of subordinate elements.
- Confirm coordination requirements between the company and adjacent units.
- Improve each Soldier's understanding of the concept of the operation, the direct fire plan, anticipated contingencies, and possible actions and reactions for various situations that may arise during the operation.
- Ensure that seconds in command are prepared to execute in their leaders' absence.

REHEARSAL CONSIDERATIONS

2-172. Rehearsals should follow the crawl-walk-run training methodology. This prepares the Infantry companies and subordinate elements for increasingly difficult conditions. The unit can conduct reduced-force or full-dress rehearsals.

Reduced-Force Rehearsals

2-173. Infantry company commanders conduct reduced-force rehearsals when time is limited or when the tactical situation affects attendance. Unit members who can participate should practice on mock-ups, sand tables, or actual terrain (usually over a smaller area than in the actual operation).

Full-Dress Rehearsals

2-174. Full dress rehearsals are the most effective, but use the most time and resources. They involve nearly every Soldier who will participate in the operations. If possible, they should be conducted under the same conditions, such as the weather, time of day, and terrain that the unit can expect in the actual operations.

REHEARSAL TYPES

2-175. Leaders may use several types of rehearsals in the same operation.

Confirmation Brief

2-176. The commander may require the platoon leaders to conduct a confirmation brief right after he issues a company OPORD or FRAGO. This is to ensure that the subordinates understand their assigned tasks, purposes, and intents.

Backbrief

2-177. The leader might require squad leaders to backbrief him once they have developed their plan. He checks to ensure it is nested with the concept of the operation, or to identify problems with synchronization.

Combined Arms Rehearsal

2-178. This is the preferred rehearsal type for Infantry units and is conducted when all subordinate OPORDs are complete. This rehearsal type involves all the elements of the unit and ensures that all subordinate plans are fully synchronized within the overall plan. Infantry company commanders can use any of several techniques to execute this type of rehearsal.

Support Rehearsal

2-179. Support rehearsals are normally conducted by a single or limited number of WFF elements such as sustainment or fire support. Infantry companies seldom conduct their own support rehearsals. However, commanders should be aware that higher headquarters might, which will affect the companies. They should consider this factor when planning their overall timelines.

PRECOMBAT CHECKS AND INSPECTIONS

2-180. Precombat checks (PCCs) and precombat inspections (PCIs) are critical to the success of any combat patrol. These checks and inspections are leader tasks and cannot be delegated below team leader level. They ensure that the Soldier is prepared to execute the required individual and collective tasks that

Troop-Leading Procedures

support the mission. Checks and inspections are part of the TLP that protect against shortfalls that could endanger Soldiers' lives and jeopardize the successful execution of a mission. PCCs and PCIs must be tailored to the specific unit and the mission requirements. Each mission and each patrol may require a separate set of checklists. Each element will have their own established set of PCCs and PCIs, but each platoon within that element should have identical checklists. A weapons squad will have a different checklist than a line squad, but each weapons squad within an organization should be the same. One of the best ways to ensure PCCs and PCIs are complete and thorough is with full-dress rehearsals. These rehearsals, run at combat speed with communication and full battle-equipment, allow the leader to envision minute details, as they will occur on the battlefield. If the operation is to be conducted at night, Soldiers should conduct full-dress rehearsals at night as well. PCCs and PCIs should include back briefs on the mission, the task and purpose of the mission, and how the Soldiers' role fits into the scheme of maneuver. The Soldiers should know the latest intelligence updates and the ROE, and be versed in MEDEVAC procedures and sustainment requirements. Table 2-9 lists sensitive items, high dollar value items, issued pieces of equipment, and supplies. This table should spur thought--it is not a final list.

ID card	T&E mechanisms	Grappling hook
ID tags	Spare barrels	Sling sets
Ammunition	Spare barrel bags	PZ marking kit
Weapons	Extraction tools	ANCD
Protective mask	Asbestos gloves	Plugger or GPS
Knives	Barrel changing handles	Handheld microphones
Flashlights	Headspace and timing gauges	NVDs
Radios and backup communication	SAW tools	Batteries and spare batteries
Communication cards	BII	Picket pounder
Nine-line MEDEVAC procedures	Oil and transmission fluids	Engineer stakes
OPORD	Antifreeze coolant	Pickets
FRAGOs	5-gallon water jugs	Concertina wire
Maps	MREs	TCP signs
Graphics, routes, objectives, LZs, and PZs	Load plans	IR lights
	Fuel cans	Glint tape
Protractors	Fuel spout	Chemical lights
Alcohol pens	Tow bars	Spare handsets
Alcohol erasers	Slave cables	Pencil with eraser
Pen and paper	Concertina wire gloves	Weapon tie downs
Tripods		
Pintles		

Table 2-9. Example PCC and PCI checklists.

Chapter 3
Movement

The purpose of tactical movement is to position units on the battlefield and prepare them for contact. This chapter focuses on the movement techniques and formations that give the company commander options for moving his unit. Included are discussions of motor or foot marches and occupation of assembly areas. Each technique and formation has advantages and disadvantages. Some are secure and slow, while others are faster, but less secure. Some work well in some terrain or tactical situations, but not so well in others. Because of the Infantry rifle company's capabilities, the commander must consider the overall movement plan, including these various advantages and disadvantages.

Commanders and leaders may adapt the movement techniques and formations in this chapter to their particular situations.

TACTICAL MOVEMENT AND ENEMY CONTACT

3-1.   Avoid confusing *movement* with *maneuver*. Maneuver is defined as "*Movement supported by fire to gain a position of advantage over the enemy.*" At company level, the two overlap considerably. Tactical movement differs from maneuver, however, because maneuver is movement while in contact, but tactical movement is movement in preparation for contact. The process by which units transition from tactical movement to maneuver is called "*actions on contact.*" Actions on contact are covered in Chapter 4, Section IV, *Actions on Contact*. Figure 3-1, page 3-2, relates movement techniques (traveling, traveling overwatch, and bounding overwatch), the possibility of enemy contact, and the transition to maneuver.

Chapter 3

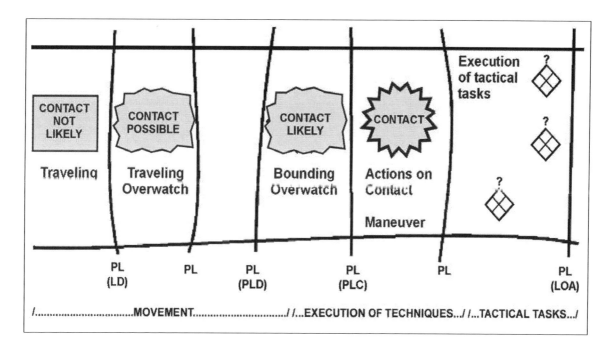

Figure 3-1. Transition from movement techniques to maneuver.

MOVEMENT TECHNIQUES

3-2. The company commander selects from the three movement techniques (traveling, traveling overwatch, and bounding overwatch) based on several battlefield factors. Figure 3-2 shows the symbols for company personnel and elements.

- The likelihood of enemy contact.
- The type of contact expected.
- The availability of an overwatch element.
- The terrain over which the moving element will pass.
- The balance of speed and security required during movement.

Movement

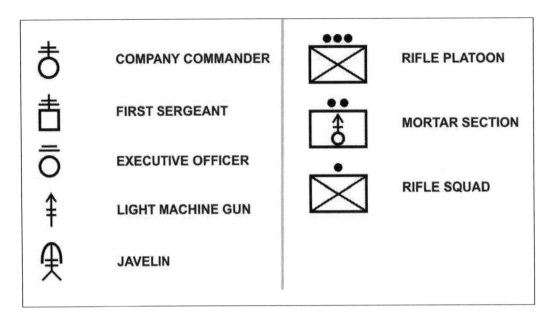

Figure 3-2. Legend of company symbols.

TRAVELING

3-3. Continuous movement characterizes the traveling technique by all company elements. It is best suited for situations in which enemy contact is unlikely and speed is important. Figure 3-3 shows the traveling technique for an Infantry rifle company.

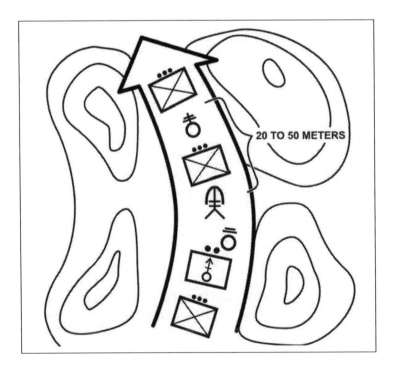

Figure 3-3. Traveling technique.

Chapter 3

TRAVELING OVERWATCH

3-4. Traveling overwatch is an extended form of traveling that provides additional security when speed is desirable but contact is possible (Figure 3-4). The lead element moves continuously. The trail element moves at various speeds and may halt periodically to overwatch movement of the lead element. Dispersion between the two elements must be based on the trail element's ability to see the lead element and to provide immediate suppressive fires in case the lead element is engaged. The intent is to maintain depth, provide flexibility, and maintain the ability to maneuver even if contact occurs, although a unit should ideally make contact while moving in bounding overwatch rather than traveling overwatch.

Note: Organization of the company (in both traveling overwatch and bounding overwatch) (consists of a lead element (also called the bounding element in bounding overwatch) and a trail (or overwatch) element. The commander constitutes these elements using varying combinations of company elements; his decisions are based on the results of his METT-TC analysis. For example, the lead element might consist of one platoon, the commander, and the FSO, overwatched by the two remaining platoons and the XO.

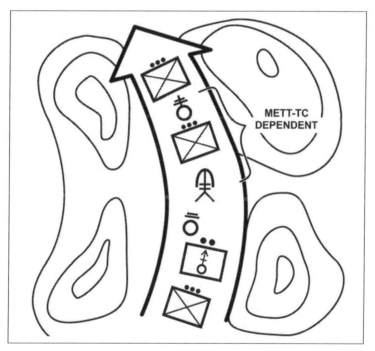

Figure 3-4. Traveling overwatch technique.

BOUNDING OVERWATCH

3-5. Bounding overwatch is used when contact is expected (Figure 3-5). It is the most secure, but slowest, movement technique. The purpose of bounding overwatch is to deploy prior to contact, giving the unit the ability to protect a bounding element by immediately suppressing an enemy force. In all types of bounding, the overwatch element is assigned sectors to scan while the bounding element uses terrain to achieve cover and concealment. The bounding element avoids masking the fires of the overwatch element; it never bounds beyond the range at which the overwatch element can effectively suppress likely or suspected enemy positions. The company can employ either of two bounding methods: alternate or successive.

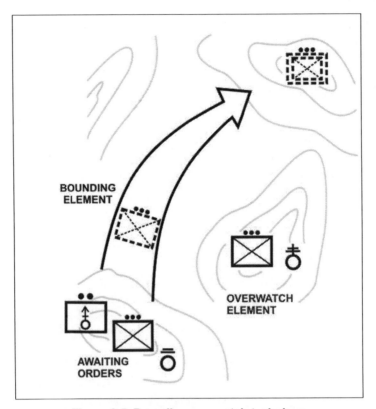

Figure 3-5. Bounding overwatch technique.

Alternate Bounds

3-6. Covered by the rear element, the lead element moves forward, halts, and assumes overwatch positions. The rear element advances past the lead element and takes up overwatch positions. This sequence continues as necessary with only one element moving at a time. This method is usually more rapid than successive bounds.

Successive Bounds

3-7. In the successive bounding method the lead element, covered by the rear element, advances and takes up overwatch positions. The rear element then advances to an overwatch position roughly abreast of the lead element and halts. The lead element then moves to the next position, and so on. Only one element moves at a time, and the rear element avoids advancing beyond the lead element. This method is easier to control and more secure than the alternate bounding method, but it is slower.

MOVEMENT FORMATIONS

3-8. The Infantry rifle company uses six basic movement formations: column, line, vee, wedge, file, and echelon right or left. These formations describe the locations of the company's platoons and sections in relation to each other. They are guides on how to form the company for movement. Each formation aids control, security, and firepower to varying degrees. The following factors should be considered in determining the best formation to use.

- Mission.

- Enemy situation.
- Terrain.
- Weather and visibility conditions.
- Speed of movement desired.
- Degree of flexibility desired.

CROSS COUNTRY MOVEMENT

3-9. When moving cross-country, the distance between Soldiers and platoons varies according to the terrain and the situation. Soldiers should constantly observe their sectors for likely enemy positions and look for cover that can be reached quickly in case of enemy contact.

Platoon Formation

3-10. The company commander may specify the platoon formations to be used within the company formation. If he does not, each platoon leader selects his platoon's formation. For example, the lead platoon leader may select a formation that permits good observation and massing of fire to the front (vee formation). The second platoon leader may select a formation that permits fast movement to overwatch positions and good flank security (wedge formation). (Squad and platoon movement formations and techniques are discussed in FM 3-21.8 (FM 7-8).)

Base Platoon

3-11. When moving in a formation, the company normally guides on the base platoon to ease control. This should be the lead platoon. In the line or the vee formation, the company commander must specify which platoon is the base platoon. The other platoons key their speed and direction on the base platoon. This permits quick changes and lets the commander control the movement of the entire company by controlling only the base platoon. Terrain features might be designated for the base platoon to guide on by using the control techniques previously described. The company commander normally locates himself within the formation where he can best see and direct the movement of the base platoon.

ALERT STATUS

3-12. One technique used to alert units for possible movement or for units to report their readiness to move is an alert status. With this technique, use a readiness condition (REDCON) system to reflect the amount of time a unit will have before it must move.

REDCON 1: Be prepared to move immediately.
REDCON 2: Be prepared to move in 15 minutes.
REDCON 3: Be prepared to move in 1 hour.
REDCON 4: Be prepared to move in 2 hours.

FORMATIONS

3-13. Formations include the column, the company line, the company wedge, the company vee, the company file, and echelon right or left.

Column

3-14. The column formation allows the company to make contact with one platoon and maneuver with the two trail platoons. It is a flexible formation, allowing easy transition to other formations. It provides good all-round security and allows fast movement. It also provides good dispersion and aids maneuver and control, especially in limited visibility. The company can deliver a limited volume of fire to the front and to the rear, but a high volume to the flanks. Figure 3-6 shows an example of the company column.

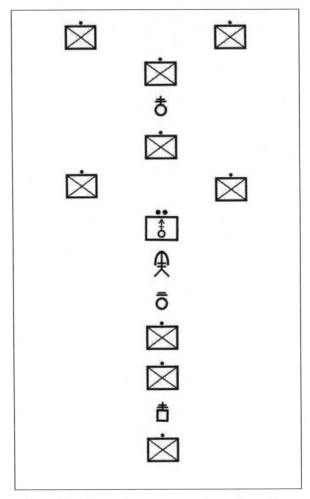

Figure 3-6. Example company column formation.

Company Line

3-15. The company line formation puts all platoons forward along the same direction of movement, and it provides for the delivery of maximum fire to the front, but less to the flanks. It is the most difficult formation to control. The company commander designates a base platoon, normally the center one, for the others to guide on. Flank and rear security is generally poor but is improved when the flank platoons use echelon formations. Figure 3-7 shows an example of the company line.

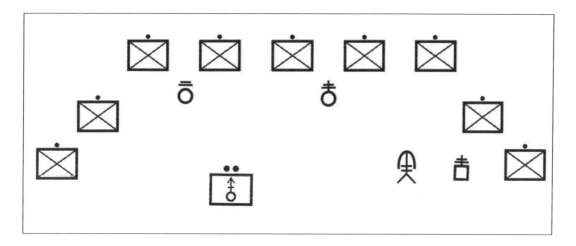

Figure 3-7. Example company line formation.

Company Wedge

3-16. The company wedge formation allows the commander to make contact with a small element and still maneuver the remaining platoons. If the company is engaged from the flank, one platoon is free to maneuver. This formation is hard to control, but it allows faster movement than the company vee formation. Figure 3-8 shows an example of the company wedge.

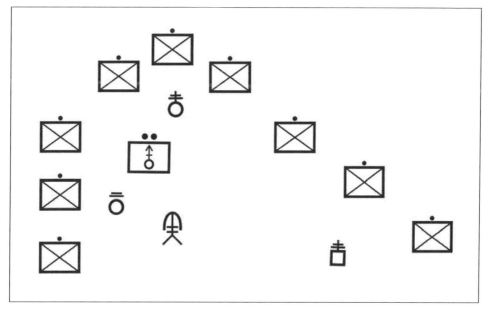

Figure 3-8. Example company wedge formation.

Company Vee

3-17. The company vee formation has two platoons forward to provide immediate fire on contact or to flank the enemy. It also has one platoon centered trailing the two forward platoons. If the company is engaged from either flank, two platoons can provide fire, and at least one platoon is free to maneuver. This formation is hard to control and slows movement. The company commander designates one of the forward platoons as the base platoon. Figure 3-9 shows an example of the company vee with all platoons in wedge.

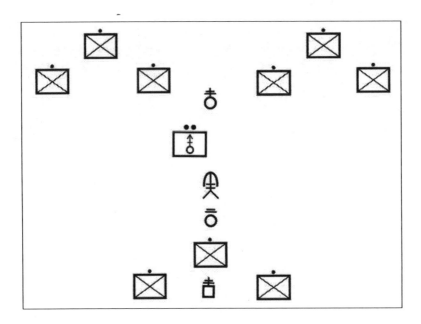

Figure 3-9. Example company vee formation.

Company File

3-18. The company file formation is the easiest formation to control. It allows rapid movement in restricted terrain and in limited visibility, and it enhances control and concealment. Light forces use a file predominantly as its movement formation during times of limited visibility. It is, however, the least secure formation and the hardest from which to maneuver. Figure 3-10 shows an example of the company file with all units in file.

Chapter 3

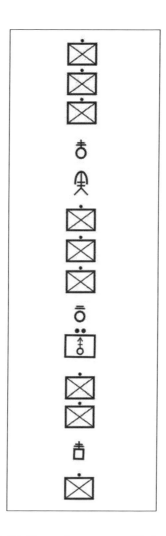

Figure 3-10. Example company file formation.

3-19. The company commander locates well forward with the lead platoon headquarters or immediately behind the lead security element. This location increases his control by putting him in position to make critical decisions. The company CP can locate farther back (behind the lead platoon) to avoid interfering with the platoon's movement and to aid communications with other elements. The 1SG (or XO) is last, or nearly last, in the company file to provide leadership and to prevent breaks in contact within the file. The company file is vulnerable to breaks in contact and should be used only when necessary and for short periods. A company may stretch out over 600 meters in a company file, with a pass time of more than 20 minutes.

Echelon Right or Left

3-20. The echelon right or echelon left formation is used if the situation is vague and the company commander anticipates enemy contact to the front or on one of the flanks. Normally, an obstacle or another friendly unit exists on the flank of the company opposite the echeloned flank, preventing enemy contact on that side. This formation provides a good volume of fire and protection to the echeloned flank, but less to the opposite flank. Figure 3-11 shows an example of the echelon right formation.

Movement

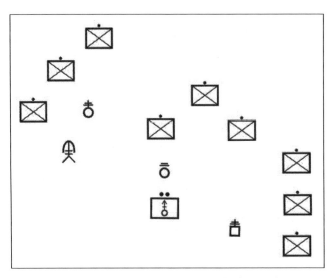

Figure 3-11. Example echelon right formation.

FORMATION SELECTION

3-21. The company commander selects the formation that provides the proper control, security, and speed. Table 3-1 compares the six movement formations.

Formation	Security	Fires	Control	Speed
Column	Good dispersion. Good all-round security.	Good to front and rear. Excellent to the flanks.	Easy to control. Flexible formation.	Fast.
Line	Excellent to the front. Poor to the flank and rear.	Excellent to the front. Poor to the flank and rear.	Difficult to control. Inflexible formation.	Slow.
Wedge	Good all-round security.	Good to the front and flanks.	Less difficult to control than the line. Flexible formation.	Faster than the line.
Vee	Better to the front.	Very good to the front.	Very difficult to control.	Slow.
File	Least secure. Effective use of concealment.	Poor.	Easy to control.	Fast.
Echelon	Good to the echeloned flank and front.	Good to the echeloned flank and front.	Difficult to control.	Slow.

Table 3-1. Comparison of movement formations.

USE OF MOVEMENT FORMATIONS

3-22. Movement should be as rapid as the terrain, mobility of the force, and enemy situation permit. The ability to gain and maintain the initiative often depends on movement being undetected by the enemy. If detected during movement, the enemy might be able to apply substantial combat power against the company. The Infantry rifle company depends heavily upon the terrain for protection from enemy fire. The company commander also protects his company during movement by ensuring the company is using proper movement formations and techniques.

Chapter 3

Fundamentals

3-23. The Infantry rifle company commander's mission analysis helps him decide how to move most effectively. When planning company movements, the commander ensures the unit is moving in a way that supports a rapid transition to maneuver. Once contact with the enemy is made, squads and platoons execute the appropriate actions on contact, and leaders begin to maneuver their units. The following fundamentals provide guidance for planning effective company movements.

Reconnoiter

3-24. All echelons reconnoiter. The enemy situation and the available planning time may limit the unit's reconnaissance, but leaders at every level seek information about the terrain and enemy. If sufficient information is still lacking, an effective technique is to send a reconnaissance element forward of the lead platoon. Even if this unit is only 15 minutes ahead of the company, it can still provide valuable information and reaction time for the company commander.

Use Terrain and Weather Effectively

3-25. One of the strengths of the Infantry rifle company is its ability to move across almost any terrain and in almost any weather conditions. The company moves on covered and concealed routes. Moving in limited visibility may provide better concealment, and the enemy might be less alert during these periods. Plan to avoid identified danger areas.

Move as Squads and Platoon

3-26. The advantages to moving the company by squads and platoons include--

- Faster movement.
- Better security. A small unit is less likely to be detected because it requires less cover and concealment.
- More dispersion. The dispersion gained by moving the company by squads and platoons makes it more difficult for the enemy to concentrate his fires against the company, especially indirect fires, CAS, and chemical agents. Subordinate units also gain room to maneuver.
- Better operations security (OPSEC). It is harder for the enemy to determine what the friendly force is doing with only isolated squad-size spot reports.

3-27. Although the advantages normally outweigh the disadvantages, when planning decentralized movements the commander should also consider the following disadvantages.

- Numerous linkups are required to regroup the company.
- May take longer to mass combat power to support a hasty attack or disengage in the event of enemy contact.

Maintain Security During Movement

3-28. Security is critical during movement since the company is extremely vulnerable to enemy direct and indirect fires. In addition to the fundamentals listed earlier, the company commander achieves security for the company by applying the following.

- Use the appropriate movement formation and technique for the conditions.
- Move as fast as the situation allows. This may degrade the enemy's ability to detect the unit and the effectiveness of his fires once he detects it.
- Ensure subordinate units correctly position security elements to the flanks, front, and rear at a distance that prevents enemy direct fire on the main body. (Normally, the company formation and movement technique provides greater security to the front; the flanks and rear must be

secured by these security elements. The company TSOP should state who is responsible for providing these security elements.)
- Enforce noise and light discipline.
- Enforce camouflage discipline (Soldiers and their equipment).
- When the situation is not clear, make contact with the smallest element possible. By making contact with a small element, the company commander maintains the ability to maneuver with the majority of his force. The Soldiers who first receive enemy fires are most likely to become casualties. They are also most likely to be suppressed and fixed by the enemy.
- When the situation is clear, the company commander must quickly mass the effects of his combat power to overwhelm the enemy.

Locations of Key Leaders and Weapons

3-29. The locations of key leaders and weapons depend on the situation, movement formation and technique, organization of the Infantry rifle company. The following paragraphs provide guidance for the company commander in deciding where these assets should locate.

Company Command Post

3-30. The company CP normally consists of the company commander, his radio operators, the fire support team (FIST), consisting of the fire support officer (FSO), fire support sergeant, and forward observer; and the CBRN sergeant along with possibly other personnel and attachments (XO, 1SG, or a security element). The company CP locates where it best supports the company commander and maintains communications with higher and subordinate units. To maintain communications, the CP may need to locate away from the CO. In this case, the XO controls the CP (or part of it) and maintains communications with higher or adjacent units while the commander locates where he can best control the company. Although the CP can move independently, it normally locates where it is secured by the other platoons and sections within the company formation.

Company Commander

3-31. The company commander locates where he can see and control the company. Normally, he positions the CP at his location, but at times, he may move separate from the CP. He might only take his company net radio operator and travel with one of his platoons. This allows him to move with a platoon without disrupting their formation. Generally, the company commander (with the CP) operates immediately behind the lead platoon.

Company Fire-Support Team

3-32. The FSO typically moves with the company commander and locates remaining team members according to METT-TC. At times, he may locate elsewhere to control indirect fires or relay calls for fire from the platoons.

Company Mortars

3-33. The company mortars locate where they can provide responsive fires in case of enemy contact. They must locate where they are provided security from the other units in the company. They are normally not last in the company formation, because they have limited capability to provide security, and because their loads often make them the slowest element in the company.

Other Attachments

3-34. The locations of other attachments depend on METT-TC. CS assets, such as engineers, are positioned where they can best support the company. For example, the engineers may follow the lead platoon where they can be more responsive.

Mechanized and Wheeled Vehicles

3-35. The Infantry rifle company may have mechanized or wheeled vehicles attached or in support of the company. These might be Bradley Fighting Vehicles, tanks, Strykers, ambulances, trucks attached for movement or resupply vehicles, and they will present certain challenges to the Infantry rifle company commander. Vehicular support greatly enhances the Infantry company's operational mobility. Woodland terrain may not support vehicular movement as well as urban. If the company secures more restrictive terrain on the flanks, it might be able to secure the roads or trails that these vehicles will move on. Several options are available to the commander for the disposition of the vehicles.

- Employ the vehicles in conjunction with the rifle platoons so that each compliments the other.
- Employ them to support the Infantry rifle platoons.
- Employ them to provide heavy direct fires or antiarmor fires.
- Leave in hide positions.
- Displace them to a secure location.

CONTROL TECHNIQUES

3-36. Using the proper formation and movement techniques helps the Infantry rifle company commander control the company, but additional control techniques are often required. The following techniques may help in controlling company movements.

GRAPHICS

3-37. Normally the battalion assigns graphic control measures to synchronize the Infantry rifle company's movement into the battalion's movement or scheme of maneuver. The company commander may need to establish additional control measures to control his units. These may include boundaries, routes, CPs, release points (RP), and TRPs on known (likely) enemy positions to control direct fires. The company commander ensures each graphic control measure is updated as needed and is easy to locate on the terrain.

RECONNAISSANCE

3-38. Prior reconnaissance aids control during movement. It provides the commander with a better idea of where movement is more difficult and where graphic control measures are needed. Elements from the company may perform this reconnaissance, however the battalion reconnaissance platoon is more likely to conduct the reconnaissance and provide the information to other organizations.

GUIDES

3-39. Guides who have already seen the terrain are the best way to provide control. When guides are not available for the entire movement, they should reconnoiter the difficult areas and guide the Infantry company through those.

NAVIGATIONAL AIDS

3-40. Even with the availability of a global positioning system (GPS), every leader should use his compass and a pace count for all moves. If possible, select routes that allow leaders to use prominent terrain to stay oriented.

LIMITED VISIBILITY MOVEMENTS

3-41. The measures already listed are the best ways to provide control for moving in limited visibility. The following measures provide extra control when moving in limited visibility.

Use Night Vision Devices

3-42. Effective limited visibility movement is possible even if there is not a sufficient quantity of night vision devices (NVDs) for every Soldier. If the Soldiers providing front, flank, and rear security use them, the entire unit can move faster. Soldiers should rotate to maintain effectiveness. Key leaders throughout the formation must also use NVDs.

Reduce Interval between Soldiers and Units

3-43. Closing up the formation allows the use of arm-and-hand signals and reduces the chance of breaks in contact. However, leaders maintain the most dispersion possible at all times. Well-trained units can operate at night as they do during the day.

Use Other Measures

3-44. Other measures include using luminous or IR tape on the back of helmets, slowing the speed of movement, using landlines to communicate or to guide units, and moving leaders closer to the front.

SECURITY DURING MOVEMENT

3-45. During company movement, each platoon is responsible for a sector, depending on its position in the formation. Each fire team and squad within the platoons has a sector, so the company has all-round security during movement (Figure 3-12).

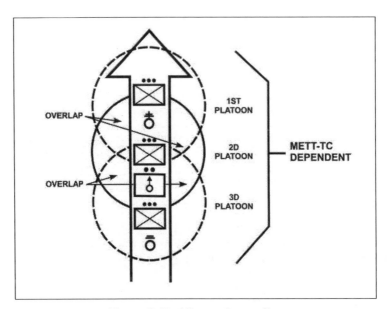

Figure 3-12. All-round security.

3-46. During short halts, Soldiers spread out and assume prone positions behind cover. They observe the same sectors they did while moving. Leaders orient machine guns and antiarmor weapons on likely enemy

AA into the position. Soldiers remain alert and keep movement to a minimum. They speak quietly and only when necessary. Soldiers with NVDs scan areas where the enemy might be concealed in limited visibility.

3-47. During long halts, the Infantry rifle company sets up a perimeter defense (see Chapter 5). The company commander chooses the most defensible terrain, which must have good cover and concealment. The company TSOP addresses the actions required during long halts.

3-48. For additional security, small ambush teams might be concealed and remain in position after a short halt. Ideally, the center platoon provides these teams, which remain in position to ambush any enemy following the company. The linkup of these teams must be coordinated and understood by all.

3-49. Before occupying a static position (objective rally point, patrol base, or perimeter defense), the Infantry commander ensures the enemy is unaware of his company's location. In addition to using the ambush teams, he may also conceal security teams in or near the tentative static position as the company passes it. The company continues movement, preferably until darkness, and then circles back to link up with the security teams, who have reconnoitered the position and guide the company into it.

MOVEMENT AS PART OF A BATTALION

3-50. The Infantry rifle company often moves as part of the battalion. The battalion commander assigns the company a position within the battalion formation, and the company commander uses the movement technique and movement formation that best suits the likelihood of enemy contact and his unit's mission. Regardless of the company's position within the battalion formation, it must be ready to make contact or to support the other elements by maneuver or by fire alone.

ROAD MARCHES AND ASSEMBLY AREAS

3-51. When the company conducts a road march as part of the battalion, the battalion staff plans the march. When the company conducts a road march alone, the company commander plans the march.

DEFINITIONS

3-52. The following definitions apply (FM 21-18 explains formulas to compute movement time).

March Unit

3-53. A unit that moves and halts at the command of a single commander is normally a platoon, but might be a company.

Serial

3-54. A group of march units under a single commander is given a number or letter designation to aid in planning and control.

Arrival Time

3-55. This is the time when the head of a column reaches a designated point or line.

Clearance Time

3-56. This is when the tail of a column passes a designated point or line.

Column Gap

3-57. This is the space, time, or distance between two consecutive elements following each other on the same route. It is stated in units of length (meters) or units of time (minutes) and is measured from the rear of one element to the front of the following element.

Vehicle Distance

3-58. This is the space between two consecutive vehicles.

Start Point

3-59. This is a well-defined point on a route where the units come under the control of the movement commander and start the move. At this point, the column is formed by the successive passing of the units.

Release Point

3-60. This is a well-defined point on a route where the elements of a column leave the control of the movement commander and return to the control of their respective commanders/leaders.

Completion Time

3-61. This is when the tail of a column passes the release point.

Critical Point

3-62. This is a point on the route of march, such as a busy intersection, which is used for reference in giving instructions. It may also designate a point on the route where interference with troop movement might occur.

Length of a Column

3-63. This is the length of roadway occupied by a column, including the gaps, measured from front to rear of the column.

Pace Setter

3-64. This is a person or vehicle in the lead element that is responsible for regulating movement speed.

Pass Time

3-65. This is the time between the passage of the first and last elements by a given point.

Rate of March

3-66. This is the average distance traveled in a given period of time (speed in kmph), including short halts or delays.

Time Distance

3-67. This is the time required for the head of a column to move between two points at a given rate of march.

Chapter 3

Traffic Density

3-68. This is the average number of vehicles that occupy 1 kilometer or 1 mile of road space; it is expressed in vehicles per kilometer or mile (VPK or VPM).

Foot Marches

3-69. The company moves prepared to fight at all times. It is normally organized into platoon-size march units for control and unit integrity. The normal march formation is the column; however, the commander may decide to use another formation based on the factors of METT-TC. (FM 21-18, *Foot Marches*, goes into more detail.)

3-70. When moving along a road, the company moves with one file on each side of the road. Do not split squads by placing a fire team in each file, because if there is contact, these teams will have a danger area between them. When moving cross-country, the company moves with two files 5 meters apart. There should be 2 to 5 meters between Soldiers and 50 meters between platoons. The normal rate of march for an 8-hour march is 4 mph. The interval and rate of a march depends on the length of the march, time allowed, likelihood of enemy contact (ground, air, or artillery), terrain and weather, condition of the Soldiers, and the weight of the Soldiers' loads.

3-71. If the company is marching to a secure area, the company vehicles (if applicable) and mortars may precede the company as a separate march unit. This permits those elements to be operational when the company arrives. If vehicles move with the company, the last vehicle should have a radio so the commander can be contacted in emergencies.

3-72. If vehicles are available, the commander may use the company's vehicles to shuttle the company. The vehicles take as many men as they can carry to the detrucking point, while the remainder of the company starts the march on foot. The vehicles unload, drive back to where they meet the marching company, and pick up another load of Soldiers. They repeat this process until the entire company is at its destination.

Motor Marches

3-73. A company must be given additional vehicles to conduct a motor march. These will normally come from the battalion or forward support company; however, the company commander is responsible for and must plan the air and ground security. He must also insure that drivers know the contingencies if attacked during assembly, loading, or movement.

3-74. The commanding officer (CO) normally organizes the platoons into march units. When moving as part of the battalion, the company is normally a serial. To provide all-round security, the CO assigns security tasks to each march unit. Some tasks might be assigned by SOP; for example, every vehicle will have an airguard with a sector of observation. When supported with vehicles armed with MK 19s or M2 machine guns, the CO positions these vehicles to provide fire support.

3-75. The formations used in a motor march are close column, open column, and infiltration. Before the move, the CO designates a maximum catch-up speed (greater than the prescribed march rate) for vehicles regaining lost distance. To control the column, the CO uses guides, escorts, and route markers. He uses radios, arm-and-hand signals, flags, and flashlights for communications.

- A close column is one in which the vehicles are spaced about 20 meters apart in daylight to increase its density and to reduce pass time. During limited visibility, they are spaced so that each driver can see the blackout markers of the vehicle to his front. This column might be used for movement through congested areas or over poorly marked routes.
- The vehicles in an open column are widely spaced as a passive defense measure, normally 75 to 100 meters apart. This permits other vehicles (not a part of the march unit) to overtake and enter the column, if necessary. It is normally used in daylight and on roads having civilian traffic. It may also be used on dusty roads to overcome the effects of the dust. Drivers do not get as tired and the chances of accidents are less than in close-column marches.

- During infiltrations, vehicles are usually dispatched singly or in small groups at irregular intervals and at a speed, that reduces traffic density. Infiltration increases control problems, but is the best passive defense against enemy observation and attack. It is used when time and road space are available and maximum security, deception, and dispersion are required.

Conduct of a Road March

3-76. The company normally moves in a column. The lead platoon (march unit) maintains the rate of march. The commander is positioned in the formation where he can best command and control the unit's movement.

3-77. Before the road march, the route should be reconnoitered and march orders issued. The march order should include a strip map (Figure 3-13). The strip map shows the assembly areas, SP, route, CPs, and RP. The CO may identify critical points on the route and post guides at those points to help control movement and to provide security. (See FM 21-18 for details on march orders.)

Chapter 3

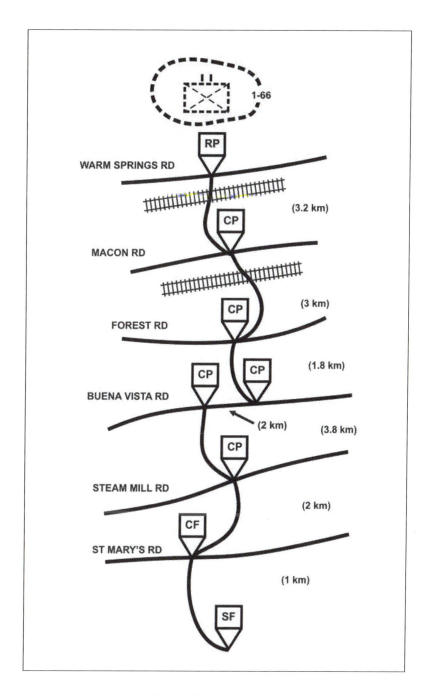

Figure 3-13. Strip map.

3-78. The battalion scout platoon may reconnoiter the route. For motor marches, the scout platoon may prepare a hasty route classification. This may include hasty bridge classifications, ford site locations and conditions, road restrictions, sharpness of curves, and the slope percentage of steep hills. (For classification symbols and their meanings, see FM 5-170.)

3-79. Arrival time at the start point is critical. The company must cross and clear the start point on time so that other units are not delayed. The CO should reconnoiter and time the route from his assembly area to the start point, so he can determine when the company must move to meet its start point time.

3-80. After crossing the start point, platoons report crossing each critical point. When moving as part of the battalion, the company commander, in turn, reports to the battalion commander when his company crosses and clears these points.

3-81. Before the company departs an assembly area, the CO should send a quartering party to the new assembly area. The XO or 1SG leads the quartering party, which may consist of the platoon sergeants, squad representatives, and the required headquarters personnel. This party provides its own security, and it follows the same route of march to the new assembly area as the company does. At the assembly area, the quartering party--

- Reconnoiters the area.
- Locates and marks or removes obstacles and mines.
- Marks platoon and squad sectors.
- Selects a position for the mortar section.
- Selects a command post location.
- Selects a company trains location.
- Provides guides for the incoming unit(s).

March Security

3-82. The CO plans for the security of the company when moving. This includes security against both air and ground threats.

- He assigns each platoon the responsibility for a security sector. For example, he may assign the lead platoon the front, the middle platoon the flanks, and the trail platoon the rear. The platoon sectors must overlap to provide all-round security.
- He plans indirect fire to support the move. He plans targets along the route as he does for all other moves. He designates warning signals and battle drills (usually SOP).
- A Stinger section may support the company from positions along the route or by moving within the company column. Each Stinger team that is on the early warning net can warn the company of an air attack. For that reason, each team should be within voice distance of someone having a radio on the company command net.
- The mortar section must be ready to go into action and fire quickly. The FO teams should be in continuous contact with the mortar and artillery fire-direction centers. The lead FO should keep the fire direction center (FDC) informed of the lead elements location.

Assembly Areas

3-83. An assembly area is a location where the company prepares for future operations. The company receives and issues orders, services and repairs vehicles and equipment, receives and issues supplies, and feeds and rests Soldiers in the assembly area. When used to prepare for an attack, the assembly area is usually well forward.

Characteristics

3-84. Cover and concealment are important if the company is to remain in the area for any length of time. Vehicles, equipment, entrances, and exits should be camouflaged to keep the enemy from detecting the location of the company. Consideration should be given to the following.

- Concealment.
- Cover from direct and indirect fire.
- Defendable terrain.
- Drainage, and a surface that will support vehicles.
- Exits and entrances, and adequate internal roads or trails.

Chapter 3

- Space for dispersion of vehicles, personnel, and equipment.
- A suitable landing site nearby for helicopters.

Planning

3-85. The CO plans for an assembly area as he does for a perimeter defense. He organizes the assembly area into a perimeter and assigns each platoon a sector of that perimeter. He also assigns positions to the anti-armor weapons and mortar section, and selects positions for the company CP and trains. The commander and the FSO plan indirect fire in and around the assembly area.

Actions In Assembly Area

3-86. Before the company moves into an assembly area, the CO sends a quartering party to reconnoiter organize it for occupation by the main body.

- When the company arrives at the RP, the platoon guides link up with their platoons and immediately lead them to their positions. The company headquarters guide links up with the headquarters personnel and leads them to their positions. The movement from the RP to the positions should be continuous.
- Once in position, the platoons establish OPs and conduct patrols to secure the area. The platoon leaders then plan the defense of their sectors. Machine gunners, and anti-armor crews prepare range cards. Fighting positions are prepared. Other defensive measures are taken as appropriate.

Communications

3-87. Wire might be the primary means of communications within the assembly area; however, it might be supplemented by messenger, radio, and prearranged signals.

Chapter 4
Offensive Operations

The Infantry company normally conducts offensive operations as part of a larger force. Offensive operations let the commander seize the initiative (choose when and where to fight), retain the initiative, and effectively exploit his company's strengths. This chapter discusses the types and characteristics, sequence of events, actions on contact, attacks, movements to contact, general and specific planning considerations, and common activities of offensive operations. This edition expands the discussion of the integration of small unmanned aerial systems (SUAS) and adds the continuum of contact.

Section	Title	Topic(s)
I	Overview	Types and characteristics of offensive operations, and forms of maneuver
II	Sequence	Order of events
III	Planning Considerations	Considerations for each WFF*
IV	Actions on Contact	All aspects of actions on contact, including the steps
V	Attacks	Types of offensive operations, including hasty and deliberate attack.
VI	Movement to Contact	Establishment and maintenance of contact with the enemy
VII	Common Activities	Warfighting actions
* Additional planning considerations are discussed where they apply.		

Section I. OVERVIEW

During any offensive operation, the commander identifies and focuses his attack on the enemy's weaknesses. He avoids attacking into enemy strengths. This section discusses the characteristics and types of offensive operations, and the various forms of maneuver.

CHARACTERISTICS

4-1. The characteristics of the offense are surprise, tempo, concentration, and audacity. Due to the nature of modern offensive operations, flexibility is included in the following discussion of the offense. For each mission, the commander decides how to apply these characteristics to focus the effects of his combat power against enemy weakness. Detailed planning is critical to achieve a synchronized and effective operation. Instead of 'fighting the plan,' commanders should exploit enemy weaknesses.

Chapter 4

SURPRISE

4-2. Units achieve surprise by striking the enemy at a time, place, or manner in which he is unprepared. Total surprise is rarely essential or attainable. Simply delaying or disrupting the enemy's reaction by attacking where he least expects is usually effective. Surprise delays the enemy's reactions, stresses his command and control, and induces psychological shock in enemy soldiers and leaders. Surprise may allow an attacker to succeed with fewer forces. The company's ability to attack in limited visibility or through restrictive terrain, to operate in small units, and to infiltrate are often key to achieve surprise. The company must exploit the effects of surprise before the enemy can recover.

TEMPO

4-3. Tempo is the rate of military action relative to the enemy. Tempo is not the same as speed. Controlling or altering tempo is essential for maintaining the initiative. Tempo promotes surprise, keeps the enemy off balance, contributes to the security of the attacking force, and prevents the defender from taking effective countermeasures. By increasing tempo, commanders maintain momentum.

4-4. When properly controlled and exploited, tempo confuses and immobilizes the defender until the attack becomes unstoppable. Leaders build tempo into operations through careful planning, synchronization, coordination, and transition to the next operation.

4-5. The company increases its tempo by using simple plans, quick decision making, decentralized control, mission orders, and rehearsed operations. The company maintains tempo by ensuring sustainment operations are well coordinated and continuous, thus preventing culmination.

CONCENTRATION

4-6. The attacker masses the effects of combat power at the decisive point to achieve the unit's purpose. Leaders concentrate the effects of their combat power, while trying not to concentrate forces.

4-7. Because the attacker often moves across terrain the enemy has prepared, he might expose himself to enemy fires. By concentrating overwhelming combat power at a weak area or system, the attacker can reduce the effectiveness of the enemy fires and the amount of time he (the attacker) is exposed to enemy fires.

4-8. The challenge for the company commander is to concentrate combat power, while reducing the enemy's ability to do the same against the friendly unit. Actions that cause the enemy to shift combat potential away from the intended decisive point yield a greater advantage, for example, moving dispersed, but concentrating at the last moment and using deception. The commander employs his Infantry capabilities to achieve overwhelming combat power at the decisive point.

AUDACITY

4-9. Audacity is a simple plan of action, boldly executed. The audacious commander develops confidence by conducting a thorough estimate. His actions, although quick and decisive, are based on a reasoned approach to the tactical situation and on his knowledge of his Soldiers, the enemy, and the terrain. He is daring and original, but he is not rash.

4-10. Audacious commanders throughout history have used the indirect approach. They maneuver to maintain a position of advantage over the enemy, seek to attack the enemy on the flank or rear, and exploit success at once, even if this briefly exposes their own flanks.

4-11. Boldness and calculated risks have always been the keystones of successful offensive operations. However, risks must be consistent with the higher commander's mission and intent. Commanders dispel uncertainty through action; they compensate for a lack of information by seizing the initiative and pressing the fight.

FLEXIBILITY

4-12. Although not a characteristic of the offense, FM 3-90 says that flexibility bears discussion. At some point in most attacks, the original plan must be adjusted to meet changes in the situation. The commander maintains flexibility at all times so he can attack identified enemy weaknesses when they are presented. The commander should avoid 'fighting the plan' and instead focus on fighting the enemy or attacking identified enemy weaknesses. Mission orders, a clear commander's intent, and competent subordinate leaders who exercise initiative ensure that proper adjustments are made.

4-13. The commander and subordinate leaders must expect uncertainties and be ready to exploit opportunities. The flexibility required often depends on the amount of reliable intelligence the commander has on the enemy.

4-14. The commander builds flexibility into his plan during the MDMP. By conducting a thorough war game and rehearsals, he develops a full appreciation for possible enemy actions. A reserve increases the company commander's flexibility.

TYPES

4-15. The four types of offensive operations, described in FM 3-90, are movement to contact (MTC), attack, exploitation, and pursuit. Companies can execute MTC and attack. Platoons generally conduct these forms of offense as part of a company or larger unit operation. Companies and platoons participate in a higher unit's exploitation or pursuit. The nature of these operations depends largely on the amount of time and enemy information available during the planning of and preparation for the operation.

ATTACK

4-16. An attack is an offensive operation that destroys enemy forces, seizes or secures terrain, or both. Movement, supported by fires, characterizes the conduct of an attack. The company will likely participate in a synchronized attack. However, a company may conduct a special purpose attack as part of, or separate from, an offensive or defensive operation. Special purpose attacks consist of ambush, spoiling attack, counterattack, raid, feint, and demonstration. (For a detailed discussion of attacks, see Section V.)

MOVEMENT TO CONTACT

4-17. An MTC is a type of offensive operation designed to develop the situation and establish or regain contact. The company may conduct an MTC on its own or as part of a larger unit's operation when the enemy situation is vague or not specific enough to conduct an attack. (For a detailed discussion of MTC, see Section VI.)

EXPLOITATION

4-18. Exploitations are conducted at the battalion level and higher. The objective of exploitation is to disorganize the enemy in depth. Exploitations seek to disintegrate enemy forces to where they have no alternative but surrender or flight. Companies and platoons may conduct movements to contact or attack as part of a higher unit's exploitation.

PURSUIT

4-19. Pursuits are normally conducted at the brigade or higher level. A pursuit typically follows a successful exploitation. Ideally, it prevents a fleeing enemy from escaping and then destroys him. Companies and platoons will participate in a larger unit's exploitation and may conduct attacks as part of the higher unit's operation.

FORMS OF MANEUVER

4-20. Each form of maneuver attacks the enemy differently. Each poses different challenges for attackers and different dangers for defenders. Maneuver places the enemy at a disadvantage through the application of friendly fires and movement. The five forms of maneuver follow.

ENVELOPMENT

4-21. Envelopment is a form of maneuver in which an attacking force seeks to avoid the principal enemy's defenses by seizing objectives to the enemy rear or flank in order to destroy him in his current positions (Figure 4-1). A successful envelopment requires discovery or creation of an assailable flank. The envelopment is the preferred form of maneuver because the enemy must fight in at least two directions and the attacking force tends to suffer fewer casualties while having the most opportunities to destroy the enemy. Envelopments focus on--

- Seizing terrain.
- Destroying specific enemy forces.
- Interdicting enemy withdrawal routes.

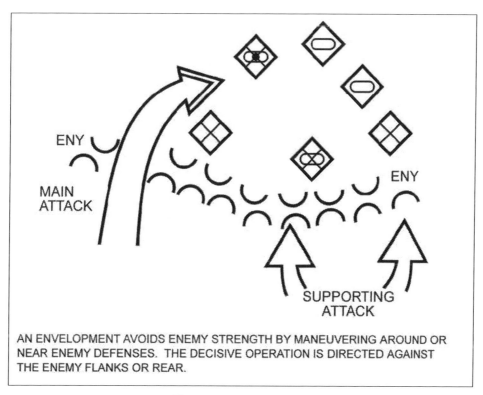

Figure 4-1. Envelopment.

TURNING MOVEMENT

4-22. Turning movement is a form of maneuver in which the attacking force avoids the enemy's principal defensive positions by seizing objectives to the enemy's rear and causing the enemy to move out of his current positions, or to divert major forces to meet the threat (Figure 4-2). For a successful turning movement, the unit trying to turn the enemy must attack something that the enemy will fight to save. This might be a supply route, artillery emplacement, or headquarters. In addition to attacking a target the enemy will fight to save, the attacking unit should be strong enough to pose a real threat to the enemy. The attacker seeks to secure key terrain deep in the enemy's rear and along his lines of communication. Faced with a major threat to his rear, the enemy is turned out of his defensive positions and forced to attack rearward.

Figure 4-2. Turning movement.

INFILTRATION

4-23. Infiltration is a form of maneuver. In an infiltration, an attacking force moves undetected into or through the enemy's main defenses, that is, an area occupied by an enemy forces. The purpose of an infiltration is to occupy a position of advantage in the enemy rear area to concentrate combat power against enemy weak points. Ideally, an infiltration exposes only small elements to enemy defensive fires (Figure 4-3). Moving and assembling forces covertly through enemy positions takes a lot of time. A successful infiltration reaches the enemy's rear without fighting through prepared positions. A company may conduct an infiltration as part of a larger unit's attack with the battalion employing another form of maneuver. The company commander may also employ maneuver by infiltration to move his platoons to locations to support the battalion's attack. A company can infiltrate--

- To attack enemy-held positions from an unexpected direction.
- To occupy a support-by-fire position to support an attack.
- To secure key terrain.

- To conduct ambushes and raids.
- To conduct a covert breach of an obstacle.

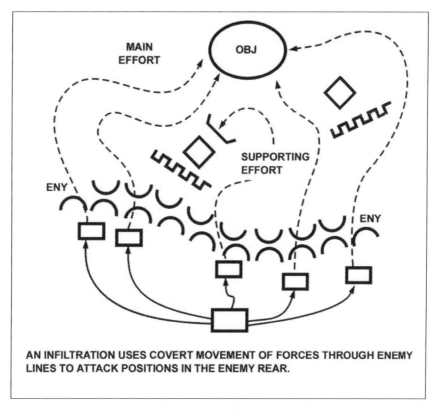

Figure 4-3. Infiltration.

PENETRATION

4-24. Penetration is a form of maneuver in which an attacking force seeks to rupture enemy defenses on a narrow front to create both assailable flanks and access to the enemy's rear (Figure 4-4). Penetration is used when enemy flanks are not assailable, when enemy defenses are overextended, when weak spots in the enemy defense are identified, and when time does not permit some other form of maneuver. A penetration normally consists of three steps.

- Breach the enemy's main defense positions.
- Widen the gap created to secure flanks by enveloping one or both of the newly exposed flanks.
- Seize the objective. As part of a larger force penetration, the company will normally isolate, suppress, fix, or destroy enemy forces; breach tactical or protective obstacles in the enemy's main defense (secure the shoulders of the penetration); or seize key terrain. A battalion may also use penetration to secure a foothold within a large built-up area.

Offensive Operations

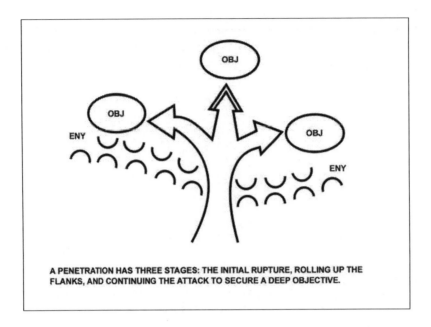

Figure 4-4. Penetration.

FRONTAL ATTACK

4-25. Frontal attack is a form of maneuver in which an attacking force seeks to destroy a weaker enemy force or fix a larger enemy force along a broad front. It is the least desirable form of maneuver, because it exposes the attacker to the concentrated fire of the defender and limits the effectiveness of the attacker's own fires. However, the frontal attack is often the best form of maneuver for an attack in which speed and simplicity are key; it helps overwhelm weak defenses, security outposts, or disorganized enemy forces (Figure 4-5).

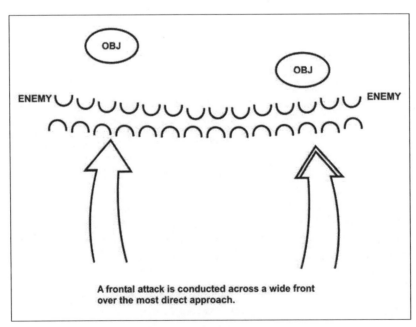

Figure 4-5. Frontal attack.

Chapter 4

Section II. SEQUENCE

This section discusses the sequence of events the company commander must consider while planning for an offensive mission. This sequence applies to many, but not to all, offensive operations.

ASSEMBLY AREA

4-26.   The commander directs and supervises mission preparations in the assembly area to prepare the company for future operations. Preparation time also allows the company to conduct precombat checks and inspections, rehearsals at all levels, and sustainment activities. See Chapter 3 for more information.

RECONNAISSANCE

4-27.   All echelons conduct reconnaissance. The enemy situation and available planning time may limit the unit's reconnaissance, but leaders at every level aggressively seek information about the terrain and enemy. Leaders must remember the benefits of having their Soldiers on the ground providing real-time information in balance with the security risks involved. The on-the-ground company reconnaissance effort reports on enemy activity in the company's AO near the LD, attack position (ATK PSN), assault position (ASLT PSN), or the company-assigned unit objective (OBJ). This provides the maneuver commander with the information needed to execute the best possible tactical plan. The commander should also consider ground reconnaissance with the use of SUASs. Most SUASs have both a daylight and limited visibility (infrared) capability and significantly aid in providing the commander with needed information on the terrain and enemy.

MOVEMENT TO LINE OF DEPARTURE

4-28.   When attacking from positions not in contact, Infantry companies often stage in rear assembly areas, road march to ATK PSN behind friendly units in contact with the enemy, conduct passage of lines, begin the attack, and then move into their AOs. The movement from the assembly area to the LD is timed so the movement to and across the LD is continuous. The lead element of the company starts crossing the LD at the attack time specified in the battalion OPORD. Before the company's movement, a patrol might be sent to reconnoiter and mark the route and check the time it takes to move to the LD. In order to be in an overwatch position, ready to support the company as it crosses LD, the support element may precede the assault and the breach elements to the LD. Mortars are moved forward to a firing position near the LD to allow maximum coverage of the objective area. The commander avoids stopping in the attack position. However, if the company is ahead of schedule or told to hold in the attack position, they occupy the attack position, post security, and wait until time to move.

MANEUVER

4-29.   Maneuver is the essence of every tactical operation. It is the use of movement in combination with fire (or fire potential) employed to achieve a position of advantage with respect to the enemy. The commander employs those techniques that avoid the enemy's strength, focus on enemy weakness, and conceals the company's true intentions. He deceives the enemy as to the location of the decisive operation, uses surprise to take advantage of his initiative in determining the time and place of his attack, and uses indirect approaches, when available, to strike the enemy from a flank or the rear. The company commander maneuvers his platoons to close with the enemy, to gain positional advantage over him, and ultimately to destroy him or force him to withdraw or capitulate.

BASE OF FIRE FORCE

4-30.   The combination of fire and movement requires a base of fire in which some elements of the company remain stationary and provide protection for the bounding forces by preventing the enemy to react to the bounding force.

4-31. The base of fire force occupies positions that afford effective cover and concealment, unobstructed observation, and clear fields of fire. Selection of the base of fire position is based on a careful study of the terrain, knowledge of enemy locations, or likely enemy locations. When the enemy situation is vague or unknown, the position is selected in such a manner as to be able to place effective fire on terrain that dominates the area the bounding force will traverse. Once in position, the base of fire is responsible for both suppressing known enemy forces and for scanning assigned sectors of observation; it identifies previously unknown enemy elements and suppresses them. The protection provided by the base of fire force allows the bounding unit to continue its movement and to retain the initiative even when under enemy observation or within range of enemy weapons.

4-32. Because maneuver is decentralized in nature, decisions on where and when to establish a base of fire must be made at the appropriate level. These decisions normally fall to a leader on a specific part of the battlefield, who knows which enemy forces can engage the bounding force, and which friendly forces are available to serve as the base of fire. At company level, these decisions might be made within the company (with the base of fire provided by a platoon), within platoons (with the base of fire provided by the weapons squad), or within squads (with a fire team as the base of fire). A detailed understanding of the terrain is critical to select appropriate positions.

BOUNDING FORCE

4-33. Movement in a maneuver situation is inherently dangerous. It is complicated by the obvious potential for harm posed by enemy weapons, and by the uncertainty posed by new terrain and other factors.

- The bounding force takes full advantage of the cover and concealment the terrain provides. Leaders enhance security by exploiting use of restrictive terrain (time dependent), intervening terrain features, inter-visibility lines, and avoidance of skylining.
- All elements involved in the maneuver maintain all-round security at all times. Elements in the bounding force scan their assigned sectors of observation continuously.
- Although METT-TC factors ultimately dictate the length of the bounds, the bounding force should not move beyond the range at which the base of fire force can effectively suppress known, likely, or suspected enemy positions. This minimizes the bounding force's exposure to enemy fires.
- In restricted and severely restricted terrain, bounds are generally much shorter than in areas that are more open.
- The bounding element remains focused on its ultimate goal of gaining a positional advantage, which it can then use to destroy the enemy by direct and indirect fires.
- When the bounding force comes under direct fire from the enemy it may become the base of fire, which allows the original base of fire force to become the maneuver force.

DEPLOYMENT

4-34. The company approaches the objective in a manner that supports its deployment prior to the assault. If a support element (from the company) is to be used, it should be positioned before the company's assault element reaches the assault position. The support element initiates its fire on the objective on order or at a specified time. Supporting indirect fires are synchronized to impact at the same time. The company should minimize time in the assault position (or not stop at all.) Movement should be as rapid as the terrain, force mobility, and enemy situation permit.

ASSAULT

4-35. The objective for the company may vary from operation to operation. In every case, the company's actions on the objective are critical and thus the focus of the commander's concept. To develop the concept, the commander starts at the decisive point. This is where the commander wants to focus his combat power. From there, he works backwards. He plans all the tasks and purposes of all the subordinate units so they shape the decisive operation's action at the decisive point. The commander's estimate will determine what other considerations must be included for the actions on the objective. During offensive operations, the unit remains enemy and effects oriented; however, based on the METT-TC factors, the company's objective might be terrain or force oriented. Terrain-oriented tasks and purposes require the company to seize, secure, or retain a designated area. These tasks and purposes may or may not require fighting through enemy forces. Force-oriented tasks and purposes require the company to destroy, suppress, or fix, enemy forces. The company's effort is focused on the enemy's actual location. The enemy might be a stationary or moving force. Actions on the objective start when the company begins placing fires on the objective; this normally occurs when the commander initiates his echelonment of fires onto the objective. Once the company achieves a position of advantage, the Infantry company conducts a rapid and violent assault to accomplish its mission, normally under the cover of darkness.

CONSOLIDATION AND REORGANIZATION

4-36. The company consolidates and reorganizes as required by the situation and mission. Consolidation is the process of organizing and strengthening a newly captured position so that it can be defended. Reorganization is the actions taken to shift internal resources within a degraded unit to increase its level of combat effectiveness. The company executes follow-on missions as directed by the battalion commander. A likely mission is to continue the attack against targets of opportunity in the objective area. Whether a raid, hasty attack, or deliberate attack, a company should posture itself and prepare for continued offensive operations to defeat local counterattacks.

Section III. PLANNING CONSIDERATIONS

The WFF are critical tactical activities the commander can use to review planning, prepare, and execute. Synchronization and coordination among the WFF are critical for success. This section discusses selected WFF and other planning considerations. For a detailed discussion of command, control, and intelligence, see Chapter 2.

INTELLIGENCE

4-37. The company commander will not have complete information about enemy intentions. Therefore, he must obtain or develop the best possible IPB products and conduct continuous ISR collection throughout the operation. He may also need to request information from the battalion staff to answer PIR.

4-38. ISR assets serve to help study terrain to determine the enemy's best area for his main defense, routes he may use for counterattacks (both enemy and own OAKOC); and confirm or deny strengths, dispositions, and likely intentions, especially where and in what strength the enemy will defend.

MOVEMENT AND MANEUVER

4-39. The battalion commander may task-organize the company with engineers as part of a breaching operation in the offense. The company commander may receive additional mobility assets such as an engineer platoon. (FM 3-34.2 discusses breaching operations in detail.)

FIRE SUPPORT

4-40. As part of the top-down fire planning system, the company commander must refine the fire plan from higher headquarters to meet his mission requirements and ensure that these refinements are incorporated into the higher headquarters plan. He incorporates the results of his METT-TC analysis and designates key observer locations and targets from the fire plan as an integral part of the company rehearsal. In addition, he works with the FSO to develop a corresponding observation plan and establishes triggers for initiating, ceasing, or shifting fires. The commander may assign responsibility for the firing of certain targets to subordinate leaders. The company commander and the FSO must have a thorough understanding of organic FSEs and traditional artillery and mortar support assets. The majority of the company's fire support is from mortar systems organic to the Infantry battalion and company. The commander employs supporting fires in the offense to achieve a variety of purposes such as--

- To suppress enemy weapons systems that inhibit movement.
- To fix or neutralize bypassed enemy elements.
- To prepare enemy positions for an assault. Preparatory fires are normally used during a deliberate attack, with fires placed on key targets before the assault begins. These indirect fires are integrated and synchronized with the company's direct fire systems to provide constant pressure on the enemy position and prevent him from reacting to, or repositioning against, the company's assaulting elements. The commander must weigh the benefits of preparatory fires against the potential loss of surprise.
- To obscure enemy observation or screen friendly maneuver. The company can take advantage of smoke in various maneuver situations, such as during a bypass or in deception operations.
- To support breaching operations. Fires are employed to obscure and suppress enemy elements that are overwatching reinforcing obstacles.
- To illuminate enemy positions. Illumination fires are always included in contingency plans for night attacks.

PROTECTION

4-41. Stinger sections, with organic vehicle support, are seldom attached to the company, but may travel with the company in order for the company to protect the air defense assets. Their security must be a consideration in planning for offensive operations. The company commander must plan for and rehearse internal air security and active air defense measures. The commander must anticipate possible contact with enemy air assets by templating enemy helicopter and fixed-wing air corridors and AA. Unit SOPs should dictate internal air security measures and active air defense measures.

SUSTAINMENT

4-42. The main purpose of sustainment in the offense is to assist maneuver elements in maintaining the momentum of the attack. Sustainment functions are performed as far forward as the tactical situation allows. Company trains normally remain one terrain feature out of direct fire range of the enemy behind the location of the company. The commander must consider the enemy situation and how it relates to the security of the company trains. If the company is conducting decentralized operations, the company trains locate where they can best support the platoons in the accomplishment of the company's mission.

COMMAND AND CONTROL

4-43. Though the company's assigned mission and objective may be the either decisive or shaping for the battalion, the commander may decide to translate, develop, and assign both decisive and shaping tasks for the platoons, sequencing his operation by "*find, fix, finish, follow through*" concepts. Typically, he will plan to make contact with the smallest element possible, deceive the enemy as to the company main or decisive effort, employ timely and synchronized fire support, and maneuver platoons to destroy the enemy and seize the company objective.

4-44. The commander will locate where he can maintain a current and accurate picture, and best control his elements as the attack progresses; this is usually with the decisive element. He is prepared to exploit unforeseen advantages and anticipates the need or requirement to shift his effort due to success or to preserve his freedom of maneuver.

Section IV. ACTIONS ON CONTACT

In both offensive and defensive operations, contact occurs when a member of the Infantry company encounters any situation that requires an active or passive response to a threat or potential threat. This section discusses the forms, circumstances, development of, times, and steps of actions on contact.

FORMS

4-45. These situations may entail one or more of the following forms of contact.

- Visual (friendly elements may or may not be observed by the enemy).
- Physical or direct fire with an enemy force.
- Indirect fire.
- With obstacles of enemy or unknown origin.
- With enemy or unknown aircraft.
- Involving CBRN conditions.
- Involving electronic warfare tactics.
- With nonhostile elements such as civilians.

CIRCUMSTANCES

4-46. Leaders at echelons, from platoon through battalion, conduct actions on contact when they or a subordinate element, recognize one of the forms of contact or receive a report of enemy contact. The company may conduct actions on contact in response to a variety of circumstances, including--

- Subordinate platoon(s) conducting actions on contact.
- Reports from the battalion or another higher unit.
- Reports from or actions of an adjacent unit.

DEVELOPMENT

4-47. To identify likely contact situations that may occur during an operation, Infantry company commanders and platoon leaders analyze the situation throughout the TLP. Commanders plan for each of the forms of contact throughout the operation. Through planning and rehearsing, they develop and refine COAs to deal with probable enemy actions. The COAs are the foundation for the company's scheme of maneuver. During the troop-leading procedures, the leaders evaluate a number of factors to determine their impact on the unit's actions on contact. For example, the commander considers how the likelihood of contact affects his choice of movement techniques and formations in order to outline procedures for the transition to more secure movement techniques before a contact situation.

TIME REQUIREMENTS

4-48. Infantry commanders must understand that properly executed actions on contact (for any of the forms of contact) require time at both platoon and company levels. To develop the situation fully, a platoon might have to execute extensive lateral movement, conduct reconnaissance by fire, or call for and adjust indirect fires. Each of these activities requires time. The commander must balance the time required for subordinate elements to conduct actions on contact with the need of the company or battalion to maintain tempo and momentum.

STEPS

4-49. The company executes actions on contact, as applicable for each form of contact, using a logical, well-organized process of decision-making and action, which actually entails four separate actions: (1) deploy and report; (2) evaluate and develop the situation; (3) choose a COA; and (4) execute the selected COA. These can be done out of sequence. In fact, some are *more likely* to be done at the same time. These actions provide an orderly framework that enables the company and its platoons to respond to initial contact and then apply sound decision-making and timely actions to complete the operation. Ideally, the company will acquire the enemy before being sighted by the enemy; the company can then initiate physical contact on its own terms by executing the designated COA.

STEP 1, DEPLOY AND REPORT

4-50. Events that occur during initial contact depend in great measure on whether the contact is expected or unexpected. Regardless of whether contact is expected or unexpected, the first step of actions on contact concludes with the unit deployed (into base of fire and maneuver forces), the enemy suppressed or destroyed (if applicable), and the commander sending a contact report to battalion headquarters. The following paragraphs examine some variables the company commander faces in expected and unexpected contact situations. The roles of platoon battle drills, SOPs, and reports are also detailed.

Expected Contact

4-51. If the commander expects contact, he will have already deployed the company by transitioning to the bounding overwatch movement technique. If the company is alert to the likely presence of the enemy, it has a better chance of establishing visual contact and then physical contact, on its own terms before being detected by the enemy. An overwatching or bounding platoon usually makes visual or physical contact that initiates the company's actions on contact. In a worst-case scenario, the platoon might be engaged by a previously undetected (but expected) enemy element. In this event, the platoon conducts a battle drill for its own survival and then initiates actions on contact.

Unexpected Contact

4-52. In some cases, the company may make unexpected contact with the enemy while using traveling or traveling overwatch. The element in contact or, if necessary, the entire company might have to deploy using battle drills to survive the initial contact.

Battle Drills

4-53. Battle drills provide automatic responses to contact situations where immediate, often violent execution is critical, both to initial survival and to ultimate success in combat. Rather than being a substitute for carefully planned COAs, drills buy time for the unit in contact, and frame the development of the situation. When enemy contact occurs, the company's platoons deploy immediately, executing the appropriate battle drills under the direction of the commander. (For additional information on dismounted platoon battle drills, see FM 3-21.8 (FM 7-8) and ARTEP 7-8-Drill.)

Maneuver Standing Operating Procedures

4-54. An effectively written, well-rehearsed maneuver SOP helps to ensure quick, predictable actions by all members of the company. The SOP, unlike platoon battle drills, allows leaders to take into account the friendly task organization, a specific enemy, and a specific type of terrain. Therefore, the SOP can assist the company in conducting actions on contact and maintaining the initiative in a number of battlefield situations.

Reports

4-55. Timely, accurate, and complete reports are essential throughout actions on contact. As part of the first step of the process, the company commander must send a contact report to the battalion as soon as possible after contact occurs. He provides subsequent reports to update the situation as necessary.

STEP 2, EVALUATE AND DEVELOP THE SITUATION

4-56. While the company deploys, the commander evaluates and develops the situation. The goal of these actions is to create conditions, which provide for the successful execution of the decisive action. The commander gathers as much information as possible, either visually or, more often, through reports from the platoon(s) in contact.

Factors

4-57. He analyzes the information to determine critical operational considerations, including these factors.

- Size of the enemy element.
- Location, composition, activity, orientation, and capabilities of the enemy force.
- Effects of obstacles and terrain.
- Probable enemy intentions.
- How to gain positional advantage over the enemy.
- Friendly situation (location, strength, and capabilities).
- Possible friendly COAs to achieve the specified end state.

Techniques

4-58. After evaluating the situation, the commander may discover that he does not have enough information to identify the necessary operational considerations. To make this determination, he must further develop the situation IAW the battalion commander's intent, using a combination of these techniques.

- Squads conducting surveillance (using binoculars and other optical aids).
- Lateral maneuver to gain additional information by viewing the enemy from another perspective.
- Indirect fire.
- Reconnaissance by fire.

Reports

4-59. Once the commander has determined the size of the enemy force the company has encountered, he sends a report to the battalion.

STEP 3, CHOOSE A COURSE OF ACTION

4-60. After developing the situation and determining that he has enough information to make a decision, the company commander selects a COA that meets the requirements of the battalion commander's intent, achieves the company's purpose, maximizes the effects of terrain, minimizes casualties, and is within the company's capabilities.

Nature of Contact

4-61. The nature of the contact (expected or unexpected) may have a significant impact on how long it takes a commander to develop and select a COA. For example, in preparing to conduct an attack, the company commander determines that the company will encounter an enemy security OP along its axis of advance. During TLP, he develops a scheme of maneuver to defeat the outpost. When the company's lead platoon makes contact with the enemy, the commander can quickly assess that this is the anticipated contact and direct the company to execute his plan. On the other hand, unexpected contact with a well-concealed enemy force may require time for development of the situation at platoon level. As the unit fights for critical information that will eventually allow the commander to make a sound decision, the company might have to employ several of the techniques for developing the situation.

Procedures for Selecting Course of Action

4-62. The company commander has several options in selecting a COA.

- If his development of the situation reveals no need for change, the company commander directs the company to execute the original plan.
- If his analysis shows that the original plan is still valid but that some refinement is necessary, the company commander informs the battalion commander (prior to execution, if possible) and issues a FRAGO to refine the plan.
- If his analysis shows that the original plan needs to be changed but the selected COA will still comply with the battalion commander's intent, the company commander informs the battalion commander (prior to execution, if possible) and issues a FRAGO to retask his subordinate elements.
- If his analysis shows that the original plan deviates from the battalion commander's intent and needs to be changed, the company commander must report the situation and, based on known information in response to an unforeseen enemy or battlefield situation, recommend an alternative COA to the battalion commander.
- If the battlefield picture is still vague, the company commander must direct the company or a platoon to continue to develop the situation. This will allow him to gather the information needed to clarify a vague battlefield picture. He then uses one of the first four options to report the situation, choose a COA, and direct further action.

STEP 4, EXECUTE SELECTED COURSE OF ACTION

4-63. In executing a COA, the company transitions to maneuver. It then continues to maneuver throughout execution (either as part of a tactical task or as an advance while in contact) to reach the point on the battlefield where it executes its tactical task. The company can employ a number of tactical tasks as COAs, any of which might be preceded and followed by additional maneuver. As execution continues, more information becomes available to the company commander. Based on the emerging details of the enemy situation, he might have to alter his COA during execution.

Section V. ATTACKS

This section discusses characteristics, types, and techniques for the hasty attack, the deliberate attack, special purpose attacks, and other information relating to attacks by an Infantry company. In the attack, the company maneuvers along lines of least resistance using the terrain for cover and concealment. This indirect approach affords the best chance to achieve surprise on the enemy force. In the attack, the company maneuvers along lines of least resistance using the terrain for cover and concealment. This indirect approach affords the best chance to achieve surprise on the enemy force.

CHARACTERISTICS

4-64. An attack is a type of offensive operation characterized by movement supported by fire. The purpose of an attack is to destroy an enemy force or to seize terrain. The attack should always try to strike the enemy where he is weakest. The company can attack independently or as part of a battalion or larger element. The two basic types of attack are the hasty attack and the deliberate attack. Figure 4-6 shows the situations under which a company conducts an attack, compares them to the amount of planning and preparation time required, and provides options for the commander to accomplish his purpose and support the higher commander's intent. All attacks, whether hasty or deliberate, depend on synchronization for success. They require planning, coordination, and time to prepare.

Attack Situations / Planning Time	Force-Oriented Moving Enemy	Force-Oriented Stationary Enemy	Terrain-Oriented
	Attack Options		
Less Time	Hasty attack to (destroy, disrupt, block) Counterattack Spoiling attack Ambush	Hasty attack to (destroy, disrupt, block) Counterattack Feint Demonstration	Hasty attack to (seize, clear, secure) Counterattack
More Time	Deliberate attack to (destroy, disrupt, block) Counterattack Spoiling attack Ambush Feint Demonstration	Deliberate attack to (destroy) Raid Counterattack Feint Demonstration	Deliberate attack to (seize, clear, secure) Counterattack

Figure 4-6. Spectrum of attacks.

4-65. The company commanders translate the mission assigned by the battalion, through analyzing the task and purpose, into specific missions for subordinate platoons and squads. To facilitate parallel planning, they immediately forward these missions, along with the appropriate portions of the battalion's plans orders, to subordinate platoons and squads. Commanders and platoon leaders must work together to develop the best plans; this requires sharing information freely between the command posts. The goal is to not simply reduce the time required to produce and distribute the plans, but, more importantly, to produce a better plan by including input from adjacent, higher, and lower elements. Also, this collaboration promotes understanding of the plan, thereby enhancing preparation and execution.

4-66. As the company plans, the enemy also has time to improve his defenses, disengage, or conduct spoiling attacks of his own. Clearly, planning must be accomplished in the shortest time possible and must accommodate the changes driven by what the enemy does.

TYPES

4-67. No clear distinction exists between deliberate and hasty attacks, because they are similar. However, the main difference between the two is the extent of planning and preparation conducted by the attacking force. Attacks range along a continuum defined at one end by FRAGOs, which direct the rapid execution of battle drills by forces immediately available. At the other end of the continuum, the company moves into a deliberate attack from a reserve position or assembly area with detailed knowledge of the enemy (a task organization designed specifically for the attack) and a fully rehearsed plan. Most attacks fall somewhere between these two ends of the continuum.

HASTY ATTACK

4-68. The commander may conduct a hasty attack during MTC, as part of a defense, or whenever he determines that the enemy is in a vulnerable position and can be quickly defeated by immediate offensive action. A hasty attack is used to--

- Exploit a tactical opportunity.
- Maintain the momentum.
- Regain the initiative.

- Prevent the enemy from regaining organization or balance.
- Gain a favorable position that might be lost with time.

4-69. Because its primary purpose is to maintain momentum or take advantage of the enemy situation, the hasty attack is normally conducted with available resources. Maintaining unrelenting pressure through hasty attacks keeps the enemy off balance and makes it difficult for him to react effectively. Attacking before the enemy can act often results in success even when the combat power ratio is not as favorable as desired. With its emphasis on agility and surprise, however, this type of attack may cause the attacking force to lose a degree of synchronization. To minimize this risk, the commander should maximize use of standard formations; well-rehearsed, thoroughly understood battle drills and SOPs; and digital tools that facilitate rapid planning and preparation. By assigning on-order and be-prepared missions to subordinate companies, as the situation warrants, the company is better able to transition into hasty attacks. The hasty attack is often the preferred option during continuous operations. It allows the commander to maintain the momentum of friendly operations while denying the enemy time to prepare his defenses and to recover from losses suffered during previous action. Hasty attacks normally result from a MTC, successful defense, or continuation of a previous attack.

Task Organization

4-70. The hasty attack is conducted using the principles of fire and movement. The controlling headquarters normally designates a base of fire force and a maneuver force. The Infantry company may also be task organized with one or more assault platoons from the Infantry battalion's Weapons company. An assault platoon has the ability to provide mobile overwatching fires with its organic M2. 50 cal heavy machine guns, Mark 19 automatic grenade launchers and TOW Improved Target Acquisition Systems (ITAS). The ITAS (and the Javelin) is also considered a close combat missile system.

Conduct

4-71. The company must first conduct actions on contact, allowing the commander to gather the information he needs to make an informed decision. The term "*hasty*" refers to limits on planning and preparation time, not to any acceleration in the conduct of actions on contact. Because the intelligence picture is vague, the commander normally needs more time, rather than less, during this process to gain adequate information about the enemy force. Execution begins with establishment of a base of fire, which then suppresses the enemy force. The maneuver force uses a combination of techniques to maintain its security as it advances in contact to a position of advantage. These techniques include, among others, the following.

- Use of internal base of fire and bounding elements.
- Use of covered and concealed routes.
- Use of indirect fires to suppress or obscure the enemy or to screen friendly movement.
- Execution of bold maneuver that initially takes the maneuver force out of enemy direct fire range.
- Once the maneuver force has gained the positional advantage, it can execute a tactical task to destroy the remaining enemy.

DELIBERATE ATTACK

4-72. The Infantry company normally conducts a deliberate attack against a strong enemy defense. As the company prepares for the attack, the enemy also continues to strengthen his position. Deliberate attacks follow a distinct period of preparation, which is used for extensive reconnaissance and intelligence collection, detailed planning, task organization of forces, preparation of troops and equipment, coordination, rehearsals, and plan refinement. The deliberate attack is a fully synchronized operation that employs every available asset against the enemy defense. It is characterized by a high volume of planned fires, use of major supporting attacks, forward positioning of resources needed to maintain momentum, and operations throughout the depth of enemy positions. Thorough preparation allows the attacking force to stage a combined arms and fully integrated attack. Likewise, however, the enemy will have more time to prepare his defensive positions and integrate fires and obstacles. The METT-TC factors dictate how thoroughly these activities are accomplished. The commander normally conducts a deliberate attack when enemy positions are too strong to be overcome by a hasty attack. In weighing his decision to take the time needed to prepare for and conduct the deliberate attack, the commander must consider the advantages that might be gained by both friendly and enemy forces.

Task Organization

4-73. The company commander normally task-organizes the company into support and assault forces for conduct of a deliberate attack. He also designates a breach force if the company may conduct a breach as part of the attack. Specific duties of these elements are covered in the discussion of a company level assault of a strongpoint and tactical tasks.

Conduct

4-74. The Infantry company's deliberate attack normally is broken into the following steps.

Attack in Zone

4-75. The attacking company advances within assault distance of the enemy position under supporting fires and uses any combination of movement techniques. Platoons advance to successive positions using available cover and concealment. The company commander may designate support by fire positions to protect friendly forces with suppressive direct fires. As the company maneuvers in zone, it employs lethal and nonlethal fires to suppress and obscure enemy positions.

Actions at Probable Line of Deployment

4-76. The probable line of deployment (PLD) is normally a phase line or CP where elements of the attacking company transition to secure movement techniques in preparation for contact with the enemy. Platoons may maneuver from the PLD to designated support-by-fire positions, assault positions, or breach or bypass sites. The PLD might be collocated with the assault position.

Actions on Objective

4-77. The assault combines the effects of overwhelming suppressive fires with the use of maneuver to gain positional advantage over the defending enemy. Fires from support forces and from indirect fire assets isolate the objective area and suppress the enemy. These fires protect the assault force as it closes with the enemy. Other measures the Infantry company may use to set the conditions for the assault include, among others, the following.

- Employment of mortar, artillery, direct fires, or a combination of these, from support-by-fire positions to destroy or isolate enemy forces on the objective and create favorable force ratios.
- Use of obscuring smoke.
- Once the conditions are set, the assault forces maneuver to close with and destroy the enemy. Other Infantry company elements continue to provide support as necessary throughout the assault.

SPECIAL PURPOSE ATTACKS

4-78. The company conducts a special purpose attack at the direction of the battalion commander. The commander's decision is based on the METT-TC factors. Special purpose attacks are subordinate forms of an attack and they include the following. As forms of attack, these share many planning, preparing, and executing considerations of the offense. Feints and demonstrations are associated with military deception operations.

- Ambush.
- Raid.
- Spoiling attack.
- Counterattack.
- Feint.
- Demonstration.

AMBUSH

4-79. An ambush is a surprise attack from concealed positions on a moving or temporarily halted enemy. It may take the form of an assault to close with and destroy the enemy, or it might be an attack by fire only, executed from concealed positions. An ambush does not require that ground be seized or held. Infantry forces normally conduct ambushes.

Purposes

4-80. Ambushes are generally executed to reduce the enemy force's overall combat effectiveness. Destruction is the primary reason for conducting an ambush. Other reasons to conduct ambushes are to harass and capture the enemy or capture enemy equipment and supplies.

Operational Considerations

4-81. The execution of an ambush is offensive in nature. However, the company might be directed to conduct an ambush in a wide variety of situations. For example, the company may stage the ambush during offensive or defensive operations (as part of battalion rear area operations), or during retrograde operations. OPSEC is critical to the success of an ambush and is a major reason the operation is normally conducted only by Infantry forces. The company must take all necessary precautions to avoid detection during movement and during the preparation of the ambush site. The company must also have a secure route of withdrawal after the ambush.

Actions

4-82. An ambush normally consists of the following steps.

- Tactical movement to the ORP.
- Reconnaissance of the ambush site.
- Establishment of ambush site security.
- Preparation of the ambush site.
- Execution of the ambush.
- Withdrawal.

Task Organization

4-83. The company normally is task-organized into assault, support, and security forces for the execution of the ambush.

Chapter 4

Support Force

4-84. The support force fixes the enemy force and prevents it from moving out of the kill zone, allowing the assault force to conduct the ambush. The support force generally uses direct fires in this role, but it can also call for indirect fires to further fix the ambushed force.

Assault Force

4-85. The assault force executes the ambush. It may employ an attack by fire, an assault, or a combination of those techniques to destroy the ambushed force.

Security Force

4-86. The security force provides protection and early warning to the ambush patrol and secures the ORP. It isolates the ambush area both to prevent the ambushed enemy force from moving out of the ambush site and to keep enemy rescue elements from reaching the site. The security force also might be responsible for securing the company's withdrawal route.

Types of Ambushes

4-87. Once the company receives an order to conduct an ambush, the commander must determine which of the two types of ambush operations is best suited to the situation and the capabilities of his company.

Point

4-88. In a point ambush, the patrol deploys to attack an enemy force in a single kill zone.

Area

4-89. In an area ambush, the patrol is deployed to conduct several related point ambushes throughout an ambush area.

RAID

4-90. A raid is a limited-objective form of attack entailing swift entry into hostile terrain. A raid operation always ends with a planned withdrawal to a friendly location upon the completion of the assigned mission. It is not intended to hold terrain. The company can conduct an independent point raid or it can participate in a battalion area raid. A point raid attacks the enemy force on a single objective; an area raid encompasses several related point raids or other related operations. Rarely will a company conduct an area raid alone.

Company Role

4-91. The company conducts raids to accomplish a number of missions, including any or all the following.

- Capture prisoners.
- Capture or destroy specific command and control locations.
- Destroy logistical areas.
- Obtain information concerning enemy locations, dispositions, strength, intentions, or methods of operation.
- Confuse the enemy or disrupt his plans.

Task Organization

4-92. Task organization of a raiding force is based on the purpose of the operation. It normally consists of the following elements.

- Support force.
- Assault force.
- Breach force.
- Security force.

Conduct of Raid

4-93. The main differences between a raid and other attack forms are the limited objectives of the raid and the associated withdrawal following completion. Raids might be conducted in daylight or darkness, within or beyond supporting distance of the parent unit. When the area to be raided is beyond supporting distance of friendly lines, the raiding party operates as a separate force. A specific objective is normally assigned to orient the raiding unit. During the withdrawal, the attacking force should use a route or axis different from that used to conduct the raid itself.

SPOILING ATTACK

4-94. A spoiling attack is a limited-objective attack to delay, disrupt, or destroy the enemy's capability to attack. Infantry commanders mount spoiling attacks from defensive postures to disrupt expected enemy attacks. A spoiling attack tries to strike the enemy when he is most vulnerable: during preparations for attack in assembly areas and attack positions, or while he is on the move prior to crossing his line of departure. In most respects, commanders conduct spoiling attacks like any other attack. They might be either hasty (when time is short) or deliberate (when the command has adequate forewarning). When the situation permits, commanders exploit a spoiling attack like any other attack.

COUNTERATTACK

4-95. A counterattack is an attack by defensive forces to exploit the success of a defense, regain the initiative, or to deny the enemy success with his attack. Commanders conduct counterattacks either with a reserve or with lightly committed forward elements. They counterattack after the enemy launches his attack, reveals his decisive operation, or creates an assailable flank. Infantry commanders conduct counterattacks much like other attacks. However, synchronizing counterattacks within the overall defensive effort requires careful timing.

4-96. Remember that timing is critical. To be decisive, the counterattack must occur when the enemy is overextended, dispersed, and disorganized during his attack. All counterattacks should be rehearsed in the same conditions that they would be conducted. Careful consideration must be given to the event that will trigger the counterattack. Once committed, the counterattack force conducts the decisive operation.

4-97. As in spoiling attacks, commanders prepare to seize the opportunity to exploit success by the entire force. However, counterattacks might be limited to movement to better terrain in order to bring fires on the enemy. Given the same forces on both sides, counterattacks can achieve greater effects than other attacks because the defender can create better conditions by rehearsing and by controlling timing.

FEINT

4-98. A feint is a form of attack used to deceive the enemy of the location or time of the actual decisive operations or main attack. Its purpose is to deceive the enemy and cause him to react in a particular way; such as by repositioning forces, committing its reserve, or shifting fires. The feint seeks direct fire contact with the enemy but avoids decisive engagement. The feint, in many ways, is identical to other attack forms. The feint is much more limited in scope than other attack forms, in part due to its extremely specific objective. The scale of the operation, however, is usually apparent only to the controlling headquarters. For the element actually conducting the feint, such as an Infantry company or battalion, execution is just as rapid and violent as in a full-scale attack.

Planning Considerations

4-99. The company normally participates in a feint as part of a larger element. Among the planning considerations for the company commander are the following.

- The higher commander's intent regarding force preservation.
- Disengagement criteria and plans.
- Assignment of limited depth and attainable objectives.
- Clear follow-on orders that ensure the feinting force is prepared to exploit the success of the main attack if necessary.

Credibility

4-100. Feints are successful only if the enemy believes that a full-scale attack operation is underway. To be believable, they must be conducted with the same violence and the same level of precision as any attack. The controlling headquarters must issue a clear task and purpose to the unit conducting the feint. This should include identification of the specific enemy action the feint is supposed to trigger (or deny), such as forcing the commitment of an enemy reserve force or preventing an enemy element from repositioning against the decisive attack. Feints are most effective under the following conditions.

- When they reinforce the enemy's expectations.
- When the attack appears to present a definite threat to the enemy.
- When the enemy has a large reserve that he has consistently committed early in the battle.
- When the attacker has several feasible COAs, any of which the enemy could mistake for the decisive operation.

DEMONSTRATION

4-101. The demonstration is an attack to deceive the enemy about the location of the decisive operation or main attack. This purpose is very similar to that of a feint, but the friendly force does not seek to make contact with the enemy. For example, the Infantry company's role might entail establishing an attack-by-fire position beyond the enemy's direct fire engagement range; the purpose would be to cause the enemy to commit a specific element simply by virtue of the positioning of the demonstration force. In preparing to participate in a demonstration as part of a larger force, the company commander should keep in mind the following planning considerations.

Limit of Advance

4-102. The limit of advance must be carefully planned so the enemy can "*see*" the demonstration force but cannot effectively engage it with direct fires. The force must also take any other security measures necessary to prevent engagement by the enemy.

Contingency Plans

4-103. The demonstration force must make contingency plans so it can respond effectively to enemy direct or indirect fires while avoiding decisive engagement.

Follow-On Orders

4-104. Clear, specific follow-on orders must ensure that the demonstration force is prepared to exploit the success of the main attack, if necessary.

OTHER ATTACK TECHNIQUES

4-105. The assault on an enemy strongpoint and a limited visibility attack are the two most demanding attacks a company conducts. The fundamentals and techniques discussed in this paragraph will help the Infantry commander plan, prepare, and conduct all attacks. The commander should try to identify and exploit enemy weakness in all attacks.

ASSAULT OF A STRONGPOINT

4-106. The most difficult objective for an Infantry force is to seize or clear an enemy strongpoint complete with obstacles and fortifications. The commander employs techniques that avoid attacking the enemy's main strength or into the enemy's main obstacle belt. Instead, he tries to identify and attack a weakness in the defense. (Most of the information contained in this section also applies to assaulting an urban strong point; however, the commander should refer to Chapter 12, *Urban Operations*, and FM 3-06.11 to ensure he considers all factors involved in urban operations.).

DECEIT OF THE ENEMY

4-107. The Infantry commander deceives the enemy to the point of the main attack; he uses surprise to take advantage of his initiative in determining the time and place for the attack. He tries to strike the enemy on exposed flanks or the rear. He tries to identify and bypass enemy obstacles. He usually conducts a deliberate attack of a strongpoint as follows.

- Reconnoiter the objective and develop the concept.
- Move to the objective.
- Isolate the objective and the selected breach site.
- Attack to seize a foothold.
- Exploit the penetration and clear the objective.

RECONNOITER AND DEVELOP THE CONCEPT

4-108. The commander reconnoiters the objective himself or has someone else do it. The reconnaissance should identify the positions on the objective (crew-served weapons, C2 locations, and vehicles), the level of preparation, the gaps in the defense, and other potential strengths or weaknesses. The Infantry commander may conduct reconnaissance of the objective to determine any changes from previous information.

4-109. The reconnaissance might be done many different ways. An effective technique is to task-organize a reconnaissance patrol with leaders from the assault, support, and breach elements. There should be sufficient personnel to establish surveillance on the objective and to secure the ORP but the party must be small enough to move undetected. The reconnaissance patrol either returns to the company's location or meets the company at a designated linkup point and guides it into the ORP, which then becomes the company's assault position. At times, the scout platoon or other battalion assets might be tasked to conduct reconnaissance in support of the company's mission.

4-110. After the commander develops his concept, he often task-organizes his unit into a breach element, a support element, an assault element, and possibly a reserve. The reserve is normally under his control and is positioned where it can best exploit the success of the attack. The reserve should not be so close that it loses flexibility during the assault. The reserve leader must know where he will locate throughout the attack.

4-111. The breach force is usually formed around an Infantry unit. Engineers, if available, are also part of the breaching element. Any mechanical or explosive breaching assets are attached to this element. The breach force makes the initial breach and passes the assault element through. It might have to organize its own assault element (to secure the breach), support element (to provide close-in suppression), and breach element (to actually breach the obstacles).

4-112. The support element is organized to provide supporting (indirect or direct) fires initially to the breach element, then to the assault element. The support element may consist of any combination of

Chapter 4

Infantry squads, the mortar section, the machine-gun teams, or M203 gunners. Their primary responsibilities are to isolate the breach point and suppress enemy forces in order to protect the assault force.

4-113. The assault element is usually one or more Infantry platoons, depending on the enemy situation (number of personnel, level of preparation, and complexity of fortifications) and the size and composition of the breach and support elements. Often, a small assault element supported by a large volume of accurate suppressive fires is effective in clearing the objective. The assault element may also need to breach enemy protective obstacles on the objective.

4-114. The commander determines the best task organization for the entire mission. It should be simple and maintain unit integrity whenever possible. Task organization should be accomplished prior to crossing the LD.

MOVE TO THE OBJECTIVE

4-115. The company approaches the objective in a manner that supports its deployment prior to the assault (Figure 4-7, page 4-26). The company may cross the LD (or depart the perimeter defense) supported by heavy suppressive direct and indirect fires, or wait until the appropriate time to initiate fires. The commander must consider the time of the movement and the ammunition available. The commander should also consider the effect of fires on his ability to achieve surprise. If so employed these fires may continue until the company reaches its assault position or final coordination line (FCL); they then shift to allow the assault on the objective. The commander can initiate indirect fires whenever he decides based on his COA. In either case, the following fundamentals should be part of this step of the attack.

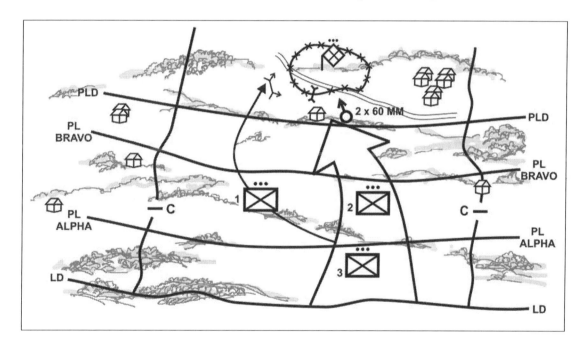

Figure 4-7. Movement to objective.

- The movement from the assembly area to the LD is timed so that movement to and across the LD is continuous. The lead element of the company starts crossing the LD at the attack time specified in the battalion OPORD. Before the company's movement, a patrol might be sent to reconnoiter and mark the route and check the time it takes to move to the LD.
- The support element may precede the assault and breach elements to the LD in order to be in an overwatch position ready to fire when the assault and breach elements cross the LD. Company mortars move forward to a firing position near the LD to provide support during movement and on the objective area. Mortars might have to displace as the operation progresses in order to

provide support in the objective area. The displacement is executed in such a manner to provide continuous indirect fire support throughout the operation.

- The commander normally avoids stopping in the attack position. However, if the company is ahead of schedule or told to hold in the attack position, it occupies the attack position, posts security, and waits until time to move (or until told to move).
- During movement from the LD to the assault position, the company makes the best use of cover, concealment, smoke, and supporting fire.
- If the company is engaged by indirect fire en route, it moves quickly out of the impact area. If it meets enemy resistance short of the objective, it returns fire at once. The leader of the platoon in contact calls for and adjusts indirect fire on the enemy. Depending on the company plan and the location and type of resistance, the platoon may bypass an enemy position that cannot affect the mission.
- If the company cannot bypass an enemy position, the company commander and the platoon leader in contact must take prompt and aggressive action. The platoon leader tries to conduct the platoon attack drill and destroy the enemy position. The commander quickly conducts an estimate of the situation and issues FRAGOs as needed to carry out his plan. He coordinates actions and fires so the company can attack the enemy with its full combat power. The commander should maneuver to assault the flank or rear of the enemy position. When it has destroyed or suppressed the enemy, the company continues toward its objective.
- The company either bypasses or breaches obstacles along the route. The company should bypass when feasible and when the terrain allows. The commander must decide the best way to overcome the obstacle without losing momentum. In selecting the scheme of maneuver, the commander normally tries to avoid COAs that require breaching of enemy obstacles. Enemy obstacles are likely to be covered by direct and indirect fires; common sense would indicate that the commander should avoid these areas. Because all forces construct defensive obstacles around their positions, however, the attacking unit must be prepared to conduct a breach if a bypass is not possible. In a battalion deliberate attack, the company might be the breach force; it may conduct breaches with its organic equipment or with attached engineer assets. The commander should consider avoiding the most obvious overwatch positions around the obstacles and enemy strongpoint since these locations will likely be covered by enemy direct and indirect fires. The company commander informs the battalion commander of obstacles that may affect units following the company. The company commander positions engineers forward to provide a rapid assessment of the obstacle.
- The support force (if any) should be in position before the company's assault force reaches the assault position. The support element initiates its fire on the objective based on a specific event, on order, or at a specified time. The commander considers stealth and surprise when deciding when to initiate direct and indirect fires. Supporting indirect fires are synchronized to impact at the same time.

ISOLATE THE OBJECTIVE AND THE SELECTED BREACH SITE

4-116. Normally, the battalion isolates the objective area to allow the company (or companies) to concentrate on the enemy strongpoint (Figure 4-8).

Chapter 4

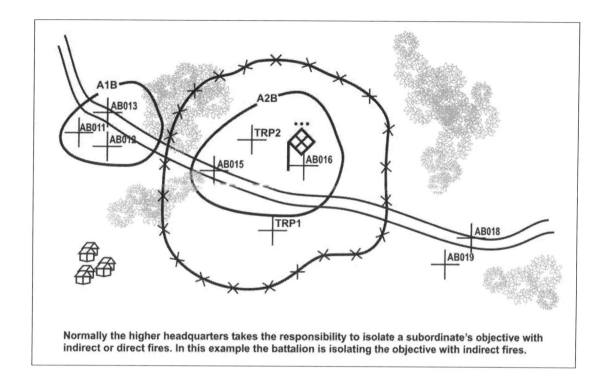

Figure 4-8. Isolation of objective.

4-117. The company may begin the isolation during the leader's reconnaissance by positioning security elements to prevent enemy movement into or out of the objective area. The commander must ensure that these units understand what actions to take in the event of enemy contact. Initially, they may only observe and report until the company deploys for the assault. At a designated time or signal, they begin active measures to isolate the objective.

4-118. Once the objective area is isolated, the commander focuses on isolation at the point of attack or breach point (if the company must breach) The commander always seeks at attack on the enemy's flank or rear, and seeks to avoid a frontal attack into enemy obstacles (this is most likely the enemy's strength.) This isolation helps prevent enemy reinforcement or repositioning at the point of attack (or breach site). It also helps in suppressing enemy weapons and positions that have observation and fire on the point of attack (or breach site.) If the precise locations of enemy weapons have not been determined, but the commander has still decided to breach, the support force concentrates on the terrain that dominates the obstacle and breach site. The support force is assigned the main responsibility for this isolation. Figure 4-9 shows the planned general area to establish a breach.

4-119. The commander masses the majority of available combat power effects at the initial penetration or breach point. He uses indirect fires to suppress or obscure adjacent enemy positions.

ATTACK TO SEIZE A FOOTHOLD

4-120. The initial penetration (or breach if necessary) of the enemy position is normally the Infantry company's initial focus. The initial assault force (or breach force) bypasses or penetrates the enemy's protective obstacles to gain a foothold and create a gap large enough for the assault element to pass through (A and B, Figure 4-9).

Offensive Operations

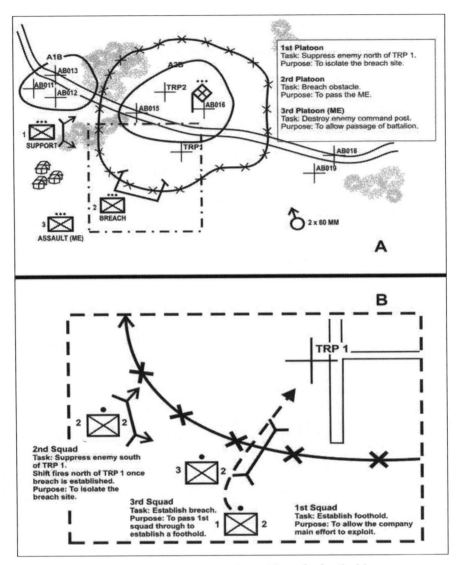

Figure 4-9. Breaching and securing of a foothold.

PREPARE

4-121. Suppress, obscure, secure, reduce, and assault (SOSRA) are the breaching fundamentals that must be applied to ensure success when breaching against a defending enemy. These fundamentals will always apply, but they may vary based on the specific battle-space situation (METT-TC). These fundamentals, and the applicable task organization, should be planned for even if a breach is not actually necessary or conducted. If the breach force can bypass obstacles and still seize the initial foothold then it will do so but the force should still be prepared to breach.

SUPPRESS

4-122. Suppression is a tactical task used to employ direct or indirect fires on enemy personnel, weapons, or equipment to prevent or degrade enemy fires and observation of friendly forces. The purpose of suppression during breaching operations is to protect forces reducing and maneuvering through an obstacle. Effective suppression is a mission-critical task performed during any breaching operation. Suppressive fires prevent the enemy from emplacing effective fire on the breach site. Successful suppression generally triggers the rest of the actions at the obstacle. Fire control measures ensure that all fires are synchronized with other actions at the obstacle. Although suppressing the enemy overwatching the obstacle is the mission of the support force, the breach force should be able to provide additional suppression against an enemy that the support force cannot effectively suppress.

OBSCURE

4-123. Obscuration must be employed to protect forces conducting obstacle reduction and the passage of assault forces. Obscuration degrades enemy observation and target acquisition and conceals friendly activities and movement. Obscuration smoke deployed on or near the enemy's position minimizes its vision. Screening smoke employed between the reduction area and the enemy conceals movement and reduction activities. It also degrades enemy ground and aerial observations. Obscuration must be carefully planned to provide maximum degradation of enemy observation and fires, but it must not significantly degrade friendly fires and control.

SECURE

4-124. Friendly forces secure the reduction area to prevent the enemy from interfering with obstacle reduction and the passage of the assault force through the lanes created during the reduction. Security must be effective against outposts and fighting positions near the obstacle and against overwatching units, as necessary. Fires must secure the far side of the obstacle, or the terrain dominating the breach site must be occupied before trying any effort to reduce the obstacle. The attacking unit's higher HQ has the responsibility to isolate the breach area by fixing adjacent units, attacking enemy reserves in depth, and providing counterfire support. Identifying the extent of the enemy's defenses is critical before selecting the appropriate technique to secure the point of breach. If the enemy controls the point of breach and cannot be adequately suppressed, the force must secure the point of breach before it can reduce the obstacle. The breach force must be resourced with enough maneuver assets to provide local security against the forces that the support force cannot sufficiently engage. Elements within the breach force that secure the reduction area may also be used to suppress the enemy once reduction is complete.

REDUCE

4-125. Reduction is the creation of lanes through or over an obstacle to allow an attacking force to pass. The number and width of lanes created varies with the enemy situation, the assault force's size and composition, and the scheme of maneuver. The lanes must allow the assault force to pass through the obstacle unimpeded. The breach force will reduce, proof (if required), mark, and report lane locations and the lane-marking method to higher HQ. Follow-on units will further reduce or clear the obstacle when required. Reduction cannot be accomplished until effective suppression and obscuration are in place, the obstacle has been identified, and the point of breach is secure.

ASSAULT

4-126. A breaching operation is not complete until--

- Friendly forces have assaulted to destroy the enemy on the far side of the obstacle that can place or observe direct and indirect fires on the reduction area.
- Battle handover with follow-on forces has occurred, unless no battle handover is planned.

PLANNING CONSIDERATIONS

4-127. In planning the breach operation, consider the following.

- The breach force moves forward by covered and concealed routes. If possible, the breach should be covert to reduce the time the breach and assault forces are exposed to enemy fire. If this is not possible or if the breaching try is compromised, the breach force moves under the suppressive fires of the support force.
- The penetration of the enemy position is made on a narrow front. The concept is to make a narrow penetration into the enemy defenses and then expand it enough to allow rapid passage of the assault force. Normally, the company concentrates all combat power at one breach point. However, it may use two breach sites if they are mutually supporting and do not result in a lack of concentration or a piecemeal assault. When using only one breach site, the company should plan an alternate site as a contingency in case the primary breach is unsuccessful.

- The support force provides effective suppression for the breach and assault force(s) to cross the killing ground. Each weapon in the support element has a specific enemy position or sector of responsibility assigned. Initially, the support force establishes fire superiority with a maximum volume of fire; then they maintain fire superiority throughout the attack. When indirect fires shift, the support force increases the rate of direct fire to maintain the suppression.
- The support force normally occupies one position to simplify control. However, at times, the support force must occupy several positions to provide effective suppression of the enemy. This might be required to prevent the masking of fires by the breach or assault force or because of the characteristics of the supporting weapons (120-mm/81-mm/60-mm mortars). The support force also often needs to reposition once the assault force begins clearing the objective. They may follow the assault force through the breach or reposition outside the enemy position.

EXPLOIT THE PENETRATION AND SEIZE THE DECISIVE POINT

4-128. After the successful breach, the assault force conducts the main attack (Figure 4-10, page 4-32). Supported by the fires of the support force and the breach force, the assault force passes through the breach. In planning the assault, consider the following points.

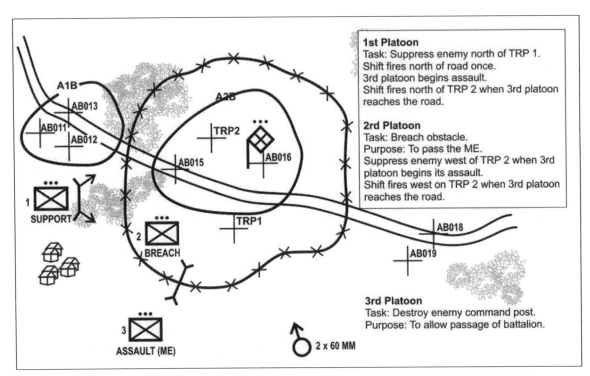

Figure 4-10. Exploitation of penetration.

4-129. The assault force must clear the enemy position as quickly as possible. If the assault force can capture or destroy the enemy's command and control facilities or other key positions and weapons, the enemy may surrender or abandon the position. If there is key terrain that allows control of the objective, this might be the decisive point for the assault force. Normally, the assault force moves within the enemy's trenches to avoid exposure to enemy fire.

4-130. The assault force must also organize into support, assault, and breach elements. As it encounters subsequent positions or bunkers, it might have to repeat the breaching operations. As in the initial breach, the breach element establishes a support-by-fire position and conducts the same sequence of breaching and assaulting to reduce the position.

Chapter 4

4-131. The designation of a reserve allows the Infantry commander to retain flexibility during the attack. The commander should be prepared to commit his reserve to exploit success and to continue the attack. The reserve may also repulse counterattacks during consolidation and reorganization.

4-132. Once an assault starts, the company maneuvers aggressively to allow the enemy less time to react. The commander monitors the situation and adjusts the plan to exploit any weakness found during the attack. If a situation develops that is beyond the capability of his company, he notifies the battalion commander. He might have to retain his position until other companies can maneuver to support him.

4-133. In moving from their assault positions, platoons advance in the formation most suitable to the terrain and situation. When the assault element must move through a narrow lane in the obstacles, it maintains dispersion and assaults through the lane by fire commands; signals should be coordinated to support this. The commander moves where he can best control his platoons and supporting fire. Indirect and direct fires of the support force shift when they endanger the advancing Soldiers.

4-134. The assaulting Soldiers clear enemy positions, secure and search prisoners, and move quickly across the objective. When they reach the far side, they take up hasty fighting positions and continue to fire at the withdrawing enemy. When the objective is secured and cleared (seized), the supporting elements and company trains are called forward.

4-135. Once it seizes the objective, the company consolidates. Reorganization, if required, is normally conducted concurrently with consolidation and consists of actions taken to prepare for follow-on operations. As with consolidation, the Infantry company commander must plan and prepare for reorganization as he conducts his TLP. He ensures that the company is prepared to take the following actions.

- Provide essential medical treatment and evacuate casualties as necessary.
- Cross-level personnel and adjust task organization as required.
- Conduct resupply operations, including rearming and refueling.
- Redistribute ammunition.
- Conduct required maintenance.

ATTACK DURING LIMITED VISIBILITY

4-136. Successful attacks in limited visibility depend on leadership, reconnaissance, training, planning, and surprise. Although these fundamentals are also key to daylight attacks, attacks in limited visibility require certain considerations and the proper application of the techniques discussed in this chapter to ensure control in the attack. Darkness, fog, heavy rain, and falling snow limit visibility. Smoke and dust from high-explosive (HE) fires also limit visibility, but their effects don't last as long. Infantry companies attack in limited visibility to--

- Achieve surprise.
- Avoid heavy losses.
- Cause panic in a weak or disorganized enemy.
- Exploit success and maintain momentum.
- Keep pressure on the enemy.

FUNDAMENTALS

4-137. The Infantry company conducts limited visibility attacks very much like daylight attacks). The fundamentals for a daylight attack, discussed earlier in this chapter, still apply for night attacks. Conducting attacks in this manner requires--

- A company that is well trained in limited visibility attacks.
- Enough natural light to employ the unit's NVDs.
- A simple, effective concept that takes advantage of the enemy's surprise and confusion.
- A successful reconnaissance of the objective area.
- Additional control measures and techniques, as needed.

CHALLENGES

4-138. When planning attacks at night, the Infantry commander must consider the increased difficulty of--

- Controlling units, Soldiers, and fires.
- Identifying and engaging targets.
- Navigating and moving.
- Distinguishing friendly and enemy Soldiers.
- Locating, treating, and evacuating casualties.
- Locating and bypassing or breaching enemy obstacles.

PLANNING CONSIDERATIONS

4-139. In planning limited visibility attacks, the Infantry commander should also consider the following.

- Feints and other deceptions might be more effective. This is true for the enemy also.
- If a small element can infiltrate the enemy position, it can be extremely effective in supporting the main attack. A small element can also covertly breach obstacles or neutralize key positions and weapons to allow the main attack to seize a foothold quickly.
- It might be possible to infiltrate the main attack inside the enemy's positions and then fight from the inside to the outside. In this case, the unit inside the position might be able to occupy defensive positions and force the enemy to attack him.

ILLUMINATION AND INDIRECT FIRE

4-140. Two basic decisions must be made for conducting limited visibility attacks: illumination on the objective and indirect fire support for the attack.

4-141. The Infantry company normally conducts nonilluminated attacks to exploit its technological and training advantage. For all night attacks, however, illumination should be readily available in case the enemy detects the attack and uses illumination, or if he possesses NVDs. Illumination may also be effective to support reorganization and consolidation after the objective is secure, particularly for casualty evacuation.

4-142. The Infantry company conducts illuminated night attacks like daylight attacks. Illumination is available from artillery, mortars, M203s, and hand-fired and aircraft flares. Permission to fire illumination is often retained by battalion because the light may affect adjacent unit operations.

4-143. Nonilluminated, nonsupported attacks offer the best chance of gaining surprise. These attacks are conducted like daylight attacks.

4-144. Illuminated, supported attacks are almost identical to daylight attacks. These attacks can be most effective when speed is essential, when there is limited time for reconnaissance, or when the enemy is weak or disorganized. When conducting these types of attacks, the attacking unit still tries to use stealth and the concealment of limited visibility to gain surprise. They then initiate fires and illumination to support the assault.

RECONNAISSANCE

4-145. Reconnaissance is critical in every attack, but especially for attacks at night. It should be conducted during daylight and down to the lowest level possible. Each unit should reconnoiter the routes on which they will move, the positions they will occupy, and the objective they are assigned. The company must balance the need for detailed information about the enemy against the risk of detection and loss of surprise.

4-146. The reconnaissance plan should establish surveillance on the objective in case the enemy repositions units and weapons, or prepares additional obstacles. Surveillance and security elements should secure critical locations, such as assault and support positions, the LD or PLD, and key routes to protect the company from enemy ambushes and spoiling attacks. These security elements assist in isolating the objective. Personnel who remain as surveillance serve as guides for the main body's movement into the assault position or PLD.

4-147. When reconnaissance is not successful due to lack of time, failure to identify critical aspects of the enemy's position, detection by the enemy, or any other reason, the commander should request a delay in the attack time to allow for further reconnaissance. If this is not possible, he should consider an illuminated or supported attack. A night attack with marginal information on the enemy's defense is very risky and difficult to conduct successfully.

4-148. The commander should also consider using SUASs for ground reconnaissance. Most SUASs have both a daylight and limited visibility (infrared) capability and significantly aid in providing the commander with needed information on the terrain and enemy. The Raven, for example, has the following characteristics:

Weight .... Just over 4 pounds
Flight endurance .... About 80 minutes
Effective operational radius .... About 10 kilometers
Flight speed .... 30 to 60 miles per hour
Operating altitude .... 100 to 1,000 feet

4-149. When beneficial to the tactical situation, SUASs may operate where they cannot be heard or where the noise signature is discernable. Even if detected, they are difficult to shoot down.

SIMPLICITY

4-150. A simple concept, particularly for the actions on the objective, also supports control during the assault. If possible, platoon and squad objectives should be small and easily identified.

4-151. Avoid developing a concept that requires the company to fight for each enemy fighting position. As in a daylight attack, identify a decisive point and focus combat power at this location. Once the decisive action is accomplished, the plan must also address any remaining enemy. If required by the higher commander's concept or for an effective consolidation, the company might have to clear all enemy forces from the objective area.

4-152. A smaller assault force maneuvering on the objective is easier to control and less likely to suffer casualties from enemy or friendly fires. The assault force must have clear signals to ensure control of all supporting fires, both direct and indirect.

4-153. The concept for a nonilluminated attack should be flexible to allow for adjustment to a daylight attack if illumination becomes appropriate due to detection by the enemy or the use of illumination by an adjacent unit. This is especially critical for a unit that plans a modified linear assault attack but might be forced to conduct an illuminated attack. A contingency plan that reorients for illumination should be prepared and issued, and every Soldier should know under what conditions to execute this plan. In some cases, such as when the unit is already deployed through the PLD and advancing on the enemy, the company might have to continue the attack as planned or try to disengage.

FIRE-CONTROL TECHNIQUES

4-154. Rehearsals are very important for achieving good fire control in limited visibility attacks. Fire-control techniques for limited visibility include--

TRACER FIRE

4-155. Leaders in the assault force fire all tracers; their men fire where the leader's tracers impact. The support force positions an automatic weapon on a tripod on the flank nearest the assault element. This weapon fires a burst of tracers every 15 seconds to indicate the near limit of the supporting fires. All other weapons in the support force keep their fires on the side of this tracer away from the assault force. The assault force signals to shift fires to the next position or to a set distance. If required, these rounds can be adjusted well over the head of the assault force to preclude casualties.

LUMINOUS TAPE OR CHEMICAL LIGHTS

4-156. Mark assault personnel to prevent fratricide. Do this in a way that avoids enemy detection. You could put luminous tape on the back of the helmet or use small infrared chemical lights (if the enemy has no NVDs). The support force should know where the lead assault force is. If individual Soldier markings do not suffice, use large chemical lights (infrared or visible). Place these on the ground or throw them in front of the assault force. When clearing a trench line, put the lights on a stick and move them with the lead element.

WEAPONS CONTROL RESTRICTIONS

4-157. Assign weapons control restrictions to reduce the risk to the assault force.

- The platoon on the right in the assault might be given weapons free to the right flank, because there are no friendly Soldiers there, but weapons tight or hold on the left because another friendly unit is located there.
- The assault force might be restricted to using only shotguns and pistols.
- The assault force might be restricted to no automatic weapons fire on the objective. This ensures that all automatic weapons in use are enemy.

OTHER TECHNIQUES

4-158. Use the following techniques to increase control during the assault.

- Not allowing flares, grenades, or smoke on the objective.
- Allowing only certain personnel with NVDs to engage targets on the objective.
- Using a magnetic azimuth for maintaining direction.
- Using mortar or artillery rounds to orient attacking units.
- Using guides.
- Reducing intervals between Soldiers and units.

SUPPORTING FIRES

4-159. Mortar, artillery, and anti-armor fires are planned for a night attack much like in a daylight attack. However, they sometimes do not fire unless the Infantry company is detected or until the company is ready to assault (based on METT-TC and the commander's scheme of maneuver.) Some weapons may fire before the attack and maintain a pattern to deceive the enemy or to help cover noise made by the company's movement. This is avoided if it will disclose the attack.

4-160. Indirect fire is difficult to adjust when visibility is poor. If doubt exists as to exact friendly locations, indirect fire is directed first at enemy positions beyond the objective and then walked onto the objective. The illumination rounds might be fired to impact on the ground, providing both light and markings on the objective. They may also be placed behind the objective and in the air, causing the enemy to be silhouetted. Once illumination begins, it should continue until the objective is secure. Sufficient ammunition must be available.

4-161. Smoke is planned to further reduce the enemy's visibility, particularly if he has night vision devices. The smoke is laid close to or on enemy positions to avoid restricting friendly movement or hindering the breaching of obstacles. Employing smoke on the objective during the assault may make it hard for assaulting Soldiers to find enemy fighting positions, but if sufficient thermal sights are available, using smoke on the objective may provide a decisive advantage for a well-trained unit.

4-162. Illumination is always planned for attacks to be conducted in limited visibility. This gives the company commander the option of calling for it and ensuring it is coordinated. The battalion commander normally controls illumination but may authorize the company commander to call for it when needed. If the company commander decides to use illumination, he should not call for it until the assault is initiated or the attack is detected. It should be placed on several locations over a wide area to confuse the enemy as to the exact location of the attack. It should also be placed beyond the objective to help assaulting Soldiers see and fire at withdrawing or counterattacking enemy Soldiers.

4-163. Illumination may also be required if the enemy uses illumination to disrupt the effectiveness of the company's NVDs. Once used, illumination must be continuous because attacking Soldiers will temporarily lose their normal night vision. Any break in illumination may also reduce the effectiveness of suppressive fire when the attackers need it most. Care must be taken to ensure that the squad and platoon leaders do not use hand flares before the commander has decided to illuminate the objective.

4-164. The thermal sights of weapons such as the ITAS and Javelin might be employed strictly for observation, if there are no targets for these weapons to engage. Positioned outside the objective area, these sights can provide critical current information. They can also assist the support force in controlling their fires or provide the assault force with reports of enemy movements on the objective.

4-165. When limited NVDs are available, they must be prioritized and employed at the most critical locations. Priorities to consider include key Soldiers in the breach force, key leaders in the assault force, other members of the assault force, and key leaders and weapons in the support force.

CONSOLIDATION AND REORGANIZATION

4-166. When it has seized the objective, the Infantry company consolidates and reorganizes. Consolidation and reorganization are the same as for a daylight attack with the following exceptions.

- Guides lead trains and support elements forward to their positions.
- The consolidation plan should be as simple as possible. Avoid changes in task organization.
- Locating and evacuating casualties and enemy prisoners of war (EPWs) takes longer. They might have to be moved to the rear of the objective and kept there until visibility improves.
- Platoon positions are closer together to ease control and improve mutual support. Position distances are adjusted as visibility improves.

MODIFIED LINEAR ASSAULT

4-167. If the company is unable to conduct a limited visibility attack, such as a daylight attack, the commander can choose from several simplified techniques. However, he should only employ these if the company is unable to fight effectively in limited visibility. The modified linear assault is a technique for conducting a nonilluminated attack. This technique is effective in controlling the fires of the assault force by maintaining a linear formation. Each Soldier assaults using individual movement techniques while remaining generally "*on line*" with the Soldier on his right and left. Each Soldier is able to engage or suppress targets to his front with fewer restrictions because there is less chance of fratricide. This technique provides extremely poor security and firepower to the flanks and poor flexibility once the assault is initiated.

MODIFICATIONS

4-168. In the true linear assault, the company deploys through its respective squad RPs, and the entire company conducts a linear assault across the objective (Figure 4-11, page 4-38). To reduce the vulnerability of the assault force, this technique is normally modified, which can be done in a number of ways depending on the situation.

Offensive Operations

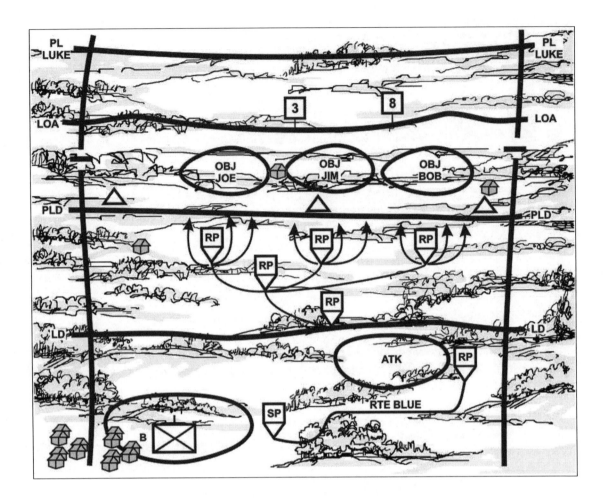

Figure 4-11. Linear assault.

4-169. The most effective modification is to establish part of the company in a support-by-fire position. The remainder of the company deploys at the PLD and conducts the assault (Figure 4-12). Machine guns, mortars, and Javelins are normally most effective in this role. M203s also might be effective if visibility is sufficient for their employment. The flank of the assault force nearest the support force must be visible to the support forwce. The fire team on this flank may mark themselves with chemical lights or glint tape to ensure visibility. If task organized with an assault platoon from the weapons company, their heavy weapons and close combat missiles can provide excellent supporting fires (see Appendix B).

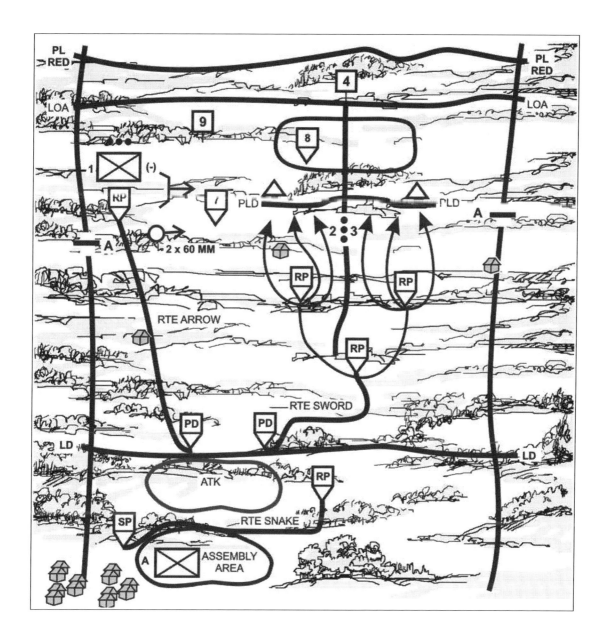

Figure 4-12. Linear assault with support element.

4-170. Other variations of this technique may include attacking on a much narrower front with a smaller assault force and having a large follow-and-support force. For example, instead of two platoons deploying at the PLD, a platoon (-) could deploy against an identified enemy weak point (Figure 4-13, page 4-37). This platoon could be tasked to bypass enemy positions to seize or destroy a critical location or facility, with the follow-and-support force reducing bypassed positions. Another variation is to assign the assault force a shallow objective to support the forward passage of the trailing unit, or to deploy through the platoon release points and then to attack in squad files. The latter is most effective when the situation supports an infiltration through the enemy defenses to seize decisive terrain or positions to the rear.

Offensive Operations

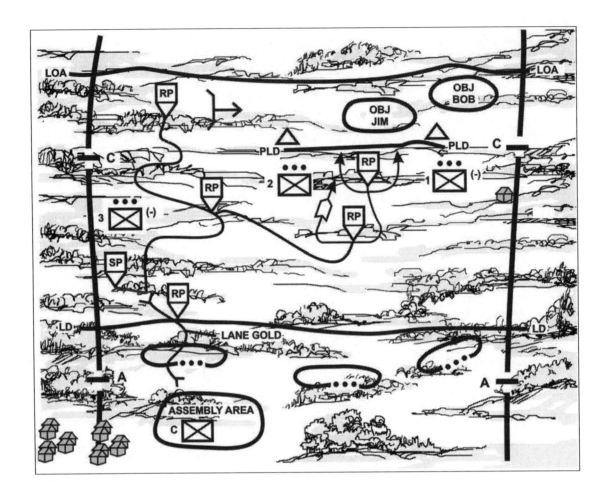

Figure 4-13. Linear assault with follow and support.

ADVANTAGES

4-171. The modified linear assault simplifies the control of supporting fires from outside the objective. By establishing support positions perpendicular to the direction of assault, the supporting fires can be employed next to the assault force and then shifted in front of them as they advance.

DISADVANTAGES

4-172. The linear formation is the biggest weakness with the modified linear assault. If the enemy is in well-prepared defensive positions, the linear formation ensures at least part of the assault force attacks through the enemy's kill zones. Also, to assault while using this technique makes it very difficult for the leader to concentrate combat power against an identified enemy weakness. Finally, if the enemy has NVDs or the assault force runs into unidentified obstacles after deploying at the PLD, fire superiority may not be achieved and the assault will rapidly come to a halt. This may result in the majority of the company being decisively engaged in the enemy killing ground.

CONDUCT OF THE ASSAULT

4-173. Although there are significant difficulties with the modified linear attack, it remains a viable technique for attack in limited visibility. It is most effective against a weak or disorganized enemy. If the enemy has NVDs or a well-prepared defense with protective obstacles, this technique should not be used. An illuminated, supported attack that is conducted as a daylight attack might be the most effective option in that situation.

4-174. Before attacking in this manner, the Infantry company should secure the PLD and provide personnel to guide the company from the LD to the PLD. Each platoon provides personnel to secure their portion of the PLD and to guide the platoon from the platoon RP. These Soldiers are briefed on the routes from the LD to the platoon RP, actions on enemy contact, time of departure, and other information needed by the patrol units to conduct their mission. They move forward to the platoon RP; then they move forward to reconnoiter and mark the platoon routes, secure their respective parts of the PLD, and observe the objective. The platoon guides go back to the platoon RP to guide their platoons to the squad RP and to the PLD.

4-175. Once the company crosses the LD, movement to the PLD is continuous. They move slowly to maintain stealth. Platoons are released at the platoon RP so they can deploy before reaching the PLD. Once their units are deployed, the platoon leaders and the support element leader notify the commander. When the company is fully deployed, the commander informs the battalion commander. On the battalion commander's order, the company moves silently forward from the PLD. The platoons guide on the base platoon.

4-176. When the attack is discovered, or on the commander's order, the support element opens fire and the platoons assault. Leaders must recognize that this technique for conducting a limited visibility attack results in a linear assault. To be successful, the assault must achieve surprise and rapidly overwhelm the defender. If the initial assault fails, control is difficult to regain. Scattered enemy fire must not be taken as a loss of surprise, and it should not be cause to start the assault.

4-177. Soldiers assault using individual movement techniques to maneuver. The support force must immediately gain fire superiority with a heavy volume of fire. Tracers are used to improve accuracy, to control fires, and to allow the assault force to see where its supporting fires are impacting. The FIST calls for indirect fire around and beyond the objective to disrupt enemy reinforcement. As the assault closes on the objective, fires are shifted beyond the limit of advance or lifted entirely. Soldiers must not go beyond the limit of advance. If the enemy discovers the attack before the company reaches the PLD, the commander may--

- Call for planned, supporting fire to suppress the enemy.
- Call for illumination (if authorized by the battalion commander) to ease control and movement.
- Continue as if it were a daylight attack by modifying the attack plan to a daylight attack.

Note: A linear assault, even a modified variation, is very risky when conducted under illumination.

Section VI. MOVEMENT TO CONTACT

This section discusses planning considerations and techniques for movements to contact; conduct of a search and attack, and conduct of an approach march.

DEFINITION

4-178. A movement to contact is an offensive operation used to develop the situation and establish and regain contact with the enemy. It is normally used when the tactical or enemy situation is vague, when the enemy has broken contact, or there is no time to reconnoiter extensively to locate the enemy. Contact results in initiation of another operation such as attack against a stationary or moving enemy force, defense, delay, or withdrawal. The fundamentals and techniques discussed here also apply to the approach phase of a hasty or deliberate attack; the main difference is the amount of enemy intelligence. In the approach phase of an attack, the enemy situation is more clear. Therefore, the company moves toward the objective in a way that avoids enemy detection and supports its deployment in the assault.

PLANNING CONSIDERATIONS

4-179. The Infantry company normally conducts MTC as part of a battalion or larger element; however, based on the METT-TC factors, it can conduct the operation independently. As an example, the company may conduct MTC prior to occupation of a screen line. Because the enemy situation is not clear, the company moves in a way that provides security and supports a rapid buildup of combat power against enemy units once they are identified. Two techniques for conducting a MTC are the search-and-attack technique and the approach-march technique. If no contact occurs, the company might be directed to conduct consolidation on the objective. The Infantry company commander analyzes the situation and selects the proper tactics to conduct the mission. He reports all information rapidly and accurately and strives to gain and maintain contact with the enemy. He retains freedom of maneuver by moving the company in a manner that--

- Ensures adequate force protection measures are always in effect.
- Makes enemy contact (ideally visual contact) with the smallest element possible (ideally, a reconnaissance and surveillance [R&S] element). The commander plans for any forms of contact to identify enemy locations.
- Rapidly develops combat power upon enemy contact.
- Provides all-round security for the unit.
- Supports the battalion concept.

SEARCH AND ATTACK

4-180. Search and attack is a technique for conducting a MTC; this technique shares many of the same characteristics of an area security mission (FM 3-0). Conducted primarily by Infantry forces and often supported by heavy forces, a commander employs this form of a MTC when the enemy is operating as small, dispersed element, or when the task is to deny the enemy the ability to move within a given area. The battalion is the echelon that normally conducts a search and attack. A brigade will assist its subordinate battalions by ensuring the availability of indirect fires and other support.

PURPOSE

4-181. A commander conducts a search and attack for one or more of the following purposes.

- Protect the force--prevent the enemy from massing to disrupt or destroy friendly military or civilian operations, equipment, property, and key facilities.
- Collect information--gain information about the enemy and the terrain to confirm the enemy COA predicted by the IPB process. Help generate SA for the company and higher headquarters.
- Destroy the enemy and render enemy units in the AO combat ineffective.
- Deny the area--prevent the enemy from operating unhindered in a given area such as in any area he is using for a base camp or for logistics support.

EXECUTION

4-182. An Infantry company is normally tasked to accomplish reconnaissance, finding, fixing, or finishing the enemy. The company normally acts within the context of the battalion concept of operation. The commander establishes control measures and communications means between any closing elements to prevent fratricide. The reconnaissance force conducts a zone reconnaissance to reconnoiter identified named areas of interest (NAIs).

4-183. Once the reconnaissance force identifies and locates the enemy force, the fixing force develops the situation, then executes one of two options based on the commander's guidance and the METT-TC factors. The first option is to block identified routes that the detected enemy can use to escape or reinforce itself. The fixing force maintains contact with the enemy and positions its forces to isolate and fix him before the finishing force attacks. The second option is to conduct an attack to fix the enemy in his current positions until the finishing force arrives. The fixing force attacks if that action meets the commander's intent and it can generate sufficient combat power against the detected enemy. Depending on the enemy's mobility and the likelihood of the reconnaissance force being compromised, the commander may need to position his fixing force before his reconnaissance force enters the AO.

4-184. If conditions are not right to use the finishing force to attack the detected enemy, the reconnaissance or the fixing force can continue to conduct reconnaissance and surveillance activities to further develop the situation. Whenever this occurs, the force maintaining surveillance must be careful to avoid detection and possible enemy ambushes.

4-185. The finishing force may move behind the reconnaissance and fixing forces, or it may locate at a pickup zone and air assault into a landing zone (LZ) near the enemy once he is located. The finishing force must be responsive enough to engage the enemy before he can break contact with the reconnaissance force or the fixing force. The battalion or brigade intelligence officer provides the commander with an estimate of the time it will take the enemy to displace from his detected location. The commander provides additional mobility assets so the finishing force can respond within that timeframe.

4-186. The commander uses his finishing force to destroy the detected and fixed enemy during a search and attack by conducting hasty or deliberate attacks, maneuvering to block enemy escape routes while another unit conducts the attack, or employing indirect fire or close air support to destroy the enemy. The commander may have his finishing force establish an area ambush and use his reconnaissance and fixing forces to drive the enemy into the ambushes.

DEVELOPMENT OF CONCEPT

4-187. Initially, the potential decisive points are identified as the most likely enemy locations. Once the enemy has been located, the specific decisive point must be determined as in any attack, and a concept must be developed for generating overwhelming combat power there. The initial concept must include the actions to finish the enemy force once they are located. At times, this part of the plan might be very general or consist only of control measures and be-prepared missions to provide flexibility and to support the rapid issuance of FRAGOs.

4-188. The commander must understand the battalion commander's concept and what freedom of action the company has to engage the enemy. At times, the company must engage and destroy all enemy forces within their capabilities. In other cases, the company must locate, follow, and report small enemy units to allow the battalion to concentrate and destroy these forces.

4-189. The commander focuses the platoons and squads on the likely enemy locations. He assigns missions IAW the battalion commander's concept. Possible operations include a zone or area reconnaissance, an ambush, or surveillance. The small-unit leaders must know what actions to take when they locate the enemy either with or without being detected. The platoon most likely to make contact is normally designated the decisive operation.

4-190. The company commander decides how the company will enter its zone or area of operations (AO), how to move once in the area, where to locate certain units or facilities, and what the requirements for contingency plans are. This includes establishing the proper graphic control measures to control the movement of the units, to provide for linkups between units, and to support the rapid concentration of the company's combat power. It also includes synchronizing the actions of the company and providing specific tasks or restraints to ensure subordinates understand what actions to take once they make contact with the enemy. The company normally enters the area or zone by moving as a company; platoons may then move toward separate objectives or areas.

4-191. The commander determines the number and size of the units that will conduct reconnaissance and combat actions against the enemy. The size of the area, the duration of the mission, the Soldier's load, and the probable size of the enemy force are key factors to this decision.

4-192. The size of the area of operations is considered in relation to how much time is available to search the area. When allocating terrain, the commander must consider how the platoons will conduct the reconnaissance and how to provide security and control. The commander may use one of the following techniques.

- Assign small AOs that keep the platoons more concentrated and help maintain control. The platoons move into the next AO on order.
- Divide the company area into zones. The commander concentrates most of the company in one zone and uses fire team or squad patrols to reconnoiter the next zone or the rest of the area. Once the company (-) has completed the reconnaissance in the initial zone, it moves into the area the small units have reconnoitered. This technique is effective when a detailed reconnaissance is required, but it also supports the seizure of the initiative through speed, stealth, and surprise. The small, dispersed units have a better chance of locating the enemy undetected. They also provide initial reconnaissance information and surveillance on which the commander focuses the remainder of the company's reconnaissance efforts.

4-193. The commander must consider how the duration of the mission affects the company's ability to conduct contingency operations. If the mission will continue for days or longer, the commander must develop a concept that allows his subordinates to maintain combat effectiveness. The concept must address the use of patrol bases and limited visibility operations. The commander must ensure that the concept provides sufficient rest to maintain his Soldiers' stealth, alertness, and security.

4-194. The duration of the mission also affects the Soldier's load, which has a tremendous impact on a search-and-attack mission. The longer the mission is expected to last, the heavier the Soldiers' loads might be to reduce the need for resupply. The ability to move with stealth and security (while close to the enemy) is hindered by heavy loads; however, resupply operations may also hinder the company's operation and allow the enemy to locate the unit by following or observing the resupply vehicles.

4-195. The company commander must determine the requirements for Soldier's load. If this causes excessive loads, he plans for resupply operations that avoid enemy detection and maintain the security of the company.

4-196. The company commander may combine techniques to reduce the risk of moving with these heavy loads. He identifies company patrol bases throughout the AO, and the company moves between these patrol bases using the approach-march technique to provide greater control and security. After securing and occupying the patrol base, the platoons leave their rucksacks and move out to conduct decentralized search-and-attack operations. A security force secures the patrol base until the units return to get their rucksacks and move to the next patrol base. Platoons can use this same technique when the risk is acceptable.

4-197. Knowing the size of the enemy units helps the company commander determine the risk to the company. He must also consider the enemy's capabilities, likely COAs, and specific weapons' capabilities. He must do so to understand the threat, and to ensure the security of his company, even during decentralized operations. He can direct specific force protection restraints such as--

- No patrols smaller than a squad.
- Platoons must be able to consolidate within 20 minutes.
- Platoons will depart their patrol bases NLT 60 minutes prior to BMNT.

LOCATE ENEMY

4-198. Information collection and analysis is normally the key to success in search and attack. During this step, the focus is on reconnaissance to locate the enemy. Generally, small units able to move quickly and with stealth are more likely to locate the enemy without detection. The company commander's concept may restrict the platoon's authority to destroy the enemy once located. It might be more important to locate and follow enemy units to identify their base camps. However, when not restricted, the unit making contact takes immediate action to destroy the enemy. If this is beyond the unit's capabilities, the platoon links up to mass its combat potential and to coordinate the attack. Considerations for reconnaissance should also include the use of SUASs. Most SUASs have both a daylight and limited visibility (infrared) capability and significantly aid in providing the commander with needed information on the terrain and enemy.

4-199. Platoons seldom receive a mission with the vague requirement to search and attack. The company commander must be more specific in stating his concept. His concept must also address the likely actions

Chapter 4

to destroy the enemy once they are located. Specific tasks may include route, area, and zone reconnaissance or surveillance tasks. Platoons may also be tasked to conduct ambushes, be prepared to conduct an attack to destroy enemy forces, provide security for another force, such as the CP or the mortar section, or act as the company reserve.

4-200. During limited visibility, reconnaissance is more difficult and potentially more dangerous. If a unit makes contact with the enemy in the dark, a hasty attack is very risky. Reconnaissance is also less effective in the dark because the unit covers less area and is unable to detect many signs of enemy activity. Although observation is reduced in limited visibility, the unit might be more likely to detect the enemy by sound or smell. Route and small-area reconnaissance tasks are more effective for limited visibility.

4-201. Ambushes are effective in limited visibility. The enemy may avoid daylight movements if aware of the company's presence in the AO. Ambushes should be set up on the enemy's likely routes or near their water and food sources. Patrol bases should integrate ambushes and OPs (with thermal sights, NVDs, and platoon early warning systems [PEWS]) into their security plans. These tasks support the seizure and maintenance of the initiative.

FIX AND FINISH THE ENEMY

4-202. These steps of a search and attack are closely related. An initial try to finish the enemy may quickly become the fixing effort for the company's attack if the enemy is too strong or the platoon is unable to achieve surprise. When the platoon's authority to destroy the enemy has been decentralized to the lowest level, the fundamentals of an attack apply at every echelon.

ACHIEVE SURPRISE

4-203. Locate the enemy without being detected. This allows more time to plan and coordinate the attack. Once detected, speed and violence in the assault may also achieve surprise, but this is rarely true against a prepared enemy defense.

LIMIT THE ENEMY'S FREEDOM OF ACTION

4-204. Fix the enemy in position. Block his routes of escape with indirect fires, maneuver forces, or both. Suppress his weapons systems, obscure his vision, and disrupt his command and control. Reconnaissance is continuous; leaders at every echelon seek out the enemy's dispositions, strengths, and weaknesses. Initially, these actions are directed toward supporting an attack by the lowest echelon. At some point, the leader of this unit must determine if he is able to achieve fire superiority and conduct the assault. If he determines he does not have sufficient combat power to complete the destruction of the enemy, the leader focuses on fixing the enemy and reconnoitering to support the attack by the next higher echelon.

MAINTAIN SECURITY

4-205. While trying to take these actions against the enemy, the enemy is trying to do the same. Do not assume the enemy is alone; there might be mutually supporting positions or units. The planned envelopment or flank attack of one enemy position may move through the kill zone of another unit, or may expose the assault force's flank to fires from undetected positions.

CONCENTRATE COMBAT POWER

4-206. Once contact is made, the plan must support the rapid concentration of combat power to fix and destroy (finish) the enemy. Leaders at each echelon plan to destroy the enemy within their capabilities. The combat potential of small units might be increased by ensuring each has the ability to request fire support.

4-207. The company commander may retain a portion of the company in reserve in order to act quickly to enemy contact by one of the small units. However, when the company is operating in a more dispersed manner, this company reserve may not be responsive enough. It might be more effective for each platoon to retain its own reserve.

4-208. If the unit or platoon cannot finish the enemy, the company commander determines how to fix or contain the enemy while concentrating his dispersed combat potential. He then develops an attack plan to destroy the enemy force. He may use the fixing force to support by fire and assault with another platoon(s), or he may use artillery and CAS to destroy him in position.

4-209. Each leader must report the results of his reconnaissance to support the company commander's planning. Leaders recommend effective support positions, good assault positions or directions of attack, and likely enemy weak points. The leader of the unit in contact should also identify good linkup points in case the preplanned points are not effective. In most cases, this leader should coordinate face-to-face with the company commander or the leader of the assault element before initiating the assault.

FOLLOW THE ENEMY

4-210. When the purpose of the operation is to locate the enemy's base camps or other fixed sites, the company concept must avoid inconclusive fights between small units. When friendly units locate small enemy units, they report and try to follow or track these units back to their base camps. Well-trained trackers familiar with the area might be able to identify and follow enemy tracks that are hours or even days old (FM 3-21.8 (FM 7-8)). The company commander must ensure that his concept does not risk the security of his force in the try to make undetected contact and track enemy units. Units tracking the enemy must be ready to react to enemy contact and avoid likely ambush situations. It also might be possible to track the enemy's movement through the AO by using stationary OPs as trail watchers to report enemy activity. Movement within the area or through the zone of attack might be conducted by the entire company or by individual platoons.

4-211. The Infantry company commander must decide where the company CP will locate. He may collocate it with the decisive operation platoon or position it in a central location where it can communicate with and move quickly to each platoon's location. A technique to support contingency operations is to rotate a reserve platoon each day to provide security for the CP and the company mortars. To prevent a serious degradation in effectiveness due to sleep loss, each platoon spends only 48 hours actively searching for the enemy and then rotates into the reserve role.

4-212. Company mortars must locate where they have security and can support the platoons. The company commander can collocate the mortars with the company CP. To overcome the difficulties of moving mortar ammunition, the company commander may direct the reserve platoon to carry the ammunition to the next firing position. Another option is to establish company patrol bases and place the mortars at these locations. The entire company can then carry the mortar ammunition; the Soldiers drop off the rounds before moving out to the platoon areas. However, the entire company must return to this location before continuing the operation through the zone.

4-213. The Soldier's load and the threat of enemy armor are two primary considerations for employing the antiarmor (shoulder-fired munitions and close-combat missile systems [Appendix B, *Employment of the TOW and Javelin Close-Combat Missile Systems*]) assets. When an armored threat exists, the company commander must provide guidance to platoon leaders on where to position the antiarmor assets. If the threat does not require antiarmor weapons, the platoons may still use some thermal sights for observation.

4-214. Contingency plans may include actions in case one platoon becomes decisively engaged or the company receives a new mission. All units should routinely report possible landing zone/pick-up zone (LZ/PZ) locations, mortar firing positions, any sign of recent enemy activity, and any sightings of civilians in the area.

APPROACH-MARCH-TECHNIQUE

4-215. The Infantry company normally uses this technique when it conducts a MTC as part of the battalion. Depending on its location in the formation and its assigned mission, the company can act as the advance guard, move as part of the battalion main body, or provide flank or rear guards for the battalion.

PLANNING CONSIDERATIONS

4-216. When planning for an approach-march MTC, the company commander needs certain information from the battalion commander. With this information, the company commander develops his scheme of maneuver and fire support plan. He provides this same information to the platoon leaders. As a minimum, he needs to know--

- The company's mission.
- The friendly and enemy situations.
- The route (axis of advance) and the desired rate of movement.
- The control measures to be used.
- The company's actions on contact.
- The fire support plan.
- The company's actions upon reaching the march objective, if one is used.

LEAD COMPANY RESPONSIBILITIES

4-217. The battalion may conduct a MTC on a single axis or on multiple axes. The lead company on an axis is responsible for--

- Protecting the battalion from a surprise attack by providing early warning of enemy positions and obstacles.
- Assisting the forward movement of the battalion by removing obstacles or finding routes around them.
- Destroying enemy forces (within its capability).
- Rapidly developing the situation once contact is made.

LEAD COMPANY MOVEMENT

4-218. The lead company or advance guard on an axis moves using traveling overwatch or bounding overwatch, depending on the enemy situation. It normally is assigned an axis of advance or a zone and a march objective on which to orient its movement. Phase lines and CPs help control movement.

- The company commander selects the movement technique and formation based on the likelihood of enemy contact and the speed of movement desired by the battalion commander. Bounding overwatch provides the best security, but traveling overwatch is faster. If the company uses traveling overwatch, the lead platoon may use bounding overwatch for added security.
- The company commander must retain the freedom to maneuver his platoons and weapons. He analyzes the terrain, anticipates where he might make contact, and plans fires on those locations. He should avoid terrain that restricts maneuver, such as draws, ravines, narrow trails, and steep slopes.

OTHER COMPANIES

4-219. A company not in the lead uses traveling or traveling overwatch. It must be ready to fire or maneuver in support of the lead company, or in another direction, or to assume the lead company's mission.

FLANK GUARD AND REAR GUARD

4-220. One company might be assigned the mission to provide forces for security of the flanks or rear of the battalion column. Normally this is the trail company of the battalion, which provides for centralized control and tactical integrity. Typically, the battalion commander specifies one platoon for each flank and one for the rear. METT-TC may call for reinforcement of certain of these elements as well. The company headquarters and remaining platoons will likely march with the main column.

REAR GUARD FORCE CONSIDERATIONS

4-221. Prevents enemy interference with the main body by stopping or delaying an attack to the rear; and prevents enemy direct fire or ground observed indirect fire from harassing the main body.

- During halts, occupies positions that enable it to protect the rear of the main body.
- Normally moves in column formation except where expected enemy action causes the need to deploy in width.

FLANK GUARD FORCE CONSIDERATIONS

4-222. Protects the main body from ground observation and surprise ground attack from the flank; Should the enemy initiate an attack, the flank guard may counterattack, defend, or delay to allow the main body to pass from the area, deploy, or maneuver.

- Responsibility might be assigned to the flank guard using a series of terrain features that block enemy likely avenues of approach; these positions must be coordinated with the protected main body unit.
- Movement of the guard force is tied to movement of the protected unit and may employ successive or alternate bounds to blocking positions.
- The flank guard must maintain close liaison with protected elements by all available means such as radio, patrols, or helicopters.
- Distances to the flank must assure mission accomplishment while trying to stay within range of the battalion mortars; normally about one terrain feature is tactically sound.

CONTACT

4-223. Once the company makes contact with the enemy, the company commander maintains that contact until ordered to do otherwise by the battalion commander. The following actions must take place at once.

4-224. When there is an unexpected contact, the platoon in contact returns fire at once and takes cover. If the enemy is unaware, the platoon making contact reports and deploys to prevent detection. Maneuver to a position of advantage to maintain the element of surprise until the company completes preparation for the hasty attack. If detected, or once the company commander decides to initiate the hasty attack, the platoon leader tries to fight through, destroying the enemy with the resources that are immediately available. His FIST should begin calling for fire. The platoon leader then reports to the company commander and continues to develop the situation. The overwatch element immediately fires at the enemy position. Trail platoons that are not able to fire take cover and wait for orders.

4-225. The squad or platoon that initially received direct fire immediately executes the attack drill. The intent is to use aggressive small-unit actions to seize the initiative rapidly and at the lowest echelon possible. The unit in contact tries to achieve fire superiority to fix or suppress the enemy with the resources that are immediately available. The unit then executes a flank attack directed against an identified enemy weakness. If this is not possible, the unit develops the situation to identify the enemy's flanks, any covered and concealed routes around the enemy position, possible supporting positions (both friendly and enemy), and any protective obstacles that the enemy has constructed. It then reports this information to the company commander.

4-226. Upon receipt of this information, the company commander determines the proper action to take. The XO reports the situation to battalion. The company commander may conduct, or direct his units to conduct, additional reconnaissance. The company FSO requests and coordinates indirect fires to support the company's maneuver. Possible reactions to contact include--

CONDUCT A HASTY ATTACK

4-227. If the company commander feels he can defeat the enemy force and an attack supports the battalion commander's concept, he conducts a hasty attack immediately, before the enemy can react.

BYPASS THE ENEMY

4-228. The company commander, with battalion permission, may bypass an enemy force. He may bypass the enemy with one platoon at a time or with the entire company at once (Figure 4-14, page 4-50). Precise criteria exist for making for a bypass. The criteria are usually conveyed as a size of enemy unit, for example, bypassing a squad-size unit without automatic weapons is authorized.

FIX OR SUPPRESS THE ENEMY

4-229. When the enemy cannot be bypassed and a hasty attack by the company is not feasible, the battalion normally directs the company to fix or suppress the enemy. This action ensures that the enemy does not have the freedom to fire or maneuver against the main body of the battalion while the battalion moves to attack the enemy. The company commander supports the battalion commander's planning by reconnoitering to identify the enemy's disposition, strengths, and weaknesses. The company identifies covered and concealed routes, good support positions, and enemy obstacles, and reports these to battalion.

ESTABLISH A HASTY DEFENSE

4-230. Although this action tends to give the initiative to the enemy force, it might provide a needed advantage. This might be required in a meeting engagement with a superior force. The company may establish a hasty defense to protect itself while the remainder of the battalion maneuvers against the enemy.

DISENGAGE

4-231. This action is not preferred unless disengaging is the only way to ensure preservation of the force. Use of indirect fires and bounding and overwatch elements is essential in disengaging from a superior force. The company may disengage while another unit maintains contact, or the company may disengage by moving back through the battalion to draw the enemy into an ambush.

TECHNIQUE CONSIDERATIONS

4-232. The battalion may direct the company's technique (search-and-attack or approach-march). If not, the Infantry company commander considers his mission and the battalion concept as he conducts his estimate to select the best technique. Normally, when operating as part of a battalion MTC, the company employs the same technique as the battalion. The following considerations may also assist the commander in developing his concept.

TIME AVAILABLE

4-233. The time available for planning, coordinating, and rehearsing may affect the decision. The approach-march technique generally requires much less time for preparation. The company may require only a brief FRAGO assigning the movement formation and technique and some simple graphic control measures to begin movement. The search-and-attack technique requires more preparation time because the platoons and squads have more planning responsibilities such as patrol base, linkups, and casualty evacuations.

SPEED OF MOVEMENT

4-234. The speed in which the company must move is a major factor. With either technique, the faster the company moves, the less effective its R&S efforts are. Thus, it becomes more likely that the enemy will initiate fires at the time and place he selects. The approach-march technique is normally more effective for quickly acting to enemy contacts.

ENEMY

4-235. The company commander considers the clarity of the enemy situation. If the enemy situation is vague then a MTC is required. Knowing where the enemy will probably locate and in what strength is key to developing a concept. The company commander considers the enemy's probable locations and strength when planning the company's movement and security needs, and he analyzes the risks for each technique. The company commander also considers the expected enemy reaction upon contact. If he expects the enemy to fight, then the approach march might be the more effective technique. If the enemy will try to avoid detection or quickly disengage, the search-and-attack technique might be the better method.

SECURITY

4-236. Preparation time, required movement speed, and the enemy situation have a direct impact on the company's security requirements. The company commander also considers the terrain, the adjacent units, the available combat support, and the present status of his unit to determine how to provide security for his company. Successful movements to contact depend on locating the enemy without being detected. This provides the company commander the initiative to develop the situation by fully coordinating and supporting the attack with all available resources.

COMBINED TECHNIQUE

4-237. An effective option might be to combine the techniques by having the lead platoon use the search-and-attack technique while the rest of the unit uses the approach-march technique. The lead platoon is assigned reconnaissance missions to find the enemy. In the example shown in Figure 4-14, the company commander assigns route reconnaissance tasks to the 2d Platoon. He assigns CPs and named areas of interest (NAIs) to focus the subordinate elements on specific locations. He can also use phase lines (PLs) to control the lead platoon by directing that PLs be crossed on order. The company's main body follows the reconnaissance at a distance that allows it to maneuver based on reports from the lead platoon. The formation and movement techniques for the main body vary but generally apply the fundamentals for the approach-march technique.

Chapter 4

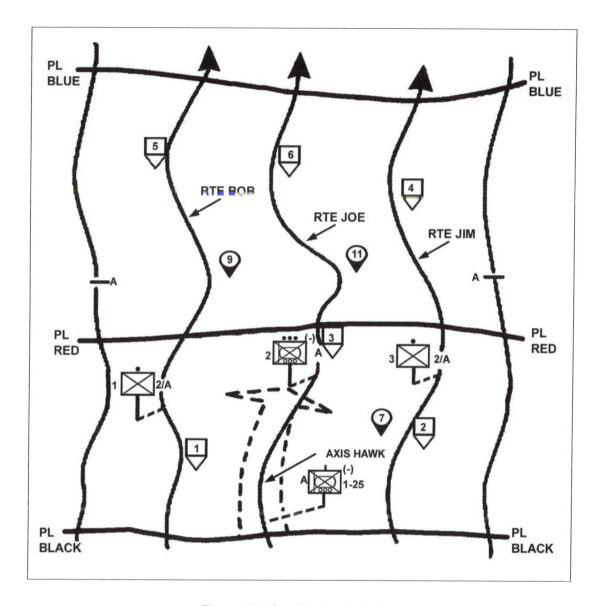

Figure 4-14. Combination technique.

Section VII. COMMON ACTIVITIES

These activities are warfighting actions the Infantry company might be called upon to perform in battle.

INFILTRATION

4-238. Infiltration is a form of maneuver used by Infantry units in many situations. During an attack, strong enemy defensive positions might be encountered. To avoid the enemy's strength, the Infantry commander may (by stealth through gaps or around enemy positions) conduct operations in the enemy's rear area. The company may infiltrate to conduct raids, ambushes, or other attacks. The company may also use infiltrations for many other types of operations, such as stay-behind and reconnaissance.

FUNDAMENTALS

4-239. By infiltrating, the Infantry company can maneuver to critical targets undetected, can achieve surprise, and can avoid the effects of enemy fires. Limited visibility, bad weather, and restrictive terrain reduce the chances of detection during an infiltration. A unit may infiltrate--

- To gather information.
- To attack the enemy at a weak point.
- To seize key terrain or destroy vital installations behind enemy positions.
- To harass and disrupt the enemy with ambushes in his rear area.
- To attack enemy reserves, fire support units, and command posts.

STEPS

4-240. The steps of an infiltration follow.

Patrol

4-241. Find gaps or weak areas in the enemy defense and locate enemy positions.

Prepare

4-242. Conduct TLP.

Infiltrate

4-243. Avoid enemy contact; move by smallest units possible.

Consolidate

4-244. Link up and prepare for actions at the objective.

Execute (Complete Mission)

4-245. Infiltration does not always require that all units move through the enemy's positions without detection or contact. Depending on the mission, the company can still complete the mission even though some of the squads make contact en route to the linkup point. Although the enemy might have some idea of what is happening, he will have difficulty estimating exactly what these small contacts mean. OPSEC might require that only key leaders have the entire plan during the infiltration step to prevent disclosure due to casualties or friendly prisoners.

CONSIDERATIONS

4-246. The Infantry company commander must prepare an infiltration plan and give units enough time for preparation and movement. The company may infiltrate by itself or as part of the battalion. In either case, movement techniques and formations are based on the likelihood of enemy contact, the terrain, the level of visibility, and the need for speed and control.

SIZE

4-247. The size of the infiltrating unit depends on the amount of time available, the amount of cover and concealment, and the enemy. Other considerations may include the need to communicate, the difficulty of navigation, and the number of infiltration routes. Generally, smaller units can move more quickly and make better use of available concealment, but they may increase the number of linkups, requiring more time.

Infiltrating by company or platoons ensures control and provides more combat power in the event of contact.

INFILTRATION LANE

4-248. The company might be assigned an infiltration lane or zone. The company commander must decide whether to move the entire company together through the company's lane or to assign each platoon a separate infiltration lane within the company lane. He also has the option to stagger the start time for each platoon on the one company lane. The infiltration lane should be wide enough to allow the infiltrating units to change their planned routes to avoid enemy contact. If the company uses a single company lane, the company commander picks a route through it and a company linkup point (Figure 4-15). If the company commander uses multiple lanes the company commander assigns each platoon a lane and a start time, picks linkup points for the platoons, and picks a company linkup point (Figure 4-16). The platoon leaders pick the routes through their lanes. In making his decision whether to use single or multiple lanes, the company commander considers several factors:

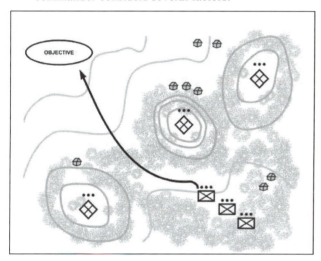

Figure 4-15. Company moving on single infiltration lane.

Figure 4-16. Company moving on multiple infiltration lanes.

Moving as a Company on a Single Lane

4-249. The advantages follow.

- Might get the company to the linkup point faster.
- Makes control easier.
- Makes navigation easier.
- Increases the chance of the entire company being detected but provides greater combat potential if detected.

Moving on Multiple Lanes or by Platoons on One Lane

4-250. The advantages follow.

- Requires linkups.
- Makes control harder.
- May make navigation more difficult.
- Decreases the chance of the entire company being detected but provides less combat potential if detected.

Routes

4-251. The routes selected must avoid enemy positions, use the best available cover and concealment, ease control and navigation, and avoid obstacles and danger areas.

- Routes should be reconnoitered without alerting the enemy. Leaders should consider using a map reconnaissance or guides, or marking the routes.
- Rally points might be selected based on the reconnaissance assets available to the commander; others are selected as the company moves along the route. If the infiltrating company is dispersed by enemy action, it rallies at the last rally point passed that is not within enemy small-arms range. The assembled unit then waits until a set number of units or Soldiers arrive at the rally point, or until a specified time, before continuing the mission. The senior man at the rally point should, in the absence of the commander, assume command and decide how best to continue the mission within the commander's intent.
- Locate the ORP as close to the objective as possible without being detected or losing security. The ORP should be large enough so that the company can deploy in it. It should be cleared before occupation.

Linkup Point

4-252. When using multiple lanes, the platoons meet at a linkup point and then move as a company to the ORP. Do not plan linkups at the ORP. If a unit misses its linkup, it moves to a contingency linkup point located away from the ORP and links up with a small element from the ORP.

Signals

4-253. Use of visual signals, such as with arms and hands, infrared devices, or flashlights with colored lenses, reduce the chance of detection. Avoid sound signals and flares. Recognition signals are critical for actions at a linkup or rally point.

4-254. Radio listening silence should be enforced, except when a unit must report its progress or when a unit detected by the enemy needs supporting fire.

4-255. Radio messages to report crossing of phase lines or CPs (if required) should be brief ? one code word. They might be transmitted without using call signs to identify units, providing each unit has separate code words.

4-256. When required, units operating out of radio contact (because of terrain or distances) can monitor or send codes only at a certain time. At these times, they set up expedient antennas or move to terrain better suited for communication.

Fire Support

4-257. Indirect fires are always planned but are used only when contact is made or when needed to support the mission. If contact is made with an enemy element, the infiltrating unit should use indirect fire to divert the enemy's attention, suppress enemy positions, and screen friendly movement as they disengage. Indirect fires may also be used to assist in navigation and to cause enemy Soldiers on security to seek cover.

Actions on Contact

4-258. When infiltrating on multiple lanes, detection of one infiltrating unit may alert the enemy and compromise the other infiltrating units. The company OPORD must state whether to continue the mission or return to friendly lines if detected by the enemy. Units following on the same lane should switch to an alternate lane. If a Soldier in the unit speaks the enemy's language, he should be positioned at or near the

Chapter 4

front of the column in case an enemy OP or patrol challenges the unit. The order also must specify what to do in the event of casualties.

Methods of Handling Casualties and Prisoners

4-259. During the infiltration, it might be extremely difficult to evacuate casualties or move prisoners without jeopardizing security. Casualties can be carried to the ORP or linkup point and evacuated when the operation has ended, or they can be consolidated and concealed with appropriate medical care and security and left for recovery and evacuation later. Moving casualties or prisoners to the ORP is dangerous when trying to avoid detection. Soldiers with medical supplies stay with any casualties left behind. The killed in action (KIA) can be concealed and recovered later when the tactical situation permits. Leave prisoners under guard at a rally point and evacuate them when the operation is over.

Rehearsals

4-260. Every Soldier must know the plan and his role in the plan. Units should rehearse their formations, their movement techniques, and their actions.

- On enemy contact.
- At rally points.
- At the linkup point.
- At the ORP.
- At danger areas.
- At the objective.

OVERWATCH

4-261. Overwatch is the component of tactical movement in which an element observes and, if necessary, provides direct fire support for a friendly moving element. Situational understanding of the tactical environment is crucial for the overwatch unit, whose objective is to prevent the enemy from surprising and effectively engaging the moving unit. The overwatch force must maintain communications with the moving element and provide early warning of enemy elements that could affect it. The overwatch must be able to support the moving element with immediate direct (to include antiarmor fires) and indirect fires; it can do this in either bounding overwatch or traveling overwatch. The key to successful overwatch is thorough scanning of gaps and dead space within the moving element's formation and on surrounding terrain. If the overwatch is unable to scan gaps and dead space and effectively engage the enemy, it must alert the moving element of the lapse in coverage. The moving element normally adjusts its movement speed, formation, or both, and initiates its own overwatch until the overwatch force completes movement to a position where it can continue the overwatch mission.

FOLLOW AND SUPPORT

4-262. Follow and support forces are employed in the offense to maintain the momentum of an operation. They do this by providing support or assistance that relieves the lead element of hindrances that could slow its advance. Follow and support missions are usually assigned when the enemy situation is vague and speed of the operation is important. The Infantry company might be task-organized to conduct follow and support missions in one of several ways.

- It can be part of a battalion with the mission of maintaining the momentum of a brigade attack.
- It can function as a separate maneuver element in support of the movement of another battalion element.
- Platoons within the company may conduct follow and support missions in support of other Infantry elements.

- Follow and support operations may require the company to conduct a variety of tactical tasks, including--
  - Conduct linkup operations with the lead element's fixing or overwatch force.
  - Destroy bypassed pockets of resistance.
  - Secure the flanks of a penetration to prevent the enemy from closing the penetration.
  - Secure lines of communications.
  - Secure bypassed key terrain.
  - Protect key installations.
  - Guard EPW.
  - Evacuate casualties.

4-263. The follow and support force receives information on the enemy or the supporting tasks from the lead element's fixing or overwatch force. The follow and support force conducts linkup with the fixing force on the ground, completes the exchange of critical tactical information, and accepts responsibility for the assigned tasks. The fixing force then rejoins the lead element, and the follow and support force executes its tasks. If enemy contact occurs, the follow and support force conducts actions on contact as outlined earlier in this chapter.

BYPASS

4-264. The company may bypass an enemy force or obstacle to maintain the momentum of the attack or for another tactical purpose. The battalion commander establishes bypass criteria.

- The Infantry company commander designates a fixing force to maintain contact with the enemy and assist the remainder of the company during the bypass. This fixing force may not come into direct fire contact with the enemy force.
- The bypassing force uses covered and concealed routes and, if possible, moves along bypass routes that are outside the enemy's direct fire range. If the situation dictates, the company can also employ smoke to obscure the enemy or to screen the bypassing force's movement. The company must conduct adequate reconnaissance of the route to confirm the feasibility of the bypass; the enemy may intentionally leave a bypass route unguarded to draw attacking forces into his kill sacks.
- Once the rest of the Infantry company clears the enemy position, the fixing platoon normally hands the enemy over to a supporting force, breaks contact, and rejoins the company. The fixing platoon might be attached to the follow-on force, but this is unlikely.

CLEARING OF AN OBJECTIVE

4-265. The company might be tasked with clearing an objective area during an attack to facilitate the movement of the remainder of the battalion or with clearing a specific part of a larger objective area. Situations in which the Infantry company may conduct the tactical task CLEAR include the following.

- Clear a defile, including high ground surrounding the defile and choke points within the defile.
- Clear a heavily wooded area.
- Clear a built-up area.
- Clear a road, trail, or other narrow corridor. This may include obstacles or other obstructions on the actual roadway, as well as surrounding wooded and built-up areas.

CLEARING PROCEDURES IN RESTRICTED TERRAIN

4-266. Clearing in restricted terrain is time-consuming and resource-intensive. During the planning process, the Infantry company commander evaluates the tactical requirements, resources, and other considerations for each of the three steps of the operation.

Chapter 4

- Approach the restricted terrain.
- Clear the area in and around the restricted area.
- Pass friendly forces, as required.

APPROACH

4-267. The approach focuses on moving combat power into restricted terrain and posturing it to begin clearing. The company commander takes the following actions.

- Establishes support-by-fire positions; destroys or suppresses any known enemy positions to allow forces to approach the restricted terrain.
- Provides additional security by incorporating suppressive indirect fires and obscuring or screening smoke.
- Provides support by fire for the dismounted Infantry. Be prepared to cover Infantry elements to the points at which they enter the restricted terrain such as high ground on either side of a defile, wooded areas on either side of a trail or road, or buildings on either side of a road in a built-up area.
- Moves dismounted Infantry elements along axes that provide the best available cover and concealment. The approach ends when the Infantry elements are prepared to conduct an attack.

4-268. The clearing begins as the Infantry squads begin their attack in and around the restricted terrain. Locations where this maneuver may take place include--

- On both sides of a defile, either along the ridgelines or high along the walls of the defile.
- Along the wood lines parallel to a road or trail.
- Around and between buildings on either side of the roadway in a built-up area.
- The following actions and considerations apply during this step.
  - Direct fire plans should cover responsibility for both horizontal and vertical observation and direct fire.
  - Infantry squads should clear a defile from the top down and should orient on objectives on the far side of the defile.
  - Engineers with manual breaching capability should move with the Infantry squads.

PASSAGE OF FRIENDLY FORCES

4-269. The Infantry company might be directed to assist the passage of another element forward to continue the clearing. When clearing is complete, the company must be prepared to take any action necessary to pass friendly forces, such as the following.

- Within the capabilities of the company, assault to destroy enemy forces and secure the far side of the restricted terrain.
- Conduct support by fire to protect the deployment of the follow-on force that is assuming the fight or to destroy or suppress any enemy elements that threaten the battalion as it exits the restricted terrain.
- Defeat any counterattacks.
- Protect the obstacle reduction effort.
- Maintain observation beyond the restricted terrain.
- Integrate indirect fires as necessary.

COMPANY AS RESERVE

4-270. The company might be held as the battalion reserve during an attack. The Infantry battalion commander commits the reserve to decisively influence the action and to maintain the momentum of the attack. To exploit the success of the other attacking Infantry/weapons companies and to achieve surprise, the reserve should attack the enemy from a new direction. Because of the various missions that the reserve might be assigned, the reserve commander must keep abreast of the tactical situation, know the missions and the tactical plans of the other companies, and be familiar with the terrain and the enemy situation in the objective area. The reserve must act quickly and effectively when committed. The reserve might be assigned one or more of the following tasks as part of its planning priorities.

- Protect the flanks or the rear of a battalion.
- Assume the mission of another company.
- Support by fire.
- Clear a position that has been overrun or bypassed.
- Attack from a new direction.
- Assist during the consolidation on an objective.
- Guard and evacuate prisoners.

Chapter 5
Defensive Operations

This chapter discusses the planning, preparing, and executing of defensive operations. These operations are temporary measures conducted to identify or create enemy weaknesses and to create the opportunity to go on the offense. However, properly conducted, defensive operations can defeat numerically superior forces. Infantry forces in the defense must use the terrain to support their maneuver and to achieve surprise. They maintain an offensive focus and seek to avoid static defenses that surrender the initiative to the enemy.

This edition adds a discussion on retrograde operations.

Section I. OVERVIEW

The immediate purpose of defensive actions is to resist, defeat, or destroy an enemy attack and gain the initiative for the offense. Defensive operations defeat an enemy attack, buy time, economize forces, or develop conditions favorable for offensive operations. Defensive actions alone are not normally decisive; frequently, they are combined with or followed by offensive action. Though the outcome of decisive combat derives from offensive actions, commanders often find that it is necessary, even advisable, to defend. Once commanders make this choice, they must set the conditions for the defense in a way that allows friendly forces to withstand and hold the enemy while they prepare to seize the initiative and return to the offense. A thorough understanding of the commander's intent is especially critical in defensive operations, which demand precise integration of combat, combat service, and sustainment elements.

TYPES

5-1. As part of defensive operations, the company may defend, delay, withdraw, or counterattack. The company may also perform security tasks. The company normally defends, as part of the battalion's defense, in the main battle area (MBA). The three types of defensive operations are--

AREA DEFENSE

5-2. Concentrates on denying the enemy access to designated terrain for a specified time, rather than the outright destruction of the enemy.

MOBILE DEFENSE

5-3. Orients on the destruction of the enemy through a decisive attack(s) by a striking force.

RETROGRADE OPERATIONS

5-4. Forced or voluntary organized movements to the rear or away from the enemy.

PURPOSE

5-5. The immediate purpose of a defensive operation is to defeat an enemy attack and gain the initiative for offensive operations. The Infantry company may also conduct the defense to achieve one or more of the following purposes.

- Gain time.
- Retain key terrain.
- Support other operations.
- Preoccupy the enemy in one area while friendly forces attack him in another.
- Erode enemy forces at a rapid rate while reinforcing friendly operations.

CHARACTERISTICS

5-6. The characteristics of the defense are also planning fundamentals for the Infantry company. These characteristics include preparation, security, disruption, massing effects, and flexibility. (FM 3-90 explains the two defensive patterns, area and mobile.)

PREPARATION

5-7. The defender arrives in the battle area before the attacker. He must take advantage of this by making the most thorough preparations for combat possible in the time available. By analyzing the factors of METT-TC, the Infantry rifle company commander gains an understanding of the tactical situation and identifies potential friendly and enemy weaknesses. He then war-games friendly and enemy options and synchronizes his concept of the operation with all available combat multipliers.

SECURITY

5-8. The goals of the company security effort are to deceive the enemy as to the location of friendly locations, strengths, and weaknesses. They also inhibit or defeat enemy reconnaissance operations. Security also provides early warning and disrupts enemy attacks early and continuously.

DISRUPTION

5-9. Defensive plans vary with the circumstances, but all defensive concepts of operation aim at disrupting the attacker's synchronization. Counterattacks, indirect fires, obstacles, and retention of key, or decisive terrain prevent the enemy from concentrating his strength against portions of the defense. Destroying enemy command and control vehicles disrupts enemy synchronization and flexibility.

MASSED EFFECTS

5-10. The successful defender concentrates combat power at the decisive time and place. Through massing effects, he can obtain a local advantage at points of decision. Offensive action and the use of surprise and deception are often the means of gaining this advantage. Concentration refers to combat power and its effects — not just numbers of Soldiers and weapons systems. To concentrate combat power, the defender may economize in some areas, retain a reserve, and maneuver to gain local superiority. Local counterattacks might be needed to maintain the integrity of the defense. Indirect fire can shift to critical points to concentrate destructive effects rapidly.

FLEXIBILITY

5-11. Flexibility is derived from sound preparation and effective C2. The defender must be agile enough to counter or avoid the attacker's blow and then strike back effectively. Flexibility results from a detailed mission analysis, an understanding of the unit's purpose, aggressive reconnaissance and security, and when applicable, organization in depth and retention or reconstitution of a reserve. Flexibility requires that the company commander "*see the battlefield*"—physically and through timely and accurate reports. Supplementary positions on secondary avenue of approach may provide additional flexibility to the company commander. After proper analysis of the terrain and enemy situation, the commander can anticipate enemy actions and be prepared to act through the positioning of maneuver units or a reserve.

Section II. SEQUENCE

Usually, as part of a larger element, the Infantry rifle company conducts defensive operations performing several integrated and overlapping activities. The following paragraphs focus on the tactical considerations and procedures involved in each activity. This discussion shows an attacking enemy that uses depth in its operations, but there will be situations where a company must defend against an enemy that does not have a doctrinal operational foundation. The Infantry company must be prepared to defend against such threats. This unconventional (insurgent or terrorist force) enemy situation requires a more flexible plan that allows for more responsive and decentralized control of combat power rather than spreading it evenly throughout the company's AO. The Infantry company may also conduct 'base-camp' or perimeter defense operations along with offensive and patrolling operations against terrorist, insurgent, or guerilla forces. (Chapter 6, *Stability Operations* discusses base defense; Chapter 8, *Tactical Enabling Operations*, includes a section on patrols and patrolling. Corresponding chapters in FM 3-21.8 (FM 7-8) and FM 3-21.20 (FM 7-20) also discuss these operations).

RECONNAISSANCE AND SECURITY OPERATIONS AND ENEMY PREPARATORY FIRES

5-12. Security forces must protect friendly MBA forces in order to allow them to prepare their defense. These security forces work in conjunction with and compliment battalion and brigade security operations. The enemy will try to discover the defensive scheme of maneuver using reconnaissance elements and attacks by forward detachments and disruption elements. He will also try to breach the battalion's tactical obstacles.

SECURITY FORCE

5-13. The goals of the battalion security force normally include providing early warning, destroying enemy reconnaissance units, and impeding and harassing enemy assault elements. The security force continues its mission until directed to displace. The battalion commander may also use security forces in his deception effort to give the *illus*ion of strength in one area while establishing the main defense in another. While conducting this type of security operation, the Infantry rifle company may simultaneously have to prepare battle positions, creating a challenging time management problem for the commander and his subordinate leaders.

GUIDES

5-14. During this activity, the Infantry company might be required to provide guides to pass the security force and might be tasked to close the passage lanes. The company may also play a role in shaping the battlefield. The battalion commander may position the company to deny likely enemy attack corridors to enhance flexibility and force enemy elements into friendly engagement areas. When it is not conducting security or preparation tasks, the company normally occupies hide positions to avoid possible chemical strikes or enemy artillery preparation.

OCCUPATION AND PREPARATION

5-15. A leader's reconnaissance is critical during this time in order for the company to conduct occupation without hesitation and begin the priorities of work. The participants in the reconnaissance are the company commander, platoon leaders, mortar section leader, FSO, leaders of any attached elements, and a security element. The goals are, but not limited to, identification of enemy avenues of approach, EAs, sectors of fire, the tentative obstacle plan, indirect fire plan, OP locations, and command post locations. The brigade and battalion establish security forces during this step, and remaining forces begin to develop EAs and prepare BPs. Operational and tactical security is critical during the occupation to ensure the company avoids detection and maintains combat power for the actual defense. Soldiers, at all levels of the company, must thoroughly understand their duties and responsibilities related to the occupation; they must be able to execute the occupation quickly and efficiently to maximize the time available for planning and preparation of the defense.

APPROACH OF ENEMY MAIN ATTACK

5-16. The company engages the enemy at a time and place where he can maximize the lethality of his direct and indirect fire systems to achieve success within his designated AO. If available, as the enemy's assault force approaches the EA, the brigade or Battalion may initiate CAS to weaken the enemy. Friendly forces occupy their actual defensive positions before the enemy reaches direct fire range; they may shift positions in response to enemy actions or other tactical factors.

Note: Long-range fires might be withheld in accordance with a higher commander's intent.

ENEMY ASSAULT

5-17. During his assault, the enemy deploys to achieve mass at a designated point, normally employing both assault and support forces. This may leave him vulnerable to the combined effects of indirect and direct fires and integrated obstacles. The enemy may employ additional forces to fix friendly elements and prevent their repositioning. Friendly counterattack forces might be committed against the enemy flank or rear, while other friendly forces may displace to alternate, supplementary, or subsequent positions in support of the commander's scheme of maneuver. All friendly forces should be prepared for the enemy to maximize employment of combat multipliers, such as dismounted Infantry operations, to create vulnerability. The enemy is also likely to use artillery, CAS, and chemical weapons to set the conditions for the assault.

COUNTERATTACK

5-18. As the enemy's momentum slows or stops, friendly forces may conduct a counterattack. The counterattack might be for offensive purposes to seize the initiative from the enemy. In some cases, however, the purpose of the counterattack is mainly defensive such as reestablishing a position or restoring control of the sector. The Infantry company may participate in the counterattack as a base-of-fire element--providing support by fire for the counterattack force--or as the actual counterattack force.

CONSOLIDATION AND REORGANIZATION

5-19. The company secures its defensive area by repositioning forces, destroying remaining enemy elements, processing EPW, and reestablishing obstacles. The company conducts all necessary sustainment functions as it prepares to continue the defense. Even when enemy forces are not actively engaging it, the Infantry company maintains awareness of the tactical situation and local security at all times. The company prepares itself for possible follow-on missions.

Section III. PLANNING CONSIDERATIONS

The WFFs are critical tactical considerations that provide a means of reviewing plans, preparation, and execution. The synchronization and coordination of activities within each WFF and among the various WFFs are critical to the success of the Infantry rifle company. This section discusses selected WFF and other planning considerations. For a detailed discussion of command, control, and intelligence, see Chapter 2.

MOVEMENT AND MANEUVER

5-20. Maneuver considerations employ direct fire weapons on the battlefield. In the defense, effective weapons positioning is critical to the company's success. Effective weapons positioning enables the company to mass fires at critical points on the battlefield and shift fires as necessary. The company commander must exploit the strengths of his weapons systems while minimizing the company's exposure to enemy observation and fires. The following paragraphs focus on tactical considerations for weapons positioning.

DEPTH AND DISPERSION

5-21. Dispersing positions laterally and in depth helps to protect the force from enemy observation and fires. The positions are established in depth, allowing sufficient maneuver space within each position to establish in-depth placement of weapons systems, and Infantry elements. Engagement areas are established to provide for the massing of fires at critical points on the battlefield. Sectors of fire are established to distribute and shift fires throughout the extent of the EA. Once the direct fire plan is determined, fighting positions are constructed in a manner to support the fire plan.

FLANK POSITIONS

5-22. Flank positions enable a defending force to fire on an attacking force moving parallel to the defender's forces. An effective flank position provides the defender with a larger and more vulnerable target while leaving the attacker unsure of the defense location. Major considerations for successful employment of a flank position are the defender's ability to secure the flank and his ability to achieve surprise by remaining undetected. Effective fire control and fratricide avoidance measures are critical considerations in the employment of flank positions. (See Chapter 9 for a more detailed discussion of direct-fire planning and control.)

DISPLACEMENT PLANNING

5-23. Disengagement and displacement allow the company to retain its flexibility and tactical agility in the defense. The ultimate goals of disengagement and displacement are to enable the company to avoid being fixed or decisively engaged by the enemy. The overarching factor in a displacement is to maintain a mobility advantage over the enemy. The commander must consider several important factors in displacement planning. These factors include, among others--

- The enemy situation, for example, an enemy attack with two battalion-size enemy units might prevent the company from disengaging.
- Disengagement criteria.
- Availability of direct fire suppression that can support disengagement by suppressing or disrupting the enemy.
- Availability of cover and concealment, indirect fires, and smoke to assist disengagement.
- Obstacle integration, including situational obstacles.
- Positioning of forces on terrain that provides an advantage to the disengaging elements such as reverse slopes or natural obstacles.
- Identification of displacement routes and times when disengagement or displacement will take place. Routes and times are rehearsed.

- The size of the friendly force that must be available to engage the enemy in support of the displacing unit.

5-24. While disengagement and displacement are valuable tactical tools, they can be extremely difficult to execute in the face of a rapidly moving enemy force. In fact, displacement in contact poses such great problems that the company commander must plan for it thoroughly and rehearse displacement before the conduct of the defense. He must then carefully evaluate the situation at the time displacement in contact becomes necessary to ensure it is feasible and will not result in unacceptable personnel or equipment losses.

DISENGAGEMENT CRITERIA

5-25. Disengagement criteria dictate to subordinate elements the circumstances, in which they will displace to alternate, supplementary, or subsequent positions. The criteria are tied to an enemy action, such as an enemy unit advancing past phase line DOG. They are also linked to the friendly situation, for example, the criteria might depend on whether artillery or an overwatch element can engage the enemy. Unique disengagement criteria are developed during the planning process for each specific situation. *They are never part of the unit's TSOP.*

DIRECT FIRE SUPPRESSION

5-26. The attacking enemy force must not be allowed to bring effective direct and indirect fires to bear on a disengaging friendly force. Direct fires from the base-of-fire element, employed to suppress or disrupt the enemy, are the most effective way to facilitate disengagement. The company may receive base of fire support from another element in the battalion. However, in most cases, the company establishes its own base-of-fire element. Having an internal base of fire requires the company commander to carefully sequence the displacement of his forces.

COVER AND CONCEALMENT

5-27. The company and subordinate platoons use covered and concealed routes when moving to alternate, supplementary, or subsequent positions. Regardless of the degree of protection the route itself affords, the company and platoons rehearse the movement prior to contact. Rehearsals increase the speed at which they can conduct the move and provide an added measure of security. The commander makes a concerted effort to allocate available time to rehearse movement in limited visibility and degraded conditions.

INDIRECT FIRES AND SMOKE

5-28. Artillery or mortar fires assist the company during disengagement. Suppressive fires slow the enemy and cause him to seek cover. Smoke obscures the enemy's vision, slow his progress, or screens the defender's movement out of the BP or along his displacement route.

OBSTACLE INTEGRATION

5-29. Obstacles are integrated with direct and indirect fires. By slowing and disrupting enemy movement, obstacles provide the defender with the time necessary for displacement and allow friendly forces to employ direct and indirect fires effectively against the enemy. The Modular Pack Mine System (MOPMS) can also be employed in support of the disengagement, to either block a key displacement route once the displacing unit has passed through it or close a lane through a tactical obstacle. The location of obstacles in support of disengagement depends on METT-TC factors. Ideally, an obstacle should be positioned far enough away from the defender that he can effectively engage enemy elements on the far side of the obstacle while remaining out of range of the enemy's massed direct fires.

MOBILITY

5-30. Mobility operations in the defense ensure the ability to reposition forces, delay, and counterattack. Initially during defensive preparations, mobility operations focus on the ability to resupply, reposition, and conduct rearward and forward passage of forces, material, and equipment. Once defensive preparations are complete, the focus normally shifts to supporting the company reserve, local counterattacks, and the higher HQ counterattack or reserve. Priorities set by the battalion may specify routes for improvement in support of such operations. Normally, most Engineer assets go to survivability and countermobility. At a set time or trigger, Engineers disengage from obstacle and survivability position construction and start preparing for focused mobility operations. The commander analyzes the scheme of maneuver, obstacle plan, and terrain to determine mobility requirements. Critical considerations may include--

- Lanes and gaps in the obstacle plan.
- Lane closure plan and subunit responsibility.
- Route reconnaissance, improvement, and maintenance.

COUNTERMOBILITY

5-31. To succeed in the defense, the company commander integrates individual obstacles into both direct and indirect fire plans. In each case, he considers his intent for each obstacle group. Obstacle intent includes the target and desired effect (clear task and purpose) and the relative location of the obstacle group. The purpose influences many aspects of the operation, from selecting and designing obstacle sites to conducting the defense. Normally, the battalion commander designates the purpose of an obstacle group. (FM 90-7 gives more information about obstacle planning, siting, and turnover.)

Tactical Obstacles

5-32. The battalion commander assigns obstacle groups. He tells the company commanders and the Engineer what he wants to do to the enemy, and then he resources the groups accordingly. Obstacle intent includes these elements.

- The target, which is the enemy force that the commander wants to affect with fires and tactical obstacles. The commander identifies the target's size, type, echelon, avenues of approach, or any combination of these.
- The obstacle effect describes how the commander wants to attack enemy maneuver with obstacles and fires. Tactical obstacles block, turn, fix, or disrupt. Obstacle effect integrates the obstacles with direct and indirect fires.
- The relative location is where the commander wants the obstacle effect to occur against the targeted enemy force. The commander initiates the obstacle integration process after he identifies where on the terrain the obstacle will most decisively affect the enemy.
- For example, the battalion commander might say,

    "Deny the enemy access to our flank by turning the northern, mechanized Infantry battalion (MIB) into our EA. Allow Companies B and C to mass their fires to destroy the enemy."

    Scatterable minefield systems and submunitions are the main means of constructing tactical obstacles. These systems, with their self- and command-destruct capabilities, are flexible, and they aid in rapid transitions between offensive and defensive operations. They do this better than conventional mines and other constructed obstacles. The force constructs conventional minefields and obstacles only for a deliberate, long-term defense. In those cases, the battalion and companies are usually augmented with assets from a divisional engineer battalion. Table 5-1 shows the symbols for each obstacle effect, and it describes the purpose and characteristics of each.

OBSTACLE EFFECT	PURPOSE	FIRES AND OBSTACLES MUST:	OBSTACLE CHARACTERISTICS
DISRUPT	• Break up enemy formations. • Interrupt the enemy's timetable and C2. • Cause premature commitment of breach assets. • Cause the enemy to piecemeal his attack.	• Cause the enemy to deploy early. • Slow part of his formation while allowing part to advance unimpeded.	• Do not require extensive resources. • Ensure obstacles are difficult to detect at long range.
FIX	• Slow an attacker within an area so he can be destroyed. • Generate the time necessary for the friendly force to disengage.	• Cause the enemy to deploy into attack formation before encountering the obstacles. • Allow the enemy to advance slowly in an EA or AO. • Make the enemy fight in multiple directions once he is in the EA or AO.	• Array obstacles in depth. • Span the entire width of the avenues of approach • Avoid making the terrain appear impenetrable.
TURN	• Force the enemy to move in the direction desired by the friendly commander.	• Prevent the enemy from bypassing or breaching the obstacle belt. • Maintain pressure on the enemy force throughout the turn. • Mass direct and indirect fires at the anchor point of the turn.	• Tie into impassable terrain at the anchor point. • Use obstacles in depth. • Provide a subtle orientation relative to the enemy's approach.
BLOCK	• Stop an attacker along a specific avenue of approach. • Prevent an attacker from passing through an AO or EA. • Stop the enemy from using an avenue of approach and force him to use another avenue of approach.	• Prevent the enemy from bypassing or penetrating through the belt. • Stop the enemy's advance. • Destroy all enemy breach efforts.	• Tie into impassable terrain. • Use complex obstacles. • Defeat the enemy's mounted and dismounted breaching effort.

Table 5-1. Obstacle effects.

Protective Obstacles

5-33. Infantry rifle companies plan and construct their own protective obstacles. For best effect, protective obstacles are tied into existing or tactical reinforcing obstacles. The company can use mines and wire, or it might receive additional materiel, including MOPMS, from the battalion Class IV or V supply point. The company might also conduct any other required coordination, such as that needed in a relief in place, to recover or destroy the obstacle, for example, MOPMS. (FM 90-7 provides detailed planning guidance for protective obstacle emplacement.)

- In planning protective obstacles, the commander evaluates the potential threat to the company's position. Then, he employs the best system for that threat. For example, MOPMS is mainly an antitank system. It is best on mounted avenues of approach, but has some antipersonnel uses. However, on dismounted avenues, wire obstacles might be more effective.
- Protective obstacles are usually located beyond hand grenade distance (40 to 100 meters) from the Soldier's fighting position, and may extend out 300 to 500 meters to tie into tactical obstacles and existing restricted terrain. As with tactical obstacles, the commander should plan protective obstacles in depth and try to maximize the effective range of his weapons.
- When planning protective obstacles, the company commander considers preparation time, the burden on the logistical system, the Soldiers' loads, and the risk of loss of surprise.

Wire Obstacles

5-34. The three types of wire obstacles are protective, tactical, and supplementary (Figure 5-1).

- Protective wire can be a complex obstacle providing all-round protection of a platoon perimeter. It might also be a simple wire obstacle on the likely dismounted avenue of approach into a squad ambush position. Command-detonated M18 Claymore mines can be integrated into the protective wire or used separately.
- Tactical wire is positioned to increase the effectiveness of the company's fires. Usually, it is positioned along the friendly side of the machine gun final protective lines (FPLs). Tactical minefields may also be integrated into these wire obstacles or used separately.
- Supplementary wire obstacles can break up the line of tactical wire. This helps prevent the enemy from locating friendly weapons (particularly the machine guns) by following the tactical wire.

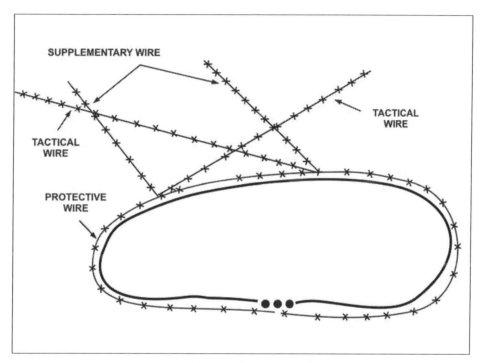

Figure 5-1. Protective wire obstacles.

Obstacle Lanes

5-35. The company might be responsible for actions related to lanes through obstacles. These duties can include marking lanes in an obstacle, reporting locations of the start and end points of each lane, operating contact points, providing guides for elements passing through the obstacle, and closing the lane.

Situational Obstacle

5-36. A situational obstacle is planned and possibly prepared before an operation, but it executes only if specific criteria are met. It gives the commander the flexibility to emplace tactical obstacles based on battlefield development.

- The commander anticipates situations that require him to modify the maneuver and fire plans to defeat the threat. He considers the use of situational obstacles to support these modifications.

Chapter 5

- By their very nature, situational obstacles must be quickly installable, but still achieve the desired effect. Therefore, SCATMINEs such as MOPMS, Hornets, and Volcanoes are the most common versions used at the company level. However, situational obstacles can consist of any type of individual obstacle.
- Commanders consider where they can employ situational obstacles. They ensure that the combination of fires and obstacles are enough to achieve the obstacle effect.
- Commanders identify execution triggers; situational obstacles are triggered based on friendly actions, enemy actions, or a combination of both.
- Finally, the commander withholds execution of a situational obstacle until the obstacle effect is required. Once committed, those assets are no longer available to support any other mission. Commanders also consider that SCATMINEs have a self-destruct (SD) time. Emplacing an obstacle too soon can cause the mines to self-destruct before the enemy arrives.

FIRE SUPPORT

5-37. For the indirect fire plan to be effective in the defense, the unit plans and executes fires in a manner that achieves the intended task and purpose of each target. Indirect fires serve a variety of purposes in the defense, including--

- Slow and disrupt enemy movement.
- Prevent the enemy from executing breaching operations.
- Destroy or delay enemy forces at obstacles using massed fires or pinpoint munitions.
- Disrupt enemy support-by-fire elements.
- Defeat attacks along Infantry avenues of approach with the use of final protective fire (FPF).
- Disrupt the enemy to allow friendly elements to disengage or conduct counterattacks.
- Obscure enemy observation or screen friendly movement during disengagement and counterattacks.
- Use smoke to separate enemy echelons or to silhouette enemy formations to facilitate direct fire engagement.
- Provide illumination as necessary.
- Execute suppression of enemy air defenses (SEAD) missions to support CAS, attack aviation, and high-payoff targets.

FIRE-SUPPORT ASSETS

5-38. In developing the fire plan, the company commander evaluates the indirect fire systems available to provide support. Considerations include tactical capabilities, weapons ranges, and available munitions. These factors help the company commander and FSO determine the best method for achieving the task and purpose of each target in the fire plan.

FIST POSITIONING

5-39. The company's fire support personnel contribute significantly to the fight. Effective positioning is critical. The company commander and FSO must select positions that provide fire support personnel with unobstructed observation of the AO and ensure survivability.

PROTECTION

5-40. Protection includes air defense, survivability, mobility, and countermobility.

AIR DEFENSE

5-41. The focus of air defense is on likely air avenues of approach for enemy fixed-wing aircraft, helicopters, and UASs. Air avenues of approach may or may not correspond with the enemy's ground avenues of approach. These systems also are frequently used to protect friendly counterattack forces against aerial observation or attack.

SURVIVABILITY

5-42. Survivability positions are prepared to protect personnel and weapon systems. Positions can be constructed and reinforced with overhead cover to provide Infantry and crew-served weapons with protection against shrapnel from air bursts. In addition, the company may use digging assets for ammunition caches at alternate, supplementary, or subsequent positions. All leaders must understand the survivability plan and priorities, and that one leader within the company is specifically designated to enforce the plan and priorities, and that completion status is accurately reported and tracked.

SUSTAINMENT

5-43. In addition to the sustainment functions required for all operations (Chapter 11, *Sustainment operations*), the IBCT rifle company commander's planning process includes the considerations highlighted in the following paragraphs.

PRE-POSITIONING AND CACHES

5-44. The commander's mission analysis may reveal that the company's ammunition needs during an upcoming operation exceed its basic load. This requires the company to pre-position ammunition caches. The caches, which might be positioned either at alternate or subsequent positions, are dug in and guarded.

POSITIONING OF COMPANY TRAINS

5-45. The company's trains normally operate one terrain feature to the rear of the company to provide immediate recovery and medical support. The company trains are established to conduct evacuation (of those wounded in action [WIA], weapons, and equipment) and resupply as required. The company trains are located in a covered and concealed position, close enough to the company to provide responsive support, but out of enemy direct fire. The 1SG or XO will position the trains and supervise sustainment operations. The commander ensures all elements know the locations of the battalion's forward support company as well as the company CCP, battalion aid station (BAS) and that casualty evacuation procedures are planned and rehearsed.

Section IV. PREPARATION AND INTEGRATION

The company commander's analysis determines the most effective measures for every mission. This section describes the techniques and planning considerations available to the company commander as he prepares his defense.

DEFENSIVE TECHNIQUES

5-46. The company normally defends using one of these basic defensive techniques.

- Defend in sector.
- Defend from a BP.

- Defend a strongpoint.
- Defend a perimeter.
- Defend in a linear defense.
- Defend in a nonlinear defense.
- Defend on a reverse slope.

5-47. The control measures for the defense are sectors, battle positions, or a combination of these measures. No set criteria exist for selecting the control measures, but Table 5-2 provides some basic considerations.

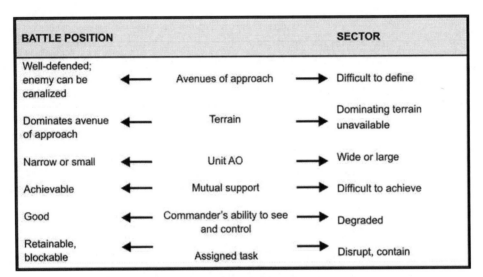

Table 5-2. Selection of control measures.

SECTOR DEFENSE

5-48. A sector is the company control measure that provides the most freedom of action to a platoon. It gives the platoon the flexibility to operate decentralized, while ensuring sufficient control to avoid confusion and synchronize the company's operation. In restricted terrain, where dismounted Infantry forces prefer to work, mutual support between the company's platoon battle positions is difficult to achieve. Seeing and controlling the fight throughout the company sector are also very difficult for the commander.

Company Disposition

5-49. The company disposition might consist of platoon sectors, a series of mutually supporting BPs, or a combination of the two (Figure 5-2). Positions are arrayed in depth. The strength of the sector comes from its flexibility. This defense normally orients on the enemy force and not on retaining terrain. It is effective because it allows the enemy to expose his flanks and critical C2 and CS assets through his own maneuver into the depth of the defense.

Defensive Operations

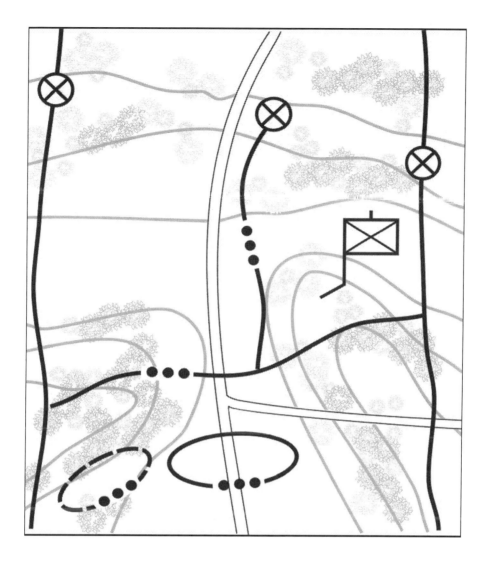

Figure 5-2. Company defense in sector, with platoon in a battle position.

Decentralization

5-50.   By assigning platoon sectors, the company may fight a defense in sector very similar to a nonlinear defense. This decentralized technique for conducting a defense in sector requires greater initiative and delegates more of the control to subordinate leaders. The small-unit actions are very similar to the nonlinear defense. When required, squads or platoons may disengage independently and move to another location within the sector to continue the fight. Considerations for the company R&S plan and employment of a reserve also are very similar to the nonlinear defense.

Platoon Battle Positions

5-51.   When fighting a company defense in sector from platoon battle positions, the concept is to defeat the attacker through the depth of his formation by confronting him with effective fires from mutually supporting BPs as he tries to maneuver around them. Infantry positions, patrols, mines, and other obstacles cover gaps that, due to terrain masking or thick vegetation, cannot be covered effectively by direct fire. Units remain in place except for local or internal movement to alternate or supplementary positions. If certain platoon positions become untenable during the battle, the company commander may withdraw the platoons according to prepared plans.

5-52. One technique is to allow the enemy to move into the EA and destroy him with massed fires. Another technique is to engage the attacker at maximum range with fires from field artillery, (and mortars) and then engage with organic antiarmor weapons positioned to deliver fires at maximum effective ranges from the flanks and rear. As the enemy closes, antiarmor weapons may move to alternate or supplementary firing positions within the BP to continue firing and avoid being bypassed.

5-53. The company defense in sector from platoon battle positions generally requires the company commander to be able to see and control the battle. It also requires good fields of fire to allow mutual support. If the terrain or the expected ECOA prevents this, the defense might be more effective if control is more decentralized and the platoons fight in sector.

5-54. A significant concern, particularly when fighting from BPs, is the enemy's ability to isolate a part of the company and then fix and destroy or bypass them. Without effective mutual support between the BPs, this is likely to occur. Even with mutual support, responsive and effective indirect fire support might be critical to defending the BPs. Without immediately available fire support, a capable enemy will quickly concentrate combat power against any BP that is identified.

BATTLE POSITION DEFENSE

5-55. A battle position is a general location orientation of forces on the ground where units defend. The platoon is located within the general area of the BP. Security elements might be located forward and to the flanks of the BP. Platoons defending from a BP may not be tied in with adjacent units; thus, the requirement for all-round security is increased. When determining the location of BPs, the commander decides first on EAs and sectors of fire; locations are then determined to support the direct fire plan. Each position must contribute to the company's accomplishment of its assigned task and purpose within the battalion commander's concept of the operation.

TYPES OF BATTLE POSITIONS AND PREPAREDNESS LEVELS

5-56. A platoon moves from its primary, alternate, supplementary, or subsequent position only with the commander's approval, or when the commander has prescribed a particular condition as a reason to move. The four types of battle positions and three levels of preparedness are--

Battle Positions

- Primary.
- Alternate.
- Supplementary.
- Subsequent.

Levels of Preparedness

- Occupied.
- Prepared but not occupied.
- Planned.

PRIMARY BATTLE POSITION

5-57. The *primary position* is the position that covers the enemy's most likely avenue of approach into the AO. It is the best position to accomplish the assigned mission such as cover an EA (Figure 5-3).

ALTERNATE BATTLE POSITION

5-52. An alternate position is a defensive position that the commander assigns to a unit for occupation when the primary position becomes untenable or unsuitable for carrying out the assigned task. (Figure 5-3). It is located so the platoon can continue to fulfill the original task such as covering the same avenue of approach or engagement area as the primary position. Alternate positions increase the defender's survivability by allowing engagement of the enemy from multiple positions. For example, a unit moves to its alternate position when the enemy brings suppressive fires on the primary position. If the alternate position is to be occupied in limited visibility, it might be forward of the primary position. The alternate position might be occupied if the platoon is driven out of the primary position by enemy fire or by assault, or it might be occupied to begin the fight to deceive the enemy of the platoon's primary position.

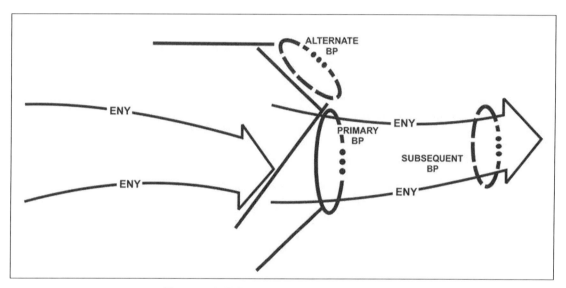

Figure 5-3. Primary and alternate positions.

SUPPLEMENTARY BATTLE POSITION

5-53. A supplementary position is a defensive position located within a unit's assigned AO that provides the best sectors of fire and defensive terrain along an avenue of approach that is not the primary avenue where the enemy is expected to attack. For example, an avenue of approach into a unit's AO from one of its flanks normally requires establishing supplementary positions to allow a unit or weapon system to engage enemy forces traveling along that avenue. It can also be assigned when the platoon must cover more than one avenue of approach (Figure 5-4, page 5-16).

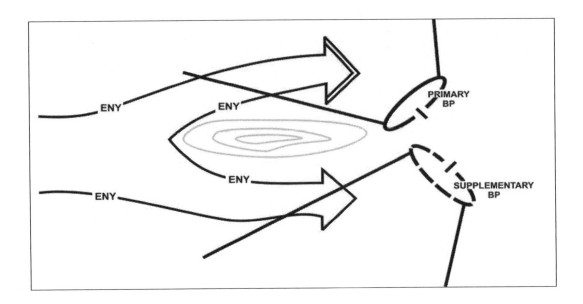

Figure 5-4. Supplementary position.

SUBSEQUENT BATTLE POSITION

5-58. A *subsequent position* is a position that a unit expects to move to during the course of battle. A defending unit may have a series of subsequent positions. Subsequent positions can also have primary, alternate, and supplementary positions associated with them.

OCCUPIED

5-59. An occupied position is one that has the unit physically is in place in the assigned position. The position is fully planned, prepared, and occupied before the "*defend no later than (NLT)*" time specified by the commander.

PREPARED BUT NOT OCCUPIED

5-60. The unit fully reconnoiters the position and the corresponding EA, marking positions in the BP and fire control measures in the EA. From the BP, the unit must accomplish all actions to enable it to execute the mission immediately on occupation. Planning, coordination, and rehearsals are required for the unit to displace to the BP and accomplish the mission. Despite time constraints, the unit digs in survivability positions, constructs fighting positions, designates target reference points (TRPs), develops direct and indirect fire plans, emplaces obstacles, clears fields of fire, and prestocks ammunition. Prepare missions are normally critical to the defense. A unit assigned such a mission must maintain security on the position and on the routes to it.

PLANNED TENTATIVE POSITIONS

5-61. The unit fully reconnoiters the EA and BP, planning tentative unit positions in the BP and establishing fire control measures in the EA. The unit coordinates and plans for defense from this position. Leaders reconnoiter, select, and mark positions, routes, and locations for security elements. They coordinate movement and other actions, such as preparing obstacles and occupation plans, with other elements of the battalion.

Defensive Operations

CENTRALIZED TECHNIQUE

5-62. Fighting from battle positions is a more centralized technique. It might also be more linear at the company level (Figure 5-5). Even so, it is not a static defense. Battle positions are positioned to achieve surprise and to allow maneuver within and between BPs. Defense from BPs is effective in concentrating combat power into an EA. It prevents the enemy from isolating one part of the company and concentrating his combat power in this area. Platoons are assigned mutual supporting battle positions that cover the enemy's likely avenues of approach. These BPs are located on terrain that provides cover and concealment and restricts vehicular movement.

SURPRISE

5-63. The commander's concept for fighting this defensive technique should concentrate on achieving surprise for each of the BPs. He does so by conducting an effective counterreconnaissance. Its purpose is to keep the enemy from finding the BPs. He initiates fires from one BP and waits for the enemy to react to this engagement before engaging from the other BPs (Figure 5-5). This confuses the enemy and disrupts his C2.

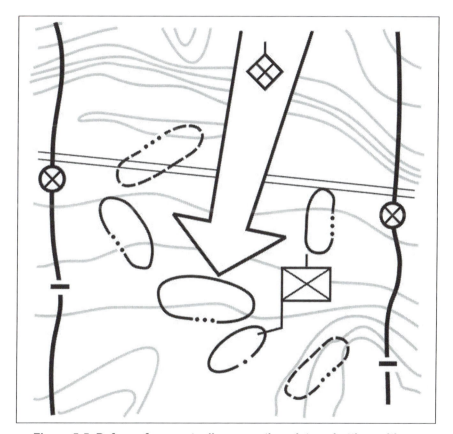

Figure 5-5. Defense from mutually supporting platoon battle positions.

- When the terrain provides a large EA and the commander's concept allows most of the enemy into the EA, the company may engage with massed fires from all platoon BPs. A disadvantage to this technique is that if there are still uncommitted enemy forces outside the EA, they will know the locations of the BPs and will try to isolate and concentrate against them. Contingency plans to disengage from these BPs and reorganize to continue the fight must be developed. This may involve displacing to alternate BPs or disengaging to conduct counterattacks against identified enemy C2, CS, or sustainment assets.

- Instead of one company EA, multiple EAs might be identified to provide flexibility to the plan (Figure 5-6, page 5-18). The plan must clearly state which platoons must reorient fires into the alternate EA and when they must do so.

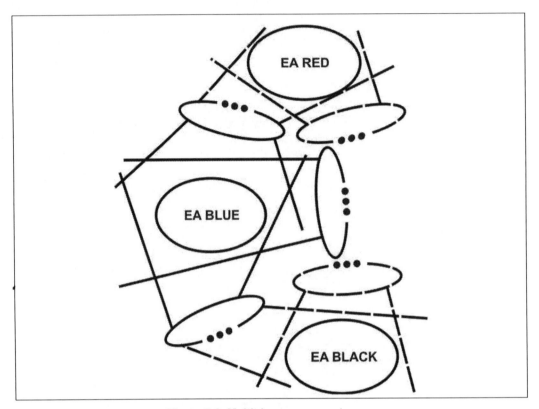

Figure 5-6. Multiple engagement areas.

STRONGPOINT DEFENSE

5-64.   A company might be directed to construct a strongpoint as part of an Infantry battalion defense (Figure 5-7). In order to do so, it is augmented with engineer support, more weapons, and sustainment resources. A strongpoint is defended until the commander directing the defense formally orders the unit out of it. The specific positioning of units in the strongpoint depends on the company commander's mission analysis. The same considerations for a perimeter defense apply, in addition to the following:

- Reinforce each individual fighting position (to include alternate and supplementary positions) to withstand small-arms fire, mortar fire, and artillery fragmentation. Stockpile food, water, ammunition, pioneer tools, and medical supplies in each fighting position.
- Support each individual fighting position with several others. Plan or construct covered and concealed routes between positions and along routes of supply and communication. Use these to support counterattack and maneuver within the strongpoint.

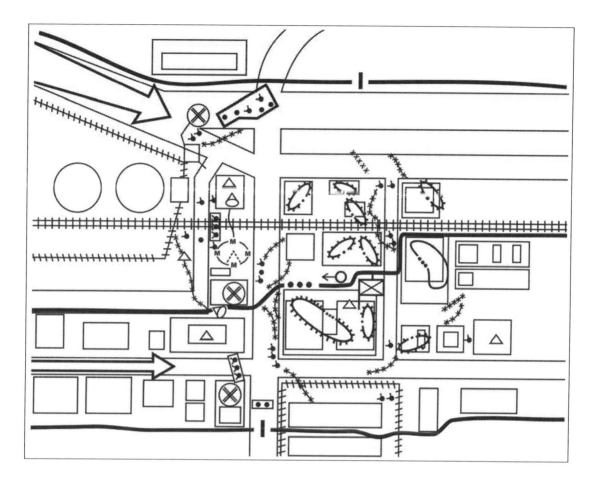

Figure 5-7. Company strongpoint.

- Divide the strongpoint into several independent, but mutually supporting, positions or sectors. If one of the positions or sectors must be evacuated or is overrun, limit the enemy penetration with obstacles and fires, and support a counterattack.
- Construct obstacles and minefields to disrupt and canalize enemy formations, to reinforce fires, and to protect the strongpoint from the assault. Place the obstacles and mines out as far as friendly units can observe them, within the strongpoint, and at points in between where they will be useful.
- Prepare range cards for each position and confirm them by fires. Plan indirect fires in detail and register them. Also, plan indirect fires for firing directly on the strongpoint using proximity fuses.
- Plan and test several means of communication within the strongpoint and to higher headquarters; possibilities include radio, wire, messenger, pyrotechnics, and other signals.
- Improve or repair the strongpoint until the unit is relieved or withdrawn. More positions can be built, tunnels and trenches dug, existing positions improved or repaired, and barriers built or fixed.

5-65. A strongpoint might be part of any defensive plan. It might be built to protect vital units or installations, as an anchor around which more mobile units maneuver, or as part of a trap designed to destroy attacking enemy forces that attack.

5-66. Mold the strongpoint to the terrain and use natural camouflage and obstacles. Existing obstacles can support formidable strongpoints, which provide cover, concealment, and obstacles. Complex and urban areas are also easily converted to strongpoints. Stone, brick, or steel buildings provide cover and concealment. Buildings, sewers, and some streets, which provide covered and concealed routes, can be rubbled to provide obstacles. Telephone systems can provide communications.

PERIMETER DEFENSE

5-67. A perimeter defense allows the defending force to orient in all directions (Figure 5-8). The perimeter defense can be employed in urban or woodland terrain. In terms of weapons emplacement, direct and indirect fire integration, and reserve employment, a commander conducting a perimeter defense considers the same factors as for a strongpoint operation. The Infantry rifle company might be called upon to execute the perimeter defense under a variety of conditions, including--

- When it must secure itself against terrorist or guerilla attacks in an urban area. This technique may also apply if the company must conserve or build combat power in order to execute offensive or patrolling operations.
- When it must hold critical terrain in areas where the defense is not tied in with adjacent units.
- When it has been bypassed and isolated by the enemy and must defend in place.
- When it conducts occupation of an independent assembly area or reserve position.
- When it begins preparation of a strongpoint.
- When it is directed to concentrate fires into two or more adjacent avenues of approach.

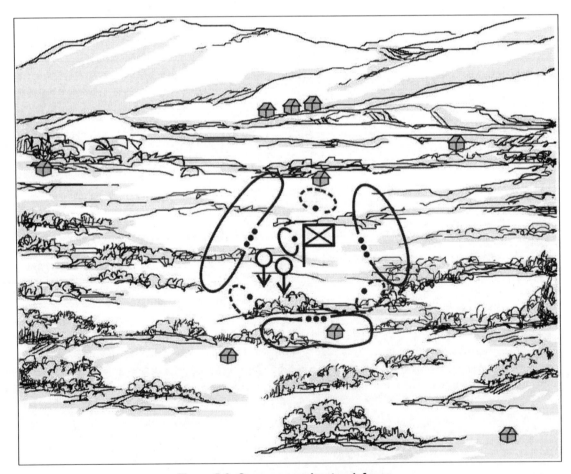

Figure 5-8. Company perimeter defense.

Defensive Operations

Preparations

5-68. The Infantry company prepares a perimeter defense when there are no friendly units adjacent to it (Figure 5-8). A perimeter defense might be used in a reserve position, in an assembly area or patrol base, on a follow-on decentralized company operation during resupply or when the company is isolated. The following actions constitute setting up a perimeter defense.

- Preparing a perimeter defense is like preparing any other position defense, but the company must disperse in a circular configuration for all-round security (the actual shape depends on the terrain). The company must be prepared to defend in all directions.
- The commander assigns the Infantry platoon to cover the most likely approach--a smaller sector than the other platoons cover. He prepares alternate and supplementary positions within the perimeter.
- If available, Javelins cover likely armor approaches. They may use hide positions and move forward to fire as the enemy appears. The commander assigns several firing positions. If there are few positions for them, they are assigned a primary position and are dug in.
- Snipers or designated shooters should cover likely or suspected enemy positions or OPs. Snipers and designated shooters should also be used to observe or overwatch areas where civilians congregate.
- Keep the mortars near the center of the perimeter so their minimum range does not restrict their ability to fire in any direction. They should dig in and have covered ammunition storage bunkers. They communicate by phone (the wire is buried). The FDC is dug in with overhead cover.
- If possible, hold one or more rifle squad in reserve. The company commander assigns a primary position to the rear of the platoon, covering the most dangerous avenues of approach. He may also assign the rifle squad supplementary positions since it must be prepared to fight in all directions.
- Prepare obstacles in depth around the perimeter.
- Plan direct and indirect fire as for any type of defense. Plan and use fire support from outside the perimeter when available.
- Counter enemy probing attacks by area fire weapons (artillery, mortars, Claymores, and grenade launchers) to avoid revealing the locations of fighting positions (ROE dependent).
- If the enemy penetrates the perimeter, the reserve destroys, and then blocks the penetration. They also cover friendly Soldiers during movement to alternate, supplementary, or subsequent positions. Even though the company's counterattack ability is limited, it must strive to restore its perimeter.
- Sustainment elements may support from within the perimeter or from another position. Supply and evacuation might be by air. Consider the availability of landing zones and drop zones (protected from enemy observation and fire) when selecting and preparing the position.

Y Variation

5-69. The Y-shaped perimeter defense is a variation of the perimeter defense that uses the terrain effectively. This defense is used when the terrain, cover and concealment, or fields of fire do not support the physical positioning of the platoons in a circular manner. The Y-shaped perimeter defense is so named because the platoon battle positions are positioned on three different axes radiating from one central point (Figure 5-9, page 5-22). It is still a perimeter defense because it is effective against an attack from any direction. The Y-shaped defense provides all-round perimeter fires without having to position Soldiers on the perimeter. It is likely to be most effective in mountainous terrain, but it may also be effective in a dense jungle environment due to limited fields of fire. All of the fundamentals of a perimeter defense previously discussed apply, with the following adjustments and special considerations.

- Although each platoon battle position has a primary orientation for its fires, each platoon must be prepared to reorient to mass fires into the EAs to its rear.

- When no most likely enemy approach is identified, or in limited visibility, each platoon may have half its Soldiers oriented into the EAs to the front and half into the EAs to the rear. Ideally, supplementary individual fighting positions are prepared to allow the Soldiers to reposition when required to mass fires into one EA.

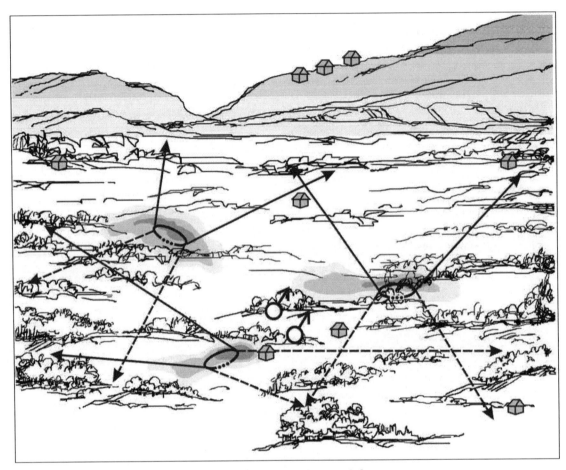

Figure 5-9. Y-shape perimeter defense.

- When a most likely enemy avenue of approach is identified, the company commander may adjust the normal platoon orientations to concentrate fires (Figure 5-10). This entails accepting risk in another area of the perimeter. The company security plan should compensate for this with additional OPs, patrols, or other measures.
- The positioning of the company CP, mortars, a reserve, or any sustainment assets is much more difficult due to a lack of depth within the perimeter.

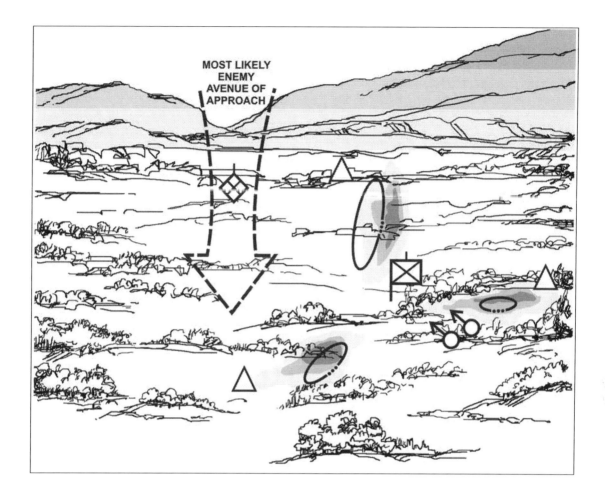

Figure 5-10. Modified Y-shape perimeter defense.

5-70. The most difficult aspect of the Y-shape perimeter defense is the fire control measures required. To safely fight this defense without casualties from friendly fire, the leaders must ensure the limits of fire for each weapon do not allow fires into the adjacent platoon position. In a mountainous environment, firing downward into the EAs may make this more simple. Some measures to consider include--

- Position machine guns near the apex of the "Y" to allow a FPL that covers the platoon front while firing away from the adjacent platoon.
- Cover the areas of the EAs closest to the apex with Claymores, other mines, or obstacles to reduce the need for direct fires in these areas.
- Identify those positions at most risk to friendly fires and prepare the fighting position to protect the Soldier from fires in this direction.
- The loss of one platoon position may threaten the loss of the entire company. To prevent this, plan and rehearse immediate counterattacks with a reserve or the least committed platoon.
- Consider allowing the enemy to penetrate well into the EAs and destroy him as in an ambush.
- Be aware that if a Y-shape defense is established on the prominent terrain feature and the enemy has the ability to mass fires, he may fix the company with direct fires and destroy it with massed indirect fires.

LINEAR DEFENSE

5-71. This technique allows interlocking and overlapping observation and fields of fire across the company's front (Figure 5-11, page 5-24). The bulk of the company's combat power is well forward. Sufficient resources must be available to provide adequate combat power across the sector to detect and stop an attack. The company relies on fighting from well-prepared mutually supporting positions. It uses a high volume of direct and indirect fires to stop the attacker. The main concern when fighting a linear defense is the lack of flexibility and the difficulty of both seizing the initiative and seeking out enemy weaknesses. When the enemy has a mobility advantage, a linear defense might be extremely risky. Obstacles, indirect fires, and contingency plans are key to this maneuver. The company depends upon surprise, well-prepared positions, and deadly accurate fires to defeat the enemy. The reserve is usually small, perhaps a squad.

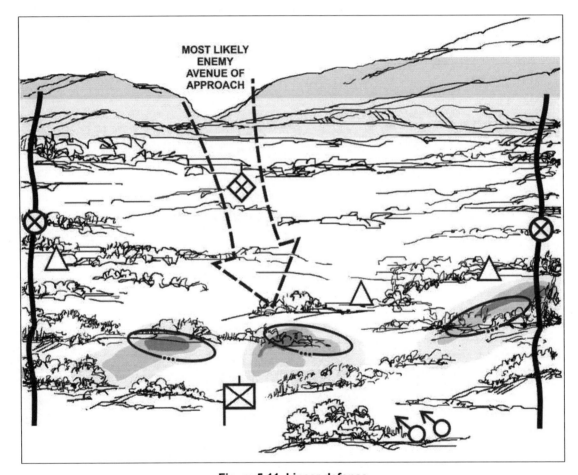

Figure 5-11. Linear defense.

Terrain Considerations

5-72. A linear defense might be used when defensible terrain is available in the forward portion of the company's sector, or to take advantage of a major linear natural obstacle. It is also used when the enemy is mainly Infantry, the company conducts a security mission such as counterinfiltration, or as directed by battalion.

Obstacles

5-73. Minefields and other obstacles are positioned and covered by fire to slow the attacker and to inflict casualties on him. Initially, engage him at long range by supporting fires (tactical air, attack helicopters, and field artillery) to disrupt the momentum of his attack. Use fires from mortars, machine guns, and small arms as he comes into range. If he penetrates the defense, block his advance with the reserve and shift fire from the forward platoons onto the enemy flanks. Then, counterattack--either by the company reserve or by the least committed platoon--with intense fires. The purpose is to destroy isolated or weakened enemy forces and regain key terrain.

Counterreconnaissance

5-74. The counterreconnaissance effort is critical when fighting a linear defense to deny the enemy the locations of the company's forward positions. If the enemy locates the forward positions, he will concentrate combat power where he desires while fixing the rest of the company to prevent their maneuver to disrupt his attack. This effort might be enhanced by initially occupying and fighting from alternate positions forward of the primary positions. This tactic enhances the security mission and deceives the enemy reconnaissance that may get through the security force.

NONLINEAR DEFENSE

5-75. The nonlinear defense is the most decentralized and dynamic defense conducted by an Infantry company. It is frequently used when operating against an enemy force that has equal or greater firepower and mobility capabilities. This type of defense is almost exclusively enemy-oriented and is not well suited for retaining terrain. To be successful, this defense depends on surprise, offensive action, and the initiative of small-unit leaders. (Figure 5-12). It is a very fluid defense with little static positioning involved.

Chapter 5

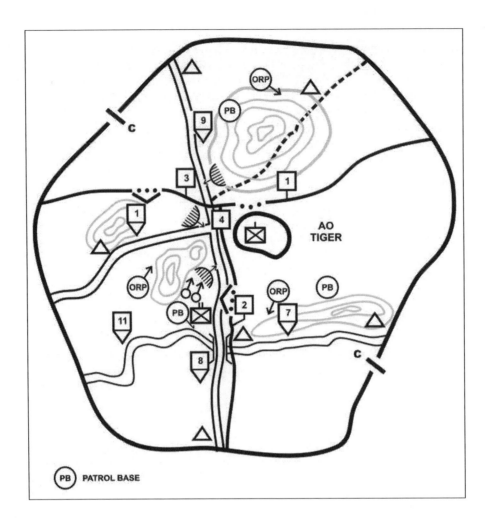

Figure 5-12. Nonlinear defense.

Company Focus

5-76. Normally, the battalion directs this defensive technique when the battalion concept does not focus the company. For example, the battalion might assign the company a sector and a mission that focuses the company on the enemy force. Mutual support is achieved solely through the linkage of purposes in the mission statements. The company commander may decide to conduct a nonlinear defense when he finds it difficult to identify a single decisive point that allows the company to concentrate combat power and achieve its purpose. Nonlinear defense may also be appropriate in terrain that prevents mutual support between platoons or against an enemy force capable of directing overwhelming firepower against identified friendly positions.

Reconnaissance and Security

5-77. The reconnaissance and security plan for this defensive technique focuses on avoiding detection by the enemy's reconnaissance assets. Operating in smaller units supports this requirement. Preparation and activity along likely reconnaissance routes must be closely controlled. Ideally, the company allows the enemy reconnaissance to move through the area before destroying him.

Platoon Sectors

5-78. The company commander assigns platoon sectors and may also identify likely ambush positions and rally points for each platoon. He identifies a decisive operation and assigns the shaping operation missions that provide mutual support and degrade the enemy's ability to generate combat power against the decisive operation. The decisive operation might be weighted by assigning priority of fires (POFs); by the allocation of mines, barrier materials, and other supplies; and by locating the company CP, CCP, and most of the caches in their vicinity.

Event-oriented Synchronization

5-79. The platoons conduct numerous squad and platoon ambushes, raids, and counterattacks, but they avoid decisive engagement. Before the enemy is able to react and concentrate against these small units, they disengage and seek out another enemy weak point. The synchronization for this defense might be event-oriented or accomplished by assigning ambush locations and initiating times or signals. The event-oriented synchronization involves identifying key enemy assets or vehicles that, if destroyed or disrupted, will have the greatest detrimental effect on the enemy.

Company Reserve

5-80. A company reserve is normally quite small. Due to the extended distances over which the company and platoons operate, the timely employment of the company reserve in a decisive action is not likely. Generally, the platoons are able to employ resources more effectively. A squad-size company reserve could be employed under the control of the 1SG as a logistics squad, for CASEVAC, or as a reaction force to shape the decisive operation.

Other Considerations

5-81. Other concerns include the difficulty of conducting resupply operations and casualty evacuation when defending in this manner. Resupply can be accomplished through pre-positioning of the critical supplies. CASEVAC requires detailed planning and battalion support. Platoon CCPs must be identified well forward to support each platoon. Litter teams moving on routes that avoid the enemy normally conduct the evacuation from these points to the company CCP. Treatment teams from the BAS should be positioned at the company collection point, particularly if casualties may need to be held until darkness for evacuation.

REVERSE SLOPE DEFENSE

5-82. An alternative to defending on the forward slope of a hill or a ridge is to defend on a reverse slope (Figure 5-13). In such a defense, the company is deployed on terrain that is masked from enemy direct fire and ground observation by the crest of a hill. Although some units and weapons might be positioned on the forward slope, the crest, or the counterslope (a forward slope of a hill to the rear of a reverse slope), most forces are on the reverse slope. The key to this defense is control of the crest by direct fire.

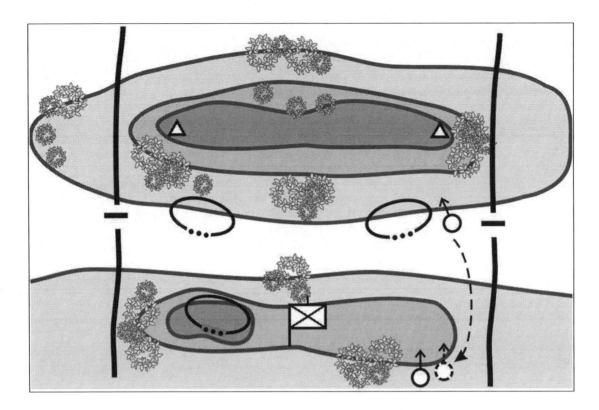

Figure 5-13. Company defense on a reverse slope.

General Considerations

5-83. These considerations generally apply when defending on a reverse slope.

- The crest protects the company from direct fire. This is a distinct advantage if the attacker has a greater weapon's range than the defender. The reverse slope defense can eliminate or reduce the standoff advantage of the attacker. It also makes enemy adjustment of his indirect fire more difficult since he cannot see his rounds impact. It keeps the enemy's second echelon from supporting his first echelon's assault.
- The enemy might be deceived and may advance to close contact before he discovers the defensive position. Therefore, the defender may gain the advantage of surprise.
- The defender can improve positions, build obstacles, and clear fields of fire without disclosing his positions.
- The defender may use dummy positions on the forward slope to deceive the enemy.
- Resupply and evacuation (when under attack) might be easier when defending on a reverse slope.
- Enemy target acquisition and jamming efforts are degraded. Enemy radar, infrared sights, and thermal viewers cannot detect Soldiers masked by a hill. Radios with a hill between them and the enemy are less vulnerable to jamming and direction finders.
- Enemy use of CAS and attack helicopters is restricted. Enemy aircraft must attack defensive positions from the flank or from the rear, which makes it easier for friendly air defense weapons to engage them.
- A counterattacking unit has more freedom of maneuver since it is masked from the enemy's direct fire.

Defensive Operations

Special Considerations

5-84. These considerations may apply when defending on a reverse slope.

- Observation of the enemy is more difficult. Soldiers in this position see forward no farther than the crest. This makes it hard to determine exactly where the enemy is as he advances, especially when visibility is poor. OPs must be placed forward of the topographic crest for early warning and long-range observation.
- Egress from the position might be more difficult.
- Fields of fire are normally short.
- Obstacles on the forward slope can be covered only with indirect fire or by units on the flanks of the company unless some weapons systems are initially placed forward.
- If the enemy gains the crest, he can assault downhill. This may give him a psychological advantage.
- If OPs are insufficient or improperly placed, the defenders might have to fight an enemy who suddenly appears in strength at close range.

Feasibility

5-85. A defense on a reverse slope might be effective when--

- The enemy has more long-range weapons than the defender.
- The forward slope has little cover and concealment.
- The forward slope is untenable because of enemy fire.
- The forward slope has been lost or not yet gained.
- Better fields of fire exist on the reverse slope.
- It adds to the surprise and deception.

Plans

5-86. The fundamentals of the defense apply to a defense on a reverse slope.

- Position forward platoons so they block enemy approaches and exploit existing obstacles. They should permit surprise fire on the crest and on the approaches around the crest. Forward fighting positions should have rear and overhead cover to protect friendly Soldiers from fratricide.
- Position OPs, including FIST personnel, on the crest or the forward slope of the defended hill. At night, increase OPs and patrol units to prevent infiltration. Machine guns might be attached to OPs.
- Position the platoon in depth or reserve where it can provide the most flexibility, support the forward platoons by fire, protect the flanks and the rear of the company, and, if necessary, counterattack. It might be positioned on the counterslope to the rear of the forward platoons if that position allows it to fire and hit the enemy when he reaches the crest of the defended hill.
- Position the company CP to the rear where it will not interfere with the reserve or supporting units. The company commander may have an OP on the forward slope or crest and another on the reverse slope or counterslope. He uses the OP on the forward slope or crest before the battle starts when he is trying to determine the enemy's intentions. During the fight, he moves the OP on the reverse slope or counterslope.
- Plan indirect fire well forward of, on, and to the flanks of the forward slope, crest, reverse slope, and counterslope. Plan indirect FPF on the crest of the hill to control the crest and stop assaults. Put the company's mortar section in defilade to the rear of the counterslope.

- Reinforce existing obstacles. Protective obstacles on the reverse slope — just down from the crest where it can be covered by fire — can slow the enemy's advance and hold him under friendly fire.
- The commander normally plans counterattacks. He plans to drive the enemy off the crest by fire, if possible. He must also be prepared to drive the enemy off by fire and movement as well.

ENGAGEMENT AREA DEVELOPMENT

5-87. The EA is where the company commander intends to contain and destroy an enemy force using the massed fires of all available weapons. The success of any engagement depends on how effectively the commander can integrate the obstacle plan, indirect fire plan, direct fire plan, and the terrain within the EA to achieve the company's tactical purpose. Beginning with evaluation of METT-TC factors, the development process covers these steps.

- Identify all likely enemy avenues of approach.
- Determine likely enemy schemes of maneuver.
- Determine where to kill the enemy.
- Emplace weapons systems.
- Plan and integrate obstacles.
- Plan and integrate indirect fires.
- Rehearse the execution of operations in the EA.

Identify Likely Enemy Avenues of Approach

5-88. The following procedures and considerations (Figure 5-14, page 5-31), apply when identifying the enemy's likely avenues of approach.

- Conduct initial reconnaissance. If possible, do this from the enemy's perspective along each avenue of approach into the sector or EA.
- Identify key and decisive terrain. This includes locations that afford positions of advantage over the enemy as well as natural obstacles and choke points that restrict forward movement.
- Determine which avenues will provide cover and concealment for the enemy while allowing him to maintain his tempo. Determine what terrain the enemy is likely to use to support each avenue.
- Evaluate lateral routes adjoining each avenue of approach.

Defensive Operations

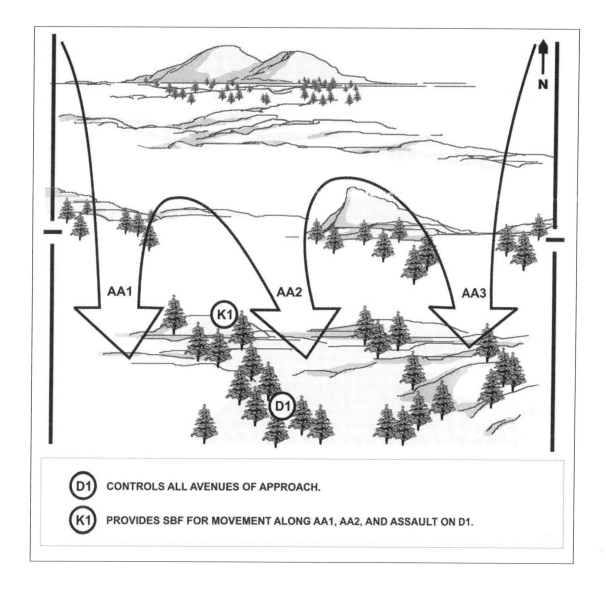

Figure 5-14. Likely enemy avenues of approach.

Determine Enemy Scheme of Maneuver

5-89. The company commander can use the following procedures and considerations (Figure 5-15) to determine the enemy's scheme of maneuver.

- Determine how the enemy will structure the attack. In what formation will he attack? How will he sequence his forces?
- Determine how the enemy will use his reconnaissance assets. Will he try to infiltrate friendly positions?
- Determine where and when the enemy will change formations and establish support-by-fire positions.
- Determine where, when, and how the enemy will conduct his assault and breaching operations. Determine likely OPs and what terrain the enemy is likely to employ for supporting fires.
- Determine where and when he will commit follow-on forces.

Chapter 5

- Determine the enemy's expected rates of movement.
- Assess the effects of his combat multipliers.
- Determine what reactions the enemy is likely to have in response to projected friendly actions.

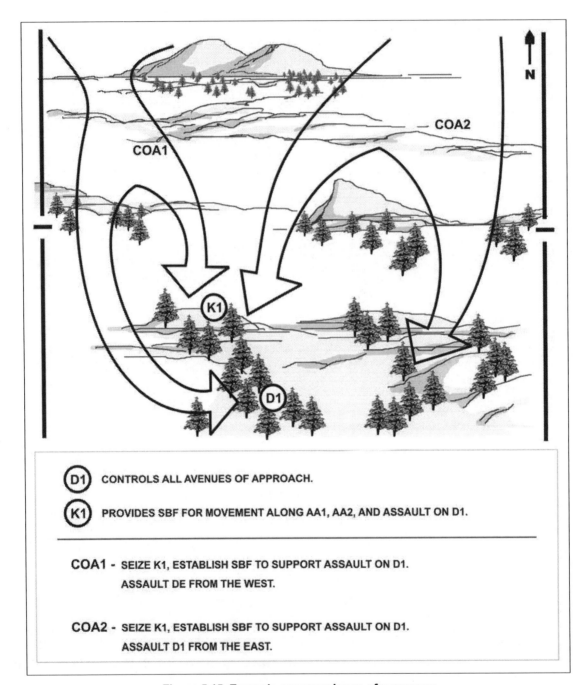

Figure 5-15. Example enemy scheme of maneuver.

Determine Where to Kill Enemy

5-90. Identify and mark where the battalion and company will engage the enemy (Figure 5-16, page 5-33).

- Identify TRPs that match the enemy's scheme of maneuver, allowing the company to identify where it will engage enemy forces through the depth of the sector.
- Identify and record the exact location of each TRP.
- Determine how many weapons systems must focus fires on each TRP to achieve the desired effects.
- Determine which platoons will mass fires on each TRP.
- Establish EAs around TRPs.
- Develop the direct fire planning measures necessary to focus fires at each TRP.

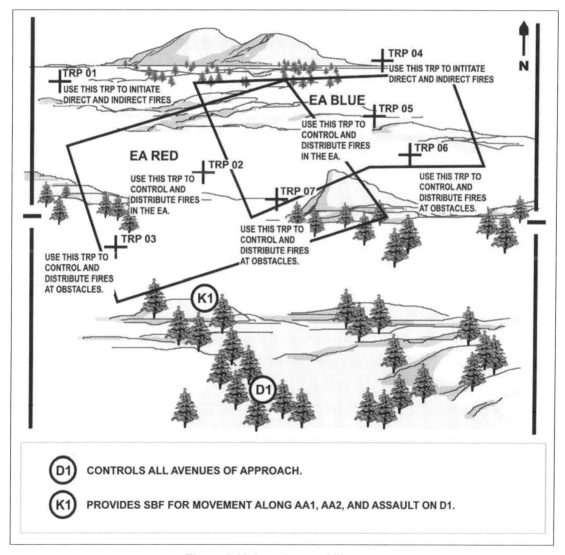

Figure 5-16. Locations to kill enemy.

> *Note:* In marking TRPs, use thermal sights to ensure visibility at the appropriate range under varying conditions, including daylight and limited visibility (darkness, smoke, dust, or other obscurants).

Emplace Weapons System

5-91. The following steps apply in selecting and improving BPs and emplacing crew-served weapons systems and Infantry positions (Figure 5-17).

- Select tentative platoon BPs. (When possible, select these while moving in the EA. Using the enemy's perspective enables the commander to assess the survivability of the positions.)
- Conduct a leader's reconnaissance of the tentative BPs.
- Traverse the EA to confirm that selected positions are tactically advantageous.
- Confirm and mark the selected BPs.
- Ensure that BPs do not conflict with those of adjacent units and that they are effectively tied in with adjacent positions.
- Select primary, alternate, supplementary, and subsequent fighting positions to achieve the desired effect for each TRP.
- Ensure that platoon leaders, platoon sergeants, and squad leaders position weapons systems so that the required number of weapons and platoons effectively covers each TRP.

Defensive Operations

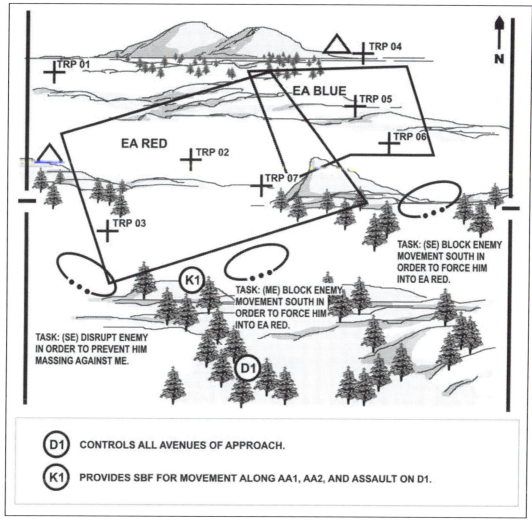

Figure 5-17. Emplacement of weapons systems.

Plan and Integrate Obstacles

5-92. The goal of obstacle planning is to support the commander's intent through optimum obstacle emplacement and integration with fires. Obstacles must allow the enemy into the EA and then contain him there. The focus at the battalion level and below is the actual integration of fires and obstacles. At the battalion level, obstacle planning is very directive and detailed and centers on obstacle groups. At the company level, obstacle planning deals with the actual sighting and emplacement of individual obstacles within the groups. The following steps apply in planning and integrating obstacles in the company defense (Figure 5-18).

- Understand obstacle group intent.
- Coordinate with the engineers.
- Site and mark individual obstacle locations.
- Combat elements should be used to provide security for the engineers as they emplace obstacles.
  - The overwatching element team marks fire control measures such as TRPs and artillery targets in the EA.

- Engineers enter the EA and move to the far side of the proposed trace of the obstacle group.
- The engineer squad/platoon leader and company commander collocate in the defensive positions covering the obstacle.
- Elements from the engineers move along the proposed trace of the obstacle group.
- From the defensive position, the leaders follow the movement of the engineers, ensuring that all points of the obstacle trace can be covered with fires.
- They maintain communications with the engineers via FM.
- The commander and engineer squad/platoon leader refines the obstacle trace, adjusting the position of individual obstacles as necessary.

- Refine direct and indirect fire control measures.
- Identify lanes and gaps.
- Report obstacle locations and gaps to higher headquarters.

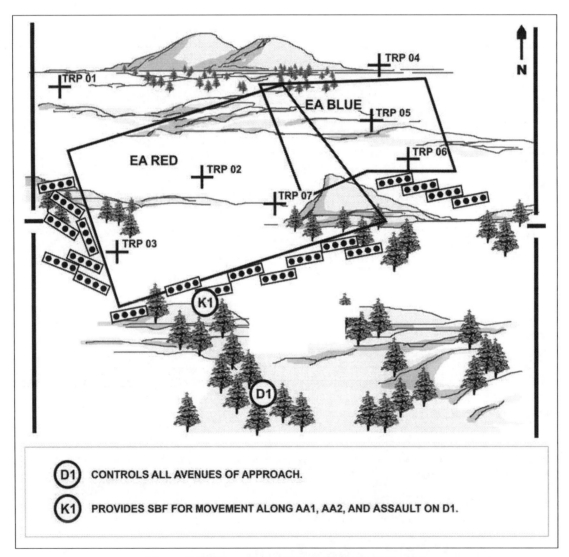

Figure 5-18. Plans for and integration of obstacles.

Plan and Integrate Indirect Fires

5-93. The following steps apply in planning and integrating indirect fires (Figure 5-19, page 5-38).

- Determine the purpose of fires and the essential fire support task (EFST) that supports it.
- Determine where the purpose can best be achieved.
- Establish the observation plan, with redundancy for each target. Observers include the FIST, as well as members of maneuver elements with fire support responsibilities such as platoon sergeants.
- Establish triggers.
- Obtain accurate target locations.
- Refine target locations to ensure coverage of obstacles.
- Adjust artillery and mortar targets.
- Plan FPFs.
- Request critical friendly zones (CFZs) for protection of maneuver elements and no-fire areas (NFAs) for protection of OPs and forward positions.

Chapter 5

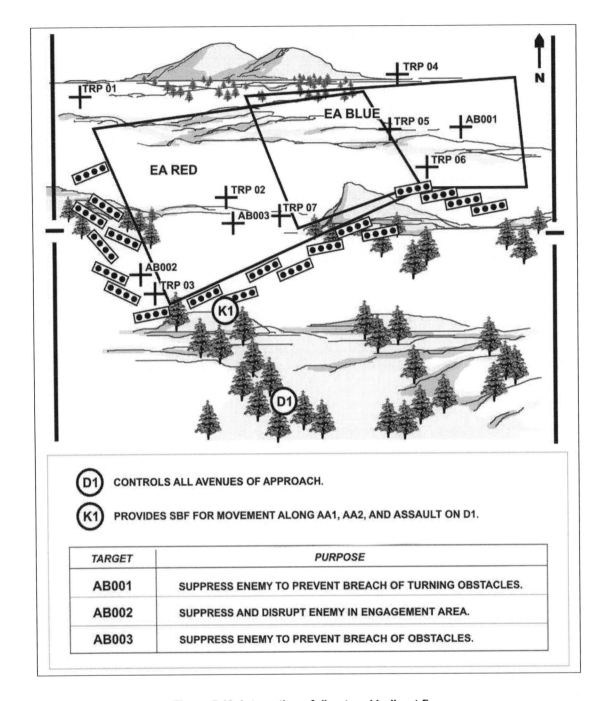

Figure 5-19. Integration of direct and indirect fires.

Conduct an Engagement Area Rehearsal

5-94. The purpose of this rehearsal is to ensure every leader and Soldier understands the plan and all elements are prepared to cover their assigned areas with direct and indirect fires. Although the company commander has several options, the most common and most effective type of rehearsal is to replicate the threat. One technique for the rehearsal in the defense is to have the company trains, under the control of the company XO, move through the EA to show the enemy force while the commander and subordinate platoons rehearse the battle from the company BP. The rehearsal should cover these actions.

- Rearward passage of security forces (as required).

- Closure of lanes (as required).
- Movement from the hide position to the BP.
- Use of fire commands, triggers, and maximum engagement lines (MELs) to initiate direct and indirect fires.
- Shifting of fires to refocus and redistribute fire effects.
- Emplacement of scatterable mine systems.
- Preparation and transmission of critical reports.
- Assessment of the effects of enemy weapons systems.
- Displacement to alternate, supplementary, or subsequent BPs.
- Cross-leveling or resupply of Class V.
- Evacuation of casualties.

Note: The company commander should coordinate the rehearsal with the battalion to ensure other units' rehearsals are not planned for the same time or location. Coordinating leads to more efficient use of planning and preparing time for all battalion units. It also eliminates the danger of misidentification of friendly forces in the rehearsal area, which could result in fratricide.

PRIORITY OF WORK

5-95. Priority of work is a set method of controlling the preparation and conduct of a defense. TSOP should describe priority of work to include individual duties. The commander changes priorities based on the situation. All leaders in the company should have a specific priority of work for their duty position. Although listed in sequence, several tasks are performed at the same time. An example priority of work sequence is as follows.

- Post local security.
- Establish the company R&S operation.
- Position Javelins, machine guns, and Soldiers; assign sectors of fire.
- Position other assets (company CP and mortars).
- Designate FPLs and FPFs.
- Clear fields of fire and prepare range cards and sector sketches.
- Adjust indirect fire FPFs. The firing unit FDC should provide a safety box that is clear of all friendly units before firing any adjusting rounds.
- Prepare fighting positions.
- Install wire communications, if applicable.
- Emplace obstacles and mines.
- Mark (or improve marking for) TRPs and direct fire-control measures.
- Improve primary fighting positions such as overhead cover.
- Prepare alternate and supplementary positions.
- Establish sleep and rest plan.
- Reconnoiter movements.
- Rehearse engagements and disengagements or displacements.
- Adjust positions and control measures as required.
- Stockpile ammunition, food, and water.
- Dig trenches between positions.
- Reconnoiter routes.
- Continue to improve positions.

Company Commander

5-96. Many of these duties can be delegated to subordinates, but the commander must ensure they are done. The commander must--

- Ensure local security and assign platoon OP responsibility.
- Conduct a leader's reconnaissance with the platoon leaders and selected personnel. Confirm or deny significant deductions or assumptions from the mission analysis. Confirm the direct fire plan, to include EAs, sectors of fire, position key weapons, and fire control measures. Designate primary, alternate, supplementary, and subsequent positions that support the direct fire plan, for platoons, sections, and supporting elements. Require platoons to conduct coordination. Integrate indirect fire plan and obstacles to support the direct fire plan. Designate the general company CP location.
- Check the company CP and brief the 1SG and XO on the situation and logistics requirements.
- Upon receipt of the platoon sector sketches, make two copies of a defensive sector sketch and a fire plan. Retain one copy and forward the other to the battalion (Figure 5-20).
- Confirm the direct fire plan and platoon positions before digging starts. Coordinate with the left and right units.
- Check with the battalion commander for any changes or updates in the orders.
- Finish the security, deception, counterattack, and obstacle plans.
- Walk the company positions after they are dug. Confirm clear fields of fire and complete coverage of the sector by all key weapons. Look at the defensive plan from an enemy point of view, both conceptually and physically.
- Check dissemination of information, interlocking fires, and dead space.
- Ensure immediate correction of deficiencies.
- Ensure EA rehearsals are conducted and obstacle locations reported.

Defensive Operations

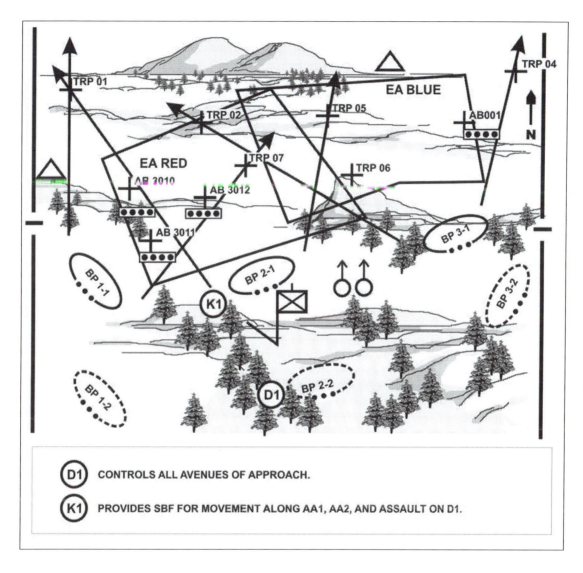

Figure 5-20. Company defensive sector sketch.

First Sergeant and Executive Officer

5-97. The first sergeant and XO must--

- Establish the company CP and ensure that wire communications link the platoons, sections, and attached elements if applicable.
- Establish casualty collection points, company logistics release points, and EPW collection points.
- Brief platoon sergeants on the company CP location, logistics plan, and routes between positions.
- Assist the company commander with the sector sketch.
- Request and allocate pioneer tools, barrier material, rations, water, and ammunition.
- Walk the positions with the company commander. Start supervising emplacement of the platoons and sections, and check range cards and sector sketches.

Fire-Support Officer

5-98. The FSO must--

- Assist the commander in planning the indirect fires to support the defense.
- Advise the commander on the status of all firing units, and on the use of smoke or illumination.
- Coordinate with the Infantry battalion FSO, firing units, and platoon leaders to ensure the fire plan is synchronized and fully understood.
- Ensure the indirect fire plan is rehearsed and understood by all.
- Ensure all FPFs are adjusted as soon as possible.
- Develop observation plan.
- Coordinate and rehearse any repositioning of observers within the company sector to ensure they can observe targets or areas of responsibility.
- Develop triggers.
- Report battlefield intelligence.
- Ensure redundancy in communications.

Mortar Section Leader

5-99. The mortar section leader must--

- Choose a tentative firing position(s) and OP(s) and complete his portion of the fire plan based on the company OPORD, coordinated with the FSO and his own analysis.
- Take part in the company leader's reconnaissance. Confirm or adjust the firing position, select OPs, and coordinate the indirect fire plan with the company FSO.
- Issue FRAGOs to the mortar squads. Conduct a section leader's reconnaissance with squad leaders. Require squad leaders to coordinate with platoons and squads for security and logistics support.
- Direct the mortar section to begin digging.
- Establish internal and external wire communications, if applicable.
- Assist the FSO in completing the fire plan and overlays.
- Register and adjust the FPF.
- Inspect the mortar position.
- Reconnoiter routes to alternate firing positions.

Senior Radio Operator (Assisted by Radio Operator)

5-100. The senior radio operator and radio operator must--

- Supervise setting up wire, radio, and voice, with the battalion, platoons, and sections.
- Organize a radio watch.
- Supervise the performance of preventive maintenance checks and services (PMCS) on the radios.
- Assist the 1SG and XO, as required. Help organize local security for the company CP, dig fighting positions, and assist in OPORD production.

Prior bullets (continued from previous page):
- Establish routine security or alert plans, radio watch, and rest plans. Brief the company commander.
- Supervise continuously and assist the commander with other duties as assigned.

Chemical, Biological, Radiological, and Nuclear NCO

5-101. The CBRN NCO must--

- Assist the commander with an updated MOPP analysis.
- Ensure that chemical detection and monitoring procedures are established and maintained.
- Coordinate for decontamination support.
- Coordinate smoke support.
- Supervise decontamination operations.
- Provide guidance on operations in CBRN conditions.

ADJACENT UNIT COORDINATION

5-102. The ultimate goal of adjacent unit coordination is to ensure unity of effort in the accomplishment of the Infantry battalion's missions. Items that adjacent units must coordinate include the following.

- Unit positions, including locations of command and control nodes.
- Locations of OPs and patrols.
- Overlapping fires (to ensure that direct fire responsibility is clearly defined).
- TRPs.
- Alternate, supplementary, and subsequent BPs.
- Indirect fire information.
- Obstacles (location and type).
- Air defense considerations, if applicable.
- Routes to be used during occupation and repositioning.
- Sustainment considerations.

Section V. RETROGRADE OPERATIONS

The retrograde is a type of defensive operation that involves organized movement away from the enemy (FM 3-0). The enemy may force these operations or a commander may execute them voluntarily. In either case, the higher commander of the force executing the operation must approve the retrograde (FM 3-90). Retrograde operations are conducted to improve a tactical situation or to prevent a worse situation from developing. Companies normally conduct retrogrades as part of a larger force but may conduct independent retrogrades (withdrawal) as required such as on a raid.

PURPOSE

5-103. Retrograde operations accomplish the following.

- Resist, exhaust, and defeat enemy forces.
- Draw the enemy into an unfavorable situation.
- Avoid contact in undesirable conditions.
- Gain time.
- Disengage a force from battle for use elsewhere in other missions.
- Reposition forces, shorten lines of communication, or conform to movements of other friendly units.
- Secure more favorable terrain.

TYPES

5-104. The three types of retrograde operations are delay, withdrawal, and retirement.

DELAY

- This operation allows the unit to trade space for time, avoiding decisive engagement and safeguarding its elements. A delay is a series of defensive and offensive actions over subsequent positions in depth. It is an economy of force operation that trades space for time. While the enemy gains access to the vacated area (space), friendly elements have time to conduct necessary operations, while retaining freedom of action and maneuver. This allows friendly forces to influence the action; they can prevent decisive engagement or postpone action to occur at a more critical time or place on the battlefield.

Types

- The two types of delay missions follow.
  - Delay in sector.
  - Delay forward of a specified line or position for a specified time.

Components of Successful Delay

- For either type of delay mission, the flow of the operation can be summarized as "*hit hard, then move.*" A successful delay has three key components.
  - The ability to stop or slow the enemy's momentum while avoiding decisive engagement.
  - The ability to degrade the enemy's combat power.
  - The ability to maintain a mobility advantage.

Delay Within a Sector

- The company might be assigned a mission to delay within a sector AO. The higher commander normally provides guidance regarding intent and desired effect on the enemy, but he minimizes restrictions regarding terrain, time, and coordination with adjacent forces. This form of a delay is normally assigned when force preservation is the highest priority and there is considerable depth to the AO.

Delay Forward of a Specified Line for a Specified Time

- The company might be assigned a mission to delay forward of a specific control measure for a specific period. This mission is assigned when the battalion must control the enemy's attack and retain specified terrain to achieve some purpose relative to another element, such as setting the conditions for a counterattack, for completion of defensive preparations, or for the movement of other forces or civilians. The focus of this delay mission is clearly on time, terrain, and enemy destruction. It carries a much higher risk for the battalion, with the likelihood of all or part of the unit becoming decisively engaged. The timing of the operation is controlled graphically by a series of phase lines with associated dates and times to define the desired delay-until period.

Culmination of Delay

- Delay missions usually conclude in one of three ways: a defense, a withdrawal, or a counterattack. Planning options should address all three possibilities.

Defensive Operations

Planning Considerations

- In preparing for the delay, the commander uses planning considerations that are identical to those for a defense in sector, varying only in their purpose. Planning for the delay must cover several areas related to hindering enemy movement and maintaining mobility. These considerations include--
  - Use of existing terrain and obstacles, enhanced by employment of reinforcing obstacles.
  - Designation of positions where the friendly force can harass or impede the enemy without risking decisive engagement itself. This applies especially to a delay in sector. When a battalion is delaying in sector, companies are normally assigned a series of specific BPs to enhance command and control across the sector. Likewise, in a company delay in sector, the commander will assign a series of specific BPs for each platoon.
  - Assessment of opportunities to conduct limited counterattacks to disrupt enemy actions.
  - Designation of high-speed avenues of withdrawal.
  - Rehearsal of operations anticipated for the delay; these may include engagement of the enemy and maneuver through the delay area.

Techniques

- In executing either a delay in sector or a time-related delay, the commander can choose from the following techniques.
  - Delay from subsequent positions or phase lines.
  - Delay from alternating positions.

Delay from Subsequent Positions or Phase Lines

- This delay technique normally is used when the sector is so wide that available forces cannot occupy more than a single line of positions. The commander must be aware of several factors that may put his unit at a disadvantage during the delay.
  - Lack of depth at any particular time.
  - The possibility of inadequate time to prepare subsequent positions.
  - Decreased security during disengagement.
  - The possibility of gaps between units.
  - When the unit receives the order to conduct the delay from its initial positions, one element, such as a company in a battalion delay or a platoon in a company delay, displaces and occupies its subsequent BP. The remainder of the unit maintains contact with the enemy until the first displacing element is in position to engage the enemy from the subsequent position. The first element then provides overwatch or base of fire as other elements displace to their subsequent positions. Figure 5-21, page 5-46, shows a company conducting a dismounted delay from subsequent positions.

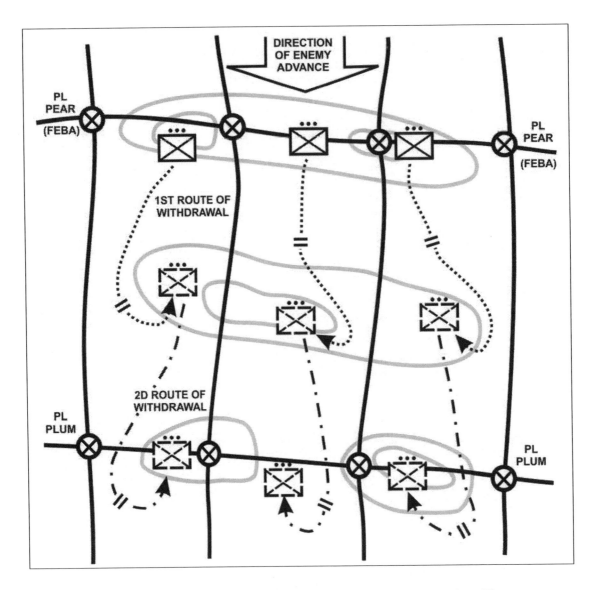

Figure 5-21 Example company dismounted delay from subsequent positions

Delay from Alternating Positions

- This method of delay might be used when the delaying element has sufficient forces to occupy more than a single line of positions (normally in a narrow sector). The delaying battalion or company arrays one or more of its subordinate elements in the initial delay positions. This first echelon then engages the enemy while the rest of the unit occupies and prepares second-echelon delay positions. The unit then alternates fighting the enemy with movement to new positions. The elements in the initial delay positions engage the enemy until ordered to displace, or until displacement criteria are met. They then displace, moving through the second-echelon delay positions to their own subsequent positions (which become the third echelon of the delay). Elements in the second echelon overwatch the displacing units' movement and assume responsibility for engaging the enemy. This sequence continues until the delay operation is completed. Figure 5-22 shows a company delay from alternating positions.

Defensive Operations

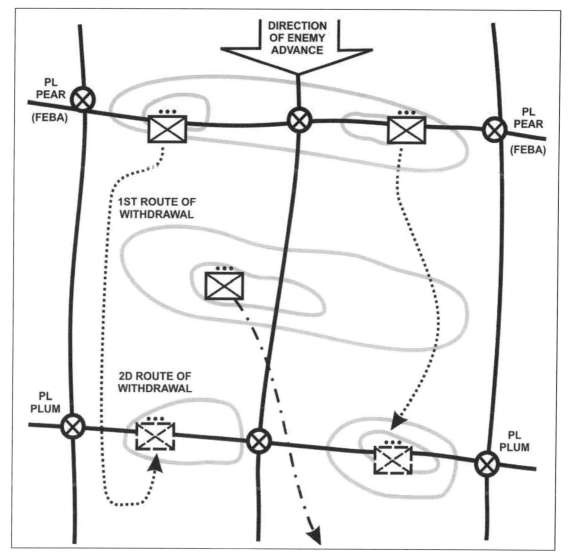

Figure 5-22. Example company delay from alternating positions.

WITHDRAWAL

5-105. The commander uses this operation to break enemy contact, especially when he needs to free the unit for a new mission. Withdrawal is a planned operation in which a force in contact disengages from an enemy force. Withdrawals may or may not be conducted under enemy pressure.

Types

5-106. The two types of withdrawals are assisted and unassisted.

Assisted

5-107. The assisting force occupies positions to the rear of the withdrawing unit and prepares to accept control of the situation. It can also assist the withdrawing unit with route reconnaissance, route maintenance, fire support, and sustainment. Both forces closely coordinate the withdrawal. After coordination, the withdrawing unit delays to a battle handover line, conducts a passage of lines, and moves to its final destination.

Unassisted

5-108. The withdrawing unit establishes routes and develops plans for the withdrawal and then establishes a security force as the rear guard while the main body withdraws. Sustainment and CS elements normally withdraw first followed by combat forces. To deceive the enemy as to the friendly movement, battalion may establish a DLIC if withdrawing under enemy pressure. As the unit withdraws, the DLIC disengages from the enemy and follows the main body to its final destination.

Phases

5-109. Withdrawals are accomplished in three overlapping phases, as follows.

Preparation

5-110. The commander dispatches quartering parties, issues WARNOs, and initiates planning. Nonessential vehicles are moved to the rear.

Disengagement

5-111. Designated elements begin movement to the rear. They break contact and conduct tactical movement to a designated assembly area or position.

Security

5-112. In this phase, a security force protects and helps the other elements as they disengage or move to their new positions. This is done either by a DLIC, which the unit itself designates in an unassisted withdrawal, or by a security force provided by the higher headquarters in an assisted withdrawal. As necessary, the security force assumes responsibility for the sector, deceives the enemy, and protects the movement of disengaged elements by providing overwatch and suppressive fires. In an assisted withdrawal, the security phase ends when the security force has assumed responsibility for the fight and the withdrawing element has completed its movement. In an unassisted withdrawal, this phase ends when the DLIC completes its disengagement and movement to the rear.

Unassisted Withdrawal

5-113. In an unassisted withdrawal, the unit conducting the withdrawal establishes the DLIC to maintain contact with the enemy or to deceive him.

Battalion Withdrawal

5-114. In a battalion withdrawal, the DLIC may consist of an element from each company (under leadership of the company XO or a platoon leader), with the battalion S-3 as the overall DLIC commander. As an alternative, a company may serve as the DLIC for the rest of the battalion. The company commander has several deployment options. He can reposition elements across the entire battalion frontage. Another option is to position the company to cover only the most dangerous enemy avenues of approach; other avenues into the sector are covered with observation from additional security elements provided by the battalion such as the reconnaissance platoon.

Defensive Operations

Company Unassisted Withdrawal

5-115. The commander has similar options in an unassisted company withdrawal. He may designate one platoon to execute the DLIC mission for the company, or he can constitute the DLIC using elements from the three rifle platoons with the XO as the DLIC commander. Figure 5-23 shows an example of an unassisted withdrawal.

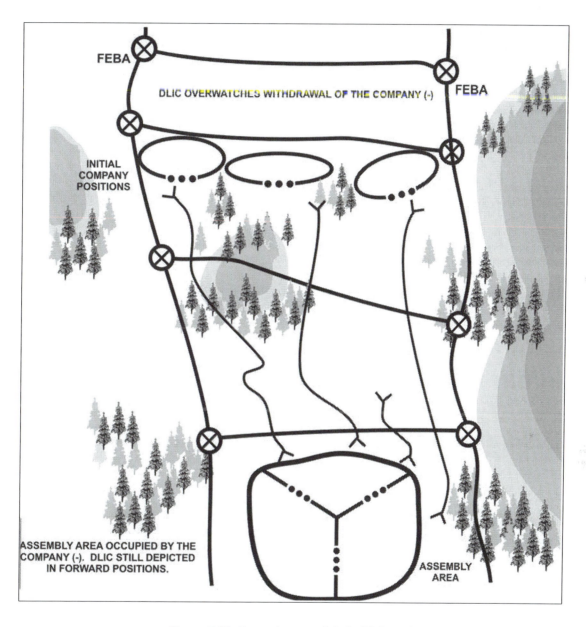

Figure 5-23. Example unassisted withdrawal.

Assisted Withdrawal

5-116. In an assisted battalion withdrawal, the higher headquarters will normally provide a security element to maintain contact with and deceive the enemy while the battalion conducts its withdrawal. Likewise, in a company withdrawal, the battalion provides the security force. The security force establishes defensive positions behind the withdrawing unit and conducts preparations for a rearward passage of lines.

The withdrawing force disengages from the enemy and conducts the rearward passage through the security force to assembly areas in the rear.

RETIREMENT

5-117. This operation is employed to move a force that is not in contact away from the enemy. Typically, the company conducts a retirement as part of a larger force while another unit's security force protects their movement. A retiring unit organizes for combat but does not anticipate interference by enemy ground forces. Triggers for a retirement may include the requirement to reposition forces for future operations or to accommodate other changes to the current CONOP. The retiring unit should move sustainment elements and supplies first, and then should move toward an assembly area that supports preparations for the next mission. Where speed and security are the most important considerations, units conduct retirements as tactical road marches.

Chapter 6
Stability Operations

This edition introduces stability operations. Stability operations encompass a range of actions that shape the political environment and respond to developing crises. Stability operations usually occur in conjunction with offensive and defensive operations. These operations are diverse, continuous, and often long-term. Stability operations may include both developmental and coercive actions. Developmental actions are aimed at enhancing a government's willingness and ability to care for its people, or simply providing humanitarian relief following a natural disaster. Coercive military actions involve the application of limited, carefully prescribed force, or the threat of force, to achieve specific objectives. Stability operations are usually nonlinear and noncontiguous, and they are often time and human intensive. Army elements might be tasked to conduct stability operations in a complex, dynamic, and often asymmetric environment, to accomplish one or more of the following purposes.

- Deter or thwart aggression.
- Reassure allies, friendly governments, agencies, or groups.
- Provide encouragement and support for a weak or faltering government.
- Stabilize an area with a restless or openly hostile population.
- Maintain or restore order.
- Satisfy treaty obligations or enforce national or international agreements and policies.
- Provide humanitarian relief outside the continental United States (CONUS) and its territories.

Before adapting anything in this chapter, the commander must consider every aspect of the situation and thoroughly analyze METT-TC.

Note: For more detailed information on stability operations, see these books--

FM 3-0, FM 3-07, FM 3-07.31, FM 7-98,
JP 3-07.2 JP 3-07.3 TC 7-98-1.

Section I. PLANNING CONSIDERATIONS

Stability operations are normally planned centrally, at the strategic and operational levels. However, tactical execution often takes the form of decentralized, small-scale, noncontiguous actions conducted over extended distances. Responsibility for making decisions on the ground falls to junior leaders. The following paragraphs examine several important considerations that influence planning and preparation for stability operations. (For a more detailed discussion of these subjects, see FM 3-07.31.) While conducting stability operations, commanders must know which phase of the COE in which they are operating. Since stability operations occur in and around populated and urban areas, commanders must also know both friendly and enemy strategic, operational, and tactical goals and purposes. The following discussions provide commanders a means of delineating warfighting functions and other key considerations as they pertain to stability operations and lean more toward the more challenging types of operations relating to combat in the urban environment:

INTELLIGENCE

6-1. Intelligence planners must consider support to situational understanding and intelligence, surveillance, and reconnaissance.

SUPPORT TO SITUATIONAL UNDERSTANDING

6-2. The commander will acquire and develop intelligence for stability operations much the same as war. The principle difference is in its focus. At lower echelons, the political, economic, linguistic, ethnic, and other factors assume greater relevance to the mission.

6-3. The many possible intelligence sources may include but are not limited to.

- The UN, governmental organizations (GO), non-governmental organizations (NGO), International Committee for the Red Cross (ICRC), special operations forces (SOF) including CA and PSYOP.
- Leaders of local disputing parties.
- Regional military and political leadership.
- Civilian populations.
- Insurgent elements.
- Police and paramilitary forces.

6-4. Units need to conduct an urban IPB to set the conditions for the overall collection plan and to improve the effectiveness of shaping operations. The minimum requirement for an urban IPB includes identifying all the routes on which the unit will move, rivers, templated IED locations (based on historical data, pattern analysis, and terrain), intervisiblity (IV) lines overlooking templated IED locations, and enemy egress routes (see FM 34-130 and FM 3-6). The civil IPB process is most effective if guided by the staff Civil-Military Officer. (See FM 3-07.31 for a complete discussion of civil IPB.) A basic and time-constrained analytical technique for the company commander may include asking the following questions.

- Who are the civilians we might encounter in the sector?
- What activities are those civilians engaged in that might affect our operations?
- How might our operations affect the civilian population?

INTELLIGENCE, SURVEILLANCE, AND RECONNAISSANCE

6-5. Many operations may be conducted quickly based on "*actionable intelligence*" received from human intelligence (HUMINT) sources. The usefulness of this intelligence may be good and the speed at which a unit can react is key to destroying or capturing and high-value target (HVT).

Stability Operations

6-6. Integrating SUAS and aviation assets into the surveillance and reconnaissance plan is highly beneficial. Aerial reconnaissance over-flights of routes helps deny the enemy use of the terrain. The enemy will likely tend to avoid an area frequently patrolled by the air. UASs have frequently provided actionable intelligence resulting in very successful targeting.

MOVEMENT AND MANEUVER

6-7. The nature of operations may require many various types of units and MOSs to conduct frequent and intensive small unit patrols both day and night. Platoons, sections, and squads will continually modify patrolling techniques to adapt to the current environment. The basic security patrol moves to or through specific areas to deter enemy aggression and reduce his activity.

6-8. During OIF stability operations, insurgents targeted U.S. forces conducting vehicle movements, mounted patrols, and convoys on a daily basis. All vehicle movements were vulnerable to attack, at almost any place and time. In this non-linear and non-contiguous environment, units continually validated the need for every vehicle movement to be planned, prepared, and executed as a combat operation. Additional relevant discussions are in Chapter 8, *Tactical Enabling Operations* and Chapter 12, *Urban Patrolling* of this field manual.

FIRE SUPPORT

6-9. Artillery, mortar, and aerial fire support may be employed in support of both decisive and shaping operations. Illumination rounds demonstrate that U. S. forces are alert and looking. Apart from employment associated with the OP, FDC, or the gun-target line, familiar FA targeting methods also relate well to information operations (IO). Therefore, the Fire Support Officers (FSOs) may also be a maneuver company IO.

PROTECTION

6-10. Almost all protection functions work much the same for stability operations as discussed throughout this manual, with only *force protection* deserving additional emphasis. Infantry commanders must implement appropriate security measures to protect the force. Offensive measures taken to protect the force are usually the most effective. Offensive measures include aggressive patrolling and offensive actions taken against identified enemy forces. Aggressive intelligence gathering and pattern analysis greatly increases the commander's situational understanding. As the commander's situational understanding increases he can more aggressively and accurately identify and deter, capture or destroy enemy elements. Defensive force protection measures include the establishment of various CPs, effective base camp security procedures, and protection against IEDs and unexploded ordnance (Appendix G, *IEDs, Suicide Bombers, UXO, and Mines*).

6-11. The Infantry company may receive security missions as part of the battalion security plan. Additional security missions result from the company commander's concept for the company defense. These missions might be oriented on friendly units, on the enemy and terrain (reconnaissance), or on the enemy's reconnaissance assets (counter-reconnaissance). Although the commander quickly establishes a security plan to keep the enemy from observing or surprising the company he must generally assume that the enemy has excellent observation and situational understanding of US units and actions in urban areas. The commander establishes this plan before moving the company into the area and maintains it continuously – the commander builds his own situational understanding to counter that of the enemy and to allow US forces to establish an effective offensive orientation. The Infantry company commander bases this plan on orders received from the battalion and on the enemy situation, terrain, and visibility conditions. The plan provides active and passive measures and counter-reconnaissance.

Chapter 6

ACTIVE MEASURES

6-12. These include OPs, stand-tos, traffic-control measures, and patrols.

- The commander develops observation and lodgement area security plans based on METT-TC. He normally incorporates snipers and designated shooter in these plans.
- If needed, he conducts a stand-to both morning and evening to help Soldiers adjust to the changing light and noise conditions. He has them prepare equipment and adjust or relocate their positions as needed.
- A variety of means exist for controlling and monitoring foot and vehicular traffic flow. Some are discussed in detail later in this chapter. The commander decides which techniques to employ, again based on METT-TC.
- The battalion or company dispatches patrols whose missions directly contribute to security, force protection, and intelligence gathering.

PASSIVE MEASURES

6-13. These measures include camouflage, movement control, light and noise discipline, and proper radiotelephone procedures.

- To ensure effective coverage, the company commander can direct platoons to cover specific areas with specific devices such as NVDs and thermal sights. He also determines how and where to employ snipers or designated marksmen.
- As much as he can, the commander varies all aspects of the operation, avoiding establishing routines that the enemy could exploit.

SUSTAINMENT

6-14. The operational environment the company faces during stability operations might be harsh, creating special sustainment considerations. The commander must remember that the enemy will target these vital and vulnerable operations and forces. He might also have to--

- Rely on local procurement of certain items, especially water and Class I.
- Plan for shortages of critical items such as repair parts, Class IV supply materials, and Class III lubricants.
- Consider special Class V supply requirements such as nonlethal munitions.

COMMAND AND CONTROL

6-15. Because of the unique requirements of stability operations, more often than not the Infantry company is task-organized to operate with a variety of units. This includes some elements with which the company does not normally work such as linguists, counterintelligence teams, PSYOP, and civil affairs teams (CATs). During platoon-level shaping operations, unless the commander is conducting battlefield circulation, the company command post will likely locate within the forward operating base (FOB). For company-level decisive or shaping operations, the commander will move and locate with his command and control element or company command post (CP).

RULES OF ENGAGEMENT

6-16. In decentralized operations, effective command guidance, a clear commander's intent, and a detailed understanding of ROE are critical at the operational and tactical levels. The ROE direct the circumstances and limitations under which US forces initiate, respond to, or continue combat engagement with forces encountered. These rules reflect the laws of war, operational concerns, and political considerations when the operational environment shifts from peace to conflict and back. The ROE must be briefed and trained to the lowest tactical level. They are established for, given to, and thoroughly understood by every Soldier in the unit. Another important consideration in development and employment of ROE is that commanders must assume that the belligerents they encounter fully understand the ROE. These enemy elements will try to use their understanding of the ROE to their own advantage and to the disadvantage of the friendly force. (See FM 3-07.31 for a more detailed discussion of ROE.)

RULES OF INTERACTION

6-17. The ROI are based on the ROE and are tailored to the specific regions, cultures, and populations affected by the operation. They provide a foundation for relating to people and groups who play critical roles in the operations. They cover an array of interpersonal communication skills such as persuasion and negotiation. They provide tools for the individual Soldier to deal with the nontraditional asymmetric threats that prevail in stability operations. These include political friction, religious and ethnic differences, unfamiliar cultures, and conflicting ideologies. Thus, the ROI help keep the Soldier out of trouble, which naturally enhances survivability. Every Soldier must know and understand the ROI. This means they must be thoroughly briefed and rehearsed.

MEDIA

6-18. The presence of the media is a reality that confronts every Soldier involved in stability operations. All leaders and Soldiers must know how to deal effectively with broadcast and print reporters and photographers. This should include an understanding of which subjects they are authorized to discuss and which subjects they must see the public affairs officer (PAO). Most current operations have media present who can immediately transmit what they see and hear. The images and words they project are powerful and can affect national policy. In our form of government, the media has the right to cover operations, and the public has a right to know. Many in the media lack a full understanding of the military, but they keep the public informed of Army operations and procedures. Therefore, there are many good things about the Army that are unknown to the public, and commanders and public affairs personnel have a responsibility to tell the Army's story. Freedom of the press does not negate the requirement for OPSEC and the accomplishment of the military mission (Appendix I, *Media Considerations*).

OPERATIONS WITH OUTSIDE AGENCIES

6-19. US Army units conduct certain stability operations in coordination with a variety of outside organizations. These include other US armed services or government agencies as well as international organizations such as private volunteer organizations (PVOs), nongovernmental organizations (NGOs), and United Nations (UN) military forces or agencies. (See Appendix F for more information.)

Chapter 6

Section II. TYPES OF OPERATIONS

Stability operations typically fall into ten broad types that are neither discrete nor mutually exclusive. For example, a force engaged in a peace operation may also find itself conducting arms control or a show of force to set the conditions for achieving an end state. This section provides an introductory discussion of stability operations (FM 3-0 and FM 3-07). Stability operations normally occur in conjunction with either offensive, defensive, or support operations (Figure 6-1).

- Peace operations (including peacekeeping, peace enforcement, and operations in support of diplomatic efforts).
- Foreign internal defense.
- Security assistance.
- Humanitarian and civic assistance.
- Support to insurgencies.
- Support to counterdrug operations.
- Combating of terrorism.
- Noncombatant evacuation operations.
- Foreign humanitarian assistance.
- Arms control.
- Shows of force.

Figure 6-1. Types of stability operations.

PEACE OPERATIONS

6-20. Peace operations encompass three general areas: operations in support of diplomatic efforts, peacekeeping, and peace enforcement.

PEACEKEEPING OPERATIONS

6-21. A peacekeeping force monitors and facilitates the implementation of cease-fires, truce negotiations, and other such agreements. In doing so, it must assure all sides in the dispute that the other involved parties are not taking advantage of settlement terms to their own benefit. The Infantry company most often observes, monitors, or supervises and aids the parties involved in the dispute. The peacekeeping force must remain entirely neutral. If it loses a reputation for impartiality, its usefulness within the peacekeeping mission is compromised.

PEACE ENFORCEMENT OPERATIONS

6-22. What characteristics distinguish peace enforcement activities from wartime operations and from other stability operations? The difference is that PEO compel compliance with international resolutions or sanctions, and they restore or maintain peace and order. They might entail combat, armed intervention, or the physical threat of armed intervention. Under the provisions of an international agreement, the battalion and its subordinate companies might have to use coercive military power to *compel* compliance with the applicable international sanctions or resolutions.

FOREIGN INTERNAL DEFENSE

6-23. Foreign internal defense means the participation, by the civilian and military agencies of a government, in any action programs taken by another government to free and protect its society from subversion, lawlessness, and insurgency (JP 1-02). The objective is to promote stability by helping the host nation establish and maintain institutions and facilities that can fill its people's needs. Army forces in foreign internal defense normally advise and assist host-nation forces conducting operations to increase their capabilities.

SECURITY ASSISTANCE

6-24. Army forces assist in providing HN security by training, advising, and assisting allied and friendly armed forces. Security assistance includes the participation of Army forces in any program through which the US provides defense articles, military training, and other defense-related services to support national policies and objectives. Security assistance can take the form of grants, loans, credit, or cash sales (JP 3-07).

HUMANITARIAN AND CIVIC ASSISTANCE

6-25. Humanitarian and civic assistance (HCA) programs help the HN populace in conjunction with military operations and exercises. Foreign humanitarian assistance (FHA) operations are also limited in scope and duration. They focus only on providing prompt aid to resolve an immediate crisis. In contrast to foreign humanitarian (only) assistance operations, HCA is planned, and provides only--

- Medical, dental, and veterinary care in rural areas of a country.
- Construction of rudimentary surface transportation systems.
- Well-drilling and construction of basic sanitation facilities.
- Basic construction and repair of public facilities.

6-26. US forces conduct FHA operations outside the borders of the US or its territories. The purpose is to relieve or reduce the results of natural or manmade disasters or other endemic conditions that pose a serious threat to life (disease, starvation) or property. The US military typically supplements the HN authorities along with other governmental agencies, nongovernmental organizations, private voluntary organizations, and unaffiliated individuals. Most FHA operations resemble civil support operations. The distinction between the two is that, inside the US and its territories, The Posse Comitatus Act (PCA) prevents the use of the military from becoming a civil police force, or guardia civil. The PCA does not apply to US forces overseas. In vague or hostile situations, FHA activities are handled as a subset of a larger stability, offensive, or defensive operation.

SUPPORT TO INSURGENCY

6-27. This type of support includes assistance provided by US forces to help a friendly nation or group that is trying to combat insurgent elements or to stage an insurgency itself. This type of stability activity is normally conducted by special operating forces.

SUPPORT TO COUNTERDRUG OPERATIONS

6-28. US military forces might be tasked for a variety of counter-drug activities, which are always conducted in conjunction with another government agency. These activities include destroying illicit drugs and disrupting or interdicting drug manufacturing, growing, processing, and smuggling operations. Counter-drug support may take the form of advisory personnel, mobile training teams, offshore training activities, and assistance in logistics, communications, and intelligence.

COMBATTING OF TERRORISM

6-29. In all types of stability operations, antiterrorism and counterterrorism activities are a continuous requirement in protecting installations, units, and individuals from the threat of terrorism. Antiterrorism focuses on defensive measures used to reduce the vulnerability of individuals and property to terrorist attacks. Counterterrorism encompasses a full range of offensive measures to prevent, deter, and respond to terrorism. (For more information on these activities, see JP 3-07.2.)

NONCOMBATANT EVACUATION

6-30. A noncombatant evacuation operation (NEO) is conducted primarily to evacuate US citizens whose lives are in danger. It can also evacuate natives and third-country nationals. An NEO involves swift insertions and temporary occupation of an objective followed by a planned withdrawal. The company uses only the force needed to protect evacuees and defend itself.

ARMS CONTROL

6-31. To prevent a conflict from escalating, an Infantry company can conduct arms-control inspections and disarm belligerents. Collecting, storing, and destroying conventional munitions and weapons systems can deter belligerents from resuming hostilities.

SHOW OF FORCE

6-32. Deploying forces abroad lends credibility to a nation's promises and commitments. Credible show-of-force operations bolster and reassure allies. Infantry companies participating in a show-of-force mission focus all preparations on the assumption that combat is probable. A show of force can quickly and unexpectedly escalate into conflict, although the intent is to avoid this.

Section III. COMPANY TASKS

Stability operations are complex and demanding. The Infantry company in a stability operation must master skills from negotiating to establishing OPs and checkpoints to escorting a convoy. The tasks and techniques in this section include lessons learned and should help the Infantry company commander implement these and other tasks. This section uses the term "*enemy forces*" to refer to guerillas, terrorists, or insurgent forces that generally try to blend into the local populations and engage in asymmetric warfare.

ESTABLISH AND OCCUPY A LODGMENT AREA OR FORWARD OPERATING BASE

6-33. A lodgment area (base camp) or forward operating base (FOB) is a well-prepared position used as a base of operations and staging area for the occupying unit (Figure 6-2, page 6-10). Like an assembly area or defensive strongpoint, the lodgment area also provides some force protection because it requires all-round security. However, several other factors distinguish a lodgment area from a less permanent position.

6-34. Due to the probability of long-term occupation, the lodgment requires a lot of preparation and logistical support. It needs shelters and facilities that can support the force and its attachments the whole time. Also, the area must be positioned and developed so the unit can effectively conduct its primary missions, such as PEO and counterterrorism, throughout its area of responsibility.

6-35. In establishing a lodgment, the Infantry company can either use existing facilities or request construction of new ones. Existing structures are immediately available, and require little or no construction support from engineers and members of the company. However, they might fall short of meeting the company's operational needs, and their proximity to other structures can pose security problems.

6-36. The company can establish and occupy a lodgment area as part of a battalion or, given enough support from battalion, as a separate element. Before preparing, constructing, and occupying the lodgment area, the company commander plans the general layout based on the following.

- Location.
- Effects of weather.
- Local traffic and pedestrian patterns.
- Observation post sites and patrol routes.
- Entry and exit procedures.
- Vehicle emplacement and orientation.
- Bunkers and fighting positions.
- Direct and indirect fires.
- Size and composition of the reserve.
- Location of possible landing and pickup zones.
- Sustainment, including--
  - Mess areas, showers, and latrines with adequate drainage.
  - Storage bunkers for Class III, IV, and V supplies.
  - Maintenance and refueling areas.
  - Aid station.
- CP site security, including--
  - Size, composition, and function of advance and reconnaissance parties.
  - Nature and condition of existing facilities such as quarters; water, sewer, and power utilities; and reinforced "*hard-stand*" areas for maintenance.
  - Proximity to structures and roadways and security implications.
- Priorities of work. The commander considers--
  - Establishment of security of the immediate area and the perimeter.
  - Establishment of initial roadblocks to limit access to the area.
- Mine clearance.
- Construction of revetments to protect vehicles, generators, communications equipment, and other facilities.
- Construction of barriers or berms around the lodgment area to limit enemy observation and to protect occupants.
- Construction of shelters for lodgment personnel.
- Construction of defensive positions.
- Construction of sanitation and personal hygiene facilities.
- Construction of hardened CP facilities.
- Continued improvement of site, for example, adding hard-wire electrical power or perimeter illumination.

Chapter 6

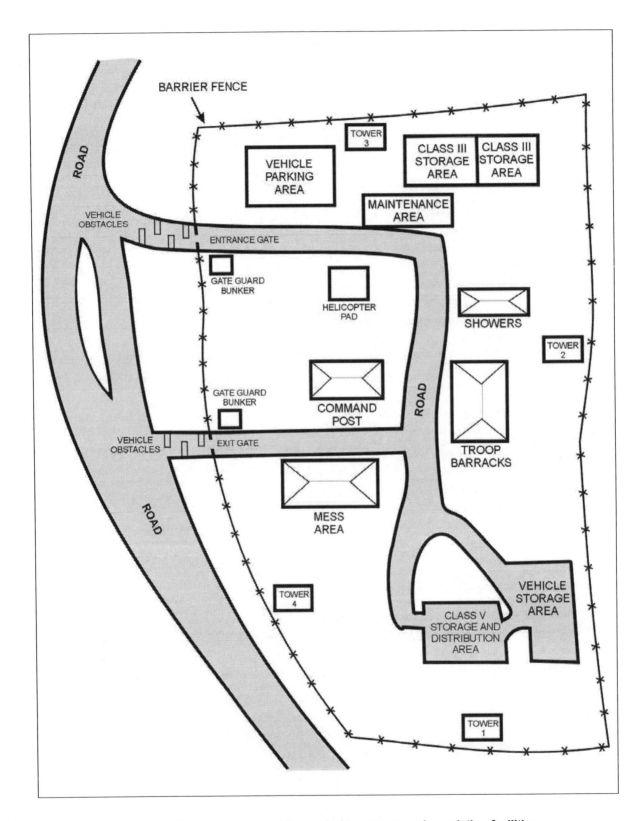

Figure 6-2. Example Infantry company lodgment area using existing facilities.

NEGOTIATE

6-37. Infantry company leaders might have to conduct negotiations. The two main types of negotiations follow.

SITUATIONAL NEGOTIATIONS

6-38. Situational negotiations allow immediate discussion and resolution of an issue or problem. For example, members of an advance guard might have to negotiate the passage of a convoy through a checkpoint.

6-39. At the company level, situational negotiations are far more common than preplanned ones. In stability operations, the commander, his subordinate leaders, and other Soldiers conduct some form of negotiations almost daily. To do this, they must thoroughly understand the ROE and ROI.

6-40. Members of the company apply this working knowledge to the process of discussing, and whenever possible, resolving issues and problems between opposing parties, which might include the company itself. The negotiator must know when he has exhausted his options under the ROE and ROI, and turn over the discussion to a higher authority. Negotiations move up through the levels of authority until the issue is resolved.

6-41. To prepare, leaders rehearse the ROE and ROI. One good way is to rehearse how to apply ROE and ROI in an example stability situation such as operating a checkpoint. This forces leaders and Soldiers to analyze the ROE and ROI while applying them in an operational environment.

PREPLANNED NEGOTIATIONS

6-42. Preplanned negotiations allow discussion and resolution of an upcoming, specific issue or problem. For example, the Infantry company commander conducts a work coordination meeting between leaders of the belligerents to determine mine-clearing responsibilities. As with situational negotiations, preplanned negotiations require leaders to know and understand the ROE and ROI. However, before a preplanned negotiation, leaders must also know every aspect of the dispute or issue. The negotiator's goal is to reach an agreement that is acceptable to both sides, and that reduces antagonism and the threat of renewed hostilities.

Identify Purpose of Negotiations

6-43. Before contacting leaders of the belligerent parties to initiate the negotiation process, the commander must familiarize himself with both the situation and the area in which his unit will operate. This includes identifying and evaluating AAs that connect the opposing forces. Results of the negotiation process, which might be lengthy and complicated, must be based on national or international agreements or accords. Negotiation topics include--

- When the sides will withdraw.
- Positions to which they will withdraw (these should preclude observation and direct fire by the opposing parties).
- What forces or elements will move during each phase of the operation.
- Pre-positioning of peace forces that can intervene in case of renewed hostilities.
- Control of heavy weapons.
- Mine clearance.
- Formal protest procedures for the belligerent parties.

Chapter 6

Establish Proper Context

6-44. The commander must earn the trust and confidence of each opposing party. This includes establishing an atmosphere (and a physical setting) that participants will judge to be both fair and safe. The commander must--

- Always conduct joint negotiations on matters that affect both parties.
- When serving as a mediator, remain neutral at all times.
- Learn as much as possible about the belligerents, the details of the dispute or issue under negotiation, and other factors such as the geography of the area and specific limitations or restrictions, including the ROE and ROI.
- Gain and keep the trust of the opposing parties by being firm, fair, and polite.
- Use tact, and remain patient and objective.
- Follow applicable local and national laws and international agreements exactly.

Prepare

6-45. Thorough, exacting preparation is another important factor in ensuring the success of the negotiation process. Company personnel--

- Negotiate sequentially, from subordinate level to senior level.
- Select and prepare a meeting place that is acceptable to all parties.
- Arrange for interpreters and adequate communications facilities, as necessary.
- Ensure that all opposing parties, as well as the negotiating company, use a common map (edition and scale).
- Coordinate all necessary movement.
- Establish local security.
- Keep higher headquarters informed throughout preparation and during the negotiations.
- Arrange to record the negotiations (use audio or video recording equipment, if available).

Negotiate

6-46. Negotiators must always strive to maintain control of the session. They must be firm, yet even-handed, in leading the discussion. At the same time, they must be flexible, with a willingness to accept recommendations from the opposing parties and from their own assistants and advisors, who--

- Exchange greetings.
- Introduce all participants by name, including negotiators and any advisors.
- Consider the use of small talk at the beginning of the session to put the participants at ease.
- Allow each side to state its case without interruptions and prejudgments.
- Record issues presented by both sides.
- If one side makes a statement that is incorrect, be prepared to produce evidence or proof to establish the facts.
- If the negotiating team or peacekeeping force has a preferred solution, present it and encourage both sides to accept it.
- Close the meeting by explaining to both sides what they have agreed to and what actions they must take. If necessary, be prepared to present this information in writing for their signatures.
- Do not negotiate or make deals in the presence of the media.
- Maintain the highest standards of conduct at all times.

MONITOR COMPLIANCE WITH AN AGREEMENT

6-47. Compliance monitoring involves observing belligerents and working with them to ensure they meet the conditions of one or more applicable agreements. Examples of the process include overseeing the separation of opposing combat elements, the withdrawal of heavy weapons from a sector, or the clearance of a minefield. Planning for compliance monitoring should cover, but is not limited to, the following considerations.

- Liaison teams, with suitable communications and transportation assets, are assigned to the headquarters of the opposing sides. Liaison personnel maintain communications with the leaders of their assigned element and talk directly to each other and to their mutual commander (the Infantry company or battalion commander).
- The commander positions himself at the point where violations are most likely to occur.
- He positions platoons and squads where they can observe the opposing parties, instructing them to assess compliance and report any violations.
- As directed, the commander keeps higher headquarters informed of all developments, including his assessment of compliance and noncompliance.

ESTABLISH OBSERVATION POSTS

6-48. Constructing and operating OPs is a high-frequency task for Infantry companies and subordinate elements whenever they must establish area security. Each OP is established for a specified time and purpose. Some OPs are overt (clearly visible) and deliberately constructed. Others are covert and designed to observe an area or target without the knowledge of the local population. Each type of OP must be integrated into supporting direct and indirect fire plans and into the overall observation plan. Based on METT-TC factors, deliberate and overt OPs may include specialized facilities such as--

- Observation towers.
- Ammunition and fuel storage areas.
- Power sources.
- Supporting helipads.
- Kitchens, sleep areas, showers, and toilets.

6-49. An OP is similar in construction to a bunker and it is supported by fighting positions, barriers, and patrols (Figure 6-3, page 6-14). Covert operations may include sniper or designated marksmen positions over-watching TAIs.

Note: If necessary, the company can also employ hasty OPs, which are similar to individual fighting positions.

Chapter 6

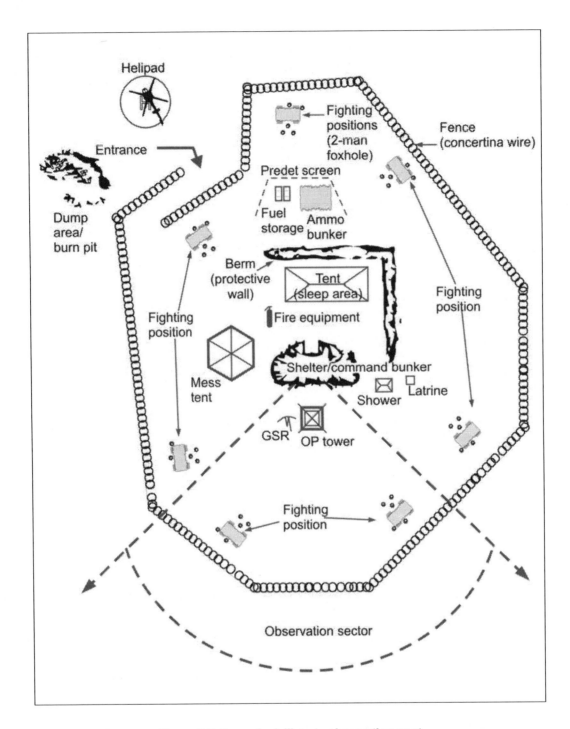

Figure 6-3. Example deliberate observation post.

ESTABLISH CHECKPOINTS

6-50. One of the main missions conducted during OIF was the vehicle or traffic checkpoint. Units considered these standard steady-state operations and through repetitive execution could perform them virtually like battle drills; clearly beneficial given the often constrained planning and preparation time at company and platoon level.

Stability Operations

Purposes

6-51. The Infantry company or a subordinate element might be directed to establish a checkpoint to achieve one or more of the following purposes.

- Obtain intelligence.
- Identify enemy combatants or seize illegal weapons.
- Disrupt enemy movement or actions.
- Deter illegal movement.
- Create an instant or temporary roadblock.
- Control movement into the area of operations or onto a specific route.
- Demonstrate the presence of US or peace forces.
- Prevent smuggling of contraband.
- Enforce the terms of peace agreements.
- Serve as an OP, patrol base, or both.

Advantages

6-52. Checkpoints offer units the following advantages regardless of the spectrum of conflict.

- They are an intimidating show of force.
1. They provide Soldiers and commanders with better situational awareness.
2. Provide reassurance to the friendly elements of the population.
3. Serve to gain the initiative for limited periods.
4. Help maintain friendly force vigilance and refine TTP.
5. Keeps the enemy off balance.

Disadvantages

6-53. Checkpoints have the following disadvantages, regardless of the spectrum of conflict.

- May create a pattern, thus giving a potential enemy the ability to gather information on TTP. May invite targeting for attack if Soldiers are undisciplined or lax.
- The level of effort required to conduct checkpoints securely and for extended periods exhausts a force rapidly.
- May incite the local populace to assist or join hostile elements.
- Their static locations can increase the potential for direct attack.

Procedures

6-54. Checkpoint layout, construction, and operating should reflect METT-TC factors, including the amount of time available for emplacing it. The following procedures and considerations normally apply.

- Position the checkpoint where it is clearly visible and where traffic cannot turn back, get off the road, or bypass the checkpoint without being observed.
- Position a combat vehicle (if available) or crew-served weapons, selected based upon METT-TC and ROE, off the road but within sight of the checkpoint. This helps deter resistance to the Soldiers operating the checkpoint. The vehicle should be in a hull-down position and protected by local security. It must be able to engage vehicles trying to break through or bypass the checkpoint.

- Place obstacles in the road to slow or canalize traffic into the search area. Traffic should enter the checkpoint single file.
- Place signs written in the host nation's language explaining what they are entering and instructions to facilitate their passing through.
- Establish a reserve if applicable.
- Establish a bypass lane for approved convoy traffic.
- If applicable, establish wire communications within the checkpoint area to connect the checkpoint bunker, combat vehicle, search area, security forces, rest area, and any other elements involved in the operation.
- Designate the search area. If possible, it should be below ground level or barrier that is protected against such threats as a booby-trapped vehicle or suicide bomber. Establish a parking area adjacent to the search area. Women are normally only checked with a metal detector or searched by female personnel. However, this depends on the ROE, the ROI, and METT-TC.
- If applicable, checkpoint personnel should include linguists.
- Properly construct and equip the checkpoint. Consider including--
  - Barrels filled with sand, concrete, or water (emplaced to slow and canalize vehicles).
  - Concertina wire (emplaced to control movement around the checkpoint).
  - Spiked vehicle chains or collapsible tire defeating devices.
  - Secure facilities for radio and wire communications with the controlling headquarters.
  - First-aid kit.
  - Sandbags for defensive positions.
  - Wood or other materials for the checkpoint bunker.
  - Binoculars, NVDs, and flashlights.
  - Long-handled mirrors (for use in inspections of vehicle undercarriages).
- Elements operating a deliberate CP may require access to specialized equipment such as--
  - Floodlights.
  - Control flags or signs in local language.
  - Barrier poles that can be raised and lowered.
  - Generators with electric wire.

Types

6-55. Some common types of checkpoints are discussed below.

Deliberate Checkpoints

6-56. These might be permanent or semi-permanent (Figure 6-4). They are typically constructed and employed to protect an operating base or well-established MSRs. Deliberate checkpoints are often used to secure the entrances to lodgment areas or base camps. They may also be used at critical intersections or along heavily traveled routes to monitor traffic and pedestrian flow. Deliberate checkpoints can be constructed so that all vehicles and personnel are checked or where only random searches occur (ROE and METT-TC dependent).

- They are useful deterrents and send a strong law and order or US presence message.
- Deliberate checkpoints and their locations are known to terrorists and insurgents. Commanders must weigh the costs to the benefits of operating deliberate checkpoints.
- Commanders must consider that deliberate checkpoints may quickly become enemy targets and US Soldiers operating deliberate checkpoints are highly visible and viable targets for enemy attack.

Stability Operations

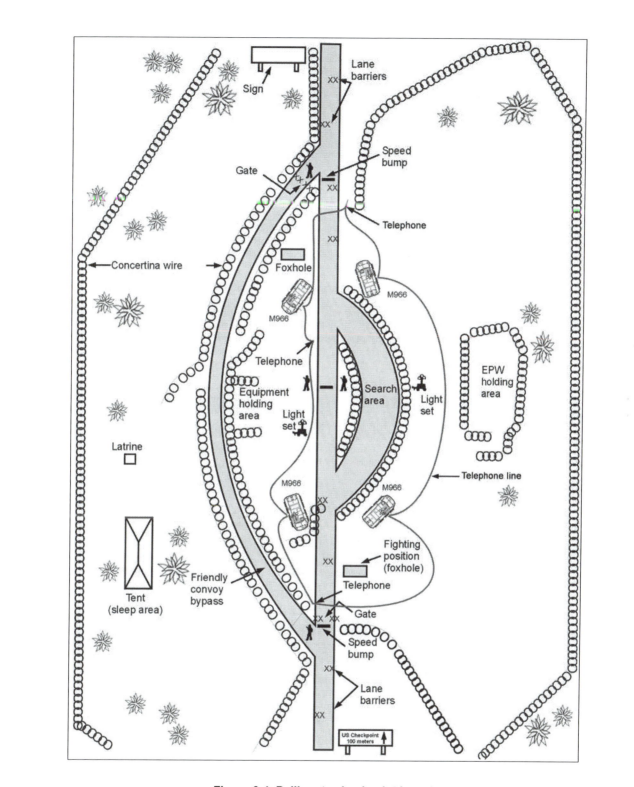

Figure 6-4. Deliberate checkpoint layout.

Chapter 6

Hasty Checkpoints

6-57. Such checkpoints are planned and used only for a short, set period. Hasty checkpoints are normally employed during the conduct of a vehicle or foot patrol. The hasty checkpoint is similar in nature to the deliberate checkpoint but only uses transportable materials.

- The hasty checkpoint is mobile and can be quickly positioned where needed.
- While more adaptable, the hasty checkpoint does not send the constant visual reminder of US presence to the local population that the deliberate checkpoint does.
- Because they can be quickly established and removed, hasty checkpoints are likely to be more effective in disrupting enemy actions. They are also less likely to be deliberately targeted by enemy forces.

Snap Checkpoints

6-58. Such checkpoints are conducted when specific intelligence indicates that a checkpoint hinders the enemy's freedom of movement at a specific time and place. Snap checkpoints are very similar to hasty checkpoints. The major difference is that hasty checkpoints are often random actions conducted as part of a patrol, whereas snap checkpoints are deliberate and based on either enemy analysis or quickly developed actionable intelligence. Snap checkpoints are normally conducted immediately and often with little to no deliberate planning; for this reason specific techniques should be develop and this specific action should be well rehearsed.

Vehicular Traffic Stop Checkpoints

6-59. Such checkpoints (Figure 6-5) are conducted by multiple sections of vehicle-equipped Infantrymen. This type of operation involves two or three sections of vehicles that patrol an area looking for a specific type of vehicle or specific personnel such as a particular model and color of car. Once this vehicle or person is identified, the vehicle or person is forced to stop and then searched. Normally the vehicle sections move single file with enough distance between the first two sections to allow civilian traffic to move between the sections (50 to 500 meters based on visibility, road conditions, and METT-TC.) If either section spots a targeted vehicle or person in a static or parked position, then the patrol cordons and searches the area, again based on METT-TC, or requests additional assistance. The patrol should move slightly slower than normal civilian traffic so that civilian traffic will pass the rear section. As civilian traffic passes the rear section, the patrol radio to the lead section if it spots a targeted vehicle. Once a targeted vehicle has moved between the two sections, both sections move abreast to effectively block the road and close the distance between themselves. They block in the targeted vehicle. The sections slowly force the targeted vehicle to pull to the side of the road and stop, and then they use normal vehicle search techniques. A third section can be employed as a reserve, as additional security, or simply as additional Soldiers.

Stability Operations

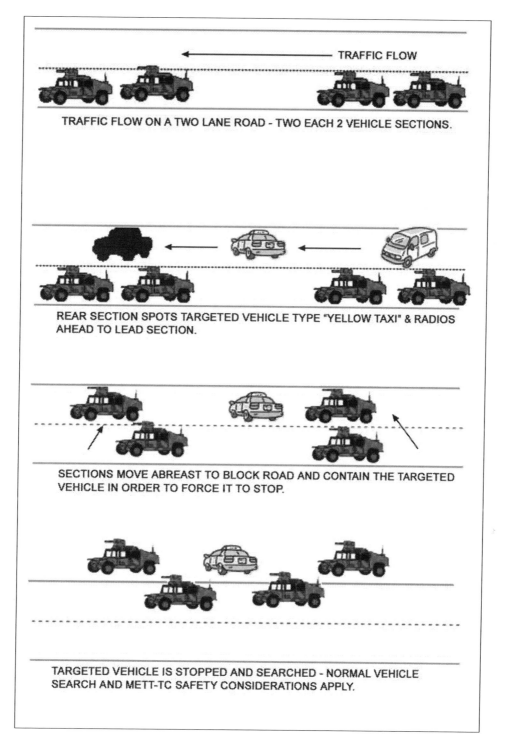

Figure 6-5. Vehicular traffic stop.

SEARCH

6-60. Searches are an important aspect of populace and resource control. The need to conduct search operations or to employ search procedures is a continuous requirement. A search can orient on people, materiel, buildings, or terrain. A search usually involves both civil police and Soldiers but may involve only Soldiers. Misuse of search authority can adversely affect the outcome of operations. Soldiers must conduct and lawfully record the seizure of contraband, evidence, intelligence material, supplies, or other minor items for their seizure to be of future legal value. Proper use of authority during searches gains the respect and support of the people. For procedures, leaders should consult available references such as unit TSOPs, theater training requirements, handbooks, or the Soldiers Manual of Common Tasks. The following discussion provides additional techniques to consider:

SEARCH A VEHICLE

6-61. Two principle types of vehicle searches include one during checkpoint operations and one for entering a forward operating base (FOB) or other secure area. Searches should be methodical yet unpredictable, efficient yet speedy, thorough yet considerate. They are used for various purposes and are conducted with varying degrees of force. Search techniques should be developed according to existing conditions and current events in theaters of operation. Therefore, the best TTP reference is the unit TSOP.

SEARCH A MALE

6-62. In all search operations, leaders must emphasize the fact that anyone in an area to be searched could be an insurgent or a sympathizer. To avoid making an enemy out of a suspect who may support the host country government, searchers must be tactful. The greatest caution is required during the initial handling of a person about to be searched. One member of the search team covers the other member, who makes the actual search. (FM 3-19.40 and STP 19-95B1-SM discuss how to correctly search people.)

SEARCH A FEMALE

6-63. The enemy can use females for all types of tasks when they think searches might be a threat. To counter this, use female searchers. If no female searchers are available, use doctors, aidmen, or members of the local populace. If male Soldiers must search females, take all possible measures to prevent any inference of sexual molestation or assault.

SEARCH A HOUSE

6-64. The object of a house search is to look for controlled items and to screen residents to determine if any are suspected insurgents or sympathizers. As far as possible, care should be taken to respect national customs. For instance, in a Muslim country, men should be allowed to take women out of the house prior to the search. The commander should have enough money to pay immediately for damages to locks and doors, and so on. A search party assigned to search an occupied building should have at least one local police officer, a protective escort for local security, and a female searcher. If inhabitants remain in the dwellings, the protective escort must isolate and secure the inhabitants during the search. Escort parties and transportation must be arranged before the search of a house. Forced entry might be necessary if a house is vacant or if an occupant refuses to allow searchers to enter. If the force searches a house containing property while its occupants are away, it should secure the house to prevent looting. Before US forces depart, the commander should arrange for the community to protect such houses until the occupants return. Several methods for controlling the inhabitants follow.

- Assemble inhabitants in a central location if they appear to be hostile. This method provides the most control, simplifies a thorough search, denies insurgents an opportunity to conceal evidence, and allows for detailed interrogation. Depending on the objective of the search, a personnel search team might be necessary in this central location. This method has the disadvantage of taking the inhabitants away from their dwellings, thus encouraging looting, which, in turn, engenders ill feelings. The security element is then responsible for controlling the inhabitants. The search element may escort individuals back to their dwellings to be present during the search or may leave them in the central location.
- Restrict inhabitants to their homes. This prohibits movement of civilians, allows them to stay in their dwellings, and discourages looting. The security element must enforce this restriction. The

disadvantages of this method are that it makes control and interrogation difficult and gives inhabitants time to conceal evidence in their homes.
- Control the heads of the households. Tell the head of each household to remain in front of the house and at the same time have everyone else in the house taken to a central location. The security element controls the group at the central location, controls the head of the household, and provides external security for the search team. During the search, the head of the household accompanies the search team through the house. This reduces looting. Plus, the head of the household sees for himself that the search team steals nothing. This is the best method for controlling the populace during a search.

PATROL

6-65. Patrolling is also a high-frequency task during stability operations. The primary advantage of the dismounted patrol is that they provide a strong presence and enable regular interface with the local population. This procedure greatly helps in gathering vital information as well as in developing the base of knowledge of the unit's AO. Planning and execution of an area security patrol and presence patrol are similar to procedures for other tactical patrols except that the patrol usually occurs in urban areas and patrol leaders must consider political implications and ROE. Figure 6-6, page 6-22 shows the use of an area security patrol, in conjunction with checkpoints and OPs, in enforcing a zone of separation between belligerent forces.

Chapter 6

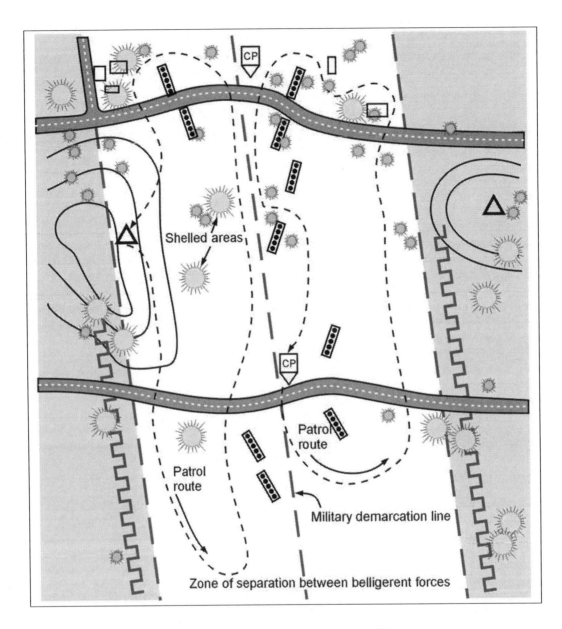

Figure 6-6. Employment of checkpoints, OPs, and patrols to enforce a zone of separation.

PRESENCE PATROLS

6-66. US forces are deployed increasingly in combat operations in urban areas and in support of stability operations missions all around the world. The Infantry company and platoons conduct a presence patrol much the same as a combat patrol, and the planning considerations are similar. The main difference is that the patrol wants to both show force and lend confidence and stability to the local population of the host nation (HN). Rarely should a commander use a presence patrol where enemy contact is likely. Presence patrols work best for some types of stability operations, for example, peace operations, humanitarian and civic assistance, NEO, foreign humanitarian assistance (FHA), or shows of force. The presence patrol is also the primary means by which the commander collects information about his AO. The presence patrol is armed, and it conducts the planning and preparation necessary for combat operations at all times. Leaders should plan the patrol based on a movement to contact model or methodology. The patrol in urban areas is planned in the same manner. Each patrol will often involve vehicle support. The platoon could be tasked to conduct mounted (if augmented with vehicles) or dismounted patrols planned by the higher headquarters to accomplish one or more of the following.

- Confirm or supervise an agreed cease-fire.
- Gain information.
- Cover gaps between OPs or checkpoints.
- Show a stability force presence.
- Reassure isolated communities.
- Inspect existing or vacated positions of former belligerents.
- Escort former belligerents or local populations through trouble spots.

PLANNING CONSIDERATIONS

6-67. Commanders must address additional considerations when planning or conducting patrols in and around urban areas. Some of these are--

- Leaders and Soldiers must plan for and rehearse actions in and around large crowds of civilians or noncombatants. Expect most civilians not to speak English which means you must have interpreters.
- All Soldiers and leaders should be briefed on information/intelligence collection priorities.
- Soldiers must be assigned responsibility to maintain all-round and high-low security that assigns responsibility for each floor and the roof of buildings.
- Leaders should consider periodically occupying rooftops during the course of the patrol to increase observation and security.
- Use of urban city maps. Navigation by grid in an urban area can be difficult. Maps that show street names, neighborhoods, and so on are much more useful.
- Ensure there is a contingency plan for dealing with large crowds of noncombatants or large hostile crowds.
- Know the numbers and locations of translators or interpreters in the patrol. Also, know the ethnicity of the translators, and how that might affect the population of the patrol area or route.
- Check recent activity or trends in the local population or urban areas.
- The speed of the patrol should normally be slow and even promoting a relaxed and confident attitude towards the local population except where the current IED threat calls for high speed movement. It also enables patrol members to concentrate on the urban environment and the population, which increases their ability to gather information. A steady pace helps stave off fatigue.

OPPORTUNITIES

6-68. Urban terrain provides multiple opportunities for attack against patrols in the stability environment. The locations of enemy firing points can be concealed by building characteristics, vehicles, civilian population, and noise. The patrol must therefore regain the initiative during an engagement by immediate and aggressive action. An attack is normally initiated on a patrol only when the attacker has an open escape route. This emphasizes the need to maneuver teams quickly in order to provide depth, and to cordon the area immediately after the initial reaction to the contact. The teams out of contact must rapidly envelop the firing point indicated by the contact team and try to close off the suspected escape route. The contact team must provide general directions or guidance to the other teams not in contact. Reacting quickly and aggressively based on limited information always beats giving the attacker a chance to escape.

Vehicle-Supported Patrols

6-69. Infantry units might find themselves conducting frequent vehicle-assisted or vehicle-mounted patrols. The same considerations that apply to any dismounted patrol apply to vehicle-mounted patrols. The commander must consider the following.

- Organize and orient vehicle gunners and commanders to maintain all-round security and, for urban areas, high-low security. Carefully consider leader locations in each vehicle and within the convoy.
- Rehearse mounted battle drills, reaction to contact, and mounting and dismounting in contact and include drivers in all rehearsals.
- Plan alternate routes, civilian traffic, and roadblocks.
- Remember that four vehicles is generally the minimum number of vehicles to conduct any operation. If one vehicle is disabled or destroyed, it can still be recovered while the third provides security. Unit SOPs determine the number of vehicles required.
- Remove outboard vehicle cargo seats that force Soldiers to face the interior of the vehicle. Replace with centerline seats that allow Soldiers to face outward. This increases observation, situational understanding, and potential firepower available.
- Position heavy crew-served weapons, such as the M2.50 caliber machine gun and the MK 19 machine gun, towards the rear of vehicle patrol convoys, where they can support the entire convoy in the event of contact.
- Harden unarmored vehicles by sandbagging or attaching improvised armor. Sandbagging vehicles seldom increases vehicle survivability, and the added weight increases maintenance problems, but it can increase crew and Soldier survivability. Add-on armor kits are preferred however, unit mechanics or welders can quickly fabricate improvised armor to protect both the vehicle engine and crew compartments if necessary.
- Plan for actions required if a vehicle breaks down and has to be repaired or recovered. Review self-recovery assets. Plan actions in case a vehicle gets stuck and cannot be recovered. Also, plan actions for catch-ups and breaks in contact.
- Establish a communications plan if not all vehicles have FM communication or in case a vehicle loses FM.
- Secure external gear to prevent theft. Inspect it to ensure it is not flammable. In the event of fire bomb or RPG attack, burning material attached to the vehicle creates a greater hazard than the initial attack.
- Plan for heavy civilian vehicle and pedestrian traffic.
- Conduct a map reconnaissance and identify likely chokepoints, ambush sites (intersections), and overpasses.
- Plan primary and alternate routes to avoid potential hazards.
- Drive offensively, unpredictably, but safely.
- Avoid stopping; it can create a potential kill zone.
- Learn the capabilities of the vehicle, including how high a vehicle can jump (curbs and other obstacles), its turning radius, its high-speed maneuverability, and its estimated width, in order to navigate narrow passageways.

6-70. Most engagements last less than one minute. Many engagements will be initiated with some type of IED or other command-detonated explosive device. Rapid maneuver against threat ensures either destruction or capture. The enemy will typically break contact and run after initially engaging with one magazine indiscriminately. Fire and maneuver is not always an option. Establish an over-watch or support-by-fire element and a maneuver element to close on the enemy (preferably from an assailable flank). An immediate cordon of the area may prevent the enemy from escaping. U.S. forces can then engage in a deliberate search of the area to capture or kill the attackers. Indigenous personnel firing weapons may not always pose a threat such as celebratory fire. Stay aware of the different situations in which weapons firing may not be threatening. Some personnel under control (PUC) are not Soldiers. They sometimes follow instructions poorly. Follow procedures for handling EPWs (secure, search, segregate, silence, safe to the rear) when handling PUCs.

ESCORT A CONVOY

6-71. This mission requires the Infantry company to provide a convoy with security and close-in protection from direct fire while on the move. Infantry forces must be augmented with additional transportation assets to carry out this mission.

COMMAND AND CONTROL

6-72. The task organization inherent in convoy escort missions makes battle command especially critical. The Infantry company commander may serve either as the convoy security commander or as overall convoy commander. In the latter role, he is responsible for the employment of his own organic combat elements and of CS and sustainment attachments and drivers of the escorted vehicles. He must incorporate all these elements into the various contingency plans developed for the operation. He must also maintain his link with the controlling TOC.

6-73. Effective SOPs and drills supplement OPORD information for the convoy. Since this is not a core mission the company commander must ensure adequate time to conduct thorough rehearsals. Also, the company conducts PCCs and PCIs, to include inspection of the escorted vehicles. The commander also coordinates with units and elements in areas through which the convoy will pass. Theater or combatant commands will provide baseline information and procedures for unit convoy TSOPs. However, Table 6-1 is an example convoy briefing checklist, which is used to provide minimum essential information to all members of a convoy.

Chapter 6

SITUATION	EXECUTION (continued)

Enemy.
- Activity in the last 48 hours.
- Threats.
- Capabilities.
Friendly.
- Units in the area or along the route.
Light and Weather Data.
- Effects of light and weather on the enemy and.
  on friendly forces.
- BMNT, sunrise, high temp, winds, sunset.
- EENT, moonrise, percent illumination.
  low temp
	- Order of movement and bumper numbers and.
	  individual manifesto.
	- Movement formation.
	- Speed and catch-up speed.
	- Interval (open areas and in built-up areas).
	- Weapons orientation, locations of key weapons.
	  systems.
	- Route.
	- Checkpoints.
	- Actions on contact.
	- Actions on breakdowns.
	- Actions at the halt (short halt and long halt)

MISSION *Task and Purpose of the Movement.* *Mission Statement*	**SUSTAINMENT** *MEDEVAC Procedures.* - Nine-line MEDEVAC request. - Location of medical support and combat. lifesavers. - Potential PZ/LZ locations. *Maintenance Procedures.* - Location of maintenance personnel. - Location and number of tow bars. - Recovery criteria. - Stranded vehicle procedures
EXECUTION *Commander's Intent.* *Endstate.* *Concept of the Operation (Concept Sketch or.* *Terrain Model).* *Task to Maneuver Units.* *Fires.* *Close Air Support.* *Coordinating Instructions.* - Timeline. - Marshal. - Rehearsals. - Convoy briefing. - Inspections. - Initiation of movement. - Rest halts. - Arrival time	
	COMMAND AND SIGNAL - Convoy commander. - Sequence of command. - Location of convoy commander. - Call signs of every vehicle/unit in the convoy. - Convoy frequency. - MEDEVAC frequency. - Alternate frequencies

Table 6-1. Example convoy briefing checklist.

6-74. Before the mission begins, the convoy commander issues a complete OPORD to all vehicle commanders in the convoy. This is vital, because the convoy may itself be task-organized from a variety of units, and because some vehicles might not have tactical radios. The order follows the standard five-paragraph OPORD format. It can emphasize these subjects.

- Inspection of convoy vehicles.
- Route of march (including a strip map for each vehicle commander).
- Order of march.
- Actions at halts (scheduled and unscheduled).
- Actions in case of vehicle breakdown.
- Actions for a break in column.
- Actions in built-up areas.
- Actions on contact, covering such situations as snipers, enemy contact (including near or far ambush), indirect fire, mine strike, and minefields.
- Riot drill.
- Refugee control drill.
- Evacuation drill.
- Actions at the delivery site.
- Chain of command.
- Guidelines and procedures for negotiating with local authorities.

- Communications and signal information.
- Tactical disposition.
- Fire support plan.

6-75. In any escort operation, the basic mission of the convoy commander (and, as applicable, the convoy security commander) is to establish and maintain security in all directions and throughout the length of the convoy. He must be prepared to move the security force to fit the situation. Several factors apply, including convoy size, organization, and composition. Sometimes, he positions the security elements, such as platoons, to the front, rear, or flanks of the convoy. He may also disperse the combat vehicles throughout the convoy body.

TASK ORGANIZATION

6-76. When sufficient escort assets are available, the convoy commander usually organizes convoy security into three distinct elements: advance guard, close-in protective group, and rear guard. He may also designate an additional reserve in the rear guard to handle contingency situations. The following paragraphs examine the role of the advance guard, of security assets accompanying the convoy main body, and of the reserve in the rear guard.

Advance Guard

6-77. The advance guard reconnoiters and proofs the convoy route. It searches for signs of enemy activity such as ambushes and obstacles. Within its capabilities, it tries to clear the route. The distance and time separation between the advance guard and the main body should be enough to give the convoy commander adequate early warning before the arrival of the vehicle column. However, the separation should be short enough that the enemy cannot interdict the route between the time the advance guard passes and the main body arrives. The advance guard should be task-organized with reconnaissance and mobility assets. As necessary, it should also include linguists.

Main Body

6-78. The commander might choose to intersperse security elements with the vehicles of the convoy main body. These can include combat elements (including the rear guard), the convoy commander, additional linguists, mobility assets, and medical and maintenance support assets. Depending on METT-TC, the convoy commander might also employ flank security.

Rear Guard

6-79. The rear guard may serve as a reserve, often called a quick reaction force (QRF). Either it moves with the convoy, or it locates at a staging area close enough to provide immediate interdiction against enemy forces. The supporting headquarters normally designates an additional reserve, consisting of an additional company or combat aviation assets, to support the convoy operation.

ACTIONS ON CONTACT

6-80. As the convoy moves to its new location, the enemy might try to harass or destroy it. This contact usually occurs in the form of an ambush, often executed at a hasty obstacle. The safety of the convoy rests on the speed and effectiveness with which escort elements can execute appropriate actions on contact. Based on the factors of METT-TC, portions of the convoy security force might be designated as a reaction force. This element performs its normal escort duties, such as conducting tactical movement or occupying an assembly area, unless enemy contact occurs. Then, it performs a reaction mission given by the convoy commander. Leaders should follow actions on contact procedures according to the unit TSOP. However, this manual also provides discussions for reacting to IEDs in Appendix G and urban patrolling react to contact in Appendix K.

Actions at an Ambush

6-81. An ambush is one of the best ways to stop a convoy. Actions in response to an ambush must be immediate, overwhelming, and decisive. This means they must be planned for and rehearsed so they can be executed by all escort and convoy elements. Procedures to consider in case of an ambush--

- The security force immediately acquires the enemy force, provides suppressive fires on known or suspected enemy positions, and tries to clear the kill zone quickly. They seek covered positions between the convoy and the enemy and suppress the enemy with the highest possible volume of fire (METT-TC and ROE dependent). They send contact reports to higher headquarters.
- The convoy vehicles, if armed, return fire only if the security force is not between the convoy and the enemy force.
- The convoy commander keeps the convoy vehicles moving on the route.
- Subordinate leaders or the convoy commander can request recovery assistance for damaged or disabled vehicles which may have to be pushed off the path or roadway.
- The convoy escort leader uses situational reports to keep the convoy security commander informed. If necessary, the convoy escort leader can then direct a reserve force from the rear guard or staging area to take action. He can also call for and adjust indirect fires.
- Once the convoy is clear of the kill zone, the convoy escort element executes one of the following COAs based on the composition of the escort and reaction forces, the commander's intent, and the strength of the enemy force.
  - Continue to suppress the enemy as the reserve moves to provide support.
  - Break contact and move out of the kill zone.
  - Assault the enemy.

Actions at an Obstacle

6-82. Obstacles are obstructions that prevent movement. They include, among others, among others, deliberate roadblocks, disabled vehicles, and large groups of demonstrators. Obstacles threaten convoy security and can canalize or stop the convoy to set up an enemy ambush. A route reconnaissance goes ahead of a convoy to identify obstacles and either breach or bypass them. Sometimes, the reconnaissance element misses an obstacle and the convoy runs into it after all. If this happens, the convoy must take actions to reduce or bypass the obstacle.

- When an obstacle is identified, the convoy escort faces two challenges: reducing or bypassing the obstacle, or maintaining protection for the convoy. Security becomes critical, and actions at the obstacle must be accomplished very quickly. The convoy commander must assume that the obstacle is overwatched and covered by enemy fires.
- To reduce the time the convoy is halted, thus reducing its vulnerability, the company should act when the convoy escort encounters point-type obstacles.

6-83. The advance guard element identifies the obstacle, and the convoy commander directs the convoy to make a short halt and establish security.

6-84. The convoy escort element overwatches the obstacle, asks the convoy commander to allow the breach force to move forward. Pay particular attention to terrain that dominates the area.

6-85. The escort maintains all-round security and provides overwatch as the breach force reconnoiters the obstacle in search of a bypass.

6-86. Once all reconnaissance is complete, the convoy commander determines if he will bypass the obstacle, breach the obstacle with the assets on hand, or breach the obstacle with reinforcing assets.

Actions During a Halt

6-87. During a short halt, the convoy escort remains at REDCON-1 status, regardless of what other convoy vehicles are doing. If the halt is for any reason other than an obstacle, the convoy escort takes the following actions.

- The convoy commander signals the short halt and transmits the order via tactical radio. Based on METT-TC factors, he directs all vehicles in the convoy to execute the designated formation or drill for the halt.
- Ideally, the convoy assumes a herringbone or coil formation. If the sides of the road are untrafficable or are mined, however, noncombat vehicles may simply pull over and establish all-round security as best they can. This allows movement of the escort vehicles through the convoy main body as necessary.
- If possible, escort vehicles are positioned up to 100 meters beyond other convoy vehicles, which are just clear of the route. Escort vehicles remain at REDCON-1 but establish local security based on the factors of METT-TC.
- When given the order to continue, convoy vehicles reestablish the movement formation, leaving space for escort vehicles. Once the convoy is in column, local security elements (if used) return to their vehicles, and the escort vehicles rejoin the column.
- When all elements are in column, the convoy resumes movement.

OPEN AND SECURE ROUTES

6-88. This task is a mobility operation normally conducted by the engineers. The Infantry company might be tasked to assist in route clearance and to provide overwatch support. Route clearance may achieve one of several tactical purposes.

- To clear a route for the initial entry of the battalion into an area of operations.
- To clear a route ahead of a planned convoy to ensure that belligerent elements have not emplaced new obstacles since the last time the route was cleared.
- To secure the route for use as a main supply route.

6-89. The planning considerations for opening and securing a route resemble those for a convoy escort operation. The company commander analyzes the route and develops contingency plans for such possibilities as likely ambush locations and sites that are likely to be mined. The size and composition of a team charged with opening and securing a route is based on METT-TC.

CONDUCT RESERVE OPERATIONS

6-90. Reserve operations in the stability environment are similar to those in other tactical operations. They too allow the Infantry company commander to plan for a variety of contingencies based on the higher unit's mission. The reserve might play a critical role in almost any stability activity or mission, including lodgment area establishment, convoy escort, and area security.

- The reserve force must be prepared at all times to execute its operations within the time limits specified by the controlling headquarters.
- The controlling headquarters can tailor the size and composition of the reserve to the mission. For a convoy mission, for example, the reserve might consist of a company.

CONTROL CROWDS

6-91. Large crowds or unlawful civil gatherings or disturbances pose a serious threat to US troops. Commanders must consider the effects of mob mentality, the willingness of enemies to manipulate media, and the ease with which a small, isolated group of Soldiers can be overwhelmed by masses of people. The police forces of each state and territory are normally responsible for controlling crowds involved in mass demonstrations, industrial, political and social disturbances, riots, and other civil disturbances. The prime role of US troops in the control of unlawful assemblies or demonstrations is to support and protect the police, innocent bystanders, and property. Therefore, this paragraph describes protective and defensive measures rather than offensive measures. For example, troops provide a firm base where police can operate, either as riot or arrest squads. The troops will only use force as a last resort to disperse the crowd or prevent its advancing past a given point or line.

6-92. Control at the scene of an incident normally falls under civil authority. The Army will act only on receiving a formal request or when danger is immediate and pressing. The command of the US military elements remains with the commander. The key to success is cooperation between military and civilian authorities. The controlling forces will not work side by side but rather, from front to rear, with one element backing up the other.

> *Note:* In the very early stages of stability operations, US forces might be the only civil or military authority present.

6-93. A military element might have to deploy without a police unit. If so, an authorized representative of the civil authority, such as a magistrate or police representative, should accompany the military.

6-94. Before going into the disturbed area, the Infantry company must try to isolate it and cut off reinforcement of dissidents. Roadblocks, checkpoints, or even a cordon can help, although complete isolation is probably unlikely.

6-95. Where possible, the company should dominate the disturbed area by unobtrusively setting up rooftop OPs or patrols before starting the street operation. If prominent rooftops are inaccessible from the ground, the unit can deploy on helicopters.

6-96. An initial deployment into an operational area has two phases: the approach, and the 'show of force.'.

6-97. The company conducts an approach march to a secure area out of view of the mob or gathering. The column formation is most suited for an approach in vehicles or on foot. The force moves one platoon behind the other on a single axis of advance, with company headquarters immediately behind the leading platoon. Barring immediate and pressing danger, a commander should avoid allowing other incidents to divert him en route to his assigned area of operations.

6-98. The original briefing, report, or order is unlikely to give the commander enough information to plan the deployment in detail before he arrives at the scene. Although rooftop OPs and patrols might have been established earlier, the commander must conduct a quick reconnaissance on arrival in the deployment area. He must make contact with the police commander or other local civil authority and plan the final deployment. The use of helicopters for crowd and route surveillance will help the commander adjust his deployment as the situation changes, and to identify the threat to the security of his forces when deployed.

6-99. Troops should be deployed outside the range of hand-thrown missiles (50 to 60 meters), but within full view of the crowd. They should deploy into the appropriate formation quickly (double time), adopting the port arms or on guard position to convey a sense of purpose to the crowd.

6-100. Such operations are likely to call for the employment of non-lethal munitions. Some examples include various size projectiles made of substances such as rubber, foam, wood, and bean-bags. (For detailed discussions of nonlethal see FM 3-22.40 and FM 3-19.15.)

Chapter 7
Civil Support Operations

During the disastrous events caused by hurricanes Katrina and Rita, US military forces executed "the largest, fastest, most comprehensive and most responsive civil support mission ever," said Paul McHale, assistant secretary of defense for homeland defense, to the Senate Committee on Homeland Security and Governmental Affairs.

McHale credited the 72,000 Active Duty, National Guard, and Reserve Soldiers who responded, particularly at a time of large-scale deployments supporting the war on terror, as a testament to the readiness, agility and professionalism of the force.

This edition adds civil support operations, which call for Army units to assist domestic civil authorities. They do this by providing essential supplies and services to control disease, to alleviate suffering, restore civil order, or to help people and communities recover from disasters. The ultimate goal is to meet the immediate crisis, and then to transfer responsibility quickly and efficiently back to the appropriate civilian authorities.

The discussions in this chapter rely on the experiences gained during recent natural disasters.

ROLES

7-1. The US active duty military is limited by the Posse Comitatus Act and other legislation in the actions it can take in within the US and its territories. Under PCA, Army forces do not conduct stability operations within the United States. Instead, the federal and state governments are responsible for those tasks, but we do conduct civil support operations. Normally, for US armed forces to conduct offensive and defensive operations inside the US and its territories, the President must do two things. First, he must identify a significant armed force that threatens the territorial integrity of the US. Second, he must declare a national emergency. Only then can jointly commanded Army forces conduct, IAW Chapter 4 (Offensive Operations) and Chapter 5 (Defensive Operations), offensive and defensive Homeland Security missions.

7-2. The Army plays different roles for offense or defense as part of Homeland Security from that of civil support. However, civil support may be conducted in circumstances that require offensive and defensive operations in order to return the affected population to a state of normalcy.

7-3. Significant legal and constitutional issues arise when military forces are committed to combat operations within the US Differences exist between the actions active duty forces, federalized reserve component forces, and National Guard forces can are authorized to perform. Commanders and leaders at all levels must ensure they clearly understand the guidelines and rules of engagement established by the President and his legal representatives.

7-4. Leaders must keep their Soldiers informed on the situation and guidance from the higher headquarters. The Infantry company will encounter various units and organizations classified by certain Titles under US Code which may effect command and control relationships. Civil authorities may need to be informed of the differences in terms of what each can and cannot do.

- National Guard (NG) in its role as the state militia can be called to respond as state active duty (SAD) under the command of the governor.

- Title 32 – National Guard on full time status can be moved from SAD to Title 32.
- Title 10 – Active duty military units including active Guard and Reserve (AGR).
- Title 14 – Pertains to the US Coast Guard.

DEFINITION

7-5. In disaster relief, the Infantry company becomes a force provider and supplements the efforts and resources of state and local governments (and possibly NGOs) within the United States.

7-6. During civil support operations, the US military responds in support of civilian agencies and may receive guidance and instructions from civil authorities through their assigned chain of command. These include responding to civil emergencies or major disasters. A presidential declaration of an emergency or disaster usually precedes the Army's commitment to civil support operations, but in cases of extreme emergency, it may follow the initial actions. Regardless of the relationship between the civil authorities and the military, the fundamental elements and responsibilities of military command do not change. A representative of the civil authority does not, except in some specific and rare circumstances, exercise command over military forces.

7-7. The US military provides civil support based on a DoD directive for military assistance to civil authorities. Such a directive normally addresses both natural and manmade disasters. It can direct military aid in civil disturbances, counterdrug and counterterror activities, law enforcement, and management of consequences associated with weapons of mass destruction (WMD).

7-8. The US Constitution mandates that the civilian government is responsible for preserving public order. However, the Constitution also allows military forces to protect federal and civilian property and functions. The Posse Comitatus Act restricts the use of the military in civilian law enforcement except in the role of supporting or technical assistance.

7-9. Federal military forces remain under the military chain of command while supporting civil law enforcement. The supported law enforcement agency coordinates Army force activities under appropriate civil laws and interagency agreements. Army and Air National Guard units that have not been federalized can assist civil authorities when active duty federal units cannot under the provisions of the Posse Comitatus Act.

TYPES OF OPERATIONS

7-10. Civil support involves using the Army to respond with a wide array of capabilities and services to aid civil authorities in the following types of actions.

- Protecting public health.
- Restoring public order.
- Assisting in disaster recovery.
- Alleviating large-scale suffering.
- Protecting critical infrastructure.

POSSIBLE TASKS

7-11. The company commander cannot predict the exact tasks his unit might have to perform during civil support operations. Infantry units must often perform nonstandard tasks during national and local emergencies. Civil support operations respond to requests for help with protection and restoration. Typically, these include riots or widespread disorder; forest and grassland fires; hazardous material releases; and floods, storms, hurricanes, tornados, and earthquakes.

7-12. State, local, and federal authorities are responsible for restoring essential services in the case of a disaster. Army forces may support their efforts. Disaster relief focuses on recovery of critical infrastructure after a natural or manmade disaster. Both humanitarian and disaster relief normally occur simultaneously.

7-13. The most common civil support tasks for Infantry companies include--

- Search and rescue of survivors.
- Recovery of human remains.
- Disposal of animal carcasses.
- Disinfection and sanitation.
- Debris and trash removal.
- Riot and civil disturbance control.
- Police augmentation.
- Food and ice distribution.
- Fuel distribution.
- Contamination containment.
- Personnel movement and control.
- Key facilities protection.
- Vital services assessment.
- Medical triage and treatment.
- Emergency fire fighting.
- Emergency flood control.
- Hazard identification.
- Water purification and distribution.
- Temporary shelter construction and administration.
- Transportation support.
- Power generation.
- Communications support.
- Clothing and blanket distribution.
- Information distribution.
- Medical evacuation.
- Operations coordination.

INTELLIGENCE

7-14. This paragraph discusses support to SA; ISR; and maps and imagery.

SUPPORT TO SITUATIONAL AWARENESS

7-15. Due to the complete destruction of communications systems prior to arrival into a devastated area, the company commander and his unit may lack any level of situational awareness. The commander must identify and conduct a face-to-face with first responders such as police, fire, or NG personnel.

7-16. The Infantry company commander must develop and disseminate PIR such as--

- What is the number and location of displaced civilians in each units sector?
- What is the number and location of civilians remaining in AO?
- Where are flooded areas?
- What is the water level at engineer test points?
- What areas currently have the most criminal activity?
- Where are criminal elements caching weapons and ammunition?

- What locations will criminals target for looting?
- Will criminal elements go beyond handguns to explosives use?
- Will civilians evaluating damage to their homes and businesses stay beyond time limits?

7-17. The Infantry company commander must also develop internal information gathering TTP relevant to disaster response situations.

INTELLIGENCE, SURVEILLANCE, AND RECONNAISSANCE

7-18. Intelligence collection and management becomes information collection and management for two reasons: the inherent statutory language preventing military intelligence collection on US citizens (DOD Directive 5240.1-R, *Procedures Governing the Activities of DOD Intelligence Components that Affect US Persons*) and the reliance on civil authorities for information regarding civil infrastructure. Tactical reconnaissance assessments become civil assessments. Centers of influence may be mobility (access to lines of communication), water and electricity distribution, telephone, access to relief agencies, and criminal activity; and tracked as indicators where there is potential to cause civil unrest or unexpected displacement of civilians.

7-19. Likely aerial collection platforms may include UASs, given adequate airspace management. It may also include observation helicopters to gain information on area trafficability, movements of displaced persons, status of flooding and drainage progress.

MAPS AND IMAGERY SUPPORT

7-20. Limited initial availability of maps and imagery may be expected. Commanders might need to obtain commercial maps or mapping software en route.

MOVEMENT AND MANEUVER

7-21. This paragraph discusses movement control, lines of communication, control points, mobility over water, and Engineer assets and munitions.

MOVEMENT CONTROL PROCEDURES

7-22. This may be delegated to task force (TF) level. If not moving with his own TF, the company commander will need to obtain procedures from the unit to which attached.

PRIMARY AND ALTERNATE LINES OF COMMUNICATIONS

7-23. The commander must determine designated primary and alternate lines of communication (LOC) as convoy routes for overall movement planning and deconfliction, C2 of forces, and ease of maintenance and recovery.

TRAFFIC AND INFORMATION CONTROL POINTS

7-24. When a disaster is severe enough to cause shortages of law enforcement officers, the military may be used for traffic control posts (TCP) or information control points.

OVER-WATER MOBILITY

7-25. If units must contend with flooded areas that cannot be forded with tactical vehicles, the commander may need to request and plan for the acquisition and employment of boating assets such as Zodiacs or coordinate for Engineer or Coast Guard support.

ENGINEER ASSETS AND MISSIONS

7-26. Response to a natural catastrophe is uniquely an engineering and logistics mission. Consequently, the Infantry company is likely to be working closely with Engineer units or require their expertise or assistance. An Engineer LNO will be located at each Emergency Operations Center (EOC).

7-27. Engineer mobility assets may provide transportation to bring food and water, open roads, build life-support centers (tent cities), provide power units for hospitals and emergency centers, initial repair to schools and local government offices, and bridging or boat assets to aid flood rescue or for emergency mobility across high-water.

FIRE SUPPORT

7-28. Fire support forces, if available, may operate to assist other critical relief functions that exploit their equipment and expertise.

PROTECTION

7-29. This paragraph discusses risk management, search and rescue, and force health protection.

RISK MANAGEMENT

7-30. During civil support operations, the commander and all company leaders and Soldiers must routinely and thoroughly integrate risk management, according to appendix A of this manual, into all missions and tasks.

SEARCH AND RESCUE (SAR)

7-31. Search and rescue may become an essential task for Infantrymen during disaster relief. Though common doctrine, procedures, and centralized control are needed for joint and inter-agency SAR operations, Army units will contribute to efforts to perform rescue by various means or may provide security during SAR operations.

FORCE HEALTH PROTECTION

7-32. Force health protection includes preventive medicine, mental health services, MEDEVAC, and decontamination.

Preventive Medicine

7-33. Preventive medicine (PM) capabilities are a must in humanitarian relief operations. In order to avoid exposing his Soldiers to unnecessary environmental and disease hazards, the commander must identify the PM assets and determine how to obtain their support.

7-34. Units ensure their Soldiers deploy with sufficient personal medications (at least a 30-day supply) to address chronic conditions. The unit medical personnel should plan to secure resupply if needed.

Mental Health

7-35. When Soldiers must become involved in the identification and recovery of human remains, the commander must identify and obtain support from mental health (MH) professionals or a Chaplain.

MEDEVAC

7-36. A lack of airspace management may significantly reduce the availability of aeromedical evacuation making ground evacuation the primary means. In any case, Infantrymen are among the most practiced at medical evacuation (MEDEVAC) procedures and will likely be needed to plan and execute those tasks.

Decontamination

7-37. Since many contaminants of both known and unknown composition may be encountered in a disaster relief scenario, another issue that may present itself in the area of medical planning is unit decontamination. Commanders must ensure the predeployment process deals with identifying decontamination element support, equipment, SOP and training.

SUSTAINMENT

7-38. For self sustainment, plan for the requirement to bring unit level logistics equipment (ULLS) for supply and maintenance management. Traditional supply requisition methods may have to be adjusted based on the environment and availability of ULLS.

7-39. The Infantry company is also likely to be heavily engaged in helping sustain a suffering civilian population by delivering food, water, fuel, and a myriad of other critical supplies. Units may anticipate the need to assist with receipt, control, and security of distribution centers. Tasks may include establishing entry, exit, staging, and distribution points. Security tasks may include performing guard duty and presence patrols.

COMMAND AND CONTROL

7-40. Although each civil support operation is different, the Army's troop-leading procedures still apply.

7-41. Civil support operations are typically joint and interagency. The potential for duplication of effort and working at cross-purposes is high. Unity of effort requires, as a minimum, common understanding of purposes and direction of all participants. Ensuring unity of effort and efficient use of resources requires constant coordination. Army forces enhance unity of effort by establishing a civil military operations center and by providing liaison elements, planning support, advisors, and technical experts to assist civil authority. In some situations, civil authority may have become so diminished that the company commander has the most effective command and control system in the area. Commanders must determine where their objectives and plans complement or conflict with those of other key agencies.

7-42. Commanders must follow the principle of providing essential support to the largest number of people as a guide prioritization and allocation. Allocate finite resources to achieve the greatest good to the largest number of people. Initial efforts usually focus on restoring civil order and vital services, which include food and water distribution, medical aid, power generation, search and rescue, and firefighting. It might be necessary to complete a lower-priority task before accomplishing a higher one. For example, Army forces might have to restore limited electrical services before restoring hospital emergency rooms and shelter operations.

7-43. Commanders must assess requirements to employ Army forces effectively. They determine how and where to apply limited assets to benefit the most people. In some cases, war fighting reconnaissance capabilities and techniques are adaptable to support operation requirements. For example, UASs can survey relief routes and locate civilian refugee groups. Standard information collection methods are reinforced and supplemented by civil affairs or dedicated disaster assessment teams as well as interagency and NGO sources. Company commanders should expect to work with many different federal, state, and local agencies to accomplish his mission. The combination of traditional and nontraditional information support allows commanders to obtain a clear understanding of the situation and adjust plans accordingly.

7-44. While the immediate goal of support operations is to relieve hardship and suffering, the ultimate goal is to create conditions necessary for civil follow-on operations. The successful handover of all activities to civil authorities and withdrawal of Army forces is a positive signal to the supported population and the Army. It indicates that the community has recovered enough for civil agencies to resume control and life is beginning to return to normal.

Chapter 8
Tactical Enabling Operations

This chapter covers tasks that the Infantry rifle company conducts, either alone or as part of a larger force, to complement or support its primary missions.

It expands the discussion of reconnaissance and breaching as tactical enabling operations and of the Company Commander's use of patrols to accomplish tactical tasks. It introduces and defines presence patrols for stability and civil support operations; point reconnaissance; tracking; and contact patrols.

Tactical enabling operations include reconnaissance, special purpose operations (linkups, reliefs in place, and passages of lines), breaches, security, and patrols. The company conducts these operations to set conditions for future operations or to support the current operations of its higher headquarters. The planning, preparation, and execution for these operations are just as important and require the same level of detail as conducting defensive or offensive operations. This edition eliminates all reference to friendly force use of non-command-detonated antipersonnel mines or booby traps.

Section I. RECONNAISSANCE

This section defines a reconnaissance and gives the categories and types offers considerations for planning; and discusses execution and techniques.

DEFINITION

8-1. Reconnaissance is any mission undertaken to get information about the activities and resources of enemy forces or the physical characteristics of a particular area, using visual observation or other methods. Successful reconnaissance is a focused collection effort, aimed at gathering timely and accurate information about the enemy and the terrain in the area of operations. Every Infantry company commander to reconnoiter to gain the information he needs to ensure the success of his mission. In addition, the company may conduct other reconnaissance operations to gather information for higher headquarters. (For a more detailed discussion of reconnaissance operations, see FM 17-95.)

CATEGORIES

8-2. The company commander develops the enemy situation through active and passive reconnaissance. Passive reconnaissance includes techniques such as map, photographic and SUAS reconnaissance and surveillance. Active methods include ground reconnaissance and reconnaissance by fire. Active reconnaissance operations are also classified as stealthy or aggressive, as discussed in the following paragraphs.

Stealthy Reconnaissance

8-3.   Stealthy reconnaissance emphasizes procedures and techniques that allow the unit to avoid detection and engagement by the enemy. It is more time-consuming than aggressive reconnaissance. To be effective, stealthy reconnaissance must rely primarily on elements that make maximum use of covered and concealed terrain. The company's primary assets for stealthy reconnaissance are its Infantry squads or SUASs.

Aggressive Reconnaissance

8-4.   Aggressive reconnaissance is characterized by the speed and manner in which the reconnaissance element develops the situation once contact is made with an enemy force. A unit conducting aggressive reconnaissance uses both direct and indirect fires and movement to develop the situation. In conducting a patrol, the unit employs the principles of tactical movement to maintain security. The patrolling element maximizes the use of cover and concealment and conducts bounding overwatch as necessary to avoid detection. (For a more detailed discussion, see Chapter 3.)

TYPES

8-5.   In addition to reconnaissance performed as part of another type of operation, three types of reconnaissances are conducted as distinct operations.

Route Reconnaissance

8-6.   A route reconnaissance is a directed effort to get detailed information on a specific route as well as on all terrain where the enemy could influence movement along that route. Route reconnaissance might be oriented on a specific area of movement, such as a road or trail, or on a more general area, like an axis of advance.

Zone Reconnaissance

8-7.   A zone reconnaissance is a directed effort to get detailed information about all routes, terrain, enemy forces, and obstacles, including areas of chemical and radiological contamination, within specified boundaries. The company normally conducts a zone reconnaissance when the enemy situation is vague, or when the company needs information about cross-country trafficability. As in route reconnaissance, the Infantry battalion commander's intent as well as METT-TC dictates the company's actions. Critical components of the operation normally include--

- Find and report all enemy forces within the zone.
- Reconnoiter specific terrain within the zone.
- Report all reconnaissance information.

8-8.   If time permits, the commander may also direct the company to--

- Reconnoiter all terrain within the zone.
- Inspect and classify all bridges.
- Locate fords or crossing sites.
- Inspect and classify all overpasses, underpasses, and culverts.
- Locate and clear all mines, obstacles, and barriers (within capability).
- Locate bypasses around built-up areas, obstacles, and contaminated areas.

Area Reconnaissance

8-9.   An area reconnaissance is a directed effort to get detailed information about the terrain or enemy activity within a prescribed area. The area can be any location that is critical to the unit's operations. Examples include easily identifiable areas covering fairly large spaces such as towns or military installations; terrain features such as ridge lines, wood lines, or choke points; or single points such as bridges or buildings. The critical tasks of an area reconnaissance are the same as those for a zone reconnaissance.

PLANNING CONSIDERATIONS

8-10. Reconnaissance planning starts with the company commander's identification of critical information requirements. The company commander then compares his CCIR list to that of the battalion commander. If the company commander identifies CCIR not covered on the battalion list, he shares them with the battalion commander and staff. The company commander requests that battalion or higher headquarter commit assets to confirm his CCIR. Based on the results of that request, the company commander can commit his forces to gather the information needed. This process begins while the unit is planning or preparing for an operation. It often continues during the conduct of the operation. Once the operation is under way, the commander continues to identify information requirements. For example, he needs to find an assailable flank or another position of advantage over an identified enemy force while the company develops the situation. In such a case, he might dispatch a platoon or section to find a flank or position where the company can effectively engage the enemy.

POSITIONING OF SUBORDINATE ELEMENTS

8-11. In conducting route, zone, or area reconnaissance, the company may employ direct and indirect fires. Based on his evaluation of METT-TC factors, the company commander establishes the roles of organic elements and support assets within his scheme of maneuver.

FOCUS OF RECONNAISSANCE

8-12. In planning for route, zone, or area reconnaissance, the company commander determines the objective of the mission, and identifies whether the reconnaissance will orient on the terrain or on the enemy force. He provides the company with clear guidance on the objective of the reconnaissance. In a force-oriented reconnaissance operation, the critical task is to find the enemy and gather information on him; terrain considerations of the route, zone, or area are a secondary concern. The company is generally able to move more quickly in a force-oriented reconnaissance than in a terrain-oriented reconnaissance.

EXECUTION

8-13. To be most effective, reconnaissance must be continuous and conducted before, during, and after operations. Before an operation, the company focuses its reconnaissance effort to confirm or deny a possible course of action (Figure 8-1, page 8-4). After an operation, the company normally reconnoiters so it can maintain contact with the enemy, collect information for upcoming operations, and provide force protection and security.

BEFORE OR AFTER AN OPERATION

8-14. Situations in which the company may conduct reconnaissance before or after an operation include--

- Reconnaissance by a quartering party of an assembly area and the associated route to it.
- Reconnaissance before an offensive operation from the assembly area to and near the LD.
- Reconnaissance by Infantry patrols to probe enemy positions for gaps prior to an attack or infiltration.
- Reconnaissance by Infantry patrols to observe enemy forward positions.
- Reconnaissance by patrols (normally Infantry and engineers) to locate bypasses around obstacle belts or to determine the best locations and methods for breaching operations.

Chapter 8

- Reconnaissance by Infantry patrols of choke points or other danger areas in advance of the remainder of the company.
- Reconnaissance of defensive positions or EAs prior to the conduct of the defense.
- Reconnaissance by patrols as part of security operations to secure friendly obstacles, clear possible enemy OPs, or cover areas not observable by stationary OPs.
- Reconnaissance to maintain contact with adjacent units.
- Reconnaissance by patrols to maintain contact with enemy elements.

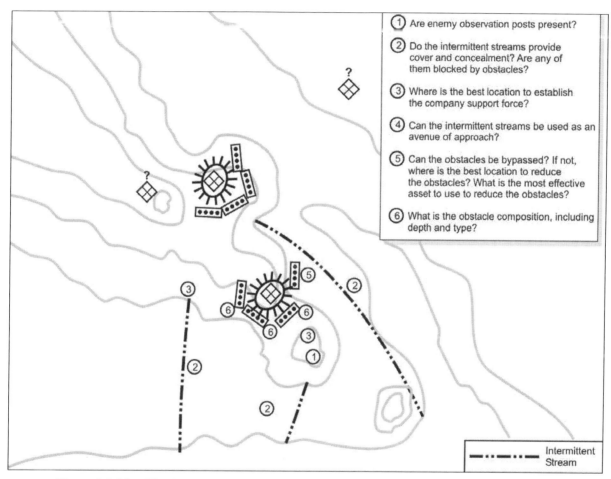

Figure 8-1. Identification of intelligence requirements and use of patrols to reconnoiter.

DURING AN OPERATION

8-15. During offensive operations, company reconnaissance normally focuses on fighting for information about the enemy and the terrain, with the primary goal of gaining an advantage over the enemy. The company conducts this type of reconnaissance during actions on contact. As the company develops the situation, the commander may dispatch patrols to identify positions of advantage or to acquire an enemy force. The information gained by the company while in contact is critical not only to the success of its own mission but also to the success of its higher headquarters. (Chapter 4, *Offensive Operations*, discusses actions on contact.)

Tactical Enabling Operations

Section II. SPECIAL PURPOSE OPERATIONS

This section defines linkup, relief in place, and passage of lines operations. It also describes when these might occur, discusses planning considerations for planning, and lists the steps to apply.

LINKUP

8-16. A linkup is an operation that entails the meeting of friendly ground forces (or their leaders or designated representatives). The company conducts linkup activities independently or as part of a larger force. Within a larger unit, the company may lead the linkup force.

SITUATIONS

8-17. Linkup may occur in, but is not limited to, the following situations.

- Advancing forces reaching an objective area previously secured by air assault, airborne, or infiltrating forces.
- Units coordinating a relief in place.
- Cross-attached units moving to join their new organization.
- A unit moving forward with a fixing force during a follow-and-support mission.
- A unit moving to assist an encircled force.
- Units converging on the same objective during the attack.
- Units conducting a passage of lines.

PLANNING CONSIDERATIONS

8-18. The plans for a linkup are detailed and cover the following.

Site Selection

8-19. Select a primary and an alternate site. These sites should be easy to find at night, have cover and concealment, and avoid the natural lines of drift. They must also be easy to defend for a short time, and must offer access and escape routes.

Recognition Signals

8-20. Far and near recognition signals help keep friendly units from firing on each other. Although units linking up exchange radio frequencies and call signs, they should avoid radio communications for short-range recognition due to possible compromise. Instead, they plan visual and voice recognition signals. They might use a sign and countersign such as a challenge and password or a number combination. Signaling means can include flashlights, chemical lights, infrared lights, or VS-17 panels. The TSOP can define near and far recognition signals.

Indirect Fires

8-21. Indirect fires are always planned, but not necessarily executed, for linkup operations. They support the movement by masking noise, deceiving the enemy of friendly intent, and distracting the enemy. Indirect fires are planned along the infiltration lanes and at the linkup sites to support in case of enemy contact.

Chapter 8

Direct Fires

8-22. Direct fire planning must include fratricide prevention. Restrictive fire lines (RFLs) control fires around the linkup site. Phase lines may serve as RFLs, which are adjusted as two forces approach each other.

Contingency Plans

8-23. The unit TSOP or the linkup annex to the OPORD must cover--

- Enemy contact before, during, and after linkup.
- Length of time to wait at the linkup site.
- Actions in case some elements fail to link up.
- Alternate linkup points and rally points.

STEPS

8-24. The linkup procedure begins as the unit moves to the linkup point. If using the radio, the unit reports its location using phase lines, checkpoints, or other control measures. Each unit sends a small contact team or element to the linkup point; the remainder of the unit stays in the linkup rally point. In a linkup, one unit occupies the linkup point as the stationary unit, while the other moves to the linkup point. The leader assigns specific duties of the contact elements and coordinates procedures for integrating the linkup units into a single linkup rally point. Full rehearsals are conducted if time permits. Figure 8-2 shows a company linkup between the 3rd platoon. The 3d platoon infiltrated early, conducted the reconnaissance of the objective, and established the ORP. The rest of the company, also shown, infiltrated later. The company (-) stops and sets up a linkup rally point about 300 meters from the linkup point. A contact team is sent to the linkup point; it locates the point and observes the area. If the unit is the first at the site, it clears the immediate area and marks the linkup point, using the agreed-upon recognition signal. It then takes up a covered and concealed position to watch the linkup point. The next unit (in this example, 3rd platoon) approaching the site repeats these actions. When its contact team arrives at the site and spots the recognition signal, they initiate the far recognition signal. The first element answers and the two elements exchange near recognition signals. The contact teams coordinate the actions required to link up the units such as to move one unit to the other unit's rally point or to continue the mission.

Tactical Enabling Operations

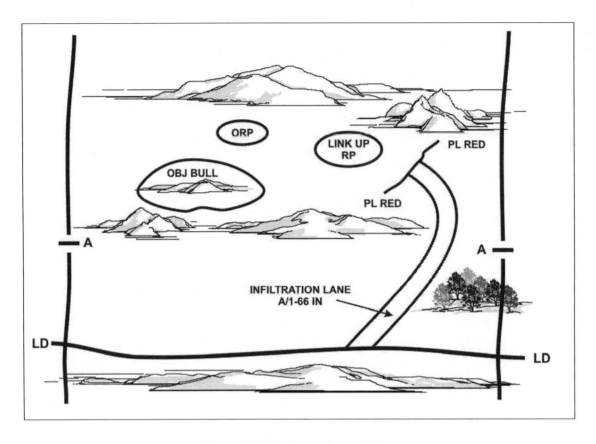

Figure 8-2. Infantry company linkup.

COORDINATION

8-25. Before initiating movement to the linkup point, the forces exchange necessary tactical information, including--

- The known enemy situation.
- Number and types of friendly units or personnel.
- Disposition of stationary forces, if either unit is stationary.
- Routes to the linkup and rally points, if used.
- Fire-control measures.
- Near recognition signal(s).
- Communications information.
- Combat support coverage.
- Sustainment responsibilities and procedures.
- Final location of the linkup point and rally point, if used.
- Any special coordination such as maneuver instructions or requests for medical support.

MOVEMENT TO LINKUP POINT AND LINKUP ITSELF

8-26. All units or elements involved in the linkup must enforce strict fire-control measures to prevent fratricide. Linkup points and RFLs must be easily recognizable by moving and converging forces. Linkup elements--

- Conduct far recognition by radio or Army Battle Command System (ABCS).

- Conduct short-range (near) recognition using visual or voice signal.
- Complete movement to the linkup point.
- Establish local security at the linkup point.
- Conduct additional coordination and linkup activities as necessary.

RELIEF IN PLACE

8-27. A relief in place is an operation in which one unit replaces another unit and assumes the relieved unit's responsibilities.

PURPOSE

8-28. The primary purpose for a relief in place operation is to maintain the combat effectiveness of committed units. A relief in place may also be conducted--

- To reorganize, reconstitute, or re-equip a unit that has sustained heavy losses.
- To rest units that have conducted sustained operations.
- To establish the security force or the DLIC during a withdrawal operation.
- To allow the relieved unit to conduct another operation.

PLANNING CONSIDERATIONS

8-29. If higher omits the time and location of the coordination meeting, the relieving unit commander contacts the relieved unit commander to coordinate them. The COs, XOs, platoon leaders, and FSOs should also attend the coordination meeting.

8-30. Each commander faces unique considerations in every tactical situation. Their respective missions, the enemy situation, and the time available are some of the factors that will affect the plan.

Command and Control

8-31. The C2 requirements during a relief are unique due to the mixing of units. To ensure effective C2, leaders must conduct detailed coordination early in the planning process (Chapter 2). The positions of key leaders and use of effective control measures will also ensure effective C2. Coordination between the relieving and the relieved units may include--

- Exchange of intelligence.
- Arrangements for reconnaissance.
- Exchange of tactical plans and sector sketches.
- Sequence and timing for each subunit's relief.
- Time or circumstance when the responsibility for the relieved unit's area of operations is transferred.
- The use of guides and liaison personnel.
- Security measures.
- Fire support.
- Transfer and exchange of equipment, supplies, ammunition, and minefields.
- Control measures.
- Exchange of frequencies, call signs, challenge and passwords, and recognition signals.

8-32. The locations of key leaders in both units are critical. The commanders and their FSOs normally collocate to best observe and control the relief. Other key leaders should be positioned to best assist the commander. They might be positioned along routes, assembly areas, points of possible congestion, or locations of greatest enemy threat.

8-33. Control measures provide control and flexibility during execution. The specific method of relief determines the number and type of control measures required. The following control measures are routine.

Assembly Areas

8-34. The relieved unit may designate platoon and company AAs to the rear of their positions. The relieving unit may also designate AAs, but should move directly into position. To avoid confusion, leaders must designate separate AAs for each unit.

Contact Points

8-35. Normally, the relieved company commander designates these. Having contact points facilitates the initial linkup between the companies. Multiple contact points might be needed to support some relief operations.

Release Points

8-36. The relieved commander normally designates the platoon release point for the relieving company. When necessary, the relieved company commander or platoon leader can designate squad release points.

Routes

8-37. All units should move along designated routes to avoid confusion and fratricide. When possible, the commander designates separate routes for the relieving and relieved units. The relieved commander ensures all movements are controlled.

RECONNAISSANCE

8-38. The relieving commander conducts a physical reconnaissance as soon as possible. The commanders and leaders of both companies should reconnoiter together, so they can coordinate their movement plans. Leaders down to squad level must reconnoiter. Only in this way can they ensure full understanding of movements, control measures, and responsibilities. Specific reconnaissance requirements may include--

- The relieved unit's disposition, to include locations of all OPs, minefields, land lines, early warning or antiintrusion devices, and crew-served weapons.
- Enemy dispositions, when the relieved unit is in contact.
- Locations for AAs, release points, contact points, and routes.
- Locations of the CP, trains, mortars, CCM, attached heavy or Stryker elements.

METHOD OF RELIEF

8-39. The method of relief is determined by the specific situation; however, regardless of the method of relief, certain actions normally occur. The relieving company occupies an AA to the rear of the relieved company or is guided along a route directly to the platoon RP. For each relieving subordinate unit, guides are provided by the relieved unit. The company CPs collocate prior to commencing the relief. The relief begins with the depth positions of the relieved company. The relieving company's trains and mortars are normally positioned before any relief begins. The relieved company's trains and mortars normally remain in position until responsibility for the area passes to the relieving company. A relief is conducted as follows.

Chapter 8

Relief of one unit at a time

8-40. This method takes longer than the others. However, it might be required when covered and concealed routes are limited, and when all platoons must use the same route. The relieving company occupies an AA to the rear of the relieved company and relieves by platoon according to the coordinated sequence (Figure 8-3). Each platoon moves forward (with guides provided by the relieved unit) to the squad release point. The squads are led to a covered and concealed location to the rear of the relieved squad's position. The necessary equipment is exchanged, and members of the relieving squads relieve the soldiers from the relieved squad. The relieved squad moves directly to the platoon AA, links up with the remainder of their platoon, and continues to the company AA. Once the entire relieving platoon assumes responsibility from the relieved platoon, the next platoon begins their relief.

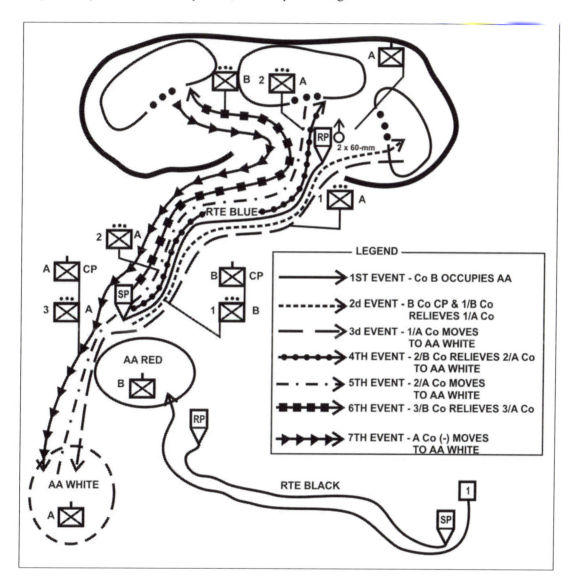

Figure 8-3. Relief in place in sequence.

Simultaneous Relief of Units

8-41. Although this method is the fastest, enemy detection is more likely, because all units move at once. This method may be appropriate when the mission requires a rapid relief, enemy detection is unlikely, and the terrain provides multiple covered and concealed routes. All relieving platoons move forward along their designated routes at the same time to the squad release points. The squad's actions are the same as previously described.

Relief by Occupation in Depth or Occupation of Adjacent Positions

8-42. This method requires the relieving unit to occupy positions to the flank or rear of the relieved unit. The relieving unit should be able to cover the relieved unit's direct-fire control measures (TRPs and EAs). This method is useful when the relieved unit has sustained chemical or nuclear contamination. It might also apply when the units involved have dissimilar TOEs such as a light unit relieving a heavy unit. The relieving unit may occupy its positions one at a time or all at once, depending on the situation. Once the relieving unit is in position, the relieved unit withdraws along designated routes.

SEQUENCE OF RELIEF

8-43. To determine the most effective sequence of relief, the commander considers the following.

Combat Effectiveness of Units

8-44. If one subordinate unit has suffered heavy losses in men or equipment, it may need to be the first relieved.

Terrain

8-45. The subordinate unit most likely to be detected during the relief should be relieved last. This allows the most relieving units to be in position before the enemy is aware of the relief operation.

Enemy

8-46. Consider relieving first the subordinate unit positioned on the most likely or most dangerous avenue of approach.

Control

8-47. When two adjacent units must share a route to conduct the relief, select a method and sequence of relief that reduces congestion and confusion. Avoid massing units in a small area.

Subsequent Mission

8-48. The subordinate unit with the most critical task may need to be relieved first. For example, a relieving subordinate unit may need to establish an OP forward of their position to provide security for the rest of the relief operation. Or when the company being relieved is moving to a LZ for an air assault operation, the platoon tasked to secure the LZ should be relieved first.

TRANSFER OF RESPONSIBILITY

8-49. The time for the transfer of responsibility must be agreed to by both commanders. Normally, this occurs once two-thirds of the relieving company are in position and have established communications and control.

TRANSFER OR EXCHANGE OF EQUIPMENT AND SUPPLIES

8-50. To simplify the relief and maintain the OPSEC, the two units might need to transfer certain equipment and supplies. These include machine gun tripods, mortar base plates and aiming stakes, camouflage nets, chemical alarms, and early warning and antiintrision devices. Supplies they should

Chapter 8

transfer can include barrier materials; excess or stockpiled supplies and ammunition; and bulky or heavy supplies that would slow the relief if the relieved unit tried to carry them out. Any prepared range cards, sector sketches, and minefield records must also be transferred to the relieving unit.

OPERATIONS SECURITY AND DECEPTION

8-51. Both units should make every effort to keep the enemy from knowing about the relief. Try to conduct the relief during limited visibility to reduce the risk of discovery by a capable threat.

8-52. The dispositions, activities, and radio traffic of the relieved unit must be maintained throughout the relief. Both companies should be on the relieved company's net. The relieved company continues routine traffic, which the relieving company monitors. Once the relief is complete and on a prearranged signal, the relieving company changes to their assigned frequency. Security activities, such as OPs and patrols, must maintain the established schedule. This might require some personnel from the relieving unit being placed under operational control (OPCON) of the relieved unit before the relief.

8-53. Additional planning and coordinating is required when a relief is conducted between a heavy or Stryker unit and an Infantry company, and when neither a relief by occupation in depth nor an occupation of adjacent positions is possible. If the relieving company is heavy or Stryker, the company should dismount, conduct the relief with the dismounts, and position the vehicles once the relieved company has withdrawn. If the relieved company is a heavy or Styker unit, the relieving company should relieve the dismounts, and then the vehicles move to the rear. The dismounts from the relieved unit may mount their vehicles or move to the rear on foot and occupy AAs until they linkup with their vehicles. If possible, the relieved company uses routine vehicle movements to reposition some vehicles to the rear before the relief. This might be possible when the unit has been using the out-of-position resupply technique. All vehicles would move to the resupply point, but only half return to their positions.

CONTINGENCY PLAN

8-54. The commanders should collocate where they can best observe and control the relief. The relieved company commander controls the relief until the conditions for the transfer of responsibility are met. If the enemy attacks before the transfer of responsibility, the subordinate units of the relieving company, which are in the area, become OPCON to the relieved company commander. If the enemy attacks after the transfer, the relieving commander assumes OPCON of all units of the relieved company still in the area. The commander should develop plans to cover these cases. Commanders must agree and understand when and which units will become OPCON. These contingency plans should address how the commanders will employ the uncommitted OPCON units. A technique for a more flexible relief plan is to designate the last relieving unit as the reserve.

CONDUCT OF RELIEF

8-55. At the time set for the start of the relief, the relieving company moves to the contact point. Once there, it makes contact with the company guide from the relieved company (Figure 8-4). The guide leads the company to the RP, where the company links up with the platoon and section guides. The platoon guides lead the platoons to their respective RPs, where the squad guides link up with their squads.

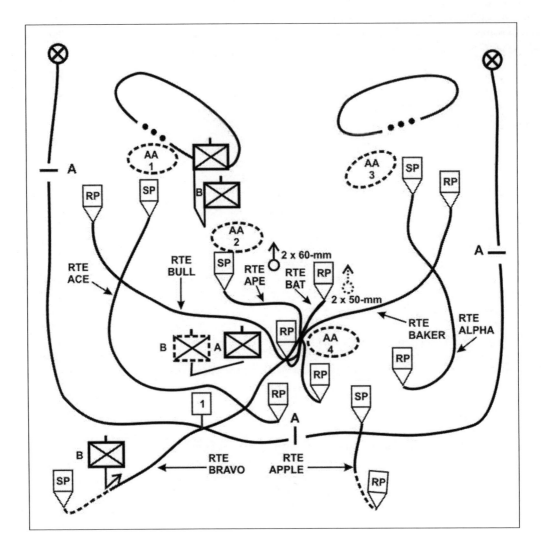

Figure 8-4. Relief in place (company graphics).

8-56. In the sequence specified in the order, each platoon conducts its relief. The platoon leader releases control of his squads, and the squad guides lead the squads to a location just to the rear of their defensive positions. The squad leaders then begin relieving a few men at a time until the relief is complete. Before each relieved soldier or leader leaves his position, he orients the relieving Soldier or leader on the position and the area around it.

8-57. As each Soldier or leader is relieved, he moves to his squad's AA. When each squad is assembled, it moves to its platoon's AA. When each platoon is assembled and its leader is relieved of his responsibility for the defense, it moves to the company AA. After the company is assembled and the transfer of responsibility is complete, the relieved commander moves his company as directed by the battalion commander.

PASSAGE OF LINES

8-58. A passage of lines is the movement of one or more units through another. This operation becomes necessary when the moving unit(s) cannot bypass the stationary unit and must pass through it. The primary purpose of the passage is to maintain the momentum of the moving elements. A passage of lines might be designated as either forward or rearward. The headquarters ordering the passage of lines is responsible for planning and coordination; however, specific coordination tasks are normally delegated to subordinate commanders.

PLANNING CONSIDERATIONS

8-59. In planning the passage of lines, the commander considers the tactical factors and procedures covered in the following paragraphs.

Passage Lanes

8-60. The passage facilitates transition to follow-on missions by using multiple lanes, or lanes wide enough to support doctrinal formations for the passing units.

Use of Deception

8-61. The company can use deception techniques, such as smoke, to enhance security during the passage.

Battle Handover

8-62. The controlling commander clearly defines the battle handover criteria and procedures for the passage. His order covers the roles of both the passing unit and the stationary unit, and of direct and indirect fires. If needed, he also gives the location of the battle handover line (BHL) as part of the unit's graphic control measures. For a forward passage, the BHL is normally the LD for the passing force. In a rearward passage, it is normally a location within the direct-fire range of the stationary force. In general, a *defensive handover* is complete when the passing unit is clear, and when the stationary unit is ready to engage the enemy. An *offensive handover* is complete when the passing unit has deployed and crossed the BHL.

Obstacles

8-63. The passing and stationary units coordinate obstacle information, to include the locations of enemy and friendly obstacles, existing lanes and bypasses, and guides for the passage.

Air Defense

8-64. Air defense coverage is imperative during the high-risk passage operation. Normally, the stationary unit provides air defense, allowing the passing unit's air defense assets to move with the passing unit.

Sustainment Responsibilities

8-65. Responsibility for sustainment actions, such as vehicle recovery or casualty evacuation in the passage lane, is clearly defined for both passing and stationary units.

Command and Control

8-66. To enhance command and control during the passage, the company collocates a command and control element, normally the company commander or XO, with a similar element from the stationary or moving unit (as applicable).

Reconnaissance Coordination

8-67. Detailed reconnaissance and coordination are critical in a passage of lines, both in its complex planning factors and to ensure that the passage goes quickly and smoothly. The company commander normally reconnoiters and coordinates the passage. At times, he may designate the XO, 1SG, or a platoon leader to conduct coordinate the following.

- Unit designation and composition.
- Type and number of personnel.
- Passing unit's arrival time(s).
- Location(s) of attack positions or assembly areas.
- Current enemy situation.
- Stationary unit's mission and plan, to include OP, patrol, and obstacle locations.
- Locations of routes, contact points, passage points, and passage lanes.
- Guide requirements.
- Order of march.
- Anticipated actions on enemy contact.
- Requirements for supporting direct and indirect fires, with location of RFL.
- Chemical, biological, radiological, or nuclear conditions.
- Available combat support assets and their locations.
- Communications information, to include frequencies and near and far recognition signals.
- Criteria for battle handover and location of the battle handover line.
- Additional procedures for the passage.

Forward Passage Of Lines

8-68. In a forward passage (Figure 8-5, page 8-16), the passing unit first moves to an assembly area or attack position to the rear of the stationary unit. Designated personnel move forward to link up with guides and confirm coordination information with the stationary unit. Guides then lead the passing elements through the passage lane. The company conducts a forward passage by employing tactical movement. It moves quickly, using appropriate dispersal and formations whenever possible, and keeping radio traffic to a minimum.

Chapter 8

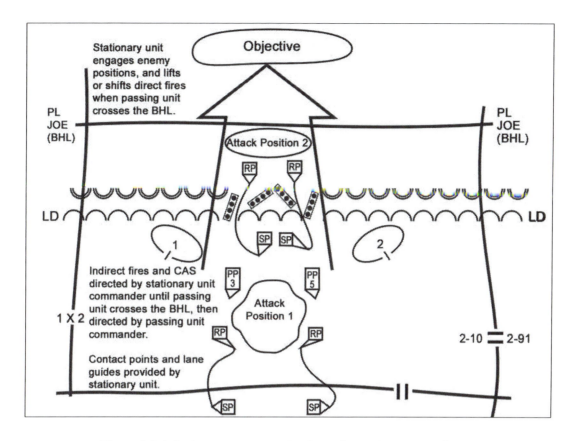

Figure 8-5. Infantry company conducting a forward passage of lines.

REARWARD PASSAGE OF LINES

8-69. Due to the increased chance of fratricide during a rearward passage, coordination of recognition signals and direct-fire restrictions is critical. While it is still beyond direct-fire range, the passing unit contacts the stationary unit, and then coordinates as previously discussed. Coordination emphasizes near recognition signals and location of the BHL. Additional fire-control measures, such as RFLs, might be used to reduce the risk of fratricide. After coordination, the passing unit continues tactical movement toward the passage lane. The passing unit is responsible for its own security until it passes the BHL. If the stationary unit provides guides, the guide meets the lead element of the passing unit and, without stopping, guides the unit to a designated location behind the stationary unit. If a guide is not provided, the passing unit moves on its own to a designated area without stopping (Figure 8-6).

Tactical Enabling Operations

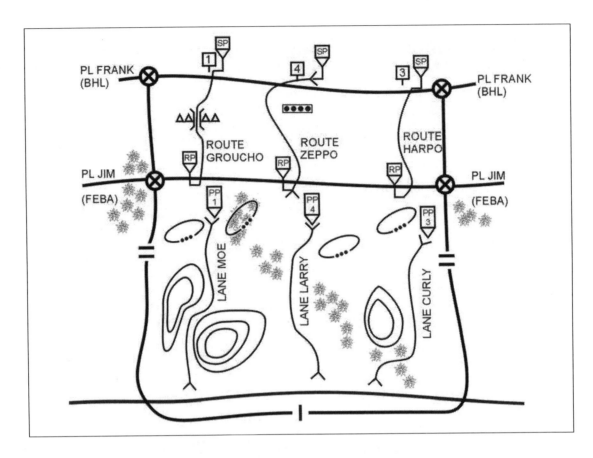

Figure 8-6. Infantry company conducting a rearward passage of lines.

Section III. SECURITY OPERATIONS

The company may conduct security operations to the front, flanks, or rear of the force. Security operations provide early and accurate warning of enemy operations. They give the protected force time and maneuver space to react to the enemy and develop the situation. This allows the commander to employ the protected force effectively. (For more on security operations, see FM 17-95.)

TYPES

8-70. The five forms of security operations are screen, guard, cover, area security, and local security. Screen, guard, and cover entail deployment of progressively higher levels of assets and provide increasing levels of security for the main body. Area security preserves a commander's freedom to move his reserves, position fire support assets, conduct command and control operations, and provide for sustainment operations. The company can conduct screen or guard operations on its own. It participates in area security missions and covering force operations only as part of a larger element. The company always provides its own local security. All forces have an inherent responsibility to provide their own local security. Local security includes OPs, local security patrols, perimeter security, and other measures taken to provide close-in security.

Chapter 8

PLANNING CONSIDERATIONS

8-71. Security operations require the commander assigning the security mission and the security force commander to address a variety of special operational factors.

AUGMENTATION OF SECURITY FORCES

8-72. When it is assigned to conduct a screen or guard mission, the company may receive additional combat, combat support, and sustainment elements. Attachments include at least--

- A scout platoon.
- An additional mortar section or platoon.

ENEMY-RELATED CONSIDERATIONS

8-73. Security operations require the company to deal with a unique set of enemy considerations. For example, the array of enemy forces (and the tactics that enemy commanders use to employ them) might differ from those for any other tactical operation the company conducts. Additional enemy considerations that could influence company security operations follow:

- The presence or absence of specific types of forces on the battlefield including--
    - Insurgent elements that might be external to the enemy force.
    - Enemy reconnaissance elements of varying strengths and capabilities, at division or brigade tactical group, or other levels.
    - Enemy security elements, such as disruption forces, including enemy stay-behind elements or other bypassed enemy elements.
- Possible locations where the enemy will employ his tactical assets, including--
    - Reconnaissance and infiltration routes.
    - OP sites for surveillance or indirect fire observers.
- Availability and anticipated employment of other enemy assets, including--
    - Surveillance devices such as radar devices or UASs.
    - Long-range rocket and artillery assets.
    - Helicopter and fixed-wing air strikes.
    - Elements capable of dismounted insertion or infiltration.
    - Mechanized forward detachments.

TIME SECURITY OPERATION IS INITIATED

8-74. The time by which the screen or guard must be set and active influences the company's method of deploying to the security area as well as the time it begins the deployment.

RECONNAISSANCE OF SECURITY AREA

8-75. The company commander uses a thorough analysis of METT-TC factors to determine the appropriate methods and techniques to accomplish this critical action.

Note: The company commander tries to personally reconnoiter the security area he expects the company to occupy, even when the operation is preceded by a zone reconnaissance by other battalion elements.

MOVEMENT TO SECURITY AREA

8-76. In deploying elements to an area for a stationary security mission, the company commander must deal with the competing requirements: to establish the security operation quickly and meet mission requirements; and to provide the necessary level of local security while doing so. The company can move to the security area using one of two basic methods: a tactical road march or a movement to contact. Either method should be preceded by a zone reconnaissance by the Infantry battalion scout platoon. The following paragraphs examine considerations and procedures for the two methods of movement.

Tactical Road March

8-77. The company conducts a tactical road march to an RP behind the security area to occupy their initial positions. This method of deployment is faster than a movement to contact, but less secure. It is appropriate when enemy contact is not expected or when time is critical.

Movement to Contact

8-78. The company moves from the LD to the security area using the appropriate movement technique based on the likelihood of enemy contact. This method is slower than a tactical road march, but it is more secure. It is appropriate when enemy contact is likely, or the situation is unclear.

LOCATION AND ORIENTATION OF SECURITY AREA

8-79. The main body commander determines the location, orientation, and depth of the security area in which he wants the security force to operate. The security force commander conducts a detailed analysis of the terrain in the security area. He then establishes his initial dispositions (usually a screen line, even for a guard mission) as far forward as possible, on terrain that allows clear observation of avenues of approach into a sector. The initial screen line is shown as a phase line. It sometimes represents the forward line of own troops (FLOT). As such, the screen line might serve as a restrictive control measure for movement. This requires the company commander to conduct all necessary coordination if he decides to establish OPs or to perform reconnaissance forward of the line.

INITIAL OP LOCATIONS

8-80. The company commander deploys OPs to ensure effective surveillance of the sector and designated NAIs. He designates initial OP locations on or behind the screen line. He provides OP personnel with specific orientation and observation guidance, including, at a minimum, the primary orientation for the surveillance effort during the conduct of the screen. Once set on the screen line, the surveillance elements report their locations. The element that occupies each OP always retains the responsibility for changing the location in accordance with tactical requirements and the commander's intent and guidance for orientation. OPs maximize stealth. Patrols might be required to cover gaps between the OPs. As required, the company commander tasks elements to conduct patrols.

WIDTH AND DEPTH OF SECURITY AREA

8-81. The company sector is defined by lateral boundaries extending out to the limit of advance (LOA) or the initial screen line. The company's ability to maintain depth through the sector decreases as the screened or guarded frontage increases.

Chapter 8

SPECIAL REQUIREMENTS AND CONSTRAINTS

8-82. The company commander specifies any additional considerations for the security operation, including at least--

- All requirements for observing NAIs, as identified by the battalion.
- Any additional tactical tasks or missions for the company and subordinate elements.
- Engagement and disengagement criteria for all company elements.

INDIRECT FIRE PLANNING

8-83. The company commander conducts indirect fire planning to integrate artillery and mortar assets into the security mission. A wide sector may require him to position mortar assets where they can provide effective coverage of the enemy's most likely axis of attack or infiltration route, as determined in his analysis of the enemy. The commander can position the mortars so that up to two thirds of their maximum range lies forward of the initial screen line. The company FSO helps the commander plan artillery fires to cover gaps in mortar coverage.

POSITIONING OF COMMAND AND CONTROL AND SUSTAINMENT ASSETS

8-84. The company commander positions himself where he can observe the most dangerous enemy axis of attack or infiltration route, with the XO positioned on the second most critical axis or route. The XO positions the company CP (if used) in depth and, normally, centered in sector. This allows the CP to provide control of initial movement, to receive reports from the screen or guard elements, and to assist the commander in more effectively facilitating command and control. Company trains are positioned behind masking terrain, but they remain close enough for rapid response. The trains are best sited along routes that afford good mobility laterally and in depth.

COORDINATION

8-85. The company commander conducts adjacent unit coordination to ensure there are no gaps in the screen or guard and to ensure smooth execution of the company's rearward passages of lines, if required. Also, he must coordinate the company's follow-on mission.

SUSTAINMENT CONSIDERATIONS

8-86. The company commander's primary consideration for sustainment during security operations is coordinating and conducting resupply of the company, especially for Class V supplies. (One technique is for the commander to pre-position Class V in successive positions.) However, in addition to normal considerations, the commander may acquire other responsibilities in this area such as arranging sustainment for a large number of attached elements or coordinating resupply for a subsequent mission. The company's support planning can be further complicated by a variety of factors. To prevent these factors from creating tactical problems, the company must receive requested logistical support, such as additional MEDEVAC vehicles, from the controlling battalion.

FOLLOW-ON MISSIONS

8-87. The complexities of security missions, combined with normal operational requirements, such as troop-leading procedures or on-the-move (OTM) planning, EA development, rest plans, and sustainment activities, can easily deprive the company commander of the time he needs for planning and preparation of follow-on missions. He addresses these competing demands in his initial mission analysis to ensure that the company and its leaders meet all requirements for current and future operations. For example, if METT-TC factors permit, the company commander can shift his focus to preparing for follow-on missions once preparations for the security mission are complete or satisfactorily underway. Another technique is to detach the XO with support personnel to prepare for follow-on missions. The XO's party can handle such operational requirements as reconnaissance, coordination, and development of follow-on EAs and BPs.

SCREEN

8-88. A screen primarily provides early warning. The screening force observes, identifies, and reports enemy actions to the main defense. A screen provides the least amount of protection for the main body of any security mission. Generally, a screening force engages and destroys enemy reconnaissance elements within its capabilities, but otherwise fights only in self-defense.

PURPOSES

8-89. A screen is appropriate to secure gaps between forces, the exposed flanks or rear of stationary and moving forces, or the front of a stationary formation. It is used when the likelihood of enemy contact is remote, the expected enemy force is small, or the friendly main body needs only a minimum amount of time, once it is warned, to act effectively. A screen is a series of OPs and patrols that ensure adequate surveillance of the assigned sector. The screen serves--

- To prevent enemy ground elements from passing through the screen undetected or unreported.
- To maintain continuous surveillance of all avenues of approach into the sector under all visibility conditions.
- To destroy or repel enemy reconnaissance elements within capability.
- To locate the lead elements of each enemy advance guard force and determine their direction of movement.
- To maintain contact with enemy forces and report any activity in sector.
- To impede and harass the enemy within capability while displacing.
- To maintain contact with the enemy main body and any enemy security forces operating on the flanks of friendly forces.

STATIONARY SCREEN

8-90. When conducting a stationary screen, the company commander analyzes infiltration routes into the sector. He assigns surveillance responsibility to the company's subordinate elements. He designates locations of OPs, which should be in-depth through the sector. Sections within the company normally operate the OPs. The commander identifies the enemy's likely axis of attack or infiltration routes. He identifies additional control measures, such as NAIs, phase lines, TRPs, or checkpoints, to assist in movement control, in tracking of enemy elements, or in confirming the enemy's course of action. The company conducts patrols to reconnoiter areas it cannot observe from OPs. Once an OP detects the enemy, the screening force normally engages with indirect fires. This disrupts the enemy and does not compromise the location of the OP. Within its capability, the screening force may destroy enemy reconnaissance assets with direct fires if indirect fires cannot accomplish the task. The screening force also impedes and harasses other enemy elements, mainly with indirect fires. If enemy pressure threatens the security of the screening force, the unit normally reports the situation and requests permission to displace to a subsequent screen line.

MOVING SCREEN

8-91. The company can conduct a moving screen to the front, flanks, or rear of the main body. The movement of the screen is keyed to time and distance factors associated with the movement of the friendly main body.

Moving Flank Screen

8-92. Responsibilities for a moving flank screen begin at the front of the main body's lead combat element and end at the rear of the protected force. In conducting a moving flank screen, the company either occupies a series of temporary OPs along a designated screen line or, if the protected force is moving too fast, continues to move while maintaining surveillance and preparing to occupy a designated screen line. The screening force may use one or more of these methods as the speed of movement of the protected force changes or contact is made. The four basic methods of controlling movement along the screened flank follow.

Alternate Bounds by Individual OP

8-93. The screening element uses this method when the protected force is advancing slowly and enemy contact is likely along the screen line. Designated elements of the screening force move to and occupy new OPs as dictated by the enemy situation and the movement of the main body. Other elements remain stationary, providing overwatch and surveillance, until the moving elements establish their new positions. These elements then move to new positions while the now-stationary elements overwatch and surveil. This sequence continues as needed. The method of alternate bounding by individual OP is secure but slow.

Alternate Bounds by Unit

8-94. The screening element uses this method when the protected force is advancing slowly and enemy contact is likely along the screen line. Designated elements of the screening force move and occupy new positions as dictated by the enemy situation and the movement of the main body. Other elements remain stationary, providing overwatch and surveillance, until the moving elements establish their new positions. These elements then move to new positions while the now-stationary elements overwatch and surveil. This sequence continues as needed. The method of alternate bounding by unit is secure but slow.

Successive Bounds

8-95. The screening element uses this method when enemy contact is possible. During this time, the main body makes frequent short halts during movement. Each platoon of the screening force occupies a designated portion of the screen line each time the main body stops. When main body movement resumes, the platoons move simultaneously, retaining their relative position as they move forward.

Continuous Marching

8-96. The screening element uses this method when the main body is advancing rapidly at a constant rate and enemy contact is not likely. The screening force maintains the same rate of movement as the main body while at the same time conducting surveillance as necessary. The screening force plans stationary screen lines along the movement route, but occupies them only as necessary to respond to enemy action.

Moving Rear Screen

8-97. The screening force may establish a moving rear screen to the rear of a main body force conducting an offensive operation, or between the enemy and the rear of a force conducting a retrograde operation. In either case, movement of the screen is keyed to the movement of the main body or to the requirements of the enemy situation. Movements to a series of phase lines normally control the operation.

GUARD

8-98. A guard force protects the friendly main body by fighting to gain time while also observing and reporting information. The guard force prevents enemy ground observation of and direct fire against the main body by reconnoitering, attacking, defending, and delaying. A guard force normally operates within the range of the main body's indirect-fire weapons. The three types of guard operations are advance guard, flank guard, and rear guard. They are conducted in support of either a stationary or a moving friendly force. The guard force normally deploys over a narrower area of operations than does a comparably sized screening force, allowing greater concentration. The guard force disrupts and delays enemy forces with both direct and indirect fires.

PURPOSES

8-99. The purposes of the guard, in addition to those listed in the earlier discussion of the screen, include--

- Destroy or repel all enemy reconnaissance elements.
- Fix or destroy enemy security elements.
- Cause the enemy main body to deploy, and then report its direction of travel to the friendly main body commander.

TYPES

8-100. The following discussion covers operational considerations for advance guards, flank guards, and rear guards.

Advance Guard

8-101. An advance guard for a stationary force is defensive in nature. The company defends or delays in accordance with the intent of the main body commander. An advance guard for a moving force is offensive in nature. The company normally conducts an offensive advance guard mission during a movement to contact as part of a battalion. The role of the advance guard is to maintain the freedom of maneuver of the main body by providing early warning of enemy activity, and by fixing or destroying enemy reconnaissance and security elements. These actions allow the main body commander to develop the situation.

Flank Guard

8-102. A flank guard protects a flank of the main body. A flank guard is similar to a flank screen except that both OPs and defensive positions are planned. The company may conduct a moving flank guard during an attack or a movement to contact. In conducting a moving flank guard, the company normally occupies a series of BPs along the protected flank. It must maintain orientation both to the front (to perform its overwatch role and to maintain its own security) and to the protected flank. It must also maintain a sufficient distance from the main body to prevent the enemy from engaging the main body with long-range direct fires before early warning can be sent.

Rear Guard

8-103. The rear guard protects the rear of the main body as well as all CS and sustainment elements within the main body. This may occur during offensive operations or during retrograde operations. Rear guards might be deployed behind either moving or stationary main bodies. The rear guard for a moving force displaces to successive BPs along phase lines or delay lines in depth as the main body moves. During retrograde operations, the rear guard normally deploys its elements across the entire sector behind the main body's forward maneuver units.

STATIONARY GUARD

8-104. As noted, a stationary guard mission is, at least initially, defensive in nature. The guard force normally employs OPs to accomplish all surveillance requirements of the guard mission. The company must be prepared to conduct actions against the enemy's main body and security elements as well as his reconnaissance forces. The following paragraphs discuss considerations for operations involving these enemy elements.

Actions against Main Body and Security Element

8-105. Once contact is made with an enemy main body or security force, the guard force attacks, defends, or delays in accordance with the enemy situation and the intent of the commander of the protected force (Chapter 5).

Actions against Reconnaissance Elements

8-106. When the company must execute counterreconnaissance tasks, it normally task-organizes into a surveillance element (normally occupying a screen line) and an attack element. Each element has specific responsibilities and must work effectively with the other to ensure success of the operation.

Surveillance Element and Surveillance Sectors

8-107. The commander assigns responsibilities for surveillance of likely avenues of approach and designated NAIs. The surveillance element is tasked with detecting, reporting, and maintaining contact with the enemy in the assigned surveillance sector. In addition, the surveillance element is responsible for passing the enemy force off to the attack element for destruction.

Attack Element

8-108. The attack element occupies hide positions, BPs, or attack-by-fire positions along likely enemy avenues of approach. Once alerted by the surveillance force, it moves into position (if necessary) and destroys the approaching enemy element. The attack element is responsible for direct fire planning and EA development in support of the commander's plan. It rehearses all necessary movement to the planned fighting positions and reports the movement times to the commander. Times of movement from hide positions to fighting positions are synchronized with the movement rates of the enemy. The time synchronization determines where the enemy must be acquired in order to provide the attack element time to move to its fighting position.

Relationship of Surveillance and Attack Elements

8-109. The company's surveillance element must track locations of any enemy moving through the sector while the attack element moves into position. Once the attack element is set and can observe the enemy, the surveillance element completes target handover. This operation requires continuous communication between the two subordinate elements conducting the handover, as well as close control by the company commander or XO. In close terrain, the surveillance and attack elements must be positioned much closer together than in open terrain. Figure 8-7 shows a company stationary guard operation.

Tactical Enabling Operations

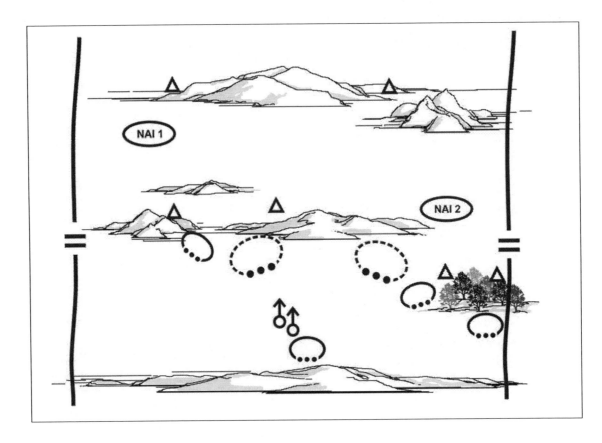

Figure 8-7. Stationary guard with OPs forward.

MOVING FLANK GUARD

8-110. Many of the considerations for a moving flank screen apply to the execution of a moving flank guard. However, unlike a moving flank screen that occupies a series of OPs, the flank guard force plans to occupy a series of defensive positions.

8-111. In conducting a moving flank guard, the company either occupies a series of temporary BPs along the protected flank or, if the protected force is moving too quickly, continues to move along the protected flank. During movement, the company maintains surveillance to the protected flank while preparing to occupy designated BPs based on enemy activity or on the movement of the protected force. The three basic methods of controlling movement along the guarded flank are--

- Alternate bounds by unit.
- Successive bounds by unit.
- Continuous marching.

Note: These are identical to the methods for controlling movement along a screened flank, except that the company and its platoons occupy designated defensive positions instead of OPs.

8-112. The lead element of a moving flank guard must accomplish three tasks. It must maintain contact with the protected force, reconnoiter the flank guard's route of advance, and reconnoiter the zone between the protected force and the flank guard's advance. The rest of the flank guard marches along the route of advance and occupies BPs to the protected flank as necessary. Figure 8-8, page 8-26, shows a company

27 July 2006 FM 3-21.10 8-25

flank guard operation during a movement to contact. One platoon is employed to provide security to the front and maintain contact with the main body. The other two platoons orient to the protected flank. Figure 8-8 also shows BPs that the platoons might occupy to respond to the approaching enemy force.

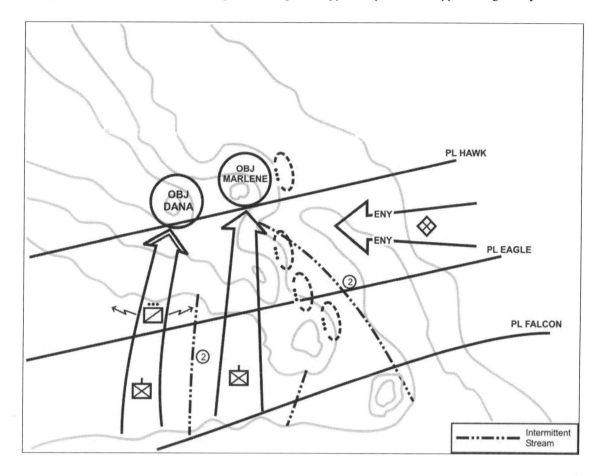

Figure 8-8. Infantry company guarding flank during movement to contact.

LOCAL SECURITY

8-113. The company is responsible for maintaining its own security at all times. It does this by deploying OPs and patrols to maintain surveillance. In addition to maintaining security for its own elements, the company may implement local security for other units as directed by the battalion commander. Examples of such situations include at least--

- Provide security for engineers as they emplace obstacles or construct survivability positions in the company BP.
- Secure LZs.
- Establish OPs to maintain surveillance of enemy infiltration and reconnaissance routes.
- Conduct patrols to cover gaps in observation and to clear possible enemy OPs from surrounding areas.

Section IV. BREACHING

Breaching operations are conducted when the company cannot bypass the obstacles with maneuver. Understanding breaching theory is the first step to understanding breaching tactics. Units should always try to bypass enemy obstacles. If the situation demands that the obstacles be reduced, then units should try to bypass the obstacles, destroy or repel the defending enemy forces, and then reduce the obstacles. Only as a last resort should commanders try to breach into an obstacle that is actively defended.

DEFINITIONS

8-114. Obstacle breaching is the use of combined tactics and techniques to advance an attacking force to the far side of an obstacle that is covered by fire. It might be the most difficult combat task a force can encounter. Breaching is a synchronized combined-arms operation under the control of a maneuver commander. The following definitions are also common to breaching operations.

TACTICAL OBSTACLE

8-115. A tactical obstacle is employed to disrupt enemy formations, turn them into a desired area, fix them in position under direct and indirect fires, or block their penetration while multiplying the effects and capabilities of firepower. Depending on the terrain, tactical obstacles are usually employed at two-thirds direct-fire maximum effective range.

PROTECTIVE OBSTACLE

8-116. A protective obstacle aids in close-in protection. They are usually employed at ranges of 30 to 100 meters to protect against hand grenades and Infantry assaults (terrain dependent).

LANE

8-117. A lane is a route through, over, or around an obstacle, which provides a passing force with safe passage.

REDUCTION

8-118. A reduction task creates and marks lanes through, over, or around an obstacle to allow the attacking force to accomplish its mission.

CLEARING

8-119. Clearing totally eliminates or neutralizes an obstacle or a portion of the obstacle. Clearing operations are not conducted under fire.

BREACH AREA

8-120. The area where a breaching operation occurs. The area must be large enough to allow the attacking unit to deploy its support force, and to extend far enough beyond the obstacle to allow follow-on forces to deploy before they leave the breach area.

BYPASS

8-121. A bypass is a tactical task that involves maneuvering around an obstacle, a position, or an enemy force to maintain the momentum of advance.

Chapter 8

POINT OF PENETRATION

8-122. A point of penetration is where the commander concentrates his efforts on the enemy's weakest point in order to seize a foothold on the far side of the objective.

POINT OF BREACH

8-123. This is where the unit tries to create a lane through the obstacle. At first, points of breach are planned locations only. Normally, the breach force determines the actual points of breach during the breaching operation.

TENETS

8-124. Successful breaching operations are characterized by the application of breaching tenets. These tenets are applied whenever an obstacle is encountered in the AO, whether during an attack or route clearance operation. The five tenets of breaching are intelligence, fundamentals, organization, mass, and synchronization.

INTELLIGENCE

8-125. In any operation where enemy obstacles can interfere with friendly maneuver, obstacle intelligence (OBSTINTEL) is always one of the information requirements and should become PIR. Examples of information that is needed to fulfill obstacle information requirements include--

- The locations of existing and reinforcing obstacles.
- The orientations and depths of obstacles.
- Locations of the weakest points of the obstacles.
- Locations of the areas that provide the best cover and concealment.
- The presence, locations, and types of wire.
- Lanes and bypasses.
- The composition of the minefield, such as buried or surface-laid antitank and antipersonnel mines or antihandling devices (AHDs), and the depths of the mines.
- Types of mines and fuzes.
- The locations of enemy indirect-fire systems that can fire into the breach area.

FUNDAMENTALS

8-126. Suppress, obscure, secure, reduce and assault (SOSRA) are the five breaching fundamentals that ensure success when breaching against a defending enemy. These fundamentals always apply, but they may vary based on the specific situation (METT-TC).

Suppress

8-127. Suppression is a tactical task used to employ direct or indirect fires. Suppression may also be an electronic attack on enemy personnel, weapons, or equipment, conducted to prevent or degrade enemy fires and observation of friendly forces. The purpose of suppression during breaching operations is to protect forces while they reduce and maneuver through an obstacle. Effective suppression is a mission-critical task performed in any breaching operation. Suppressive fires in sufficient volume secure the reduction area. Successful suppression generally triggers the rest of the actions at the obstacle. Fire-control measures ensure that all fires are synchronized with other actions at the obstacle. Although suppressing the enemy overwatching the obstacle is the mission of the support force, the breach force should be able to provide additional suppression against an enemy that the support force cannot effectively suppress.

Tactical Enabling Operations

Obscure

8-128. Obscuration protects forces conducting obstacle reduction and the passage of assault forces. Obscuration hampers enemy observation and target acquisition, and it conceals friendly activities and movement. Obscuration smoke deployed on or near the enemy's position minimizes its vision. Screening smoke employed between the reduction area and the enemy conceals movement and reduction activities. It also degrades enemy ground and aerial observations. Obscuration must be carefully planned to provide maximum degradation of enemy observation and fires, but it must not significantly degrade friendly fires and control.

Secure

8-129. Friendly forces secure the reduction area to prevent the enemy from interfering with obstacle reduction and the passage of the assault force through the lanes created during the reduction. Security must be effective against outposts and fighting positions near the obstacle and against overwatching units, as necessary. The far side of the obstacle is secured by fires, or is occupied before any effort to reduce it. The attacking unit's higher HQ isolates the breach area. It fixes adjacent units, attacks enemy reserves in depth, and provides counterfire support. Identifying the extent of the enemy's defenses is critical before selecting the best way to secure the point of breach. If the enemy controls the point of breach, and cannot be adequately suppressed, then the force must secure the point of breach before it can reduce the obstacle. The breach force must have enough maneuver assets to provide local security against the forces that the support force cannot sufficiently engage. Elements within the breach force that secure the reduction area may also be used to suppress the enemy once reduction is complete.

Reduce

8-130. Reduction is the creation of lanes through or over an obstacle to allow an attacking force to pass. The number and width of lanes created varies with the enemy situation, the assault force's size and composition, and the scheme of maneuver. The lanes must allow the assault force to rapidly pass through the obstacle. The breach force will reduce, proof (if required), mark, and report lane locations and the lane-marking method to higher HQ. Follow-on units will further reduce or clear the obstacle when required. Reduction cannot be accomplished until effective suppression and obscuration are in place, the obstacle has been identified, and the point of breach is secure.

Assault

8-131. A breaching operation is not complete until--

- Friendly forces have assaulted to destroy the enemy on the far side of the obstacle that can place or observe direct and indirect fires on the reduction area.
- Battle handover with follow-on forces has occurred, unless no battle handover is planned.

ORGANIZATION

8-132. A commander organizes friendly forces to accomplish the five breaching fundamentals quickly and effectively. This requires him to organize support, breach, and assault forces with the necessary assets to accomplish their roles (Table 8-1).

Breaching Organization	Breaching Fundamentals	Responsibilities
Support force	Suppress. Obscure	Suppress enemy direct-fire systems covering the reduction area. Control obscuring smoke. Prevent enemy forces from repositioning or counterattacking to place direct fires on the breach force.
Breach force	Suppress (provide additional suppression). Obscure (provide additional obscuration in the reduction area). Secure (provide local security). Reduce	Create and mark the necessary lanes in an obstacle. Secure the near and far sides of an obstacle. Defeat forces that can place immediate direct fires on the reduction area. Report the lane status and location.
Assault force	Assault. Suppress (if necessary)	Destroy any enemy on the far side of an obstacle that can place direct fires on the reduction area. Assist the support force with suppression if the enemy is not effectively suppressed. Be prepared to breach follow-on or protective obstacles after passing through the reduction area.

Table 8-1. Relationship between breaching organization and fundamentals.

Support Force

8-133. The support force's primary responsibility is to eliminate the enemy's ability to interfere with a breaching operation. Suppression depends on the commander massing enough direct fires to protect the breach force. The support force must--

- Isolate the reduction area with fires.
- Mass and control direct and indirect fires to suppress the enemy and to neutralize any weapons that can fire on the breach force.
- Control obscuring smoke to prevent enemy-observed direct and indirect fires.

8-134. The support force should be provided with assets to reduce the impact of unexpected obstacles or scatterable minefields on their approach to and occupation of support-by-fire (SBF) positions. Failure to provide reduction assets can greatly affect the synchronization of the entire breaching operation. As a technique, a unit may create a reserve that supports the decisive operation throughout the operation. Initially, the reserve can support the support force until it seizes SBF positions. Then, the reserve shifts support to the breach or assault force. If possible, the support force should follow a covered or concealed route to the SBF position, take up its assigned sectors of fire and observation, and begin to engage the enemy. It might have to adjust its direct-fire plan.

8-135. Observation is critical. Artillery observers with the support force may initially bring indirect fires on enemy positions to fix and suppress the enemy. The support force adjusts the indirect fire-delivered obscuring smoke to protect the breach and assault forces as they approach the reduction area. When resourcing the support force, consider possible personnel and equipment losses as it fights its way into its SBF position. To increase the survivability of the support force, the commander may request a CFZ in support of the support force once it occupies the SBF positions. A CFZ is an area, usually a friendly unit or location, which the maneuver commander designates as critical to protect an asset whose loss would seriously jeopardize the mission. Covered by a radar sector, the CFZ supports counterfire operations by providing the most responsive submission of targets to the fire support system when rounds impact inside the CFZ.

Breach Force

8-136. The breach force helps in the passage of the assault force by creating, proofing (if necessary), and marking lanes. The breach force might be a combined-arms force. It includes reduction assets, enough maneuver forces to provide additional suppression, and local security and engineers (if available). The breach force applies portions of the following breaching fundamentals as it reduces an obstacle.

Suppress

8-137. The breach force must be provided with enough maneuver forces to provide additional suppression against various threats, including enemy direct-fire systems and counterattacking or repositioning forces. Enemy direct-fire systems that cannot be effectively observed and suppressed by the support force are engaged by other forces, indirect fire, or air fire support.

Obscure

8-138. The breach force may employ quick building artillery or mortar-delivered smoke, followed by vehicle-mounted smoke systems or smoke pots. It uses these for self-defense and lane coverage during the passage of the assault force.

Secure

8-139. The breach force secures itself from threat forces that are providing close-in protection of the obstacle. The breach force also secures the lanes through the tactical obstacles, once they are created, to allow safe passage of the assault force.

Reduce

8-140. The breach force performs its primary mission of reducing the obstacle. To support the development of a plan to reduce the obstacle, the composition of the obstacle system must be an information requirement.

Assault

8-141. The assault force assaults through both the point of breach and the breach force to reach the far side of an obstacle and seize the reduction area.

8-142. The breach force has two subordinate elements: security and reduction. The security element is mostly maneuver forces. It normally provides additional suppression, obscuration, and local security. The reduction element normally reduces the obstacle.

8-143. The breach force must be able to deploy and begin reducing the obstacle as soon as enemy fires are suppressed and effective obscuration achieved. It can expect enemy artillery fires within a matter of minutes. If available, CFZs should be activated at the point of breach, before the commitment of the breach force.

8-144. After the breach force has reduced the obstacle and passed the assault force through, the breach force might have to hand over the lane to follow-on units. At a minimum, the lanes must be marked and their locations and conditions reported to both higher HQ and follow-on units, IAW the unit's SOP.

8-145. Achieving necessary mass for the assault requires the breach force to open lane(s) through the obstacle for rapid passage and the buildup of forces on the far side. The size of the assault force determines the number of lanes initially created. A dismounted assault force normally requires one lane for each leading assault platoon. The tactical situation might require additional lanes to pass a large assault force quickly through the obstacle to achieve sufficient combat-power ratio.

Chapter 8

Assault Force

8-146. The principle of *mass* influences the selection of the point of breach, the task organization of the support, breach, and assault forces, and the integration of engineers in movement or attack formations.

- The need to generate enough mass in the maneuver space available strongly influences which echelon can breach the obstacle. A company cannot simultaneously mass sufficient fires, reduce the obstacle, and assault the defending position unless the obstacle is simple and defended by no more than one platoon.
- The assault force's primary mission is to destroy the enemy and seize terrain on the far side of the obstacle. The purpose is to prevent the enemy from placing direct fires on the created lanes. The assault force might be tasked to assist the support force with suppression while the breach force reduces the obstacle.
- The assault force must be sufficient in size to seize the point of penetration. Breach and assault assets may maneuver as a single force when conducting lower-level breaching operations.
- When a small enemy force defends the obstacle, assault and breach force's missions might be combined. This simplifies C2 and provides more immediate combat power for security and suppression.
- Fire-control measures are essential, since support and breach forces might be firing on the enemy when the assault force is committed. Suppression of overwatching enemy positions must continue, and other enemy forces must remain fixed by fires until the enemy has been destroyed. The assault force must assume control for direct fires on the assault objective as support and breach force's fires are shifted or cease.

MASS

8-147. Breaching is conducted by rapidly concentrate efforts at one point to reduce the obstacle and penetrate the defense. Massed combat power is directed against the enemy's weakness. The location selected for breaching depends largely on the weakness in the enemy's defense, that is, where it has the least covering fires. If friendly forces cannot find a natural weakness, they create one by fixing the majority of the enemy force and isolating a small portion for attack.

SYNCHRONIZATION

8-148. Breaching operations require precise synchronization of the breaching fundamentals by support, breach, and assault forces. Failure to synchronize effective suppression and obscuration with obstacle reduction and assault can cause rapid, devastating losses of friendly troops in the obstacle or the enemy's EA. The commander ensures synchronization through proper planning and force preparation. Fundamentals to achieve synchronization are detailed reverse planning, clear subunit instructions, effective C2, and well-rehearsed forces.

Detailed Reverse Planning

8-149. Synchronizing a breach begins by using the reverse-planning process to ensure that actions at obstacles support actions on the objective. Planning the breach without regard to actions on the objective may cause the operation to fail. During COA development, the commander analyzes the relative combat power and compares enemy and friendly strengths and weaknesses. The commander decides how he must attack the objective to accomplish his mission.

Effective Command and Control

8-150. Effective C2 is paramount to mission success. It is integrated into the plan by maneuver and fire-control measures and the positioning of key leaders to see the battle space. Maneuver control measures enable the commander to convey his intent, scheme of maneuver, and subunit instructions graphically. Relating subunit actions to the terrain is critical to successful execution. Key leaders must be able to see the battle space to make informed decisions. This is most critical in breaching operations. The commander positions himself where he can best control the engagement. Since effective suppression is the most critical event in a breaching operation, the commander can position himself with the support force, or he can go where he can observe the effects of the suppression effort. This enables him to personally influence fire control and facilitate the necessary cross talk between breach and assault forces.

Well-Rehearsed Forces

8-151. The most effective synchronization tool available to the commander is the rehearsal. The inherent complexity of the breaching operation makes rehearsals at every level essential to success. The commander must give his subordinates time to plan how they will execute their assigned missions and time to rehearse it with their units.

Section V. PATROLS

This section introduces and defines presence patrols for stability, reconstruction, and civil support operations. It also introduces and defines point reconnaissance, tracking, and contact patrols.

DEFINITION

8-152. A patrol is a detachment sent out by a larger unit to conduct a combat, reconnaissance, or security mission. A patrol's organization is temporary and specifically matched to the immediate task. Because a patrol is an organization, not a mission, it is not correct to speak of giving a unit a mission to "*Patrol*."

8-153. Commanders sends a patrol out from the main body to conduct a specific tactical task with an associated purpose. Upon completion of that task, the patrol leader reports to the commander and describes the events that took place, the status of the patrol's members and equipment, and any observations.

8-154. If a patrol is made up of a single unit, such as a rifle squad sent out on a reconnaissance patrol, the squad leader is responsible. If a patrol is made up of mixed elements from several units, then the senior officer or NCO is designated as the patrol leader. This temporary title defines his role and responsibilities for that mission. The patrol leader may designate an assistant, normally the next senior man in the patrol, and any subordinate element leaders he requires.

8-155. A patrol can consist of a unit as small as a fire team. Squad- and platoon-size patrols are normal. Sometimes, for larger combat tasks, normally for a raid, the patrol can be a company (-).

8-156. The leader of any patrol, regardless of the type or the tactical task assigned, has an inherent responsibility to prepare and plan for possible enemy contact while on the mission. Patrols are never administrative. They are always assigned a tactical mission. On his return to the main body, the patrol leader must always report to the commander. He then describes the patrol's actions, observations, and condition.

TYPES

8-157. The planned action determines the type of patrol. The two main types of patrols are combat and reconnaissance. Regardless of the type of patrol, the unit needs a clear task and purpose.

COMBAT PATROL

8-158. A combat patrol provides security and harasses, destroys, or captures enemy troops, equipment, or installations. When the commander gives a unit the mission to send out a combat patrol, he intends for the patrol to make contact with the enemy and engage in close combat. A combat patrol always tries to escape detection while moving, but of course discloses their location to the enemy in a sudden, violent attack. For this reason, the patrol normally carries a significant amount of weapons and ammunition. It may carry specialized munitions. A combat patrol collects and reports any information gathered during the mission, whether related to the combat task or not. The three types of combat patrols are--

Raid Patrol

8-159. A raid is a surprise attack against a position or installation for a specific purpose *other than* seizing and holding the terrain. It is conducted to destroy a position or installation, to destroy or capture enemy soldiers or equipment, or to free prisoners. A raid patrol retains terrain just long enough to accomplish the intent of the raid. A raid always ends with a withdrawal off the objective and a return to the main body.

Ambush Patrol

8-160. An ambush is a surprise attack from a concealed position on a moving or temporarily halted target. It can include an assault to close with and destroy the target, or it can include only an attack by fire. An ambush need not seize or hold ground.

Security Patrol

8-161. A security patrol is sent out from a unit location during a halt, when the unit is stationary, to search the local area, to detect any enemy forces near the main body, and to engage and destroy them within the capability of the patrol. This type of combat patrol is normally sent out by units operating in close terrain with limited fields of observation and fire. Although this type of combat patrol seeks to make direct enemy contact and to destroy enemy forces within its capability, the patrol should try to avoid decisive engagement. A security patrol detects and disrupts enemy forces that are conducting reconnaissance of the main body or that are massing to conduct an attack. Security patrols are normally away from the main body of the unit for limited periods, returning frequently to coordinate and rest. They do not operate beyond the range of communications and supporting fires from the main body, especially mortar fires.

RECONNAISSANCE PATROL

8-162. A reconnaissance patrol collects information or confirms or disproves the accuracy of information previously gained. The intent for this type of patrol is to avoid enemy contact and accomplish its tactical task without engaging in close combat. With one exception (presence patrols), reconnaissance patrols always try to accomplish their mission without being detected or observed. Because detection cannot always be avoided, a reconnaissance patrol carries the necessary arms and equipment to protect itself and break contact with the enemy. A reconnaissance patrol travels light, that is, with as few personnel and as little arms, ammunition, and equipment as possible. This increases stealth and cross-country mobility in close terrain. Regardless of how the patrol is armed and equipped, the leader always plans for the worst case: contact. The types of reconnaissance patrols are--

Route Reconnaissance Patrols

8-163. This kind of patrol obtains detailed information about a specified route, and about all terrain where the enemy could influence movement along that route.

Area Reconnaissance Patrols

8-164. This kind of patrol focuses only on obtaining detailed information about the terrain or enemy activity within a prescribed area.

Tactical Enabling Operations

Zone Reconnaissance Patrols

8-165. This kind of patrol is a directed effort to obtain detailed information on all routes, obstacles, terrain, and enemy forces within a zone defined by boundaries.

Point Reconnaissance Patrols

8-166. This patrol goes straight to a specific location and determines the situation there. As soon as it does so, it either reports the information by radio or returns to the larger unit to report. This patrol can obtain, verify, confirm, or deny extremely specific information for the commander. These patrols are often used in stability, reconstruction, or civil support operations. For example, in a reconstruction operation, the commander might send such a patrol to determine the exact situation at a specific sewage pumping station, that is, has it begun operation? Does it have all the equipment and personnel it needs? Have the necessary repairs have been completed?

Leader's Reconnaissance Patrols

8-167. This patrol reconnoiters the objective just before an attack or, in the case of a point reconnaissance, prior to sending elements forward to locations where they will observe. It confirms the condition of the objective. It gives each subordinate leader a clear picture of the terrain where he will move, and of the part of the objective he must seize or observe. The patrol can consist of the unit commander or representative, the leaders of major subordinate elements, and, sometimes, a few security personnel and unit guides. This patrol gets back to the main body as quickly as possible. The commander can use the following aid to help in remembering a five-point contingency:

G..........Going--as in, where is the leader going?
O..........Others--what others are going with him?
T ....Time (duration)--how long will the leader be gone?
W..........What do we do if the leader fails to return?
A ....Actions--what actions do the departing reconnaissance element and the main body staying in the ORP take on contact ?

Presence Patrols

8-168. This patrol is used in stability or civil support operations. It has many purposes, but it should always see and be seen. Its main goal is to gather information about the conditions in the unit's AO. To do this, the patrol gathers critical (as determined by the commander) information, both specific and general. The patrol seeks out this information, and then observes and reports.

8-169. In addition to reconnaissance tasks, presence patrols demonstrate to the local populace the presence and intent of the US forces. That is, the patrol clearly demonstrates the determination, competency, confidence, concern, and sometimes the overwhelming power of the force to all who observe it, including local and national media. A presence patrol is planned for the possibility of enemy contact, even though this is not their intent. Rarely should a commander use a presence patrol where enemy contact is likely. Presence patrols work best for some types of stability operations, for example, peace operations, humanitarian and civic assistance, NEO, FHA, or shows of force. To accomplish its secondary purpose, to be seen, a presence patrol reconnoiters overtly. It takes deliberate steps to visibly reinforce the impression that the commander wants to convey to the populace. Where the patrol goes, what it does there, how it handles its weapons, what equipment and vehicles it uses, and how it interacts with the populace are all part of that impression. Before sending out a presence patrol, the commander should carefully consider what he wants to convey, and then clearly describe his intent to the patrol leader. When the presence patrol returns to the main body, the commander thoroughly debriefs it not only for hard information, but also for the patrol leader's impressions of the effects of the patrol on the populace. This allows the commander to see to modify the actions of subsequent patrols.

Tracking Patrols

8-170. A tracking patrol is normally a squad-size, possibly smaller, element. It is tasked to follow the trail of a specific enemy unit in order to determine its composition, final destination, and actions en route. Members of the patrol look for subtle signs left by the enemy as he moves. As they track, they gather information about the enemy unit, the route it took, and the surrounding terrain. Normally, a tracking patrol avoids direct fire contact with the tracked unit, but not always. Tracking patrols often use tracker dog teams to help them maintain the track.

Contact Patrols

8-171. A contact patrol is a special reconnaissance patrol sent from one unit to physically contact and coordinate with another. Modern technology has reduced, but not eliminated, the need for contact patrols. Now, they are most often used when a US force must contact a non-US coalition partner who lacks compatible communications or position-reporting equipment. Contact patrols may either go to the other unit's position, or the units can meet at a designated contact point. The leader of a contact patrol provides the other unit with information about the location, situation, and intentions of his own unit. He obtains and reports the same information about the contacted unit back to his own unit. The contact patrol also observes and reports pertinent information about the area between the two units.

COMPANY COMMANDER INVOLVEMENT

8-172. The company commander may be involved in patrolling in one of three ways. He may lead a company-size patrol; he may provide small patrols from his company (as directed by battalion); or he may send out patrols on his on initiative to support his company's operations. The company routinely conducts patrols as part of the company and battalion R&S plans. When he receives a mission from battalion to send out a patrol, the company commander--

- Obtains all necessary enemy information from the S-2 and other sources.
- Issues warning orders to the platoon he chooses for the patrol.
- Initiates appropriate troop-leading procedures.
- Coordinates and develops a detailed plan.
- Ensures the unit is prepared and properly organized and equipped for the mission.
- Assists the patrol leader with preparations, coordination, and final inspections before the patrol departs.
- Ensures that the patrol is debriefed upon its return.

8-173. When the company commander plans to use a patrol to support a company operation, he identifies its mission, organization, key time(s) and places(s) for departure and return, and (possibly) its routes. Depending on the mission, he may assign the task, give his intent, and allow the platoon leader to plan the patrol. He assists in planning fire support, logistic support, and communications.

ORGANIZATION

8-174. The commander decides what elements, teams, weapons, equipment, and men or units are needed for his mission. However, he should use his unit's normal organization (squads and platoons) and chain of command (squad and platoon leaders) as much as possible to meet these needs. For example, a combat patrol may be organized as follows:

- Company headquarters serves as the patrol headquarters.
- 1st Platoon serves as the assault element.
- 2d Platoon serves as the security element.
- 3d Platoon and weapons platoon comprise the support element.

8-175. When task-organizing a company patrol, the company commander only selects as many personnel as he needs. For example, if the security element only requires three security teams, the CO should task the platoon for a security element headquarters and three fire teams. A patrol generally consists of a patrol headquarters and the elements needed for the mission.

HEADQUARTERS

8-176. The headquarters of a company-size patrol may consist of the same number of men as a regular company headquarters. However, regardless of a patrol's size, the commander tailors the headquarters to meet mission needs. The patrol headquarters has the same responsibilities as any other command element.

ELEMENTS

8-177. In an area reconnaissance (Figure 8-9, page 8-37), a patrol has a reconnaissance element and a security element.

8-178. In a zone reconnaissance, a patrol has several reconnaissance elements (Figure 8-10, page 8-38). Each provides its own security.

8-179. A combat patrol normally has an assault element, a security element, and a support element (Figure 8-11, page 8-38). At times, the support element is omitted and instead combined with the assault element, or a reserve element might be required.

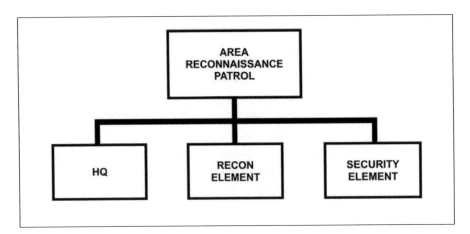

Figure 8-9. Area reconnaissance patrol.

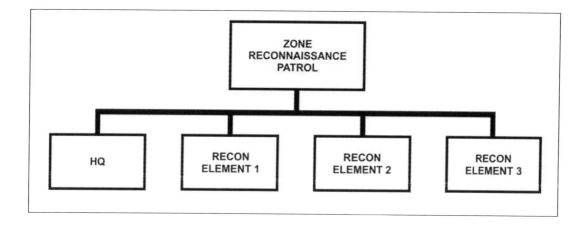

Figure 8-10. Zone reconnaissance patrol.

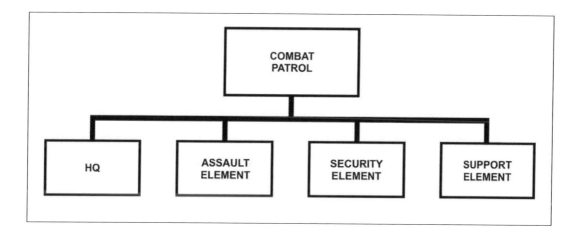

Figure 8-11. Combat patrol.

TEAMS

8-180. Each element of a patrol may be further organized into the teams needed to perform various tasks (Figure 8-12, page 8-39).

Reconnaissance Patrol Elements

8-181. Reconnaissance patrol elements may be organized into several reconnaissance teams for an area reconnaissance, or into R&S teams for a zone reconnaissance. R&S teams must provide their own security while reconnoitering.

Security Elements

8-182. Security elements are organized into the number of security teams needed to secure the objective area.

Combat Patrol Elements

8-183. Combat patrol elements are also organized into the teams needed for various tasks (assault, security, support, and special purpose).

- Two or more assault elements are organized when the assault element leader cannot directly control all of the assault element. This may be the case when the objective is to be assaulted from more than one location.
- Security teams are organized as needed to secure and or isolate the objective area.
- Two or more support teams are organized when the support element leader cannot directly control all of the weapons of the support element. This may be the case when there are many supporting weapons, or they are too far apart for direct control; by the element leader.
- Special-purpose teams may also be organized for missions involving the use of scout dogs, demolitions, litters for wounded, and EPW handling.

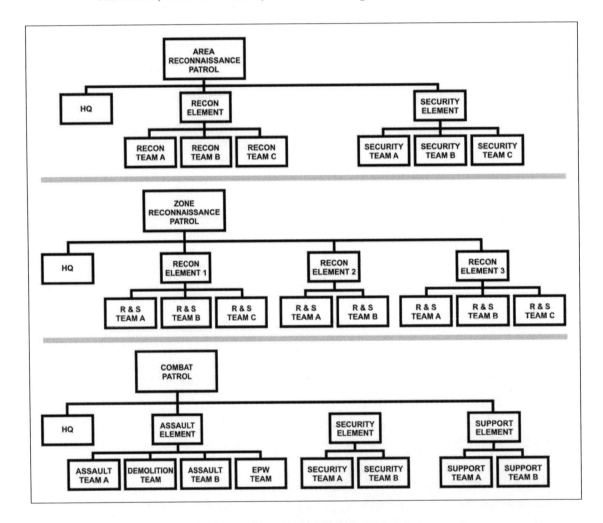

Figure 8-12. Organization of elements.

RAID

8-184. A raid is a surprise attack against a position or installation for a specific purpose other than seizing the terrain. It is conducted to destroy a position or installation, to destroy or capture enemy soldiers or equipment, or to free friendly prisoners. Since the purpose does not include holding terrain, the operation must include a planned withdrawal.

KEY CHARACTERISTICS

8-185. Surprise, firepower, and violence are the key characteristics for a successful raid. Surprise is best achieved by attacking when the enemy least expects an attack, when visibility is poor, and from an unexpected direction. Firepower is concentrated at critical points to suppress and kill the enemy. Violence is best achieved by gaining surprise, by using massed fire, and by attacking aggressively.

PLANNING CONSIDERATIONS

8-186. Although the planning process (Chapter 2) for the attack also applies to a raid, some differences exist. A raid is normally conducted in enemy controlled territory, often against an enemy of equal or greater strength. The plan must ensure that the unit retains the element of surprise and avoids detection prior to initiating the assault. An extraction or withdrawal plan must also be developed and coordinated to ensure the unit's survival after they successfully accomplish the actions on the objective. The fire support plan might be complex, depending on the depth of the raid. It can include a greater than usual reliance on artillery, CAS, AC-130 gunships, and attack helicopters. Finally, a raid often requires more detailed intelligence of the objective area. This may be obtained from higher units, or the company might have to develop this information through reconnaissance.

ACTIONS ON OBJECTIVE

8-187. Raids are normally conducted on an objective that is a valuable asset to the enemy. The enemy often has extra forces in position to react to any threat. The assault element must conduct a rapid and precise assault into and through the objective. The element must spend as little time as possible on the objective. Task organization should include only the personnel and teams who are essential to complete the assigned mission. This is particularly important during limited visibility, to reduce confusion and friendly casualties. Also, the assault must be thoroughly rehearsed to ensure precise execution.

PREPARATION

To achieve the surprise, violence, and speed of execution required, the unit's preparation is crucial to the success of the operation. The following requirements are key to the success of a raid mission.

Maximum Use of Intelligence Information

The gathering and disseminating of information must be continuous, and the information provided to the raid force, even while they are en route to the target area. To ensure mission accomplishment, the unit must be kept informed of the latest enemy developments in the objective area to prevent being surprised.

Plan Development

The reverse planning sequence and the planning process discussed in Chapter 2 will assist in conducting the detailed planning required for a raid. The plan must address the following phases.

PHASE 1

8-188. The unit is inserted or it infiltrates into the objective area.

PHASE 2

8-189. The objective area is then sealed off from outside support or reinforcement, to include the enemy air threat.

PHASE 3

8-190. Any enemy force at or near the objective is overcome by surprise and violent attack, using all available firepower for shock effect.

PHASE 4

8-191. The mission is accomplished quickly before any surviving enemy can recover or be reinforced.

PHASE 5

8-192. The unit quickly withdraws from the objective area and is extracted, or it infiltrates to link up with friendly units or to conduct a new mission.

Coordination

8-193. Coordination is normally conducted through the battalion headquarters. At times, the company may coordinate directly with adjacent, supporting, or host nation/allied forces.

Rehearsals

8-194. Rehearsals validate all aspects of planning for the raid and ensure precision in execution. They allow changes to be made in the plan before it is carried out. Full-scale rehearsals should be conducted under the most realistic conditions possible.

FAVORABLE CONDITIONS

8-195. A successful raid is ensured by--

- Launching the raid at an unexpected time or place by using limited visibility, and by moving over terrain the enemy might think impassable.
- Avoiding detection through proper movement techniques and skillful camouflage and concealment, to include using the natural cover of the terrain.
- Timing the operation as closely as possible.
- Using all available support, both organic and nonorganic, to include use of special weapons such as Air Force laser and GPS-guided bombs.
- Performing quick, violent, precise, and audacious actions that focus full combat power at the decisive time and place.
- Disengaging quickly upon mission completion.
- Withdrawing swiftly using planned routes and including a deception plan.

FUNCTIONS

8-196. Four functions are normally performed by the unit when conducting a raid. Each supplement is organized and equipped to do a specific part of the overall mission. Depending upon the specific mission, nature of the target, enemy situation, and terrain, the functions are as follows.

Command Group

8-197. The command group controls movement to and actions at the objective. This unit normally consists of the company commander, other subordinate leaders, and communications to support these leaders.

Chapter 8

Security Element

8-198. The security element, whose organization is determined by the mission of the raid force, size and type of the enemy force, and its mobility and state of alert, terrain, avenues of approach into the area, and the time needed to seal off the objective area. The security element may--

- Secure the objective rally point.
- Give early warning of and interdict approaching enemy forces.
- Block avenues of approach into the objective areas.
- Prevent enemy escape from the objective area.
- Provide overwatch for the units at the objective and suppressive fires to assist in their withdrawal.
- Provide short-range air defense fires.

Support Element

8-199. The support element provides the heavy volume of fire needed to neutralize the enemy or objective. Because fires from this unit are violent and devastating, they must be closely controlled to ensure the precision needed. On order or as planned, fires are shifted, lifted, or both to cover the maneuver of the assault element by suppressing enemy fire from the objective or aerial fires.

8-200. If an enemy quick-reaction force moves toward the objective area, the support element may also be given specific locations to cover by fire in support of the security element. These may include routes to and from the objective site, key terrain features, or installations adjacent to the main objective. Once the assault has been completed, or on order from the raid force commander, the support element displaces to the next planned position. Organization of the support element is determined by the following.

- Size of the objective, the geography of surrounding area, and the enemy threat in the area. This element should be able to neutralize the objective and to lift or shift fires either when the assault is launched or when so ordered by the raid force commander.
- Mission of the assault unit.
- Suitable firing positions.
- Size and nature of the enemy force in the objective area and those enemy forces capable of reinforcement at the objective.
- Fire support from other units.

Assault Element

8-201. The assault element seizes and secures the objective and protects demolition teams, search teams, prisoner-of-war teams, and other teams.

- The organization of the assault element is always tailored to the mission. Each objective must be examined carefully. The element's mission is to overcome resistance, secure the objective, and destroy the installation or equipment. Other specialized teams might also be needed. For example, sniper teams could be needed to remove key sentries. To capture prisoners, liberate personnel, and seize or destroy equipment, the assault element could be organized into assault teams, prisoner teams, search teams, medical teams, demolition teams, or breach teams.
- To destroy a point target or installation in a heavily defended area where the USAF cannot get close enough to be effective, the assault element might be organized with one small team equipped with laser target designators. From covered and concealed positions, members of this team could then guide USAF delivery of laser-guided munitions from a safe distance.

CONDUCT OF A RAID

8-202. The unit moves to the ORP, secures it, and sends out a leaders' reconnaissance. Once the final plan is confirmed, elements and teams then move to their positions. After the raid, the patrol unit reassembles at the ORP, and then it moves a safe distance away to recognize and disseminate information. It then returns to friendly lines or continues the mission.

Security Element

8-203. The teams of the security element move to positions (Figure 8-13, page 8-44) where they can secure the ORP, warn of enemy approach, and block avenues of approach into the objective area. They also situate themselves where they can prevent enemy escape from the objective area and perform any combination of these tasks within their capability. As the assault element and support element move into position, the security element keeps the leader informed of all enemy action. It fires only if detected, or on the leader's order. Once the assault starts, the security element prevents enemy entry into, or escape from, the objective area. When the assault is over, the security element covers the withdrawal of the unit to the ORP. It withdraws on order or on a planned signal.

Support Element

8-204. The support element moves into position before the assault element (Figure 8-14, page 8-45). From its position, it suppresses the objective and shifts its fire when the assault starts. It normally covers the withdrawal of the assault element from the immediate area of the objective. It withdraws on order or on signal.

Assault Element

8-205. The assault element deploys close enough to the objective to permit immediate assault if detected by the enemy. As supporting fire is shifted, lifted, or both, the assault element attacks and secures the objective. It protects demolition teams, search teams and other special teams while they work. On order, the assault element withdraws to the ORP. The assault element should be as small as possible and conduct thorough rehearsals to avoid confusion on the objective.

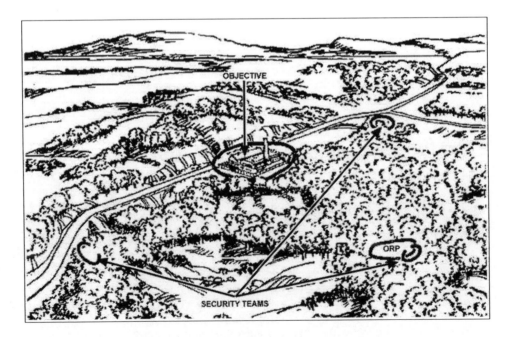

Figure 8-13. Security elements move into position.

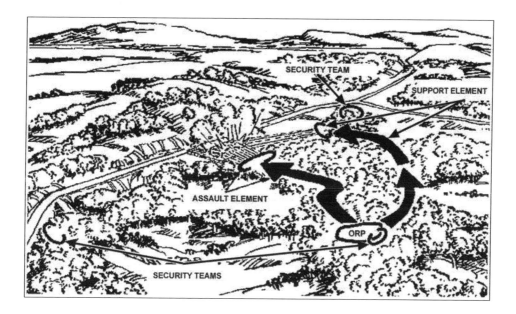

Figure 8-14. Support and assault elements move into position.

AMBUSH

8-206. An ambush is a surprise attack from a concealed position on a moving or temporarily halted target. The ambush can include an assault to close with and destroy the target, or it can consist of fire only. An ambush need not seize and hold ground. The company plans, prepares, and conducts ambush patrols the same as a platoon. An ambush is a useful tactic because small, well-trained, disciplined forces with limited weapons and equipment can destroy much larger enemy forces. Also, an ambush reduces the enemy's overall combat effectiveness by destroying and harassing his forces. Enemy morale and effectiveness suffer heavily at little cost to the unit executing the ambush.

EXECUTION

8-207. A successful ambush must be executed with precision, violence, speed, and audacity. For success, ambush operations must emphasize the following.

Surprise

8-208. Surprise, more than any other single aspect, enhances the value of an ambush. Surprise increases the potential for inflicting damage on the enemy with less risk to the unit.

Coordinated Firepower and Shock Effect

8-209. Coordinated firepower is used for maximum shock effect.

- Massive volumes of accurate fire, explosives, and mines, coupled with an aggressive attack, break the enemy's spirit to fight back. Surprise increases shock effect and the chances for success. Shock effect can cover unexpected defects in an ambush such as ambushing a much larger force than expected.
- All weapons must be sited with interlocking fires in the kill zone and along likely avenues of entrance or exit. Mortars should be used if the terrain permits. Tripods and traversing and elevating mechanisms are normally used with machine guns to lock in fires. All riflemen use firing stakes to mark left and right limits, and elevation stakes. There is a tendency to shoot

high in an ambush--especially at night. The M203 grenade launchers are sited to cover the dead space and routes of escape.

Control

8-210. Control is essential; leaders must have contact with all members of their unit to alert them to the oncoming enemy.

- Leaders should not move around the ambush site during this crucial period. A method used to alert members can be to tie strings or vines to Soldiers' legs or arms. A series of light tugs can alert all members of the ambush to the enemy's presence.
- The leader must initiate the ambush with a casualty-producing device. A bank of Claymore mines on a double-ring main is an excellent device to spring an ambush. Other good techniques are to use a machine gun or Javelins, if vehicles are in the kill zone. Whistles or pyrotechnics are not used, because they would give the enemy time to react.

8-211. As soon as the enemy is hit, he reacts. The ambush force has only a few seconds to destroy the enemy before the enemy recovers from the initial shock and leaves the kill zone, either by directly counterattacking or by withdrawing. Subsequent fires and other banks of Claymore mines must be planned.

8-212. The leader initiates the ambush except when a member of the ambush knows he has been discovered. He then has the authority to execute--with killing fire, not by yelling.

8-213. The cease-fire must be controlled by the leader. A whistle or other device may be used to get attention and then cease-fire is signaled.

Security

8-214. The flanks and rear of an ambush site are open to counterattack. Flank and rear security may be enhanced by--

- Echeloning in depth.
- Designating sectors of observation.
- Positioning RSTA devices.
- Enforcing noise and light discipline.
- Preparing a good withdrawal plan.
- Securing routes of withdrawal.
- Executing with speed and violence.
- Positioning a security force to seal off the ambush area.
- Preparing effective camouflage.

Simplicity

8-215. A simple, direct plan improves the chance of success. The ambush plan must be clear yet concise to offer the greatest likelihood of success. For example--

- Mission statements for security, support, and assault elements must be clear, concise, and direct.
- Tasks to be performed by the ambush elements should be easy to understand.
- Simple contingency plans.
- Routes into positions and withdrawal routes should not cross. They should be the shortest, most secure routes.

Chapter 8

Training and Self-Discipline

8-216. All advantages must be exploited. Discipline must be strict. There must be no sleeping, talking, eating, or smoking in the ambush site. If an ambush is to be set up for long periods, then the elements of the ambush must be pulled back to the ORP at set times for rest. Extended ambushes of 24, 36, or 48 hours require six- or eight-hour shifts. It may take a company to man an extended platoon ambush position. Tired troops cannot man an ambush well; they cannot perform vigorous operations all day and be alert on an ambush all night.

ORGANIZATION

8-217. A unit conducting an ambush must be task-organized to perform the following functions: assault, security, and support. The ambush forces should be task-organized according to the TOE--by platoons, squads, and fire teams. The TOE should not be changed to create smaller elements for an ambush. The TOE formations may be reinforced with machine gun or recoilless rifle teams, or a 60-mm mortar squad.

Assault

8-218. The elements assigned the assault mission either move directly into their positions or move through a release point. The mission may include any combination of the following actions.

- Conduct the main assault.
- Halt an enemy's motorized column or any moving target.
- Kill or capture personnel.
- Recover supplies and equipment.
- Destroy vehicles and supplies.

8-219. The leader decides how and when to use search teams. When Soldiers leave the security of their well-chosen, concealed ambush position, they are subject to the fires of the enemy who may also be hidden and ready. Always assume there is hidden enemy--the ambush will not kill them all. Either night vision devices or white-light flashlight should be used to make a quick search. A white light flashlight is faster if loss of night vision is not critical. If the return fire from the enemy is great or if the ambush missed the main body, then the leader may choose to break contact and leave without searching the kill zone.

Security

8-220. The elements assigned the mission of security may move to their positions directly or by way of a release point. Their missions may include any or all of the following actions.

- Secure flanks, rear, or ORP.
- Provide early warning.
- Seal off the kill zone to prevent the enemy from escaping or reinforcing.
- Assist in executing the ambush.
- Cover withdrawal of main ambush force.

Support

8-221. The units assigned a support mission provide fires that may include employment of--

- Heavy automatic weapon fires.
- Antitank fires.
- Mortar fires.
- Mines.
- Flame munitions.

AMBUSH SITE

8-222. When choosing an ambush site, all sources of information must be used to enhance surprise, exploit the enemy's weak points, and take advantage of the terrain. Emphasis is on--

- Natural cover and concealment for the ambush force.
- Concealed, easily accessible routes of entry and withdrawal.
- Good observation and fields of fire.
- Limited enemy escape routes.
- Limited enemy reinforcement ability.
- Nearby assembly or rendezvous area.
- Terrain that will canalize enemy into kill zones, and natural obstacles to keep him there.

Take Advantage of Terrain

8-223. Emphasize exploiting all natural cover and concealment afforded by the terrain. Site the ambush and individual positions based on the terrain rather than trying to adapt the terrain to a fixed geometric design.

Restrict Enemy Movement

8-224. Restricting enemy movement by natural or man-made obstacles should also be planned.

TYPES OF AMBUSHES

8-225. Ambushes have two basic categories - area ambush and point ambush.

Area Ambush

8-226. An area ambush may be set up by platoons, companies, or battalions. It is used to interdict enemy movement in a given area or inflict casualties on his forces. An area ambush consists of a series of point ambushes. The size and locations of the ambushes are dictated by the METT-T analysis.

- Companies may conduct area ambushes independently or as part of a battalion area ambush. The company may receive very specific guidance or only an area of operations and a mission statement. The CO may develop a very detailed concept with a central ambush supported by smaller ambushes for security/isolation. Or, the CO may assign platoon areas of operation and allow decentralized execution.
- Considerations in selecting point ambush sites as part of a company area ambush include--
  - Ensuring fires from one ambush force do not endanger other ambush units.
  - The enemy's likely course of action, both before and after the ambush is initiated.
  - The withdrawal/linkup plan after completing the ambush mission.
- The CO must establish clear criteria to each ambush site leader on when to initiate fires.

Point Ambush

8-227. Point ambushes are set at the most ideal location to inflict damage on the enemy. Such ambushes must be able to handle being hit by the enemy force from more than one direction. The ambush site should enable the unit to execute an ambush in two or three main directions. The other directions must be covered by security that gives early warning of enemy attack.

8-228. The mechanical ambush is a special type of point ambush. It consists of Claymore mines set in series with a double-ring main. It is command detonated. Soldiers prepare to engage the enemy with direct fire after the mechanical ambush detonates. Mechanical ambushes are an effective way to interdict a large area using a small force. If the mechanical ambush is effective and our Soldiers do not reveal their presence, the enemy is confused. This has a devastating effect on his morale and effectiveness.

EXECUTION OF AMBUSH

8-229. Stealth and security are important factors; the following are various ways to accomplish these factors.

- Position security teams and early warning detection devices first.
- Use the best route to main ambush position consistent with security.
- Quickly occupy the ambush position and set up communications and signaling devices.
- Position key weapons (automatic and antiarmor).
- Rig Claymore mines and trip flares.
- Ensure that all weapons are correctly positioned. Assign sectors of fire to provide mutual support and cover dead space.

Camouflage

8-230. During mission preparation, each man camouflages himself and his equipment, and secures his equipment to prevent noise. At the ambush site, prepare positions with minimum change in the natural appearance of the site. Conceal all resulting debris to prevent any evidence of occupation.

Movement

8-231. Keep movement to a minimum. Closely control the number of men moving at a time. Keep every Soldier as quiet as possible, especially at night. Enforce light discipline rigidly at night and forbid smoking.

Signals

8-232. Change audible and visual signals, such as whistles or pyrotechnics, often to avoid setting patterns and alerting the enemy. Three or four simple signals are needed to execute the ambush. Signals are used--

- To provide early warning of an enemy approach. A signal by the security force to alert the patrol leader to the correct direction of enemy approach may be given. This includes arm-and-hand signals, radio, or field telephone.
- To initiate the ambush. This may be the detonation of mines or explosives. Fire is then delivered at once in the heaviest, most accurate volume possible. Properly timed and delivered fires add to the achievement of surprise, as well as to the destruction of the target.
- To lift or shift fires if the kill zone is to be assaulted. Voice commands, whistles, or pyrotechnics maybe used. When the kill zone is assaulted, the lifting or shifting of fires must be as precise as when starting the ambush. Otherwise, the assault is delayed and the enemy has a chance to recover and react.
- To withdraw. The signal for withdrawal can be voice commands, whistles, or pyrotechnics.

Objective Rally Point

8-233. Locate the ORP far enough from the ambush site so that it will not be overrun if the enemy manages to attack the ambushers. Withdrawal routes should provide cover and concealment for the unit and hinder enemy pursuit; they are a main consideration in the selection of the ambush site. They may be the key to survival after executing the ambush. On signal, the ambush force quickly (but quietly) withdraws to the ORP. If the force is pursued, they can withdraw by bounds, and use grenades or hasty ambushes to delay pursuing forces.

Ambush Variety

8-234. Use more than one ambush method. If one method is used predominantly, the enemy will develop an effective defense against it and will be affected less by the shock of the ambush since he knows what to expect. No single method will fit all combinations of terrain, equipment, weather, and enemy capabilities. Use a variety of signals as well, both audible and visual. Use weapons fire, mines, and RSTA when possible and vary signals to avoid compromise.

Swift Action

8-235. Speed in the execution of the ambush and the withdrawal should prevent enemy reaction forces from engaging the ambush force. Speed is often a shield against casualties and failure. If there is contact with reaction forces, speed may enhance quick disengagement.

SUCCESSFUL AMBUSH

8-236. Emphasize the following to succeed.

- Intelligence, which ensures the enemy is ambushed at a time and place when he least expects or is least prepared to fight.
- Detailed planning, thorough training, and rehearsing of all elements in all phases of the ambush. This ensures maximum shock effect through swift, precise execution.
- All available night vision and detection devices, when appropriate.
- All available firepower.
- Speed, which aids in achieving surprise and enhancing the security of the force.
- Cover, concealment, and overall protection afforded by the terrain when moving or when occupying ambush positions.

PATROL BASE

8-237. A patrol base is a position set up when the patrol unit halts for an extended period. When the unit must halt for a long time in a place not protected by friendly troops, it takes active and passive security measures. The time the patrol base may be occupied depends on the need for secrecy. It should be occupied only as long as necessary, but not for more than 24 hours--except in an emergency. The unit should not use the same patrol base more than once. The considerations for a perimeter defense apply for establishing a company patrol base.

Chapter 9
Direct Fire Control

Suppressing or destroying the enemy with direct fires is fundamental to success in close combat. Effective direct fires are the unique contribution of maneuver forces to the combined arms team, and fire and movement are complementary components of maneuver. The Infantry company commander must effectively plan to focus, distribute, and shift the overwhelming mass of his direct fire capability at critical locations and times to succeed on the battlefield. Effective and efficient fire control means that the company acquires the enemy and masses the effects of direct fires to achieve decisive results in the close fight.

This edition introduces a discussion of direct fire control and distribution including principles, processes, planning, and control.

Section I. FIRE-CONTROL PRINCIPLES

When planning and executing direct fires, the Infantry company commander and subordinate leaders must know how to apply several fundamental principles. The purpose of these direct fire control principles is not to restrict the actions of subordinates, but to help the company accomplish the primary goal of any direct fire engagement: to eliminate the enemy by *acquiring first and shooting first*. Applied correctly, these principles give subordinates the freedom to respond rapidly upon acquisition of the enemy. This discussion focuses on the following principles.

- Mass the effects of fire.
- Destroy the greatest threat first.
- Avoid target overkill.
- Employ the best weapon for the target.
- Minimize friendly exposure.
- Plan and implement fratricide avoidance measures.
- Plan for extreme limited visibility conditions.
- Develop contingencies for diminished capabilities.

MASS EFFECTS OF FIRE

9-1. The Infantry company must mass its direct fires to achieve decisive results. Massing entails focusing direct fires at critical points and distributing the effects. Random application of fires is unlikely to have a decisive effect. For example, concentrating the company's fires at a single target may ensure its destruction or suppression; however, that fire control option will fail to achieve the decisive effect on the remainder of the enemy formation or position.

DESTROY GREATEST THREAT FIRST

9-2. The order in which the Infantry company engages enemy forces is in direct relation to the danger these forces present. The threat posed by the enemy depends on his weapons, range, and positioning.

Presented with multiple targets, a unit must initially concentrate direct fires to destroy the greatest threat, and then distribute fires over the remainder of the enemy force.

AVOID TARGET OVERKILL

9-3. Use only the amount of fire required to achieve necessary effects. Target overkill wastes ammunition and is not tactically sound. To the other extreme, the company cannot have every weapon engage a different target because the requirement to destroy the greatest threats first remains paramount.

EMPLOY BEST WEAPON FOR TARGET

9-4. Using the appropriate weapon for the target increases the probability of rapid enemy destruction or suppression; at the same time, it conserves ammunition. The Infantry company has many weapons with which to engage the enemy. Target type, range, and exposure are key factors in determining the weapon and ammunition that should be employed, as are weapons and ammunition availability and desired target effects. The company commander arrays his forces based on the terrain, enemy, and desired effects of all of his available direct fires.

MINIMIZE FRIENDLY EXPOSURE

9-5. Units increase their survivability by exposing themselves to the enemy only to the extent necessary to engage him effectively. Natural or manmade defilade provides the best cover from ATGMs and other large caliber direct fire munitions. Dismounted Infantry minimize their exposure by constantly seeking effective available cover, trying to engage the enemy from the flank, remaining dispersed, firing from multiple positions, and limiting engagement times.

PLAN AND IMPLEMENT FRATRICIDE AVOIDANCE MEASURES

9-6. The company commander must work proactively to reduce the risk of fratricide and noncombatant casualties. He must plan and use the numerous tools to assist him in this effort: identification training for combat vehicles and aircraft, the unit's weapons safety posture, the weapons control status (WCS), and recognition markings. Knowledge and employment of applicable ROE are the primary means of preventing noncombatant casualties.

PLAN FOR EXTREME LIMITED VISIBILITY CONDITIONS

9-7. At night, limited visibility fire control equipment enables the Infantry company to engage enemy forces at nearly the same ranges that are applicable during the day. However, obscurants such as dense fog, heavy rain, heavy smoke, and blowing sand can reduce the capabilities of thermal and IR equipment. The company commander develops contingencies for limited visibility conditions. Although a decrease in acquisition capabilities has little effect on area fire, point target engagements are likely to occur at decreased ranges. Firing positions, whether offensive or defensive, typically must be adjusted closer to the area or point where the commander intends to focus fires. Another alternative is the use of visual or IR illumination when there is insufficient ambient light for passive light intensification devices.

PLAN FOR DIMINISHED CAPABILITIES

9-8. Leaders initially develop plans based on their units' maximum capabilities; they make backup plans for implementation in the event of casualties, weapon damage, or failure. While leaders cannot anticipate or plan for every situation, they develop plans for what they view as the most probable occurrences. Building redundancy into these plans, such as having two systems observe the same sector, is an invaluable asset when the situation (and the number of available systems) permits. Designating alternate sectors of fire and supplementary firing positions provides a means of shifting fires if adjacent elements become unable to fire.

Section II. FIRE-CONTROL PROCESS

To bring direct fires against an enemy force successfully, commanders and leaders continuously apply the four steps of the fire control process. At the heart of this process are two critical actions: rapid, accurate target acquisition and the massing of fires to achieve decisive effects on the target. Target acquisition consists of detecting, identifying, and locating the enemy in sufficient detail to permit the effective employment of weapons. Massing entails focusing fires at critical points and then distributing the fires for optimum effect. The four steps are--

- Identify probable enemy locations and determine the enemy scheme of maneuver.
- Determine where and how to mass (focus and distribute) fire effects.
- Orient forces to speed target acquisition.
- Shift fires to refocus or redistribute their effects.

IDENTIFY PROBABLE ENEMY LOCATIONS AND DETERMINE ENEMY SCHEME OF MANEUVER

9-9. The Infantry company commander plans and executes direct fires based on his analysis of the factors of METT-TC. In particular, his analysis of the terrain and the enemy force are essential and aid him in visualizing how the enemy will attack or defend a particular piece of terrain. A defending enemy's defensive position or an attacking enemy's support position is normally driven by terrain. Typically, there are limited points on a piece of terrain that provide both good fields of fire and adequate cover for a defender. Similarly, an attacking enemy will have only a limited selection of avenues of approach that provide adequate cover and concealment. The company commander's understanding of the impact of a specific piece of terrain on maneuver assist him in identifying probable enemy locations and likely avenues of approach both before and during the fight. Figure 9-1, page 9-4, shows the commander's analysis of enemy locations and scheme of maneuver. He uses any or all of the following products or techniques in developing and updating the analysis.

- A SITEMP provided by the battalion.
- A SPOTREP or contact report on enemy locations and activities.
- Reconnaissance of the area of operations.

Chapter 9

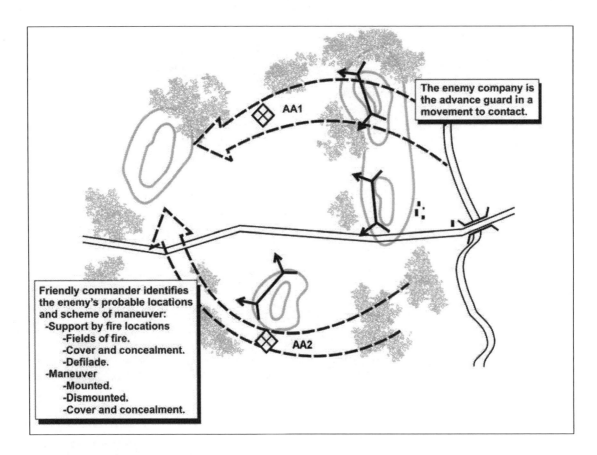

Figure 9-1. Identification of probable enemy locations and determination of enemy scheme of maneuver.

DETERMINE WHERE AND HOW TO MASS FIRES

9-10. To achieve decisive effects, the Infantry company masses direct fires. Effective massing requires the company commander both to focus the fires of subordinate elements and to distribute the effects of those fires. Based on his analysis and his concept of the operation, the company commander identifies points where he wants to or must focus the company's direct fires. Most often, he has identified these locations as probable enemy positions or points along likely enemy avenues of approach where the company can mass direct fires. Because the platoons may not initially be oriented on the point where the commander wants to mass direct fires, he may issue a fire command to focus the fires. At the same time, the company commander must use direct fire control measures to distribute the direct fires of his subordinate elements effectively, fires that are now focused on the same point. Figure 9-2 shows how the commander masses fires against the enemy.

Direct Fire Control

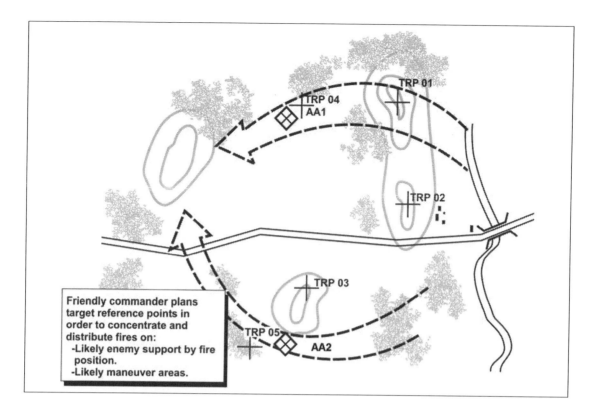

Figure 9-2. Determination of where and how to mass (focus and distribute) fire effects to kill enemy.

ORIENT FORCES TO SPEED TARGET ACQUISITION

9-11. To engage the enemy with direct fires effectively, the Infantry company must rapidly and accurately acquire enemy elements. Orienting the company on probable enemy locations and on likely enemy avenues of approach will speed target acquisition. Conversely, failure to orient the company slows acquisition, which greatly increases the chance that enemy forces can engage first. The clock direction orientation method, which is prescribed in most unit SOPs, is good for achieving all-round security, but it does not ensure that friendly forces are most effectively oriented to detect the enemy. To achieve this critical orientation, the commander typically designates TRPs on a recognizable permanent feature on or near a probable enemy location or avenues of approach and orients his platoons using directions of fire or sectors of fire. Figure 9-3, page 9-6, shows how the company commander orients the company for quick, effective acquisition of the enemy force.

Chapter 9

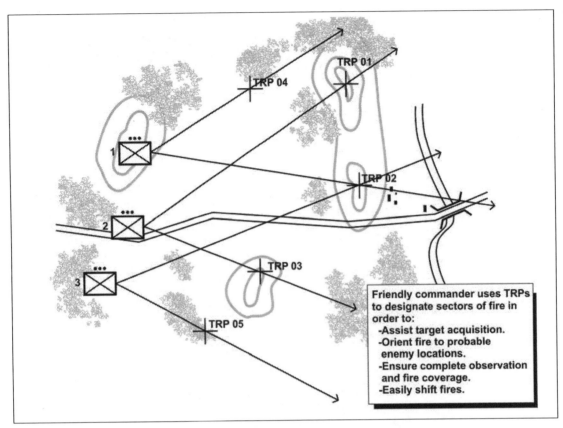

Figure 9-3. Orientation of forces to speed target acquisition.

SHIFT FIRES TO REFOCUS AND REDISTRIBUTE

9-12. As the engagement proceeds, leaders shift direct fires to refocus and redistribute the effects based on evolving friendly and enemy information. Figure 9-4 provides an example of shifting to refocus and redistribute fires. The Infantry company commander and his subordinate leaders apply the same techniques and considerations that they used earlier to focus and distribute fires, including fire control measures. A variety of situations dictate shifting of fires, including--

- Appearance of an enemy force posing a greater threat than the one currently being engaged.
- Extensive destruction of the enemy force being engaged, creating the possibility of target overkill.
- Destruction of friendly elements that are engaging the enemy force.
- Change in the ammunition status of friendly elements that are engaging the enemy force.
- Maneuver of enemy or friendly forces resulting in terrain masking.
- Increased fratricide risk as a maneuvering friendly element closes with the enemy force being engaged.

Direct Fire Control

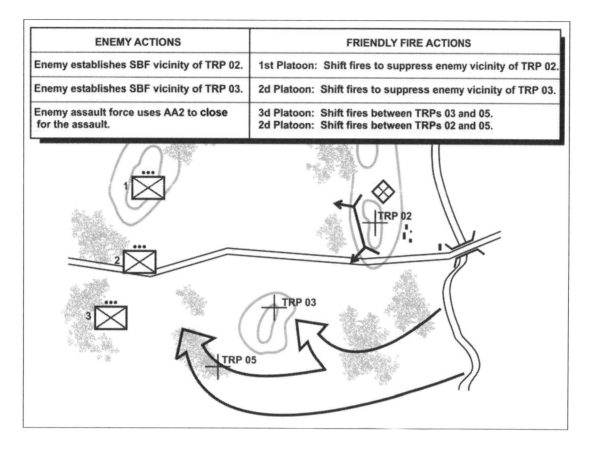

Figure 9-4. Shifting of fires to refocus and redistribute them.

Section III. PLANNING CONSIDERATIONS

The Infantry company commander plans direct fires as part of the troop-leading procedures. Determining where and how the company can and will mass fires are essential steps as the commander develops his concept of the operation.

OVERVIEW

9-13. After identifying probable (or known) enemy locations, the Infantry commander determines points or areas where he will focus his combat power. His situational understanding (SU), or vision, of where and how the enemy will attack or defend helps him determine the volume of fires he must focus at particular points to have a decisive effect. In addition, if he intends to mass the direct fires of more than one platoon, he must establish a means for distributing those fires effectively.

- Based on where and how he wants to focus and distribute direct fires, the commander can establish the weapons ready postures for company elements as well as triggers for initiating fires. He must evaluate the risk of fratricide and establish controls to prevent it. Fratricide prevention measures include designation of recognition markings, weapons control status (WCS), and weapons safety posture.
- Having determined where and how he will mass and distribute direct fires, the company commander orients platoons so they can rapidly and accurately acquire the enemy. The commander anticipates how the enemy will fight. He gains this anticipation through a detailed war-game of the selected course of action. With this war game, he determines probable

requirements for refocusing and redistributing fires and for establishing other necessary controls. Also during the troop-leading procedures, the company commander plans and rehearses direct fires (and the fire-control process) based on his analysis.

- The company commander continues to apply planning procedures and considerations throughout execution. When necessary, he must also apply effective direct fire SOPs.

STANDING OPERATING PROCEDURES

9-14. A well-rehearsed direct fire SOP enhances direct fire planning and ensures quick, predictable actions by all members of the company. The Infantry company commander bases the various elements of the SOP on the capabilities of his force and on anticipated conditions and situations. SOP elements should include standard means for focusing fires, distributing their effects, orienting forces, and preventing fratricide. The commander should adjust the direct fire SOP whenever changes to anticipated and actual factors of METT-TC become apparent.

FOCUS FIRES

9-15. One technique is to establish a standard respective position for TRPs in relation to friendly elements and then to consistently number the TRPs such as from left to right. This allows leaders to quickly determine and communicate the location of the TRPs.

DISTRIBUTE FIRES

9-16. Two useful means of distributing the effects of the company's direct fires are engagement priorities and target array. Engagement priorities, by type of enemy vehicle or weapon, are assigned for each type of friendly weapon system. The target array technique helps in distribution by assigning specific friendly elements to engage enemy elements of approximately similar capabilities.

ORIENT FORCES

9-17. A standard means of orienting friendly forces is to assign a primary direction of fire, using a TRP, to orient each element on a probable (or known) enemy position or likely avenue of approach. To provide all-round security, the SOP can supplement the primary direction of fire with sectors using a friendly based quadrant. The following sample SOP elements show the use of these techniques.

- The front (center) platoon's primary direction of fire is TRP 2 (center) until otherwise specified; the platoon is responsible for the front two quadrants.
- The left flank platoon's primary direction of fire is TRP 1 (left) until otherwise specified; the platoon is responsible for the left two friendly quadrants (overlapping with the center platoon).
- The right flank platoon's primary direction of fire is TRP 3 (right) until otherwise specified; the platoon is responsible for the right two friendly quadrants (overlapping with the center platoon).

PREVENT FRATRICIDE

9-18. The SOP must address the most critical requirement of fratricide prevention. It must direct subordinate leaders to inform the commander, adjacent elements, and subordinates whenever a friendly force is moving or preparing to move. One technique is to establish a standing WCS of WEAPONS TIGHT, which requires positive enemy identification prior to engagement. The SOP must also cover means for identifying dismounted Infantry squads and other friendly dismounted elements. Techniques include using arm bands, medical heat pads, or an IR light source, as well as detonating a smoke grenade of a designated color at the appropriate time.

Direct Fire Control

Section IV. CONTROL

Acquiring the enemy is a precursor to direct fire engagement. He must expect the enemy to use covered and concealed routes effectively when attacking and to make best use of flanking and concealed positions in the defense. As a result, the company may not have the luxury of a fully exposed enemy that it can easily see. The acquisition of the enemy often depends on visual recognition of very subtle indicators such as exposed antennas, reflections from the vision blocks of enemy vehicles, small dust clouds, or smoke from vehicle engines or ATGM or tank fires. Because of the difficulty of target acquisition, the company commander must develop surveillance plans to assist the company in acquiring the enemy.

MEASURES

9-19. Fire control measures are the means by which the Infantry company commander or his subordinate leaders control direct fires. Application of these concepts, procedures, and techniques helps the unit acquire the enemy, focus fires on him, distribute the effects of the fires, and prevent fratricide. At the same time, no single measure is enough to control fires effectively. At company level, fire control measures are effective only if the entire unit has a common understanding of what they mean and how to employ them. Table 9-1 lists terrain-based and threat-based fire control measures.

Terrain-Based Fire-Control Measures	Threat-Based Fire-Control Measures
Target reference point Engagement area Sector of fire Direction of fire Terrain-based quadrant Friendly based quadrant Maximum engagement line Restrictive fire line Final protective line	Fire patterns Target array Engagement priorities Weapons ready posture Engagement criteria Weapons control status Rules of engagement Weapons safety posture Engagement techniques

Table 9-1. Common fire-control measures.

TERRAIN-BASED FIRE-CONTROL MEASURES

9-20. The Infantry company commander uses terrain-based fire-control measures to focus and control fires on a particular point, line, or area rather than on a specific enemy element. The following paragraphs describe the techniques associated with this type of control measure.

Target Reference Point

9-21. A target reference point is an easily recognizable point on the ground that leaders use to orient friendly forces and to focus and control direct fires. In addition, when TRPs are designated as indirect fire targets, they can be used in calling for and adjusting indirect fires. Leaders designate TRPs at probable (or known) enemy locations and along likely avenues of approach. TRPs, natural or manmade, can be established sites such as hills or buildings. They can also be expedient, temporary features designated as TRPs on the spot such as a burning enemy vehicle or smoke generated by an artillery round. While not ideal, TRPs can also be made by the unit with items such as engineer pickets with visible or IR chem-lights or IR strobe in the recessed grove of the picket. Ideally, TRPs should be permanent features and visible in three observation modes (unaided, passive-IR, and thermal) so all forces can identify them, for example---

- Prominent hill mass.
- Distinctive building.
- Observable enemy position.

- Destroyed vehicle.
- Ground-burst illumination.
- Smoke round.
- Laser point.

Engagement Area

9-22. This fire control measure is an area along an enemy avenue of approach where the company commander intends to mass the direct fires of available weapons to destroy an enemy force. The size and shape of the EA is determined by the degree of relatively unobstructed intervisibility available to the unit's weapons systems in their firing positions and by the maximum range of those weapons. Typically, company commanders delineate responsibility within the EA by assigning each platoon a sector of fire.

Sector of Fire

9-23. A sector of fire is a defined area that must be covered by direct fire. It is used to distribute fires within an EA. Leaders assign sectors of fire to subordinate elements, crew-served weapons, and individual Soldiers to ensure coverage of an area of responsibility. They may also limit the sector of fire of an element or weapon to prevent accidental engagement of an adjacent unit. In assigning sectors of fire, commanders and subordinate leaders consider the number and types of weapons available. They also consider acquisition system type and field of view in determining the width of a sector of fire. For example, while unaided vision has a wide field of view, its ability to detect and identify targets at distant ranges and in limited visibility conditions is restricted. Conversely, most fire control acquisition systems have greater detection and identification ranges than the unaided eye, but their field of view is narrow. Means of designating sectors of fire include--

- Target reference points.
- Clock direction.
- Terrain-based quadrants.
- Friendly based quadrants.

Direction of Fire

9-24. A direction of fire is an orientation or point used to assign responsibility for a particular area on the battlefield that must be covered by direct fire. Leaders designate directions of fire for the purpose of acquisition or engagement by crew-served weapons, or individual Soldiers. Direction of fire is most commonly employed when assigning sectors of fire would be difficult or impossible because of limited time or insufficient reference points. Means of designating a direction of fire include--

- Closest target reference point.
- Clock direction.
- Cardinal direction.
- Tracer on target.
- Infrared laser pointer.

Quadrants

9-25. Quadrants are subdivisions of an area created by superimposing imaginary perpendicular axes over the terrain to create four separate areas, or quadrants. Quadrants can be based on the terrain, on friendly forces, or on the enemy formation. The technique in which quadrants are based on the enemy formation is usually referred to as the target array and is covered in threat-based fire control measures. The method of identifying quadrants is established in the unit SOP, but the unit must take care to avoid confusion when using quadrants based on terrain, friendly forces, and enemy formations simultaneously.

Terrain-Based Quadrant

9-26. A terrain-based quadrant entails use of a TRP, either existing or constructed, to designate the center point of the axes that divide the area into four quadrants. This technique can be employed in both offensive and defensive operations. In the offense, the company commander designates the center of the quadrant using an existing feature or by creating a reference point such as using a ground burst illumination round, a smoke marking round, or a fire ignited by incendiary or tracer rounds. The axes delineating the quadrants run parallel and perpendicular to the direction of movement. In the defense, the company commander designates the center of the quadrant using an existing or constructed TRP. In Figure 9-5, the quadrants are marked using the letter "Q" and a number (Q1 to Q4); quadrant numbers are in the same relative positions as on military map sheets (from Q1 as the upper left quadrant clockwise to Q4 as the lower left quadrant).

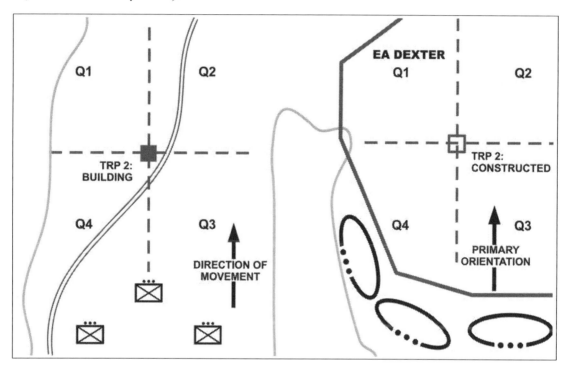

Figure 9-5. Terrain-based quadrants.

Friendly Based Quadrant

9-27. The friendly based quadrant technique entails superimposing quadrants over the unit's formation. The center point is based on the center of the formation, and the axes run parallel and perpendicular to the general direction of travel. For rapid orientation, the friendly quadrant technique might be better than the clock direction method because different elements of a large formation are rarely oriented in the same exact direction and because the relative dispersion of friendly forces causes parallax to the target. Figure 9-6, page 9-12, shows use of friendly based quadrants.

Chapter 9

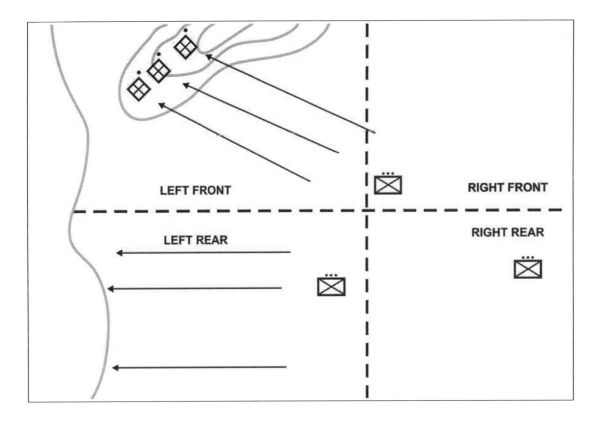

Figure 9-6. Friendly based quadrants.

Maximum Engagement Line

9-28. A MEL is the linear image of the farthest limit of effective fire for a weapon or unit. This line is determined both by the weapon's or unit's maximum effective range, and by the effects of terrain. For example, slope, vegetation, structures, and other features provide cover and concealment that may prevent the weapon from engaging out to the maximum effective range. A MEL serves several purposes. The company commander may use it to prevent engaging beyond the maximum effective range of their weapons, to define criteria for the establishment of triggers, and to delineate the maximum extent of battle space on the sector sketch.

Restrictive Fire Line

9-29. An RFL is a line established between converging friendly forces (one or both might be moving) that prohibit fires and effects across the line without coordination with the affected force. In the offense, the company commander may designate an RFL to prevent a base of fire platoon from firing into the area where an assaulting platoon is maneuvering. In the defense, the company commander may establish an RFL to prevent the unit from engaging an Infantry squad positioned in restricted terrain on the flank of an enemy avenue of approach.

Final Protective Line

9-30. The FPL is a line of fire established where an enemy assault is to be halted by the interlocking fires of all available weapons. The unit reinforces this line with protective, tactical, and supplemental obstacles and with FPFs whenever possible. Initiation of the FPF is the signal for elements and individual Soldiers to shift fires to their assigned portion of the FPL.

Direct Fire Control

THREAT-BASED FIRE-CONTROL MEASURES

9-31. The Infantry company commander uses threat-based fire control measures to focus and control direct fires by directing the unit to engage a specific enemy element rather than to fire on a point or area. The following paragraphs describe the techniques associated with this type of fire control measure.

Fire Patterns

9-32. Fire patterns are a threat-based fire control measure designed to distribute the fires of a unit simultaneously among multiple, similar targets. They are most often used by platoons to distribute fires across an enemy formation. Leaders designate and adjust fire patterns based on terrain and the enemy formation. The basic fire patterns are frontal fire, cross fire, and depth fire.

Frontal Fire

9-33. Leaders may initiate frontal fire (Figure 9-7) when targets are arrayed in front of the unit in a lateral configuration. Weapons systems engage targets to their respective fronts. For example, the left flank weapon engages the left-most target, and the right flank weapon engages the right-most target. As the unit destroys targets, weapons shift fires toward the center of the enemy formation and from near too far.

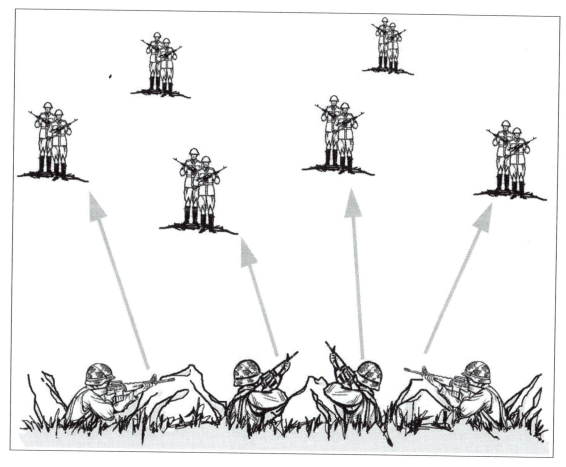

Figure 9-7. Frontal fire.

Chapter 9

Cross Fire

9-34. Leaders initiate cross fire (Figure 9-8) when targets are arrayed laterally across the unit's front in a manner that permits diagonal fires at the enemy's flank or when obstructions prevent unit weapons from firing frontally. Right flank weapons engage the left-most targets, and left flank weapons engage the right-most targets. Firing diagonally across an EA provides more flank shots, thus increasing the chance of kills. It also reduces the possibility of the enemy detecting friendly elements. As the unit destroys targets, weapons shift fires toward the center of the enemy formation.

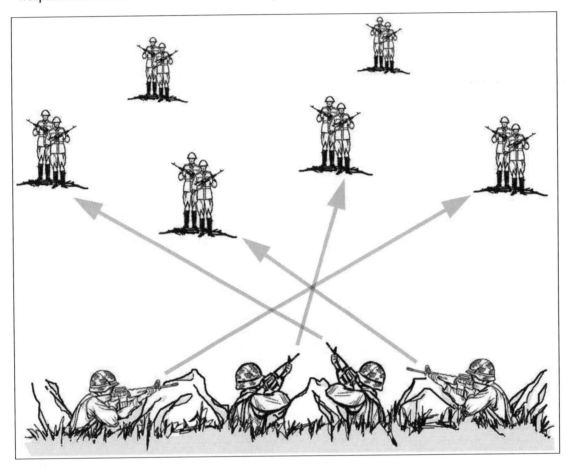

Figure 9-8. Cross fire.

Depth Fire

9-35. Leaders initiate depth fire (Figure 9-9) when targets are dispersed in depth, perpendicular to the unit. Center weapons engage the closest targets, and flank weapons engage deeper targets. As the unit destroys targets, weapons shift fires toward the center of the enemy formation.

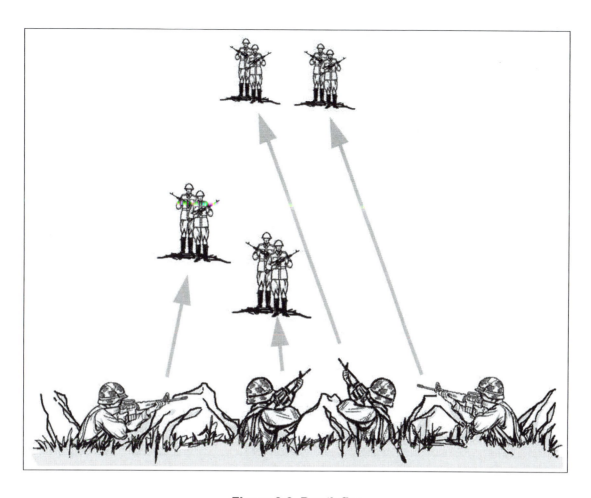

Figure 9-9. Depth fire.

Target Array

9-36. Target array permits the company commander to distribute fires when the enemy force is concentrated and terrain-based controls are inadequate. This threat-based distribution measure is similar to the quadrant method mentioned in terrain-based fire control measures. The company commander creates the target array by superimposing a quadrant pattern over an enemy formation. The pattern centers on the enemy formation, with the axes running parallel and perpendicular to the enemy's direction of travel. Quadrants are described using their relative locations. Figure 9-10, page 9-16 shows examples of the target array technique.

Chapter 9

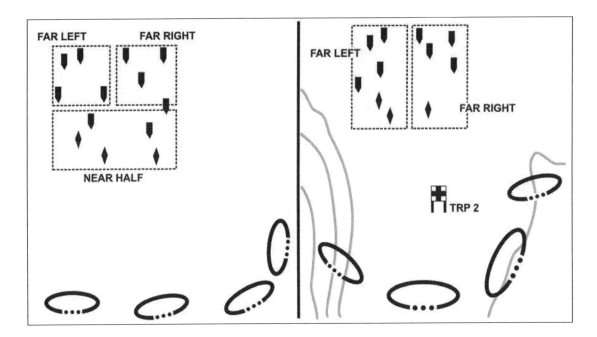

Figure 9-10. Example target arrays.

Engagement Priorities

9-37. Engagement priorities entail the sequential ordering of targets to be engaged. They serve one or more of the following critical fire control functions.

Prioritize Targets

9-38. In concert with his concept of the operation, the company commander determines which target types provide the greatest threat to the company and sets these as engagement priorities. For example, he may decide that destroying enemy engineer assets is the best way to prevent the enemy from breaching an obstacle.

Employ Best Weapons for Target

9-39. Establishing engagement priorities for specific friendly systems increases the effectiveness with which the unit employs its weapons. As an example, the engagement priority for the Javelin could be enemy fortifications first, then enemy armored vehicles.

Distribute Unit's Fires

9-40. Establishing different priorities for similar friendly systems helps to prevent overkill and achieve effective distribution of fires. For example, the company commander may designate the enemy fortifications as the initial priority for one Infantry platoon while making the enemy vehicles the priority for another Infantry platoon. This decreases the chance of multiple Javelins being fired against two enemy vehicles while ignoring the dangers posed by the fortifications.

Weapons Ready Posture

9-41. The weapons ready posture is a means by which leaders use the tactical information available to specify the ammunition and range for the most probable engagement. Ammunition selection depends on the target type, but the leader may adjust it based on engagement priorities, desired effects, and effective range. Range selection depends on the anticipated engagement range, and it is affected by terrain intervisibility, weather, and light conditions. Within the company, the weapons ready posture affects the types and quantities of ammunition and where they are located or cached.

9-42. For Infantry squads, weapons ready posture is the selected ammunition and indexed range for individual and crew-served weapons. For example, an M203 grenadier whose most likely engagement is to cover dead space at 200 meters from his position might load high-explosive, dual-purpose (HEDP) rounds and set 200 meters on his quadrant sight. To prepare for an engagement in a wooded area where engagement ranges are extremely short, an AT4 might be more appropriate than a Javelin.

Engagement Criteria

9-43. Engagement criteria are a specific set of conditions that specify the circumstances in which subordinate elements are to engage. This is often referred to as a trigger. The circumstances can be based on a friendly or an enemy event. For example, the engagement criteria for a friendly platoon to initiate engagement could be three or more enemy combat vehicles passing or crossing a given point or line. This line can be any natural or manmade linear feature, such as a road, ridgeline, or stream, or a line perpendicular to the unit's orientation, delineated by one or more reference points.

Weapons Control Status

9-44. The three levels of weapons control status (WCS) outline the conditions, based on target identification criteria, under which friendly elements may engage. The commander sets and adjusts the weapons control status based on friendly and enemy disposition. In general, a more restrictive WCS relates to a higher probability of fratricide. The three levels, in descending order of restriction, are--

> WEAPONS HOLD (Engage only if engaged or ordered to engage)
> WEAPONS TIGHT (Engage only targets positively identified as enemy)
> WEAPONS FREE (Engage any targets not positively identified as friendly)

Note: As an example, the company commander may establish the WCS as WEAPONS HOLD when friendly forces are conducting a passage of lines. As the passage progresses and the likelihood of enemy presence rises, the WCS might be lowered. In such a case, the company commander might be able to set a WEAPONS FREE status when he knows there are no friendly elements near the passage. This permits his elements to engage targets at extended ranges, even when distinguishing targets accurately is difficult under battlefield conditions. The WCS is extremely important for forces using combat identification systems: establishing the WCS as WEAPONS FREE permits leaders to engage an unknown target when they fail to get a friendly response.

Rules of Engagement

9-45. The ROE specify the circumstances and limitations under which forces may engage. ROE include definitions of combatant and noncombatant elements and prescribe the treatment of noncombatants. Factors influencing ROE are national command policy, the mission and commander's intent, the operational environment, and the law of war. ROE always recognize a Soldier's right of self-defense; at the same time, they clearly define circumstances in which he may fire.

Chapter 9

Weapons Safety Posture

9-46. Weapons safety posture is an ammunition handling instruction that allows the company commander to control the safety of his unit's weapons precisely. Leaders supervise the weapons safety posture and Soldier adherence to it, minimizing the risk of accidental discharge and fratricide. Table 9-2 shows the company's procedures and considerations in using the four weapons safety postures, which are listed in ascending order of restriction: AMMUNITION LOADED, AMMUNITION LOCKED, AMMUNITION PREPARED, and WEAPONS CLEARED.

Note: In setting and adjusting the weapons safety posture, the company commander must weigh the need to prevent accidental discharges against the requirement for immediate action based on the enemy threat. If the possibility of direct contact with the enemy is high, the company commander may establish the weapons safety posture as AMMUNITION LOADED. If the requirement for action is less immediate, he may lower the posture to AMMUNITION LOCKED or AMMUNITION PREPARED. Also, the company commander may designate different weapons safety postures for different elements of the unit.

Weapons Safety Postures	Infantry Squad Weapons and Ammunition
Ammunition loaded	• M4 rounds chambered. • M240B and M249 ammunition on feed tray; bolt locked to rear. • M203 launcher loaded. • Weapons on safe.
Ammunition locked	• Magazines locked into M4s. • M240B and M249 ammunition on feed tray; bolt locked forward. • M203 launcher unloaded.
Ammunition prepared	• Magazines, ammunition boxes, launcher grenades, and hand grenades prepared but stowed in pouches or vests.
Weapons cleared	• Magazines, ammunition boxes, and launcher grenades removed; all weapons cleared.

Table 9-2. Weapons safety posture levels.

Engagement Techniques

9-47. Engagement techniques are effects-oriented direct fire distribution measures. The following engagement techniques are the most common in Infantry company operations.

Point Fire

9-48. Point fire concentrates the effects of a unit's fire against a specific, identified target such as a vehicle, machine gun bunker, or ATGM position. When leaders direct point fire, all of the unit's weapons engage the target, firing until it is destroyed or the required time of suppression has expired. Employing converging fires from dispersed positions makes point fire more effective because the unit engages the target from multiple directions. The unit may initiate an engagement using point fire against the most dangerous threat, then revert to area fire against other, less threatening point targets. (Use of point fire has been rare because a unit seldom encounters a single, clearly identified enemy weapon.)

Area Fire

9-49. Area fire involves distributing the effects of a unit's direct fires over an area in which enemy positions are numerous or are not obvious. If the area is large, leaders assign sectors of fire to subordinate elements using a terrain-based distribution method such as the quadrant technique. Typically, the primary purpose of the area fire is suppression; however, sustaining effective suppression requires judicious control of the rate of fire.

Alternating Fire

9-50. In alternating fire, pairs of elements continuously engage the same point or area target one at a time. For example, a company team may alternate fires of two platoons; a tank platoon may alternate the fires of its sections, or an Infantry platoon may alternate the fires of a pair of machine guns. Alternating fire permits the unit to maintain suppression for a longer duration than simultaneous fire. It also forces the enemy to acquire and engage alternating points of fire.

Sequential Fire

9-51. In sequential fire, the subordinate elements of a unit engage the same point or area target one after another in an arranged sequence. Sequential fire can also help preserve ammunition, as when an Infantry platoon waits to see the effects of the first Javelin before firing another. Also, sequential fire permits elements that have already fired to pass on information they have learned from the engagement. An example would be an Infantryman who missed an armored vehicle with AT4 fires passing range and lead information to the next Soldier preparing to engage the same armored vehicle with an AT4.

Simultaneous Fire

9-52. Units employ simultaneous fire, also referred to as volley fire, to mass the effects of their fires rapidly or to gain immediate fire superiority. For example, a unit may initiate a support-by-fire operation with simultaneous fire, then change to alternating or sequential fire to maintain suppression. Simultaneous fire is also employed to negate the low probability of hit and kill of certain antiarmor weapons. As an example, an Infantry squad may employ volley fire with its AT4s to ensure rapid destruction of a BMP that is engaging a friendly position.

Observed Fire

9-53. Observed fire is normally used when the company is in concealed defensive positions with extended engagement ranges. It can be employed between elements of the company, such as an Infantry platoon observing while the machine gun section fires; or it can be employed between machine guns in the section. The company commander or platoon leader directs one element to engage. The remaining elements or vehicles observe fires and prepare to engage on order in case the engaging element consistently misses its targets, experiences a malfunction, or runs low on ammunition. Observed fire allows for mutual observation and assistance while protecting the location of the observing elements.

Time of Suppression

9-54. Time of suppression is the period, specified by the company commander, during which an enemy position or force is to be suppressed. Suppression time typically depends on the time it will take a supported element to maneuver. Normally, a unit suppresses an enemy position using the sustained rate of fire of its weapons. In planning for sustained suppression, leaders must consider several factors: the estimated time of suppression, the size of the area being suppressed, the type of enemy force to be suppressed, range to the target, rates of fire, and available ammunition quantities.

Reconnaissance by Fire

9-55. Reconnaissance by fire is the process of engaging possible enemy locations to elicit a tactical response such as return fire or movement. This response permits the company commander and his subordinate leaders to make accurate target acquisition and then to mass fires against the enemy element. Typically, the company commander directs a platoon to conduct the reconnaissance by fire. For example, he may direct an overwatching platoon to conduct the reconnaissance by fire against a probable enemy position before initiating movement by a bounding element.

COMMANDS

9-56. Fire commands are oral orders issued by the Infantry company commander and his subordinate leaders to focus and distribute fires as required to achieve the desired effects against an enemy force. Fire commands allow leaders, in the already confusing environment of close combat, to articulate their firing instructions rapidly and concisely using a standard format. Unit fire commands include these elements:

ALERT

9-57. The alert specifies the units that are directed to fire. It does not require the leader who initiates the command to identify himself. Examples of the alert element (call signs and code words based on unit SOP) include--

> *GUIDONS (all subordinate elements)*
> *RED (1st platoon only)*

WEAPON OR AMMUNITION (OPTIONAL)

9-58. This element identifies the weapon and ammunition to be employed by the alerted units. Leaders may designate the type and number of rounds to limit expenditure of ammunition. Examples of this element include--

> *JAVELIN*
> *MACHINE GUN*

TARGET DESCRIPTION

9-59. Target description designates which enemy forces are to be engaged. Leaders may use the description to focus fires or achieve distribution. Example target descriptions include--

> *TROOPS IN TRENCH*
> *BUNKER*
> *PCs*

ORIENTATION

9-60. This element identifies the location of the target. The location of the target might be designated in any of several ways, for example--

> *Closest TRP   TRP 13*
> *Clock direction   ONE O'CLOCK*
> *Terrain quadrant   QUADRANT ONE*
> *Friendly quadrant LEFT FRONT*
> *Target array   FRONT HALF*
> *Tracer on target   ON MY TRACER*
> *Laser pointer   ON MY POINTER*

RANGE (OPTIONAL)

9-61. The range element identifies the distance to the target. Announcing range is not necessary for systems that have range finders or that employ command-guided or self-guided munitions. For systems that require manual range settings, leaders have a variety of means for determining range, including--

- Predetermined ranges to TRPs or phase lines.
- Handheld rangefinders.
- Range stadia.
- Mil reticles.

CONTROL (OPTIONAL)

9-62. The company commander may use this optional element to direct desired target effects, distribution methods, or engagement techniques. Subordinate leaders may include the control element to supplement the company commander's instructions and achieve effective distribution. Examples of information specified in the control element include--

Target array   FRONT HALF
Fire pattern   FRONTAL
Terrain quadrant   QUADRANT ONE
Engagement priorities   M203 ENGAGE BUNKERS; MACHINE GUNS   ENGAGE TROOPS
Engagement technique ALTERNATING
Target effect   AREA

EXECUTION

9-63. The execution element specifies when direct fires should be initiated. The company commander may engage immediately, delay initiation, or delegate authority to engage. Examples of this element include--

FIRE
AT MY COMMAND
AT YOUR COMMAND
AT PHASE LINE ORANGE

Chapter 10
Maneuver Support

For a unit to achieve its full combat potential, the commander must effectively integrate all available combat arms and combat support assets. This chapter focuses on the elements that the Infantry company is most likely to work with such as fire support, engineers, air defense, and CBRN. It also briefly describes available support from Army aviation units.

This edition expands the discussion of aerial fire support to the rifle company, including close air support (CAS), and joint tactical air control (JTAC) procedures. It provides checklists such as the CAS 9-line briefing.

Section I. COMMAND AND SUPPORT RELATIONSHIPS

The Infantry company commander must understand the command and support relationships established between his company and supporting units.

COMMAND RELATIONSHIPS

10-1. Command responsibility and authority are established routinely through the following standard relationships, also shown in Figure 10-1, page 10-3. To simplify these definitions, the gaining organization is referred to as "*the organization*" and the organic, assigned, attached, OPCON, or TACON unit simply as "*the unit*." The term "*parent unit*" is self-explanatory.

ORGANIC

10-2. This unit is assigned to and forms an essential part of an Army organization. It (the unit) is listed in the organization's table of organization and equipment (TOE) or table of distribution and allowances (TDA).

ASSIGNED

10-3. This is a unit permanently placed in an organization and completely controlled and administered by the organization.

ATTACHED

10-4. This is a unit temporarily assigned to an organization other than its parent unit. The organization's commander has the same command and control over the attached unit as he does with his organic units. This control is limited only as caveated in the attachment order. The organization is responsible for the unit's logistics, training, and operations. The parent unit normally retains responsibility for transfers and promotions; nonjudicial punishment; administrative actions such as SIDPERS and unit strength reporting; and uniform code of military justice actions. Having an attached unit increases an organization's administrative and logistical burden.

Operational Control

10-5. An OPCON unit is one placed under the control of an organization for one or more specific operations. The organization controls *all* aspects of the unit's military operations. The relationship is limited by function, time, or location. The organization is not necessarily responsible for administration, logistics, discipline, internal organization, or training. Otherwise, the commander's relationship with OPCON units is the same as with organic or attached subordinate units.

Tactical Control

10-6. A TACON unit is normally only subject to the authority of the organization with regard to specified local direction of movement and maneuver of forces to accomplish a task. In this command relationship, a combatant commander delegates limited authority to the organization to direct the tactical use of a combat unit. TACON is often the command relationship established between forces of different nations in a multinational force.

SUPPORT RELATIONSHIPS

10-7. Support is the action of an element or unit that aids, protects, complements, or sustains another unit IAW an order requiring such support. A supporting unit helps another unit but is not under the command of that unit. The commander's relationship with supporting units follows (Figure 10-1).

- He ensures that the supporting unit establishes liaison and communications with his unit.
- He keeps the supporting unit informed of the situation and the support needed.
- He communicates with the leader of the supporting unit on employment considerations.
- His request for support is honored by the supporting unit as an order. The supporting unit leader refers any conflicts or problems to his parent unit commander. However, the request in question is honored until the conflict is resolved.

Direct Support

10-8. A unit in direct support (DS) of another organization remains under the command of its parent unit. However, although the supporting unit answers the supported unit's requests directly, the organization's commander may not reallocate, reassign, or task-organize the DS force.

General Support

10-9. A unit in general support (GS) to another organization remains under the control of the parent unit. A GS unit supports the organization as a whole, not any specific subunit. Therefore, subunit commanders must request support from the GS unit through their own parent unit.

Reinforcing Support

10-10. Reinforcing support is tactical (fire) augmentation of one artillery unit by another in overall support of the maneuver force. Coordination for support is normally provided through the DS artillery unit.

General Support Reinforcing

10-11. General support-reinforcing (GS-R) artillery is a tactical artillery mission where an artillery unit has the mission of supporting the force as a whole. It also provides reinforcing fires for another artillery unit. Coordination for support is the responsibility of the DS artillery unit.

Maneuver Support

IF RELATIONSHIP IS:		INHERENT RESPONSIBILITIES ARE:						
		Has Command Relationship with:	May Be Task Organized by:	Receives CSS from:	Assigned Position or AO By:	Provides Liaison To:	Establishes/ Maintains Communications with:	Has Priorities Established by:	Gaining Unit Can Impose Further Command or Support Relationship of:
COMMAND	Attached	Gaining unit	Gaining unit	Gaining unit	Gaining unit	As required by gaining unit	Unit to which attached	Gaining unit	Attached; OPCON; TACON; GS; GSR; R; DS
	OPCON	Gaining unit	Parent unit and gaining unit; gaining unit may pass OPCON to lower HQ. Note 1	Parent unit	Gaining unit	As required by gaining unit	As required by gaining unit and parent unit	Gaining unit	OPCON; TACON; GS; GSR; R; DS
	TACON	Gaining unit	Parent unit	Parent unit	Gaining unit	As required by gaining unit	As required by gaining unit and parent unit	Gaining unit	GS; GSR; R; DS
	Assigned	Parent unit	Parent unit	Parent unit	Gaining unit	As required by parent unit	As required by parent unit	Parent unit	Not Applicable
SUPPORT	Direct Support (DS)	Parent unit	Parent unit	Parent unit	Supported unit	Supported unit	Parent unit; Supported unit	Supported unit	Note 2
	Reinforcing (R)	Parent unit	Parent unit	Parent unit	Reinforced unit	Reinforced unit	Parent unit; reinforced unit	Reinforced unit; then parent unit	Not Applicable
	General Support Reinforcing (GSR)	Parent unit	Parent unit	Parent unit	Parent unit	Reinforced unit and as required by parent unit	Reinforced unit and as required by parent unit	Parent unit; then reinforced unit	Not Applicable
	General Support (GS)	Parent unit	Parent unit	Parent unit	Parent unit	As required by parent unit	As required by parent unit	Parent unit	Not Applicable

NOTE 1. In NATO, the gaining unit may not task organize a multinational unit (see TACON).
NOTE 2. Commanders of units in DS may further assign support relationships between their subordinate units and elements of the supported unit after coordination with the supported commander.

Figure 10-1. Command and support relationships.

Section II. FIRE SUPPORT

Fire support is the collective and coordinated use of indirect fire weapons and armed aircraft in support of the battle plan. Fire support assets include mortars, field artillery cannons and rockets, and CAS. Desired effects from fire support assets can be achieved through a combination of both lethal and nonlethal means. The integration of fire support assets is critical to the success of the company. The Infantry battalion FSO plans fires (in coordination with the plans developed by the Infantry battalion S-3) to support the Infantry battalion commander's concept of the operation. The company FSO plans company fire support, and the Infantry company commander approves his plan. Fire support planning is the process of analyzing, allocating, and scheduling fire support assets.

Chapter 10

INDIRECT FIRE CAPABILITIES

10-12. Indirect fires affect an enemy force much more than do the fires of the Infantry company's organic weapons. Indirect fire assets offer the most destructive, accurate, and flexible combat multiplier available to the Infantry company commander in immediate support of his operations. Table 10-1 lists the capabilities of the indirect fire systems that might support the Infantry company.

Capabilities of the Indirect Fire System						
CALIBER	60-mm	81-mm	120-mm	105-mm	155-mm	155-mm
MODEL	M224	M252	M285	M119	M198	M109A6
MAX RANGE (HE)(m)	3,490	5,608	7,200	14,000	24,000	24,000
PLANNING RANGE (m)	(2/3 max)	(2/3 max)	(2/3 max)	11,500	14,600	14,600
PROJECTILE	HE, WP, illum, IR illum	HE, WP, illum, RP, IR illum	HE, SMK, illum, IR illum	HE M760 illum, HEP-T, APICM, chem, RAP	HE, WP, illum, smk, chem, nuc, RAP, FASCAM, CPHD, AP/DPICM	HE, WP, illum, smk, chem, nuc, RAP, FASCAM, CPHD, AP/DPICM
MAX RATE OF FIRE	30 RPM for 4 min	30 RPM for 2 min	16 RPM for 1 min	6 RPM for 3 min	4 RPM for 3 min	4 RPM for 3 min
SUSTAINED RATE OF FIRE (rd/min)	20	15	4	3	2	1
MINIMUM RANGE (m)	70	83	200	Direct fire	Direct fire	Direct fire
FUZES	PD, VT, time, dly, MO	PD, VT, time, dly, MO	PD, VT, time, dly, MO	PD, VT, MTSQ, CP, MT, dly	PD, VT, CP, MT, MTSQ, dly	PD, VT, CP, MT, MTSQ, dly

LEGEND

AP	armor piercing	MO	multioption (VT, PD, dly)
APICM	antipersonnel improved conventional munitions	MT	mechanical time
chem	chemical	MTSQ	mechanical time super quick
CP	concrete piercing	nuc	nuclear
CPHD	Copperhead	PD	point detonating
dly	delay	RAP	rocket assisted projectile
DPICM	dual purpose improved conventional munitions	RD	round
FASCAM	family of scatterable mines	RP	red phosphorus
HE	high explosive	RPM	rounds per minute
HEP-T	high explosive plastic--tracer	smk	smoke
illum	illumination	time	adjustable time delay
IR	infrared	VT	variable time
min	minute	WP	white phosphorus

Table 10-1. Indirect fire capabilities.

FIRE-SUPPORT TEAM

10-13. For artillery and mortar support, the FIST personnel are the observers or "*eyes*" for the maneuver company. Each FIST has a four-Soldier headquarters consisting of an FSO, a fire support sergeant, a fire support specialist, and a radio operator. Normally the commander sends FO teams to the platoons. An FO team has an FO and a radio operator. The mission of the fire support team is to provide fire support for the supported maneuver company. To do this the team must provide fire support planning, fire support coordination, target location, calls for indirect fire, battlefield information reporting, and emergency control of CAS.

FIRE-SUPPORT PLANS AND COORDINATION

10-14. At all levels, leaders plan fire support and maneuver concurrently. Infantry battalions typically plan fire support from the top down, and refine plans from the bottom up. The commander develops guidance for fire support tasks, purposes, and effects. The fire support planner determines the method for accomplishing each task. Individual units then incorporate assigned tasks into their fire support plans. In addition, units tasked to initiate fires refine and rehearse their assigned tasks. The company commander refines his unit's assigned portion of the battalion fire support plan, ensuring that the designated targets will achieve the intended purpose. He also conducts rehearsals to prepare for the mission and, as specified in the plan, directs the company to execute its assigned targets.

TERMS AND DEFINITIONS

10-15. Leaders must understand basic fire support terms to effectively plan and employ fire support assets.

Fire support plans

10-16. Fire support planning is the continual process of analyzing, allocating, and scheduling fire support. The goal of fire support planning is to effectively integrate fire support into battle plans to optimize combat power. It is performed as part of the TLP.

Fire support Coordination

10-17. Fire support coordination is the continual process of implementing fire support plans and managing the fire support assets that are available to a maneuver force.

Fire Plans

10-18. Fire planning is the continual process of selecting and prearranging fires on particular targets to support a phase of the commander's plan.

Essential Fire-Support Task

10-19. An EFST is a task that a fire support element must accomplish in order to support a combined-arms operation. Failure to achieve an EFST may require the commander to alter his tactical or operational plan. A fully developed EFST has a task, purpose, method, and effects (TPME). The *task* describes the targeting objective, for example, delay, disrupt, limit, or destroy. Fires must achieve these on an enemy formation's function or capability. The *purpose* describes why the task contributes to maneuver. The *method* describes how the task is accomplished by assigning responsibility to observers or units and delivery assets and providing amplifying information or restrictions. Typically, the method is described by covering three categories: priority, allocation, and restrictions. The effects statement describes how successful accomplishment of the task will be measured, quantitatively.

Scheme of Fires

10-20. The scheme of fires is the detailed, logical sequence of targets and fire support events the fire support element uses to find and attack high-payoff targets (HPTs). It details how to execute the fire support plan in accordance with the time and space of the battlefield to accomplish the commander's EFSTs. The products of the fire support annex: fire support execution matrix (FSEM), target list or overlay, or a target synchronization matrix (TSM): articulate the scheme of fires.

Linking Tasks and Maneuver Purpose

10-21. A clearly defined purpose and commander's intent enables the maneuver commander to articulate precisely how he wants fire support to affect the enemy during different phases of the battle. This, in turn, allows fire support planners to develop a fire support plan that effectively supports the intended purpose. The planners can determine each required task (in terms of effects on target), the best method for accomplishing each task (in terms of a fire support asset and its fire capabilities), and a means of quantifying accomplishment. A carefully developed method of fire is equally valuable during execution of the fire support mission; it helps not only the firing elements but also the observers who are responsible for monitoring the effects of the indirect fires. With a clear understanding of the intended target effects, fire support assets and observers can work together effectively, planning and adjusting the fires as necessary to achieve the desired effects on the enemy. The following paragraphs describe several types of targeting objectives associated with fire support tasks and provide examples of how the Infantry company commander might link a target task to a specific maneuver purpose in his order.

Delay

10-22. The friendly force uses indirect fires to cause a particular function or action to occur later than the enemy desires. For example, the commander might direct delaying fires this way, "*Delay the repositioning of the enemy's reserve, allowing B Company to consolidate on Objective Bob.*"

Disrupt

10-23. Disrupting fires are employed to break apart the enemy's formation; to interrupt or delay his tempo and operational timetable; to cause premature commitment of his forces; or, otherwise, to force him to stage his attack piecemeal. The commander might direct, for example, "*Disrupt the easternmost lead motorized battalion to prevent the enemy from massing two battalions against Alpha and Charlie.*"

Limit

10-24. Indirect fires help prevent the enemy from executing an action or function where he wants it to occur. The commander might direct, for example, "*Limit the ability of the enemy's advance guard to establish a firing line on the ridge line to the flank of the battalion axis of advance to prevent the enemy from fixing the battalion main body.*"

Destroy

10-25. The friendly force uses indirect fires to render an enemy formation ineffective. For example, the commander might direct destroying fires this way, "*Destroy enemy platoon on Objective Harry in order to allow the decisive operation to assault Objective Tom.*"

Divert

10-26. Diverting fires are used to cause the enemy to modify his course or route of attack. The commander might direct, for example, "*Divert the enemy's combined arms reserve. Counterattack to EA Dog to facilitate its destruction by Delta.*"

Screen

10-27. Screening fires entail the use of smoke to mask friendly installations, positions, or maneuver. Normally, they are conducted for a specified event or a specified period. The commander might direct, for example, "*Screen the movement of the counterattack force (B Company) along Route Red to attack by fire (ABF) position 21 to prevent the remnants of the enemy battalion from engaging the company.*"

Obscure

10-28. Smoke is placed between enemy forces and friendly forces or directly on enemy positions to confuse and disorient the enemy's direct fire gunners and artillery FOs. Obscuration fires are normally conducted for a specified event or a specified period. The commander might direct, for example, "*Obscure the northernmost company to protect our breach force until the breach site is secured.*"

FINAL PROTECTIVE FIRE PLANS

10-29. FPF planning is designed to create a final barrier, or "*steel curtain*", to prevent the enemy from moving across defensive lines. These are fires of last resort and take priority over all other fires. The employment of FPFs presents several potential problems. They are linear fires, with coverage dependent on the firing sheaf of the fire support asset(s). In addition, while an FPF may create a barrier against penetration by enemy Infantry, armored vehicles may simply button up and move through the fires into the friendly defensive position. FPFs are planned targets and thus must have a clearly defined purpose. FPF planning is normally delegated to the Infantry company that is allocated the support.

TARGET REFINEMENT

10-30. The Infantry company commander is responsible for the employment of indirect fires in his zone or sector. The most critical aspect of this responsibility is target refinement, in which he makes necessary changes to the fire support plan to ensure that targets accomplish the commander's intended battlefield purpose. Rather than merely executing targets without regard to the actual enemy situation, the company commander and FSO must be ready to adjust existing targets or to nominate new targets that allow engagement of specific enemy forces.

10-31. Necessary refinements usually emerge when the Infantry company commander war-games as part of step 6 (complete the plan) of the TLP. The war-gaming process allows him to identify required additions, deletions, and adjustments to the Infantry battalion fire support plan. The company FSO then submits the refinements to the battalion FSE for inclusion in the scheme of fires for the operation. (This is normally only the first step of target refinement, with the commander and FSO making further adjustments as the enemy situation becomes clearer.)

10-32. As a specific requirement in defensive planning, the company commander must focus on target refinement for the ground he will "*own*" during the operation. This usually takes place as part of engagement area development. The commander makes appropriate adjustments to the targets based on refinements to the SITEMP such as the actual positions of obstacles and enemy direct fire systems.

10-33. Because fire support is planned from the top down, cutoff times for target nomination and target refinement are normally specified in the battalion OPORD. Commanders must ensure that nominations and refinements meet these deadlines to provide fire support planners with sufficient time to develop execution plans.

FIRE SUPPORT PREPARATION

10-34. As noted, although the Infantry battalion and brigade commanders establish target tasks and purposes and allocate appropriate fire support assets, the Infantry company commander is the one who must ensure execution of assigned targets. In turn, successful execution demands thorough preparation, focusing on areas covered in the following paragraphs.

Observation Plan

10-35. In developing the observation plan, the commander ensures that both primary and alternate observers cover all targets and determine whether the desired target effects have been achieved. The plan provides clear, precise guidance for the observers. Perhaps the most important aspect of the plan is positioning. An observer's positions must allow him to see the trigger for initiating fires as well as the target area and the enemy force on which the target is oriented; this is done to help the observer determine if the target effects have been achieved. The commander must also consider other aspects of observer capabilities, including the Bradley Fire Support Team (BFIST) when operating with HBCT elements or Long Range Advanced Scout Surveillance System (LRAS) when available. As another example, the ground/vehicle laser locator designator (G/VLLD), or similar device, provides first round fire-for-effect capability; without it, observers might have to use adjust-fire techniques that take longer and are more difficult to implement. The observation plan must also include contingency plans that cover limited visibility conditions and backup communications.

Note: In addition to providing the specific guidance outlined in the observation plan, the commander ensures that each observer understands the target task and purpose for which he is responsible. For example, observers must understand that once the first round impacts, the original target location is of no consequence; rather, they must orient on the targeted enemy force to ensure that fires achieve the intended battlefield purpose.

Rehearsals

10-36. The Infantry company commander is responsible for involving his FSO in company-and battalion-level rehearsals, making the company available for any separate fire support rehearsals, and for rehearsing the company's FOs in the execution of targets. He uses rehearsals to ensure that the company's primary and backup communications systems adequately support the plan.

Target Adjustment

10-37. In the defense, the commander confirms target location by adjusting fires as part of engagement area development.

Tactical and Technical Triggers

10-38. The two types of triggers associated with a target are tactical and technical. The company commander develops a tactical trigger for each target and then he or his FSO develop the technical trigger. A tactical trigger is the maneuver related event or action that causes the commander to initiate fires. This event can be friendly or enemy based. The tactical trigger is usually determined during COA development. The technical trigger is the mathematically derived solution for firing the indirect fires based on the tactical trigger to ensure that the indirect fires arrive at the correct time and location to achieve the desired effects.

Note: Triggers can be marked using techniques similar to those for marking TRPs.

10-39. When selecting the tactical trigger the commander must ensure that either he, or the designated observer, is able to observe the enemy forces or event that is the tactical trigger if it is enemy driven; for example, *"When enemy forces occupy their defensive positions vic Objective Brown."* The tactical trigger may also be friendly event or time driven; for example, *"When Charlie company crosses PL Bowen"* or *"at 0900."*

10-40. Several factors govern the selection and positioning of the technical trigger. Critical factors are the enemy's likely locations or rate of travel, and the time required for the enemy force to move from the technical trigger to the target area. Using this information, the commander can then select the technical trigger location based on the following considerations:

- The amount of time required to initiate the call for fire.
- The time needed by the fire support element to prepare for and fire the mission.
- The time required to clear the fires.
- Any built-in or planned delays in the firing sequence.
- The time of flight of the indirect fire rounds.
- Possible adjustment times.

10-41. The company commander can use an estimated rate of enemy movement, along with the information in Table 10-2, to complete the process of determining the location of the technical trigger in relation to the target area. Table 10-2 lists the response time required by field artillery assets to prepare for and fire various types of support missions. Trigger lines or points (used in this method) are usually employed as technical triggers to synchronize the effects of direct fires, countermobility efforts, and indirect fires in time and space, rather than try to engage moving targets based on mathematical calculations.

Grid or polar mission (unplanned)	5 to 7 minutes
Preplanned mission	3 minutes
Preplanned priority mission	1 to 2 minutes
NOTE: These are approximate times (based on ARTEP standards) needed to process and execute calls for fire on normal artillery targets. Special missions may take longer.	

Table 10-2. Artillery response times.

Ceasing or Shifting of Fires

10-42. As in trigger planning for the initiation of fires, the commander must establish triggers for ceasing or shifting fires based on battlefield events such as the movement of enemy or friendly forces. One technique is the use of a minimum safe line (MSL) when a friendly element, such as a breach force, is moving toward an area of indirect fires. As the element approaches the MSL, observers call for fires shift or cease, allowing the friendly force to move safely in the danger area.

Clearance of Fire

10-43. The maneuver commander has the final authority to approve (clear) fires and their effects within his zone or sector. Although he may delegate authority to coordinate and clear fires to his FSO, the ultimate responsibility belongs to the Infantry company commander. Normally, the FSO helps the commander by making recommendations on the clearance of fires.

Fires Support Execution Matrix

10-44. As a tool in fires support planning and execution, the company commander may develop a graphic summary outlining the critical elements of the fire support plan and the company's role. The commander incorporates this information into his own execution matrix or into a separate fire support execution matrix, similar to the battalion's fire support execution matrix as shown in Table 10-3, page 10-10. The company fire support execution matrix is similar and should include, as a minimum, the following information for each target.

- Target number and type, to include FPF designation.
- Allocated fire support asset and munitions type.
- Observer and backup observer.
- Trigger.
- Target purpose.
- Target grid.
- Priority of fire.
- Priority targets.
- Fire support coordination measures (FSCMs).

Event Support Data	Event I LD to SBF O1	Event II Set conditions for breach from SBF 01	Event III B Co Breach	Event IV C Co Assault
TARGET/ GRID	AE0001 (PK 10184938).	AE0002 (PK 09005031).	O/O shift AE0001 to AE0003 (PK 10204810) and lift AE0002.	O/O lift AE0003.
ASSET	155-mm HE.	Mortar smoke.	155-mm.	155-mm.
OBSERVER/ BACKUP	Scout platoon will initially call for and adjust fires; FSO adjusts upon arrival at SBF; 1st platoon leader is backup.	FSO (primary)/ 1st platoon leader (backup).	AE0003: FSO (primary)/ 2d platoon leader (backup).	FSO (primary)/ 3d platoon leader (backup).
TRIGGER	Weapons company crosses PL Lynx.	On-call at SBF.	B Company crosses PL Lion.	C Company completes consolidation on OBJ Bob.
PURPOSE	Disrupt enemy on OBJ Bob to facilitate maneuver of A Company to SBF position.	Obscure enemy to prevent interference with B Company's breach.	Disrupt MRB reserve to protect the assault force (C Company).	Protect the assault force (C Company).

Table 10-3. Example battalion fire support execution matrix.

MANEUVER COMMANDER'S INTENT

10-45. The Infantry company commander ensures the FSO clearly understands the intent and desired effects for maneuver and fire support. He identifies the role of fire support in the scheme of maneuver (when, where, what, and why) by explaining in detail the concept of the operation, scheme of maneuver, and tasks and desired effects for fire support to the FSO.

- Providing this level of guidance is not easy. Artillery fires are not instantaneous, and planning must allow for this lag time. It takes several minutes to process targets of opportunity and deliver fires in the target area. While war-gaming the maneuver, the company commander refines the critical targets or EAs, priority of targets, priority of engagement, sequence of fires, and results desired. He then can see when and how to synchronize direct and indirect fires to destroy the enemy and protect the force.
- The company commander normally designates the company's decisive operation to have priority of fires. This prioritizes requests when two or more units want fires at the same time. He also designates where to place obscuration or illumination, suppressive fires, and preparation fires.

> *Note:* An element can still request fires even if it has not been allocated priority of fires.

PLANNING PROCESS

10-46. While the Infantry company commander develops and refines the tactical plan, he also develops the fire support plan. The FSO concurrently helps develop and refine fire support. Targets are placed in the fire support planning channels as soon as possible, so they can be processed at the battalion FSE or battery FDC (Figure 10-2). Regardless of the planning method used, the company fire support plan includes--

- Target number and location.
- A description of the expected target.
- Primary and alternate persons responsible for shooting each target.
- The amount of effect required and purpose.
- Radio frequency and call sign to use in requesting fires.
- When to engage the target.
- Priority of fires and shifting of priority.
- Size, location, code word, and emergency signal to begin FPF.

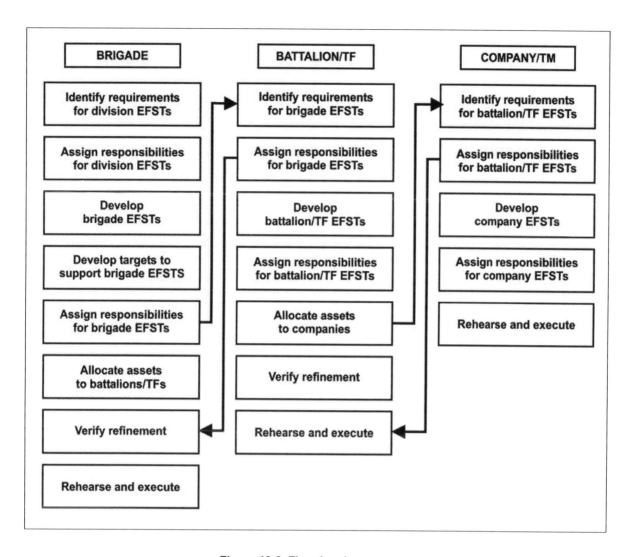

Figure 10-2. Fire planning process.

10-47. The company FSO does most of the technical aspects of the company fire support planning; however, he may receive targets and target information from platoon leaders and the battalion FSO. The company commander and FSO should not plan too many targets.

- The number of targets planned by the company and included in the formal fire support plan depends upon the company's priority for fire support and the number of targets allocated to them. The total number of targets in the fire support plan or the battalion mortar plan might be constrained. An excessive number of targets tends to dilute the focus of fire planning and can lead to increases in response time.
- Informal planning continues with target locations being recorded on terrain sketches or the FSO's map or being stored in the buffer group of the Advanced Field Artillery Tactical Data System (AFATDS) for quick reference and transmission. Fire planning for the company mortars complements these plans; the primary constraint is normally ammunition availability and the rapid resupply ability. Care must be taken to ensure that planning focuses on the critical fire support requirements identified by the company commander.

10-48. The company FSO completes the indirect fire plan and briefs the company commander. The company commander may alter the plan or approve it as is, but he makes the final decision. After the company commander approves the plan, the FSO makes sure the targets are passed to the battalion FSE where the fire plans are integrated into the battalion scheme of maneuver.

10-49. The commander and the FSO ensure platoon leaders are thoroughly familiar with the indirect fire plan. The FSO also provides target overlays to the platoon leaders, FOs, and the commander. He may also disseminate the company fire support plan as a target list and a fire support execution matrix. The FSO does this in sufficient time to allow subordinates to brief their platoons and sections. (A good plan given with the company order is better than a perfect plan handed out at the line of departure.)

FIRES PARAGRAPH

10-50. The fires subparagraph in the CONOP describes the concept of fires that, along with the scheme of maneuver, communicates how the force as a whole will achieve the commander's intent. It clearly describes the logical sequence of EFSTs and how they contribute to the CONOP. The overall paragraph organization should mirror that of the scheme of maneuver paragraph. If the maneuver paragraph is phased or otherwise organized, the fires paragraph takes on the same organization.

TASK, PURPOSE, METHOD, AND EFFECTS

10-51. The internal format for the fires paragraph uses the four subcategories of TPME. Within each phase of an operation, each EFST will be described in the sequence of planned execution using TPME. The fires paragraph must clearly and concisely state what fires will accomplish in the operation. Describe each category and subcategory as follows.

Task

10-52. Task describes the targeting objective fires must achieve against a specific enemy formation's function or capability. These formations are HPTs or contain one or more HPT. Task is normally expressed in terms of objective, formation, and function.

Objective

10-53. Clearly describes the targeting objectives that must be achieved. Use terms such as destroy, disrupt, delay, or limit to describe the effects required.

Formation

10-54. This is a specific element or subelement of the enemy. This can specify a specific vehicle type or target category as long as the element or sub element is clear.

Function

10-55. This is a capability of the formation that is needed for it (the enemy formation) to achieve its primary task and purpose.

Purpose

10-56. Purpose describes the maneuver or operational reason for the task. This identifies (as specifically as possible) the friendly maneuver formation that will benefit from the targeting objective and also describe in space and time what the objective will accomplish.

> **EXAMPLE**
>
> **Task:** Disrupt the ability of the motorized Infantry platoon at point of penetration to place effective direct fire against the breach force.
>
> | Objective | Formation | Function |
> |---|---|---|
> | Disrupt | The motorized Infantry platoon at point of penetration | To prevent the enemy from placing effective direct fire against the breach force. |
>
> **Purpose:** To allow an Infantry rifle company to breach the obstacle without becoming decisively engaged by the motorized Infantry platoon at the point of penetration.

Method

10-57. Method describes how the task and purpose will be achieved. It ties the "*detect*" function to the "*deliver*" function in time and space and describes how to accomplish the task. Method is normally described in terms of priority, allocation, and restriction.

Priority

10-58. For detection assets, it assigns priorities for finding NAIs, targeted areas of interest (TAIs), EAs, or HPTs. For delivery assets, it assigns the priority of the HPT against which that system will mostly be used.

Allocation

10-59. For both detection and deliver assets, it describes the allocation of assets to accomplish the EFST.

Restriction

10-60. Describes constraints, either requiring or prohibiting a particular action. Considerations include ammunition restrictions and FSCMs. The method subparagraph includes the following information.

- Priority of fires.
- Observers (primary and alternate).
- Triggers (tactical and technical).
- Target allocation.
- Priority targets.
- Close air support allocations.
- Final protective fires.
- Restrictions.
- Special munitions.
- Intelligence and electronic warfare assets.
- Any other instructions.

> **EXAMPLE METHOD**
>
> FA POF to 1st platoon, mortar POF to 2d platoon. Primary observer for AB 1000 (motorized Infantry platoon at point of penetration) is 1st platoon from OP 1, NFA 1. Alternate observer is company FIST, NFA 3 ... no DPICM within 300 meters NP177368 ... airspace coordination area (ACA) Lion in effect with CAS at initial point.

Effects

10-61. Effects try to quantify the successful accomplishment of the task. They provide a guide to determine when the task is completed. One measure is to determine if the purpose has been met. If multiple delivery assets are involved, it helps clarify what each must accomplish. Effects determination also provides the basis for the assess function of targeting and contributes to the decision of whether to reattack the target.

> **EXAMPLE EFFECTS**
>
> No hostile fire on the breach force from enemy motorized Infantry platoon until at least the assault force has passed through. Twenty-five percent of vehicles and fifty percent of enemy motorized Infantry platoon destroyed.

Note: At battalion and below, a formal written OPORD may not be produced. A fire support plan at this level might be an operations overlay with written instructions, an FSEM, and a target list or overlay.

FIRE-SUPPORT EXECUTION MATRIX

10-62. Battalion fire support plans might be distributed in matrix format. The fire support execution matrix is a concise, effective tool showing the many factors of a detailed plan. It aids the company FSO and the commander in understanding how the indirect fire plan supports the scheme of maneuver. It explains what aspects of the fire support plan each element is responsible for, and at what time during the battle these aspects apply. (For more on the battalion fire support matrix, see FM 3-21.20).

10-63. The advantage of the matrix is that it reduces the plan to one page and simplifies execution. The company fire support execution matrix also directs execution responsibilities and reduces the possibility that planned fires will not be executed. The company commander is responsible for disseminating the fire plan. The commander and his key subordinate leaders must understand the categories of targets, and how to engage those targets to create the desired result.

10-64. Figure 10-3 shows an example completed fire support execution matrix for a company deliberate attack. In the AA, a field artillery FPF is allocated for 1st and 2d platoons; 3d platoon has been allocated a mortar FPF; 2d platoon has priority of mortar fires from the LD to Checkpoint 7. From Checkpoint 7 to Objective Green, 3d platoon has been allocated a mortar priority target and has designated it as CA3017; 2d platoon is backup for execution. 1st platoon has been allocated a mortar FPF; 2d and 3d platoons have been allocated field artillery FPFs. At company level, information in each box of the matrix includes the following.

- Priorities of indirect fire support to a platoon appear in the upper left corner of the appropriate box (FA).
- If a unit is allocated an FPF, the type of indirect fire means responsible for firing appears next to the indicator (FA FPF or MORT FPF).

Chapter 10

- The target number of priority targets allocated to a platoon appear in the box preceded by the target, followed by the target number (MORT PRI TGT CA3014).
- If the company FSO is responsible for initiating specific fires, the target number, group, or series designation is listed in the box for the FSO (CA3012). Specific guidelines concerning fires not included on the target list are included in that box.
- Alternate element responsible for the execution of specific fires is listed in the lower right hand corner of the box (2d platoon). If fires have not been initiated when they were supposed to have been, that unit initiates them (unless ordered not to).
- Each fire support measure to be placed in effect, followed by a word designated for the measure, is shown in the box (CFL CHUCK). For airspace coordination areas, the time for the arrival of the planned CAS or attack helicopters is listed (ACA 1400Z).
- Other factors that apply to a certain platoon during a specific time might be included in the appropriate box. General guidance is issued in the written portion of the operation order.

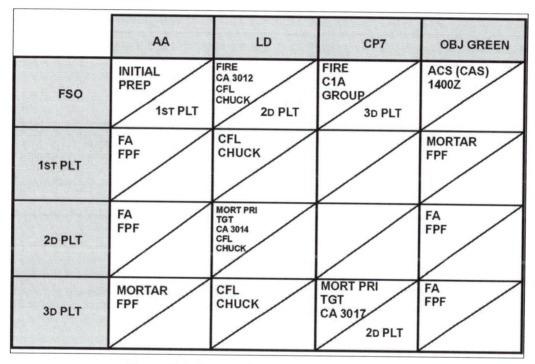

Figure 10-3. Example fire support execution matrix.

FINAL PROTECTIVE FIRES

10-65. These are immediately available planned fires that block enemy movement, especially dismounted Infantry approaching defensive lines or areas. These areas are integrated with defensive plans. The pattern of FPF plans may be varied to suit the tactical situation. They are drawn to scale on the target overlay. The size of the FPF is determined by the number and type of weapons used to fire on it (Figure 10-4). The company commander is responsible for the precise location of FPFs. The company FSO--

- Reports the desired location of the FPF to the supporting FDC.
- Adjusts indirect fire on the desired location, by weapon.
- Transmits the call to fire FPF to the supporting FDC.

10-66. The leader (normally the company commander or a platoon leader) in whose area the FPF is located has the authority to call for the FPF. The FPF has the highest priority of any target assigned to a fire support means. The FPF is only fired when required to repel the enemy's assault. Premature firing wastes ammunition and allows the enemy to avoid the impact area.

WEAPONS	SIZE (METERS)
60-mm Mortar (2 tubes)	60 x 30
81-mm Mortar (4 tubes)	100 x 40
120-mm Mortar (2 tubes)	120 x 60
120-mm Mortar (4 tubes)	240 x 60
155-mm Howitzer (4 guns)	200 x 50
155-mm Howitzer (6 guns)	300 x 50
155-mm Howitzer (8 guns)	400 x 50

Figure 10-4. Dimensions of final protective fires.

SPECIAL MUNITIONS

10-67. Obscuration fires use smoke and WP ammunition to degrade the enemy by obscuring his view of the battlefield. (High explosive ammunition may also obscure his view with dust and fires, but the unit should not rely on it as the primary means.) Because smoke is subject to changes in wind direction and terrain contours, its use must be coordinated with other friendly units affected by the operation. Used properly, obscuration fires can--

- Slow enemy vehicles to blackout speeds.
- Obscure the vision of enemy direct fire weapon crews.
- Reduce accuracy of enemy-observed fires by obscuring OPs and CPs.
- Cause confusion and apprehension among enemy Soldiers.
- Limit the effectiveness of the enemy's visual command and control signals.

10-68. Screening fires are closely related to obscuration fires; they also involve the use of smoke and WP. However, screening fires mask friendly maneuver elements to disguise the nature of their operations. Screening fires may assist in consolidation by placing smoke in areas beyond the objective. They may also be used to deceive the enemy to believe that a unit is maneuvering when it is not. Screening fires require the same precautions as obscuration fires.

10-69. Special munitions might be used for illumination, which might be scheduled or on-call. Use friendly direct fire weapons and adjustment of indirect fires to illuminate areas of suspected enemy movement or to orient moving units.

SMOKE SUPPORT

10-70. Internal smoke capabilities consist of company mortars and smoke pots. Smoke pots are the commander's primary means of producing small-area screening smoke. An external smoke platoon is required for long-term, large-area obscuration. If attached, the smoke platoon can provide both hasty smoke and large-area smoke support for tactical operations in the main battle area.

OBSERVER POSITIONS

10-71. To ensure that indirect fire can be called on a specific target, observers are designated and in the proper position. As the company plans indirect fire targets to support the operation and passes these down to the platoon, specific observers are positioned to observe the target and the associated trigger line or TRP. Any Soldier can perform this function as long as he understands the mission and has the communications capability and training. Once the target has been passed to the platoon or included by the platoon in the fire support plan, the platoon leader must position the observer and make sure he understands the following in precise terms.

- The nature and description of the target he is expected to engage.
- The terminal effects required (destroy, delay, disrupt, limit, and so on) and purpose.
- The communications means, radio net, call signs, and FDC to be called.
- When or under what circumstances targets are to be engaged.
- The relative priority of targets.
- The method of engagement and method of control to be used in the call for fire.
- Purpose and location of target; observers (primary and alternate); trigger; communications; and the resource providing the fires.

REHEARSALS AND EXECUTION

10-72. Once the company has developed and coordinated the fire support plan, it rehearses the plan. As the company rehearses the maneuver, it rehearses the fire plan. The target list is executed as the maneuver is conducted; fires are requested (though not actually executed by the firing units) just as they would be during the operation. Under ideal circumstances, an FPF can be adjusted during the rehearsal. Rehearsals on the terrain reveal any problems in visibility, communications, and coordination of the fire support plan. Conduct rehearsals under degraded conditions (at night and in MOPP4) to make sure the company can execute the plan in all circumstances.

- If time or conditions do not permit full-scale rehearsals, key leaders can meet, preferably at a good vantage point, and brief back the plan. They can use a sand table to show it on the terrain. Each participant explains what he does, where he does it, and how he plans to overcome key-leader casualties. The fire support plan execution is integral to this process and is rehearsed in exactly the same way.
- The company executes the fire plan as it conducts the operation. It fires targets as required and makes adjustments based on enemy reactions. Priority targets are cancelled as friendly units pass them or they are no longer relevant to the maneuver.

COMMUNICATIONS

10-73. The FSO can monitor three of four possible radio voice nets and three digital nets (Figure 10-5). The company's mission and priority determine the specific nets.

Maneuver Support

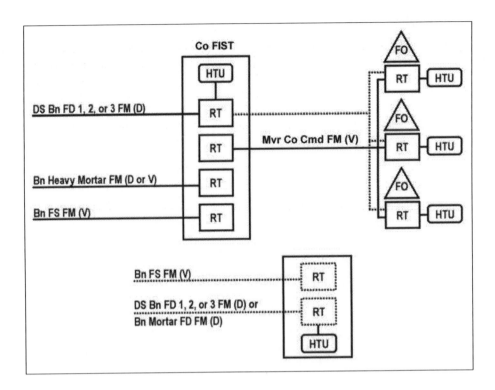

Figure 10-5. Company FIST communications.

COMPANY COMMAND NET FM (VOICE) AND FBCB2 (DIGITAL)

10-74. Platoon leaders, the XO, and attachments use this net to send reports, receive instructions, and request fires. Any Strikers, Bradleys, or tanks attached to the company monitor this net. This net also allows the FIST to monitor company operations and links it to the company commander, platoon leaders, and observers for planning and coordination. The company headquarters is the net control station (NCS).

BATTALION MORTAR FIRE DIRECTION NET FM (VOICE)

10-75. Observers may use this net to request fires of the battalion mortar platoon. Other stations on the net include the FIST headquarters and the battalion FSE. The battalion mortar platoon is the NCS.

BATTALION MORTAR FIRE DIRECTION NET (DIGITAL)

10-76. As necessary, the FIST sends fire missions to the supporting mortar platoon or section using this net.

COMPANY MORTAR NET (VOICE)

10-77. Observers or the company FSO use this net to request fire from the company mortars.

DIRECT SUPPORT BATTALION FIRE DIRECTION NET FM (VOICE) AND DIGITAL

10-78. This net is used for FA fire direction. The FIST uses this net to relay calls for fire through the battalion FSE to supporting artillery assets. The direct support battalion FDC is the NCS. When a Striker is present, it uses this net to request FA fires. The battery FDC and battalion FSE are also on this net.

QUICKFIRE CHANNEL

10-79. A QuickFire channel is established to link an observer (or other target executor) directly with a weapon system (Figure 10-6). QuickFire channels might be either voice or digital nets. QuickFire channels within a brigade combat team are normally established on FA or mortar nets. These channels are designed to expedite calls for fire against HPTs or to trigger preplanned fires. QuickFire channels may also be used to execute fires for critical operations or phases of the operation and to link an observer with a battery or platoon FDC for counterreconnaissance fires. Copperhead missions can best be executed by using QuickFire channels. The fire support coordinator or FSOs establish QuickFire channels and procedures based on the commander's intent and the CONOP.

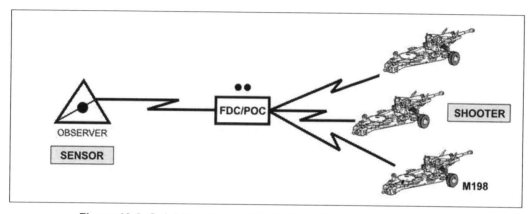

Figure 10-6. QuickFire channel *illu*stration of sensor-to-shooter link.

INDIRECT FIRES IN CLOSE SUPPORT

10-80. Effective indirect fire support often requires artillery and mortar fires near friendly Infantry Soldiers. A safe integration of fires and maneuver this close demands careful planning, coordination, and knowledge of the supporting weapons. These close supporting fires are most commonly FPFs in a defensive operation or suppression or obscuration fires to support an assault on an enemy position. When planning these fires, the company commander considers.

EFFECTS

10-81. In the defense, this might be to destroy enemy Soldiers and to degrade the effectiveness of enemy vehicles by causing them to fight buttoned-up. In the attack, the suppression/obscuration of enemy positions to allow the breach and seizure of a foothold on the objective is probably the desired effect.

ACCURACY

10-82. Many variables affect the accuracy of the weapon. The FSO has the technical knowledge to assist the company commander. Artillery and mortars are area weapons systems, which means that every round fired from the same tube impacts in an area around the target or aiming point. This dispersion is greater in length than in width. The weather conditions (wind, temperature, and humidity), the condition of the weapon, and the proficiency of the crew also affect the accuracy.

PROTECTION

10-83. If the unit is in well-prepared defensive positions with overhead cover, an FPF can be adjusted very close, just beyond bursting range. If required, the company commander can even call for artillery fires right on his company position using proximity or time fuzes for airbursts. Calling for close indirect fires is more dangerous during an attack. The commander considers the terrain, the breach site, and the enemy positions to determine how close to adjust his supporting indirect fires.

INTEGRATION OF SUPPRESSIVE FIRES

10-84. When integrating indirect suppressive fires to support the breach and assault, the following points are key.

- The danger increases with the size of the weapons. Use artillery to isolate the objective; use mortars on enemy positions away from the breach site; and use the 60-mm mortars, M203s, and direct fire weapons for close suppression.
- Assaulting perpendicular to the gun target (GT) line increases the probability of safety. If the rounds are coming over the head of the assault element, the margin of safety is reduced.
- Company mortars firing direct lay or direct alignment are the most responsive system. They are able to observe the rounds' impact and adjust accordingly. The safest method is to fire the 60-mm mortar with a bipod.
- Ideally, the firing units register prior to firing close-support missions. If not, the first rounds fired might be off target by a considerable distance. Once the firing units are adjusted on a target, then any shifts from that target are much more reliable.

TIMES AND CONTROLS

10-85. The final requirement for integrating these fires is to establish timings and control to ensure these targets are initiated, adjusted, and shifted properly. If possible, the company FSO should locate where he can observe these targets (possibly with the support element). A detailed execution matrix that assigns responsibility for each target to the leader or observer who is in the best position to control them should be developed. These Soldiers must know when each target, series, or group is fired, what effect is desired on which enemy positions, and when to lift or shift the fires. Consider the use of pyrotechnic or other signals to ensure communication.

ECHELONMENT OF FIRES

10-86. Company commanders will often find themselves as the observer (and executor) of battalion fires. Understanding the concept of echelonment of fires is critical for the indirect fire plan to be effectively synchronized with the maneuver plan. The purpose of echeloning fires is to maintain constant fires on a target while using the optimum delivery system up to the point of its risk-estimate distance (RED) in combat operations or minimum safe distance (MSD) in training (Table 10-4). Echeloning fires provides protection for friendly forces as they move to and assault an objective, allowing them to close with minimal casualties. It prevents the enemy from observing and engaging the assault by forcing the enemy to take cover, allowing the friendly force to continue the advance unimpeded. (Appendix A in this book; FM 3-90.2; and Appendix E in FM 3-09.32 all provide more information on RED.)

10-87. The concept behind echeloning fires is to begin attacking targets on or around the objective using the weapons system with the largest RED-combat (or MSD-training). As the maneuver unit closes the distance, that is, crosses the RED line for that specific munition en route to the objective, the fires cease, shift, or switch to a different system such as to the 81- or 60-mm mortar. This triggers the engagement of the targets by the delivery system with the next largest RED-combat (or MSD-training). The length of time to engage the targets is based on the rate of the friendly force's movement between the RED-combat (or MSD-training) trigger lines. The process continues until the system with the least RED-combat (or MSD-training) ceases fires and the maneuver unit is close enough to eliminate the enemy with direct fires or make its final assault and clear the objective.

10-88. The RED for combat (or MSD training) take into account the bursting radius of particular munitions and the characteristics of the delivery system. It associates this combination with a probability of incapacitation for Soldiers at a given range. The RED-combat (or MSD-training) is defined as the minimum distance friendly troops can approach the effects of friendly fires without 0.1 percent or more probability of incapacitation. A commander may maneuver their units into the RED-combat area based on the mission. However, he is making a command decision to accept the additional risk to friendly forces.

> **WARNING**
>
> Risk estimate distances are for combat use and do not represent the maximum fragmentation envelopes of the weapons listed. Risk estimate distances are not minimum safe distances for peacetime training use.

10-89. The casualty criterion is the 5-minute assault criterion for a prone Soldier. Physical incapacitation means that a Soldier is physically unable to function in an assault within a 5-minute period after an attack. A probability of incapacitation (PI) value of 0.1 percent can be interpreted as being less than or equal to one chance in one thousand and a PI value of 10% is one chance in ten (Table 10-4 and Table 10-5).

System	Description	Risk Estimate Distances (Meters)					
		10% PI			0.1% PI		
		1/3 range	2/3 range	Max range	1/3 range	2/3 range	Max range
M224	60-mm mortar	60	65	65	100	150	175
M252	81-mm mortar	75	80	80	165	185	230
M120/121	120-mm mortar	100	100	100	150	300	400
M102/M119	105-mm howitzer	85	85	90	175	200	275
M109/M198	155-mm howitzer	100	100	125	200	280	450
	155-mm DPICM	150	180	200	280	300	475

Table 10-4. Risk estimate distances for mortars and cannon artillery.

Item	Description	Risk Estimate Distance (meters)	
		10% PI	0.1% PI
MK-82 LD	500 lb. bomb	250	425
MK-82 HD	500 lb. bomb (retarded)	100	375
MK-82 LGB	500 lb. bomb (GBU-12)	250[1]	425[1]
MK-83 HD/LD	1,000 lb. bomb	275	475
MK-83 LGB	1,000 lb. bomb (GBU-16)	275[1]	475[1]
MK-84 HD/LD	2,000 lb. bomb	325	500
MK-84 LGB	2,000 lb. bomb (GBU-10/24)	225[1]	500[1]
MK-20[2]	Rockeye	150	225
MK-77	500 lb. Napalm	100	150
CBU-55/77[2]	Fuel-air explosive (FAE)	[1]	[1]
CBU-52[2]	CBU (all types)	275	450
CBU-58/71[2,3]	CBU (all types)	350	525
CBU-87[2]	CBU (all types)	175	275
CBU-89/78[3]	CBU (all types)	175	275
2.75" folding fin aircraft rocket (FFAR)	Rocket with various warheads	160	200
5" Zuni	Rocket with various warheads	150	200
SUU-11	7.62-mm minigun	[1]	[1]
M-4, M-12, SUU-23, M-61	20-mm Gatling gun	100	150
GAU-12	25-mm gun	100	150
GPU-5a, GAU-8	30-mm Gatling gun	100	150
AGM-65[5]	Maverick (TV, IIR, laser-guided)	25	100
MK-1/MK-21	Walleye II (1,000 lb. TV-guided bomb)	275	500
MK-5/MK-23	Walleye II (2,400 lb. TV-guided bomb)	[1]	[1]
AC-130	105-mm cannon.	80[4]	200[2]
	40-/25-/20-mm gun	35	125

[1] Risk-estimate distances are to be determined. For LGBs, the values shown are for weapons that do not guide and that follow a ballistic trajectory similar to general purpose bombs. This does not apply to GBU-24 bombs, because GBU-24s do not follow a ballistic trajectory.

[2] Not recommended for use near troops in contact.

[3] CBU-71/CBU84 bombs contain time-delay fuzes, which detonate at random times after impact. CBU-89 bombs are antitank and antipersonnel mines and are not recommended for use near troops in contact.

[4] AC-130 estimates are based on worse case scenarios. The 105-mm round described is the M-1 HE round with M-731 proximity fuze. Other fuzing would result in smaller distances. These figures are accurate throughout the firing orbit. The use of no-fire headings has no benefits for reducing risk-estimate distances and should not be used in contingency situations.

[5] The data listed applies only to AGM-65 A, B, C, and D models. AGM-65 E and G models contain a larger warhead. Risk-estimate distances are unknown.

Table 10-5. Risk estimate distances for aircraft-delivered ordnance.

10-90. Using echelonment of fires within the specified RED-combat (or MSD-training) for a delivery system requires the unit to assume some risks. The maneuver commander determines, by delivery system, how close he will allow fires to fall in proximity to his forces. The maneuver commander makes the decision for this risk level, but he relies heavily on the FSO's expertise. The commander considers the effects of terrain and weather, the experience of the observers, and communication systems involved. While this planning is normally accomplished at the battalion level, the company FSO has input and should be familiar with the process.

EXECUTION CONSIDERATIONS

10-91. When the lead elements of the battalion task force approach the designated phase line or control measure en route to the objective, the FSO begins the preparation. Lead element observers or company team FSOs track movement rates and confirm them for the battalion task force FSO. The battalion task force FSO may need to adjust the plan during execution based on unforeseen changes to anticipated movement rates (Figure 10-7, Figure 10-8, Figure 10-9, Figure 10-10, and Figure 10-11, this page thru page 10-27).

- As the unit continues its movement toward the objective, the first delivery system engages its targets. It maintains fires on the targets until the unit crosses the next phase line that corresponds to the RED-combat (or MSD-training) of the weapon.
- To maintain constant fires on the targets the unit starts the next asset before the previous asset lifts. This ensures no break in fires, enabling the friendly forces' approach to continue unimpeded. However, if the unit rate of march changes, the fire support system must remain flexible to the changes.
- The FSO lifts and engages with each asset at the prescribed triggers, initiating the fires from the system with the largest RED to the smallest. Once the maneuver element reaches the final phase line to lift all fires on the objective, the FSO shifts to targets beyond the objective.

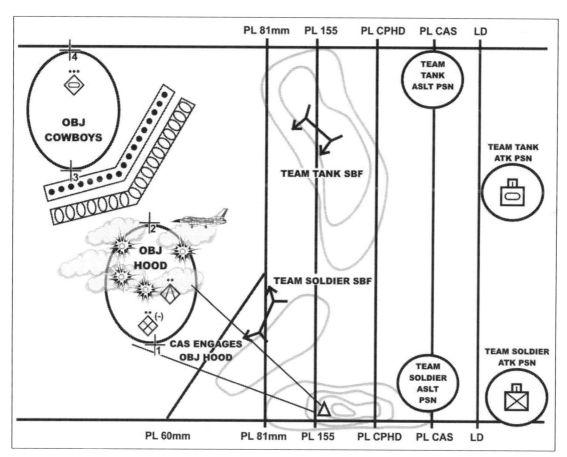

Figure 10-7. Beginning of close air support.

Maneuver Support

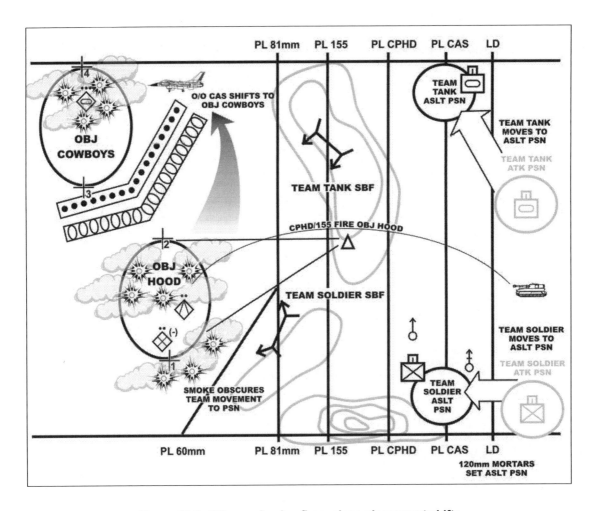

Figure 10-8. 155-mm shaping fires, close air support shifts.

Chapter 10

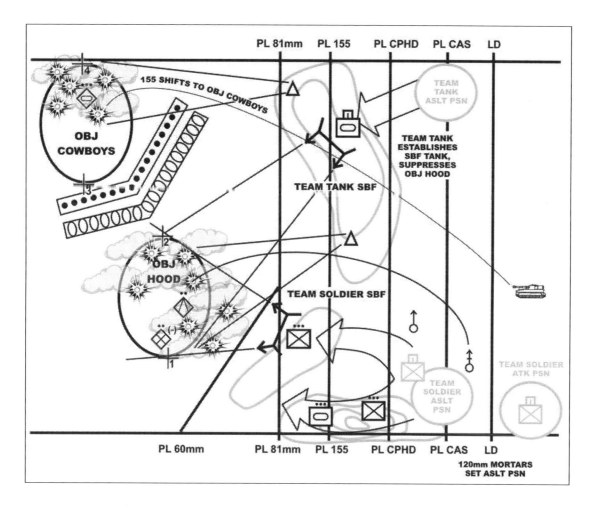

Figure 10-9. 155-mm shift, 81-mm, and supporting fires.

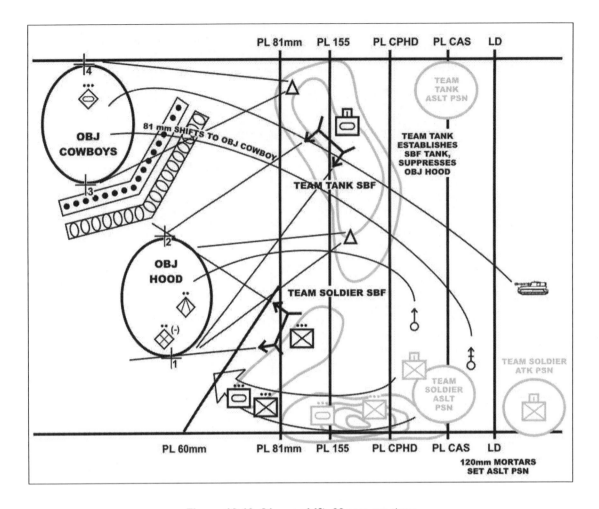

Figure 10-10. 81-mm shift, 60-mm mortars.

Chapter 10

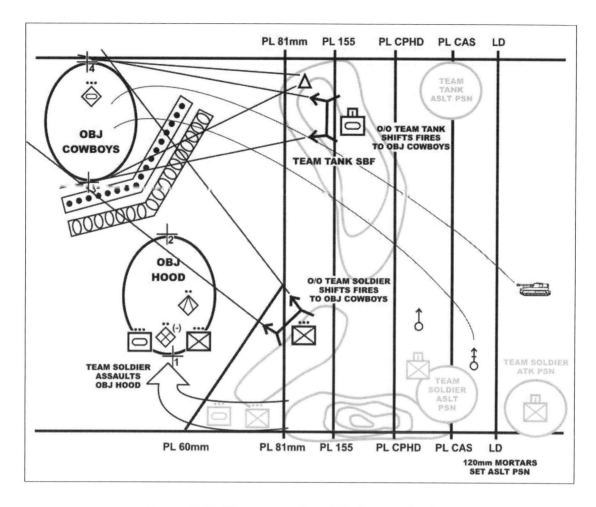

Figure 10-11. 60-mm cease fire, shift of supporting fires.

MORTARS

10-92. Mortars, located in both the battalion and company level, act as both a killer of enemy forces and as an enhancer of friendly mobility. They provide the maneuver commander with responsive, organic indirect fire support at a higher rate of fire than field artillery units. Using mortars to suppress the enemy inhibits his fire and movement while allowing friendly forces to gain a tactical mobility advantage. (Table 10-6 shows the different ranges and types of ammunition for the company mortar systems.)

- Mortars provide the maneuver commander with immediately available, responsive indirect fires in support of combat operations, and reinforce direct fires during close combat.
- Mortars are integrated with field artillery assets in an echelonment of fires. Echelonment of fires is a technique for the integration and synchronization of maneuver and fires. It is the execution of a schedule of fires fired from the highest caliber to the lowest caliber weapon based on the risk estimate distances as the maneuver force moves toward an objective. The echelonment of fires helps ensure that ground forces are able to move to an objective without losing momentum, and sets the conditions for the direct firefight and reduces the risk of friendly casualties.
- In the offense, mortars establish conditions for the maneuver elements in conducting their combat operations. They assist in suppressing and fixing the enemy and provide close support fires during the assault. Additionally they provide smoke for screening and obscuring friendly

movements. Heavy mortars can penetrate buildings and destroy enemy field fortifications, preparing the way for the dismounted assault force.
- In the defense, mortars can force the enemy to button up, obscure his ability to employ supporting fires, deny his use of defilade terrain, break up enemy concentrations and formations, and separate enemy dismounted Infantry from their armored personnel carriers (APCs) and accompanying tanks. They can destroy synchronization, reduce enemy mobility, and canalize enemy units into engagement areas.
- The mortar section leader works closely with the company commander and his FSO to maximize mortar fires and ensure the mortars are integrated into the echelonment of fires. Figure 10-12, Figure 10-13, and Figure 10-14, page 10-29, show the mortars most likely to support company operations.

Size/ Nomenclature	Model	Type	Min Range	Max Range	Diameter of Illum	ROF
60-mm M224	M720/M9 98	HE	70	[1]3,500		30 rpm for 4 minutes,[2] then 20 rpm sustained
	M722	WP	70	3,500	500	
	M721	illum	200	3,500	300	
	M302A1	WP	35	1,830		
	M83A3	illum	725	950		
	M494A	HE	45	1,830		
81-mm M252	M821	HE	83	5,608		30 rpm for 2 minutes, then 15 rpm sustained
	M889	HE	83	5,608	1,200	
	M819	RP	300	4,875		
	M853	illum	300	5,100		
120-mm M121	M57	HE	200	7,200		15 rpm for 1 minute, then 4 rpm sustained
	M68	WP	200	7,200	1,500	
	M91	illum	200	7,100	1,500	
	M933	HE (PD)	200	7,200		
	M934	HE (MOF)	200	7,200		
	M929	WP	200	7,200		
	M930	illum	200	7,200		

[1] Bipod mounted, charge 4 (maximum range handheld is 1,300 meters).
[2] Charge 2 and over. 30 rpm can be sustained with charge 0 or 1.

Table 10-6. Mortar ammunition characteristics.

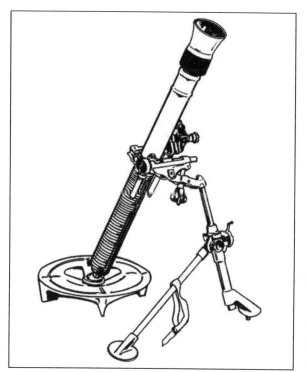

Figure 10-12. 60-mm mortar.

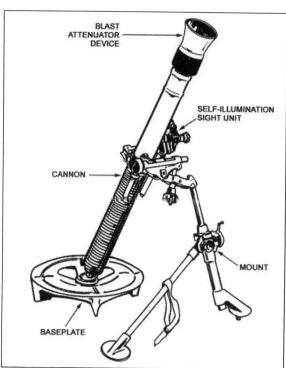

Figure 10-13. 81-mm mortar.

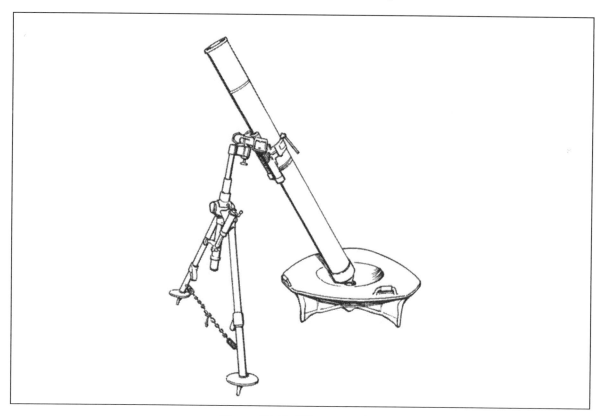

Figure 10-14. 120-mm mortar.

MORTAR POSITIONS

10-93. Based on the mission, terrain, and company commander's guidance, the company mortar section leader reconnoiters and selects mortar firing positions. A company mortar section position should--

- Allow firing on targets throughout the company's sector or zone, or the supported platoon's sector or zone. In the offense, one half to two thirds of the range of the mortars should be forward of the lead platoon.
- Be in defilade to protect the mortars from enemy observation and direct fire. Places such as the reverse slope of a hill, a deep ditch, the rear of a building, and the rear of a stone wall are well suited for mortar positions. The reverse slope of a hill may protect mortars from some indirect fire.
- Have concealment from air and ground observation. Vegetation is best for breaking up silhouettes. Mortars are positioned in defilade where natural camouflage conceals them. When the location of the firing position provides little concealment, consider the use of a hide position, which provides good cover and concealment and allows the mortar crews to occupy their firing positions quickly when required.
- Have overhead and mask clearance. Overhead clearance is checked by setting the sight at maximum elevation and looking along the mortar tube. Mask clearance is checked the same way, but at minimum elevation.
- Have solid ground that supports vehicle movement and precludes excessive settling of base plates. On soft ground, put sandbags under base plates to reduce settling.
- Have 25 to 30 meters between 60-mm mortars and 35 to 40 meters between 81-mm and 120-mm mortars. This reduces the chances of having more than one mortar hit by one enemy round. It also provides proper sheaf dispersion without plotting for each gun.
- Have routes in and out. These routes should ease resupply and displacement.
- Be secure. The section might have to provide its own local security. Being near other friendly units improves security.
- Avoid overhead fire of friendly Soldiers when possible.

10-94. The FDC might be in voice-distance of the squads; however, telephone wire should be laid from the FDC to each squad for security purposes and because battle noise might be so intense, that the squads cannot hear the commands. The mortar section has a very limited capability to secure itself. Normally, it collocates with other elements or has a security element attached. Mortar crews prepare mortar positions to protect themselves and to serve as firing positions for the mortars. The crews construct the positions with sandbags, ammunition boxes, earth, or any other available materials. (FM 7-90 describes these dug-in positions.)

MORTAR EMPLOYMENT

10-95. In a movement to contact, the mortar section usually supports the Infantry company with priority of fire to the lead platoon. The section normally displaces one squad at a time so that at least one squad is always in position and ready to fire. The section's displacement is based on the company's movement. The leader keeps the company commander informed of the location and status of his weapons and ammunition. The observers report their locations to the FDC.

- In an attack, sections prepare initial firing positions and may stockpile ammunition. They occupy positions at the last moment before the attack. The section must remain ready throughout the attack to respond to calls for fire and displace, if necessary.
- In the defense, mortars are farther to the rear than in the offense. The company commander plans his mortar section's final protective fire on a dangerous, dismounted enemy AA. Extra ammunition is stockpiled (if feasible). The mortars have some security when behind forward troops, but they still prepare to defend their positions.

Chapter 10

- To avoid being suppressed, a number of mortar positions are designated, prepared (if feasible), and occupied as required during the battle.
- In a withdrawal not under enemy pressure, one or more mortars might be left in position to support the DLIC.
- No matter where the platoon or section is located, it does everything it can for its own security. It might be able to post one or two security elements on the most dangerous approaches. It also uses early warning devices. If attacked, the security elements give warning, kill as many of the enemy as they can, and then rejoin their squads. The rest of the unit defends them from the dug-in mortar positions. The company reserve might be employed for a counterattack or to improve the security and defense of the mortar section.
- Targets are planned around the section's position so other mortars or artillery can provide support.

MORTAR DISPLACEMENT

10-96. Mortars displace to provide continuous support and to evade suppression, whether the company is attacking or defending. The displacement plan and the position of the mortar section in the company formation are responsive to the commander, and should not disrupt the maneuver elements but they should provide the mortar section with local security. It also allows the mortars to go into action quickly using the desired method of engagement and should provide ammunition resupply for the mortars. The displacement plan flows logically from other decisions made by the company commander, the company FSO, and the mortar section leader. If the company commander determines that operations (offensive or defensive) will move slowly enough to stay within mortar range and that continuous indirect fires must be available, he may order the mortars to displace to a suitable support position before the company moves out. In this event, he may not move them again until the company reaches its next position. The choices available for displacement are displacement by section and displacement by squad.

DISPLACEMENT BY SECTION

10-97. The whole section displaces at the same time. This allows the section to mass fires and the section sergeant to maintain control of his section. Moving as a section maximizes the limited FDC capability. It is also the fastest method of displacement. While the section is moving, its fire support is not immediately available unless it is positioned to fire using the direct lay or direct alignment methods or by conducting a hip shoot. Using any of these methods, the mortar section can only be available with minimum delay.

DISPLACEMENT BY SQUAD

10-98. This method allows continuous coverage of at least part of the company's sector. However limited communications equipment may still make it difficult to provide continuous indirect fire coverage even when displacing by squad. However, the company can attach one squad to each of two bounding platoons, so that, while using the direct lay or direct-alignment methods, one squad is always in overwatch of the company's movement. This may allow increased mortar coverage of the company sector during decentralized operations. It reduces the difficulty of transporting the mortar ammunition, and might also be the best way to infiltrate the mortars. Each platoon carries the ammunition for the attached gun squad.

10-99. The company commander also decides whether to move the mortars as a separate element in the company formation or to attach each gun squad to a subordinate element.

Attached

10-100. The mortars are attached to a subordinate element when the situation requires task organization, such as on a patrol or with the company support element; or when the mortars need additional control, security, and load-carrying capacity such as during an infiltration.

Separate

10-101. The mortars move as a separate element in the company formation when the commander wishes to control them directly and keep them together for massed use. When the mortars move as an element, they can displace by section or by squad.

MORTAR ENGAGEMENTS

10-102. Various engagement methods exist: direct lay and direct alignment (no FDC required), conventional indirect fire, and hip shoot. The primary methods of engagement for the 60-mm mortar are direct lay and direct alignment.

DIRECT LAY

10-103. This method is used when the gunner can see the target. The mortar might be handheld or bipod-mounted. An initial fire command is required to designate the target and (if desired) specify the shell-fuze combination and number of rounds. The gunner then adjusts fire and fires for effect without additional instructions (Table 10-7).

Advantages of Direct Lay	*Disadvantages of Direct Lay*
Can engage target immediately in handheld mode (the mortar weighs only 18 pounds and is therefore highly portable).	Requires the mortar crew to be relatively close to the enemy and therefore susceptible to direct and indirect fires.
Can be used by relatively untrained gunners such as cross-trained Infantrymen.	Is less effective at night (the gunner cannot engage when he cannot see).
Does not require an FDC.	

Table 10-7. Advantages and disadvantages of direct lay.

DIRECT ALIGNMENT

10-104. This method allows the mortar crew to fire from full defilade positions without an FDC. It requires that an observer be within 100 meters of the gun-target line and, if possible, within 100 meters of the guns. Direct alignment can only be used when handheld or bipod-mounted, although bipod-mounted is much more accurate (Table 10-8).

Chapter 10

Advantages of Direct Alignment	Disadvantages of Direct Alignment
Can engage target more quickly than the methods requiring an FDC. Allows crew more protection than direct lay. Does not require an FDC.	Is slightly slower than direct lay. Requires the mortar crew to be relatively close to the enemy and therefore vulnerable to indirect fires or assault. Requires a well-trained observer to be within 100 meters of the gun-target line (preferably within 100 meters of the guns). Requires observers to be in direct communication with the gun crew by voice, arm-and-hand signal, landline, or radio. Requires gun to be relaid to engage each different target.

Table 10-8. Advantages and disadvantages of direct alignment.

CONVENTIONAL INDIRECT FIRE

10-105. This method is used when the mortars have been laid for direction and an FDC established with positions plotted on the M16 plotting board or the mortar ballistic computer (MBC). In this situation (for the 60-mm mortar), the section leader operates the MBC or the M19 plotting board and the radio as the FDC (Table 10-9).

Advantages of Conventional Indirect Fire	Disadvantages of Conventional Indirect Fire
Can fire accurately at any target within range as long as an observer who can communicate with the FDC observes the target. Can accurately engage plotted targets in limited visibility. Can locate well away from enemy direct fires.	Requires an FDC (there is no designated FDC in the light Infantry mortar section). Is not as responsive as direct lay.

Table 10-9. Advantages and disadvantages of conventional indirect fire.

HIP SHOOT

10-106. When a call for fire is received during movement and the target cannot be engaged by either the direct lay or direct alignment method, a hip shoot is initiated. A hip shoot is a hasty occupation of a firing position; it requires both an FDC and an observer. The section leader normally acts as the FDC (60 mm only). The observer's corrections might be sent over the radio or by a wire net. The platoon or section leader must determine an azimuth of fire by map inspection. He then gives this direction to the mortar squads. The second squad leader uses the M2 compass (for the 60-mm section) to lay the base mortar. The section leader uses the MBC, the graphical firing scale, or the firing tables to determine the appropriate elevation and charge. He uses either the MBC or the M19 plotting board to refine the firing data based on the observer's corrections. The section leader may use the aiming-point deflection method, depending upon the terrain. The second mortar is laid either by sight-to-sight or M2 compass (Table 10-10).

Advantages of Hip Shoot	Disadvantages of Hip Shoot
Allows fire support when other methods of engagement are not usable.	Requires an FDC (there is no designated FDC in the light Infantry mortar section).
Is able to move at the same time as the unit and still provide adequate fires.	Is the slowest method of fire and the least accurate.

Table 10-10. Advantages and disadvantages of hip shoot.

AIR FIRE SUPPORT

10-107. Infantry company operations might be supported by attack aircraft including Army helicopters or ground attack fighters of the Air Force, Navy, or Marines. However, next to Army aviation, the Air Force most commonly provides sorties for the close-in fight. This type air power is typically close air support (CAS) but can also be joint air attack team (JAAT) operations. Though JAAT missions might be flown in or near the company AO, they are more complex than pure CAS, requiring higher level C2. Therefore, this discussion primarily addresses CAS. Attack helicopter operations are discussed in Appendix D, *Aviation Support for Ground Operations*.

Close Air Support

10-108. Tactical air control party (TACP) personnel are provided by the air force to maneuver units for control of CAS. Controllers might be assigned down to the IBCT company level to direct CAS missions and to perform terminal control. (FM 3-09.32 discusses this in detail, including complete procedures and checklists.)

Requests

10-109. CAS requests might be initiated at any level. The two types of CAS request are preplanned and immediate. Preplanned CAS is an air strike on a target that can be anticipated sufficiently in advance to permit detailed mission coordination and planning. These missions are categorized as scheduled or on call. A scheduled mission is executed at a specific time. An on call mission involves aircraft placed in a ground/air alert status and preloaded with ordnance for a particular target or type of target. Immediate CAS is an air strike on a target of opportunity that was not identified or requested sufficiently in advance to permit detailed mission coordination or planning.

Request Procedures

10-110. Requests for preplanned CAS missions are submitted to the task force FSE. The commander, ALO, FSO, and S-3 or S-3 Air evaluates requests, consolidate them and, if approved, assign a priority and precedence. The S-3 or S-3 Air then forwards approved requests to brigade. Immediate CAS requests are forwarded to the task force command post by the most expeditious means available. The ALO, FSO, and S-3 or S-3 Air considers each request. Approved requests are transmitted by the TACP to the air support operations center (ASOC). The ASOC coordinates with the senior ground HQ, which approves the request. The TACP at each intermediate HQ monitors the request and informs the S-3 or S-3 Air, the ALO, and the FSO or FSCOORD. Silence by an intermediate TACP indicates approval by the associated HQ.

Chapter 10

Joint Terminal Attack Controller

10-111. A JTAC is a qualified (certified) service member who, from a forward position directs the action of combat aircraft engaged in close air support and other offensive air operations. A qualified and current joint terminal attack controller is recognized across the Department of Defense as capable and authorized to perform terminal attack control. Terminal attack control is the authority to control the maneuver of and grant weapons release authority to attacking aircraft. Based on a risk assessment, the supported commander will weigh the benefits and liabilities of authorizing a particular type of terminal attack. JTACs will broadcast the type of control (1, 2, and 3) upon aircraft check-in.

Execution Procedures

10-112. The JTAC must coordinate with ground maneuver forces and obtain required CAS information before building the CAS briefing. The memory tool "*TTFACOR*," shown in Figure 10-15, ensures that JTACs coordinate the minimum information required for a CAS mission. Part of execution includes coordination and deconfliction considerations as pre-execution measures. The following discussion includes procedures that exceed minimum requirements in the event that CAS must be controlled by a non-JTAC qualified person. Under such circumstances, the controller must identify himself as "*non-JTAC qualified*" upon aircraft check-in:

> **CAS TTFACOR Pre-Execution Information Checklist:** *
> - Target commander's intent, valid, hostile target ID, coordinates.
> - Threat intelligence updates or pilot (weather) report (PIREPS), SEAD.
> - Friendlies update or confirm location, troops in contact, danger close.
> - Artillery ACA activation, SEAD coordination.
> - Control Cdr's Approval, risk assessment, type control.
> - Ordnance CAS, ordnance type, effects.
> - Restrictions, artillery, weapons effects, friendlies, collateral damage.
>
> *Also useful as TACP to FAC(A) and situation update briefing guide.

Figure 10-15. TTFACOR technique.

Tactical Operations Center Coordination

10-113. Facilitating coordination at the TOC involves the CAS battle drill. Either the ALO, as part of the fire support element (FSE), or the JTAC uses this procedure to ensure CAS is integrated with surface fires and maneuver forces to meet the commander's intent. Coordination includes the TOC staff, who rehearse the battle drill so they can execute it quickly. (FM 3-09.32, *JFIRE*, lists the procedures for a CAS battle drill. These procedures also apply to any non-CAS missions that require terminal attack control, but that do not require detailed integration with artillery or ground force assets.) The following format applies to "*CAS check-in*" (aircraft transmits to controller). The net control agency initiates the authentication statements (examples follow this paragraph) in the "*CAS check in*" brief. The NCA can shorten their brief to save time or to enhance security. To do this, they might say, for example, "*as fragged*" or "*with exception*":

```
Aircraft: ................................................"This is (controller call sign)(aircraft call sign)
Identification and mission number: ......... ""
Number and type of aircraft: .................. ""
Position and Altitude: ............................. ""
Ordnance: ............................................... "" (Fusing, laser code)
Play time: ................................................ ""
**Abort code: .......................................... ""
**Remarks: ............................................. "" (NVG, LST, special mission items)
Flight lead will establish abort code
**Optional entry
```

10-114. The following format provides a situation update to inbound aircraft:

Situation update # (JTAC to fighter)
Target and general enemy situation
Activity
Friendly situation
Artillery activity
Clearance authority
Ordnance requested
Restrictions and remarks
Localized SEAD efforts (suppression and EW)
Hazards (WX, terrain, obstructions)

10-115. A situation update is normally given once, when a fighter first checks in. Higher echelons, such as division or brigade, may assign an alphanumeric tracking number to ease subsequent check-ins at lower echelons. For example, "*Icebox 21, Hog Flight checking in as fragged with situation update Hotel.*" This briefing should be broad in scope. More specific information is passed in the nine-line briefing. Situation update might be passed to supporting airborne platforms (JSTARS) to speed information flow. Figure 10-16 shows an example format for a nine-line, close air support briefing.

Chapter 10

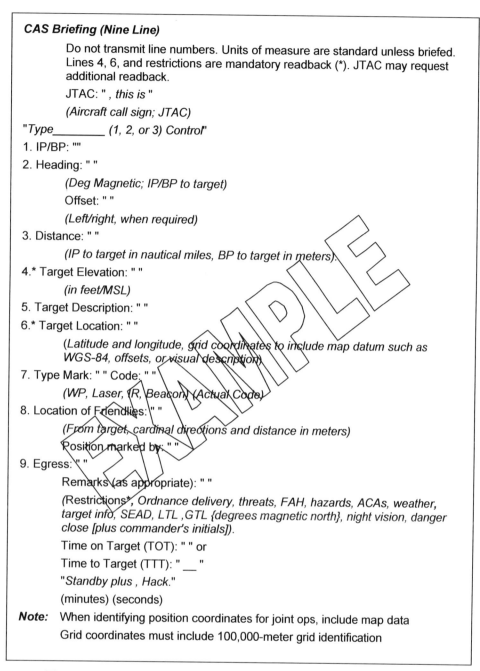

Figure 10-16. Example format for a nine-line close air support briefing.

Joint Air Attack Team

10-116. JAAT is a method of integrating rotary-wing and fixed-wing aircraft to locate and attack high-priority targets and targets of opportunity. JAAT is a method of employment, not a mission. JAAT fires are integrated mutually supportive, and synergistic, not simply deconflicted. The land force commander typically determines when to employ a JAAT but any commander (air, land, or maritime) may request one. JAAT can be employed anywhere on the battlefield across the spectrum of operations. CAS procedures may/may not be required depending on the proximity of friendly forces and requirement for detailed integration.

Maneuver Support

JAAT Planning

10-117. A mission commander will be designated for JAAT operations. The mission commander should be the element with the highest situational awareness and ability to provide command and control. JAAT can be accomplished with minimum coordination, if the participants are trained and proficient. Maximum JAAT synergy occurs when the JAAT mission commander at the tactical level, normally an Air Mission Commander (AMC), possesses the authority to coordinate attack execution directly with the other team members. In non-CAS JAAT application, direct attack coordination is more efficient because there is no requirement for JTAC/FAC(A) control. When JAAT is employed where CAS procedures are required, Type 2 or 3 (Table 10-11) control options offer increased control flexibility that can preserve JAAT synergy if the tactical risk assessment allows.

Type	Results of Risk Assessment	JTAC Observes Target and Aircraft	Timely and Accurate Target Data Provided
1	Commander assesses a high risk of fratricide to friendlies or noncombatants	Required	By JTAC (Inherent to Type 1 control)
2	Lower risk to friendlies or noncombatants but JTAC maintains control of individual attacks	Not Required.	By Observer or through other JTAC sensors*
3	Commander assesses the lowest risk of fratricide to friendlies or noncombatants. JTAC may provide blanket clearance.	Not. Required.	By JTAC or Observer or by aircrew if targets comply with prescribed guidance**

*   Observer: Scout, COLT, FIST, UAS, SOF, or assets that provide real-time targeting information.
** Supporting commander delegates weapons release to JTAC for all types of control. JTAC will provide "*cleared hot*" as appropriate for each attack in Types 1 and 2 control and "*cleared to engage*" for Type 3.

Table 10-11. Close air support types for terminal attack attributes.

Section III. ENGINEERS

The engineer company is tailored to fight as part of the combined arms team in the IBCT. It focuses on mobility but also provides limited countermobility and survivability engineer support. Only one engineer company is organic to the IBCT.

ORGANIZATION

10-118. The engineer company can be augmented according to the mission, with units from combat support brigades at echelons above the IBCT. Augmentation provides additional engineer capability and functions.

Chapter 10

ENGINEER COMPANY

10-119. The IBCT engineer company is assigned and executes engineer missions that are identified by the BCT commander. Their employment depends on the BCT commander's analysis of METT-TC. The engineer company commander may receive augmentation from other engineer units. He directs his unit in the execution of mission support to the BCT. The engineer company is self-sufficient for mobility purposes. Figure 10-17 shows an example of IBCT engineer company organization.

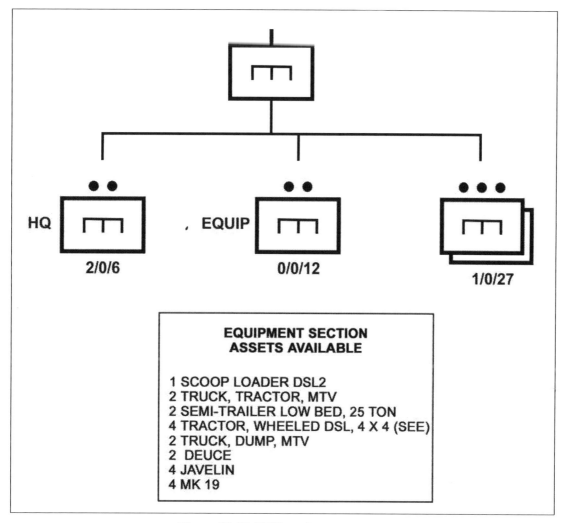

Figure 10-17. IBCT engineer company.

ENGINEER PLATOON

10-120. An engineer platoon (Sappers) might be task-organized to a battalion or company, based on the BCT commander's analysis of METT-TC. The engineer platoon can be employed to accomplish almost any engineer mission. However, the engineer platoon lacks organic sustainment assets and has minimal C2 depth and combat systems. Thus, it will most likely require augmentation or external support to conduct continuous operations over a sustained period of time (more than 48 hrs). The engineer platoon might also require some augmentation to conduct combined-arms tasks such as breaching operations. The engineer platoon may receive augmentation from its engineer company or other units as required.

SAPPER SQUAD

10-121. A sapper squad might be task organized to a company. It executes engineer tasks to support the company mission. Task organization is based on the battalion commander's analyses of METT-TC. The squad is the smallest engineer element that can be employed with its own organic C2 assets and as such can accomplish tasks such as reconnaissance, manual breaching, demolitions, or route clearance as part of a platoon or company mission. The sapper engineer may receive augmentation of engineer equipment such as a small emplacement excavator (SEE) or other specialized engineer equipment based on METT-TC. Figure 10-18 shows an example of a sapper squad.

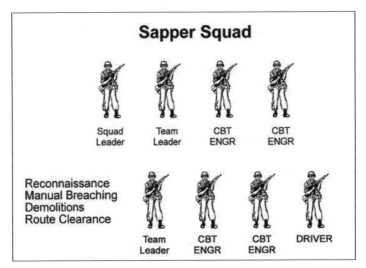

Figure 10-18. Example sapper squad.

MISSIONS

10-122. The tactical missions of engineers correspond to those of IBCT Infantry units. Engineer units can operate in restrictive terrain such as forests, jungles, mountains, and urban areas. Because of their austere nature, IBCT engineers have limited tactical mobility. To compensate for this, IBCT engineers train to operate in a decentralized manner. Like their supported maneuver force, they are very well suited to operate under conditions of limited visibility.

10-123. The mobility, countermobility, and survivability tasks for Engineer fall within the WFF of *Maneuver and Movement* or *Protection*. (Table 10-12 shows the tasks included in each of these categories). An engineer platoon or squad might be attached to a company depending on METT-TC. Engineers also conduct reconnaissance, evaluate obstacles, and employ demolitions.

10-124. The Engineer Corps is also responsible for general construction and topographic and geospatial mapping. Combat engineers must also be prepared to fight as Infantry.

Chapter 10

Mobility	Countermobility	Survivability
Breach obstacles.	Construct obstacles to turn, fix, block, or disrupt enemy forces.	Construct crew-served weapons and vehicle fighting positions.
Clear minefields.		
Clear routes.		
Cross expedient gaps.		
Construct combat roads or trails.		

Table 10-12. Engineer missions.

CAPABILITIES

10-125. The IBCT engineer's focus is mobility. They are expert in supporting infiltrations, air assaults, parachute assaults, ambushes, and raids. In this role, the engineer may conduct covert breaches, route reconnaissance, and obstacle reduction. They may also identify potential enemy counterattack routes to establish countermobility measures, such as scatterable mines (SCATMINEs), to protect the force. IBCT engineers train in Infantry skills and are able to move undetected when close to the enemy.

Weapon Characteristics

10-126. Weapons listed in the following paragraphs show munitions, which might be used or encountered in the maneuver company area. Commanders should be aware of the safety zones for the particular system they wish to employ:

Antipersonnel Obstacle Breaching System

10-127. The antipersonnel obstacle breaching system (APOBS) is a man-portable device that can quickly create a footpath through AP mines and wire entanglements. The APOBS is normally employed by combat engineers or Infantry Soldiers. The APOBS provides a lightweight, self-contained, two-Soldier, portable line charge that is rocket-propelled over AP obstacles from a standoff position away from the edge of the obstacle. For dismounted operations, the APOBS is carried in 25-kilogram backpacks by no more than two Soldiers for a maximum of 2 kilometers. One backpack assembly consists of the rocket-motor launch mechanism, containing a 25-meter line-charge segment and 60 attached grenades. The other backpack assembly contains a 20-meter line-charge segment and 48 attached grenades. The total weight of the APOBS is about 54 kilograms. It can breach a footpath that is about 0.6 by 45 meters and is fired from a 25-meter standoff.

Selectable Lightweight Attack Munition

10-128. The SLAM is a multipurpose munition with an antitamper feature. The SLAM is compact and weighs only 1 kilogram, so it is easily portable. The SLAM is intended for use against APCs, parked aircraft, wheeled or tracked vehicles; stationary targets such as electrical transformers, small fuel-storage tanks (less than 10,000-gallon); and ammunition storage facilities. The EFP warhead can penetrate 40 millimeters of homogeneous steel. The SLAM has two models—one is self-neutralizing (M2) and the other is self-destructing (M4).

- The M2 is solid green and has no labels, brands, or other distinguishing marks. This device is used by SOF and is not available to other units.
- The M4 is green with a black warhead (EFP) face. Units designated as Ranger, airborne, air assault, crisis response, and rapid deployment normally use this device.

- The SLAM has four possible employment methods—bottom attack, side attack, timed demolition, and command detonation.

Demolitions

10-129. Engineers use different types of demolitions to accomplish their missions. (FM 5-250 covers the types of demolitions used to support the light Infantry mission.)

Bangalore Torpedoes

10-130. Consist of 10 tube assemblies, 10 connecting sleeves, and 1 nose sleeve. The tube assemblies, or torpedoes are steel tubes 5 ft long and 2 1/8 inches in diameter. The main charge is 10 ½ pounds of composition B4 explosive. The primary use of the Bangalore is for clearing paths through wire obstacles and heavy undergrowth. It will clear a 3 to 4 meter path through wire obstacles.

M183 Satchel Charge

10-131. Consists of 16 M112(C-4) charges and 4 priming assemblies. It has a total explosive weight of 20 lbs. The M183 assembly is used primarily for breaching obstacles or demolishing structures when large demolition charges are required. The charge also is effective on smaller obstacles such as small dragon's teeth.

M112 Charge

10-132. Consists of 1.25 lbs of C-4 packed in an olive drab Mylar- film container with a pressure-sensitive adhesive tape on one surface. The M112 is primarily used for cutting and breaching. Because of its ability to cut and be shaped, the M112 is ideally suited for cutting irregularly shaped targets such as steel. The adhesive backing allows you to place the charge on any relatively flat surface.

Modernized Demolition Initiator

10-133. The MDI is a new family of nonelectric blasting caps and associated items. MDI components simplify initiation systems and improve reliability and safety. The components include the M11 high-strength blasting cap, the M12 and M13 low strength blasting cap, and the M14 high strength cap (time delay).

Detonating Cord

10-134. Consists of a core of HE (6.4 pounds of PETN per 1,000 feet) wrapped in a reinforced and waterproof olive drab coating. Detonating cord can be used to prime and detonate single or multiple explosive charges simultaneously. Detonating cord can be used in conjunction with the MDI components.

Scatterable Mines

10-135. Scatterable mines might be used to support the commander's intent by turning, fixing, disrupting, and blocking the enemy; however, they are used, they must be planned and coordinated to fit into the overall obstacle plan. The different types of mine systems and their emplacement authority are described in Table 10-13.

Scatterable Mine System	Emplacement Authority
Ground- or artillery-delivered, with SD time greater than 48 hours (long duration).	The corps commander may delegate emplacement authority to division level, which may further delegate to brigade level.
Ground- or artillery-delivered, with SD time of 48 hours or less (short duration).	The corps commander may delegate emplacement authority to division level, which may further delegate to brigade level (which may further delegate to battalion level).
Aircraft-delivered (Gator), regardless of SD time.	Emplacement authority is normally at corps, theater, or army command level, depending on who has air-tasking authority.
Helicopter-delivered (Volcano), regardless of SD time.	Emplacement authority is normally delegated no lower than the commander who has command authority over the emplacing aircraft.
MOPMS when used strictly for a protective minefield.	Emplacement authority is usually granted to the company or base commander. Commanders at higher levels restrict MOPMS use only as necessary to support their operations.

Table 10-13. Emplacement authority.

Area-Denial Artillery Munitions (Field Artillery Delivered)

10-136. The wedge-shaped ADAM is a bounding-fragmentation mine that deploys up to seven tension-activated trip wires 6 meters away from the mine. After ground impact, trip wires are released and the mine is fully armed. The lethal casualty radius is between 6 and 10 meters.

Remote Antiarmor Mines (Field Artillery Delivered)

10-137. The RAAM mine has a cylindrical shape and provides a full-width or catastrophic kill (K-kill). Using a magnetically influenced fuze, the mine projects a bi-directional, shaped-charge warhead through the crew compartment of a vehicle.

Multiple Delivery Mine System, or Volcano (Ground or Air Delivered)

10-138. The Volcano is a scatterable mine system that can be mounted on a cargo truck or UH-60A Blackhawk helicopter. It can rapidly produce tactical minefields with a linear frontage of up to 1,100 meters and a depth of 120 meters. The system can be employed to reinforce existing obstacles; close lanes, gaps, and defiles; provide flank protection for advancing units. The Volcano dispenses mines with 4-hour, 48-hour, and 15-day self-destruct (SD) times. The SD times are field-selectable before dispensing and do not require a change or modification to the mine canister. Reload time (not including movement time to the reload site) for an experienced four-Soldier crew is about 20 minutes. The average time to emplace one ground Volcano load (160 canisters) is 10 minutes.

Modular Pack Mine System (Man-Portable)

10-139. The MOPMS is a man-portable, 162-pound, suitcase-shaped mine dispenser. It contains 21 mines (17 antitank mines and 4 antipersonnel mines) and propels them in a 35-meter, 180-degree semicircle from the container. Mines are dispensed on command using the M71 remote control unit (RCU) or an electronic initiating device such as the M34 blasting machine. When dispensed, an explosive propelling charge at the bottom of each tube expels mines through the container roof. The company can use MOPMS to create a protective minefield or to close lanes in tactical obstacles. The safety zone around one container is 55 meters to the front and sides and 20 meters to the rear. MOPMS has duration of four hours, which can be extended up to three times (a total of 16 hours). Once mines are dispensed, they cannot be recovered or reused. If mines are not dispensed, the container might be disarmed and recovered for later use. The RCU can also self-destruct mines on command, allowing a unit to counterattack or withdraw through the minefield. The RCU can control up to 15 MOPMS containers or groups of MOPMS containers from a distance of 300 to 1,000 meters.

Hornet (Man-Portable)

10-140. The M93 Hornet, a wide area mine, introduces an entirely new obstacle concept to the combined-arms company. The Hornet is a top-attack special munition that type-categorizes, reports, and engages individual vehicles. It is an antitank and antivehicular off-route munition made of lightweight material (35 pounds) that one person can carry and employ. The Hornet is a nonrecoverable munition that can destroy vehicles by using sound and motion detection methods. It will automatically search, detect, recognize, and engage moving targets by using top attack at a standoff distance up to 100 meters. It can be a stand-alone tactical obstacle or can reinforce other conventional obstacles. It disrupts and delays the enemy, allowing long-range, precision weapons to engage more effectively. (This feature is particularly effective in non-LOS engagements.) It is employed by combat engineers, rangers, and SOF. The RCU is a hand-held encoding unit that interfaces with the Hornet when the remote mode is selected at the time of employment. After encoding, the RCU can be used to arm the Hornet, reset its SD times, or destroy it. The maximum operating distance for the RCU is 2 kilometers.

Special Engineer Vehicles

10-141. Engineer earth-moving capabilities are key to survivability missions. Additional considerations for survivability planning include command and control of digging assets, site security, sustainment (fuel, maintenance, and Class I), and movement times between BPs. The commander should start the survivability effort as soon as practical. He may employ blade assets to support systems such as mortars, C2, and key weapons before the bulk of his combat systems are ready for survivability support. The commander should establish a NLT time or a directed time to be ready for survivability. This helps prevent waste of blade time. Companies prepare their area for the arrival of the blades by marking positions, identifying leaders to supervise position construction, and designating guides for the blade movement between positions.

Small Emplacement Excavator

10-142. The SEE has a backhoe, bucket loader, a handled hydraulic rock drill, a chain saw, and a pavement breaker, among others. It can dig positions for individual, crew-served, and antitank weapons or for Stinger missile teams. It can also be employed to dig in ammunition prestock positions.

Deployable Universal Combat Earthmover

10-143. The DEUCE is a high-speed, high-mobility earth-moving system that can perform clearing, leveling, and excavation operations in support of all tactical engineer missions. It can drop by parachute and travels up to 30 MPH on rubber tracks. The enclosed operator compartment and controls provide numerous advantages over the D5 dozer it replaces.

Chapter 10

Scoop Loader

10-144. The 2.5 cubic yard scoop loader features a quick-coupler mechanism to attach and detach the multipurpose bucket. Tactical delivery means include airdrop or low altitude parachute extraction and some models are of sectionalized design for helicopter lift. Employment capabilities include loading trucks, performing excavations, or other similar engineer operations.

Section IV. AIR DEFENSE ARTILLERY

Air defense assets may operate in and around the Infantry company AO. However, the company is unlikely to receive task-organized air defense assets. Therefore, the company conducts its own air defense operations. It relies on disciplined, passive air defense measures and the ability to engage aerial platforms actively with organic weapons systems. Troops should be familiar with air defense assets, capabilities, operational procedures, as well as self-defense measures.

SYSTEMS, ORGANIZATION, AND CAPABILITIES

10-145. The man-portable Stinger and the HMMWV-mounted Stinger (then called the "*Avenger*") (Figure 10-19) might be used in and adjacent to the company AO. A maneuver battalion might be task organized with an air defense platoon equipped with four Avengers.

Figure 10-19. Stinger, man-portable and mounted (as "*Avenger*") on a HMMWV.

EMPLOYMENT

10-146. In offensive situations, man-portable Stingers and Avengers accompany the main attack. They may maneuver with the battalion's lead companies, orienting on low-altitude air avenues of approach. When the unit is moving or in a situation that entails short halts, the Stinger gunners can dismount to provide air defense when the unit reaches the objective or pauses during the attack. In the defense, man-portable Stinger and the HMMWV-mounted Stinger (then called the "*Avenger*") (Figure 10-19) might be used in and adjacent to the company AO. A maneuver battalion might be task-organized with an air defense platoon equipped with four Avengers.

WEAPONS CONTROL STATUS

10-147. The WCS describes the relative degree of control in effect for air defense fires. It applies to all weapons systems. The WCS is coordinated between the airspace controlling agency and brigade and disseminated when required.

LEVELS OF CONTROL

10-148. The three levels of control are--

Weapons Free

10-149. At this least restrictive level of control, crews can fire at any air target not positively identified as friendly.

Weapons Tight

10-150. At this moderate level of control, crews can fire only at air targets positively identified as hostile according to the prevailing hostile criteria.

Weapons Hold

10-151. At this most restrictive level of control, crews may fire *only* in self-defense or in response to a formal order.

EARLY WARNING PROCEDURES

10-152. Air defense warnings (ADWs) include--

> *RED: Air or missile attack imminent or in progress*
> *YELLOW: Air or missile attack probable*
> *WHITE: Air or missile attack not likely*

10-153. While air defense warnings cover the probability of hostile air action over the entire theater of war or operations, local air defense warnings describe with certainty the air threat for a specific part of the battlefield. Air defense units use these local warnings to alert Army units to the state of the air threat in terms of "*right here, right now.*" The three local air defense warning levels are--

> *DYNAMITE:* *Air platforms are inbound or are attacking locally now.*
> *LOOKOUT:* *Air platforms are in the area of interest, but are not threatening They might be inbound, but there is time to react.*
> *SNOWMAN:* *No air platforms pose a threat at this time.*

Note: The area air defense commander routinely issues air defense warnings for dissemination throughout the theater of war or operations. These warnings describe the general state of the probable air threat and apply to the entire area.

Chapter 10

REACTION PROCEDURES

10-154. Reaction procedures include both passive and active air defense measures.

Passive Air Defense

10-155. Passive air defense consists of all measures taken to prevent the enemy from detecting or locating the unit, to minimize the target acquisition capability of enemy aircraft, and to limit damage to the unit if it comes under air attack. One advantage the company can exploit is that target detection and acquisition are difficult for crews of high-performance aircraft. In most cases, enemy pilots must be able to see and identify a target before they can launch an attack.

Guidelines

10-156. The Infantry company should follow these guidelines to avoid detection or limit damage.

- When stopped, occupy positions that offer cover and concealment, dig in, and camouflage. When moving, use covered and concealed routes.
- Disperse as much as possible to make detection and attack more difficult.
- If moving when an enemy aircraft attacks, disperse and seek covered and concealed positions.
- Do not fire on a hostile fixed-wing aircraft unless the aircraft has identified friendly elements. Premature engagement compromises friendly positions.
- Designate air guards for every position; establish and maintain all-round security.
- Establish an air warning system in the unit SOP, including both visual and audible signals.

Procedures

10-157. When the company observes fixed-wing aircraft, helicopters, or UASs that could influence its mission, it initially takes passive air defense measures unless the situation requires immediate active measures. This reaction normally takes the form of each platoon's "*react to air attack*" battle drill. However, if needed, the commander can initiate specific passive measures (discussed previously in this section).

> *Note*: Passive air defense also includes the company's preparations for conducting active air defense measures.

10-158. Passive air defense has three steps.

- Step 1... Alert the company with a contact report.
- Step 2... Deploy or take the appropriate actions. If the company is not in the direct path of an attacking aircraft, the commander or platoon leaders order Soldiers to seek cover and concealment. They may also be ordered to continue moving as part of the battalion.
- Step 3... Prepare to engage.

Active Air Defense

10-159. If the commander determines that the company is being targeted by or is in the direct path of attacking aircraft, he may initiate active air defense procedures, including "*react to air attack*" drills by the platoons. This decision must be weighed against the possibility of exposing his positions to threat aircraft that might not have already seen them. If engagement is necessary, they use a technique known as volume of fire. This technique is based on the premise that the more bullets a unit can put in the sky, the greater the chance the enemy will fly into them. Even if these fires do not hit the enemy, a "*wall of lead*" in the sky can intimidate enemy pilots, causing them to break off their attack, or it may cause inaccuracy in their ordnance delivery. This technique may involve a designated leader firing a magazine of tracer ammunition for other shooters to follow. The Soldiers maintains the aiming point, not the lead distance. (Figure 10-20).

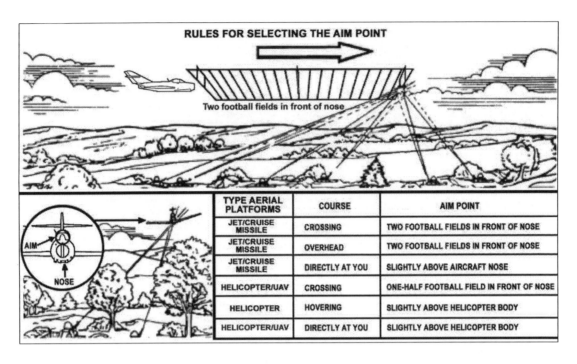

Figure 10-20. Machine-gun aim points against helicopters and high-performance aircraft.

CHEMICAL, BIOLOGICAL, RADIOLOGICAL, OR NUCLEAR SUPPORT

10-160. Chemical, biological, radiological, and nuclear (CBRN) weapons can cause casualties, destroy or disable equipment, restrict the use of terrain, and disrupt operations. They might be used separately or in combination to supplement conventional weapons. The company must be prepared to fight on a CBRN-contaminated battlefield. The CO designates principal CBRN defense trainers and advisors on CBRN defense operations and CBRN equipment maintenance. These trainers include a CBRN defense officer, a chemical NCO (MOS 74D), and an enlisted alternate. The CO ensures all personnel in his command can operate and perform maintenance on all organic CBRN equipment. CBRN assets within the Infantry company are limited. To survive on a contaminated battlefield, the company must practice the fundamentals of CBRN defense, avoidance, protection, and decontamination. (For more on CBRN, see Appendix H or FM 3-11.)

ARMY AVIATION

10-161. Army aviation is an asset at echelons above the BCT and might be requested by the Infantry battalion. Requests from the Infantry company go through the battalion. Army aviation can be used for command and control, reconnaissance, MEDEVAC, movement of troops and supplies, direct fire support, and as maneuver units. Air assault helicopter operations deliver assault elements of the Infantry company to locations on or near tactical objectives. (See Appendix D for a detailed discussion on Army aviation support.)

OTHER ATTACK HELICOPTERS

10-162. Other Attack helicopters are employed as integral parts of the joint and combined arms team and might be provided by US Marine or Navy assets. They are aerial attack systems also suited for situations calling for a quick response if available. Previously discussed JFIRE procedures also apply in employing them.

Chapter 11
Sustainment Operations

The role of sustainment support in any military unit is to sustain the force for continuous combat operations. Within the Infantry battalion, sustainment at the company level is provided by the support and medical platoons in conjunction with the battalion XO, HHC commander, S-1, and S-4. At the Infantry company level, the company commander has ultimate responsibility for sustainment. The XO and the 1SG are the company's primary sustainment operators; they work closely with the battalion staff to ensure they receive the required support for the company's assigned operations.

This edition updates the discussion of sustainment, incorporates new terms, and better addresses principles and TTP for unit trains, resupply, health service support, and weapons replacement operations.

Section I. PLANNING CONSIDERATIONS

Infantry company commanders, as well as the battalion S-4, make plans and key decisions concerning sustainment. The battalion S-4, company XO, company 1SG, company supply sergeant, platoon sergeants, and squad leaders implement these plans. Platoon leaders plan and relay support requirements for mission accomplishment to the company headquarters where it is consolidated and passed on to the battalion. Unit SOPs address planning, implementation, and responsibilities in detail and standardize as many routine sustainment operations as possible.

OVERVIEW

11-1. The Infantry company plans, prepares, and executes its portion of the sustainment plan. Concurrent with other operational planning, the company develops and refines its sustainment plan during troop-leading procedures. Rehearsals are normally conducted at both battalion and company levels to ensure a smooth, continuous flow of materiel and services. The company's sustainment responsibilities follow.

- Determine requirements.
- Report status.
- Request support.
- Receive support.
- Distribute.

11-2. Force health protection (FHP) in the Infantry company is a critical sustainment function. Medical support comes from these life saving elements and measures.

- Self aid.
- Buddy aid.
- Combat life saver.

- Combat medics.
- Treatment team, that is, support from the battalion aid station.

COMPANY RESPONSIBILITIES

11-3. In sustainment operations, roles differ slightly from other operations.

COMMANDER

11-4. The commander ensures that his sustainment operations meet the tactical plan. He will--

- Assure sustainment operations sustain his company's fighting potential.
- Identify special requirements for the mission.
- Integrate and synchronize sustainment activities into the tactical plan.
- Provide guidance to the operators.

EXECUTIVE OFFICER

11-5. The XO coordinates and supervises the company's logistical effort. During planning, he receives status reports from the platoon leaders, platoon sergeants, and 1SG. He then reviews the tactical plan with the company commander to determine company sustainment requirements, and coordinates these needs with the battalion S-4. During execution, as determined by the company commander, the XO locates at the second most important place on the battlefield. At times, this is where he can best supervise sustainment operations. The XO also performs the following functions.

- Determines the location of the company's resupply point based on data developed during operational planning and the war gaming process.
- Selects resupply method according to METT-TC.
- Tailgate.
- Service station.
- Maintains logistics status (LOGSTAT).
- Receives LOGSTAT from platoons.
- Completes company rollup and forwards to the combat trains command post (CTCP).
- Along with the 1SG, ensures that the company executes sustainment according to the battalion plan and SOP.
- Ensures his unit sustainment requirements are met.

FIRST SERGEANT

11-6. In addition to his tactical responsibilities, the 1SG is a key player in sustaining the company. He is also key in the execution of the company's plan and may supervise the company trains based upon the commander's intent and the factors of METT-TC. He may assist the XO with LOGSTAT management and in preparing paragraph 4 of the OPORD. He normally supervises the evacuation of casualties, EPW, and damaged equipment in addition to supervising company resupply activities and monitoring company maintenance activities. The 1SG orients new replacements and assigns them to squads and platoons IAW the company commander's guidance. He assures proper tracking of casualties between battalion, platoon leadership, and the senior trauma specialist; and oversees the NCO chain performing sustainment functions and tasks IAW the company SOP. The 1SG may also perform the following functions:

- Conduct sustainment rehearsals at the company level and integration with maneuver rehearsals.
- Perform C2 over company medic and oversees the evacuation plan from platoon to company CCP.
- Maintain the company battle roster.

Sustainment Operations

SUPPLY SERGEANT

11-7. The supply sergeant is the company representative for resupply to the company and based upon METT-TC may locate in either the combat trains or battalion field trains He assembles the logistics package (LOGPAC) and moves with the LOGPAC forward to the company. He coordinates the company's sustainment requirements with the support platoon leader and the Infantry battalion S-4. The supply sergeant may control the MEDEVAC vehicle when it is unable to remain forward with the company. He monitors the tactical situation and adjusts the sustainment plan as appropriate to meet the tactical plan and the company commander's guidance. He may assist the commander by establishing caches. He forecasts the company's consumption of food; water; ammunition; petroleum, oils, and lubricants (POL); and batteries; based on the operation. The supply sergeant also performs the following sustainment functions.

- Coordinate with the battalion S-4 for resupply of Classes I, III, and V.
- Maintain individual supply and clothing records.
- Requisition Class II resupply as needed.
- Request Class IV and Class VII equipment and supplies.
- Coordinate for maintenance support from the Forward Support Company maintenance section to include turn in and pick up maintenance documents, routine Class IX supplies, and recoverable materials.
- Pick up replacement personnel and, if necessary, deliver them to the 1SG.
- Coordinate for receipt and evacuation of human remains and personal effects.
- Transport, guard, and transfer EPW as required.
- Accompany the LOGPAC to the logistics release point (LRP).
- Guide the LOGPAC to the company resupply point.
- Accompany the LOGPAC along with EPW and damaged vehicles (if applicable) back to the BSA.
- Coordinate with the battalion S-1 section to turn in and pick up mail and personnel action documents.
- Collect hazardous material (HAZMAT) and transport it to collection points as part of LOGPAC procedures.
- Maintain and provide supplies for company field sanitation activities.

PLATOON SERGEANT

11-8. Each PSG in the company performs the following sustainment functions.

- Ensure Soldiers perform proper maintenance on all assigned equipment.
- Compile and submit all personnel and logistics status reports for the platoon as directed or in accordance with SOP.
- Collect each equipment inspection and maintenance form (DA Form 2404, *Equipment Inspection or Maintenance Worksheet* or DA Form 5988-E, *Maintenance Request Register*) within the platoon.
- Obtain supplies and equipment (all classes except Class VIII) and mail from the supply sergeant and ensures proper distribution within the platoon.

SENIOR TRAUMA SPECIALIST/SENIOR COMPANY MEDIC

11-9. The senior trauma specialist or senior company medic is attached to the rifle company to provide emergency medical treatment for sick, injured, or wounded company personnel. Emergency medical treatment procedures performed by the trauma specialist might include opening airways, starting intravenous fluids, controlling hemorrhages, preventing or treating shock, splinting suspected or confirmed fractures, and relieving pain. The emergency medicine performed by the trauma specialist is supervised by the battalion surgeon or physician's assistant (PA). The senior trauma specialist/company medic must--

Chapter 11

- Oversee and provide guidance to each platoon medic as required.
- Triage injured, wounded, or ill friendly and enemy personnel for priority of evacuation as they arrive at the company CCP.
- Oversee sick-call screening for the company.
- Request and coordinate the evacuation of sick, injured, or wounded personnel under the direction of the company 1SG.
- Assist in the training of the company personnel on first aid (self-aid and buddy-aid) and combat lifesavers in enhanced first-aid procedures.
- Requisition Class VIII supplies from the BAS for the company according to the TSOP.
- Assist the commander with medical planning, advise on higher headquarters' plan, and recommend locations for company CCPs.
- Monitor the tactical situation, and anticipate and coordinate HSS requirements and Class VIII resupply as necessary.
- Advise the company commander and 1SG on mass casualty operations.
- Keep the 1SG informed on the status of casualties, and coordinate with him for additional HSS requirements.

Section II. SOLDIER'S LOAD

The Soldier's load is of crucial concern to the leader. How much do Soldiers carry, how far, and in what configuration? These critical mission considerations require command emphasis and inspection. Army research shows that a Soldier can carry 30 percent of his body weight and retain much of his agility, stamina, alertness, and mobility. For the average Soldier, who weighs 160 pounds, this means carrying 48 pounds. Success and survival in company operations demand that Soldiers retain these capabilities. When it cannot move with stealth, agility, and alertness, the unit is at risk. For each pound over 30 percent of his body weight, the Soldier loses function. When his load exceeds 45 percent of his body weight, or 72 pounds for the average Soldier, his functional ability drops rapidly, and his chances of becoming a casualty increase. Research also shows that training can only improve load-carrying capability by 10 to 20 percent--at best. Commanders must ensure that Soldiers carry no more than 30 percent of their body weight when in contact or when contact is expected. At other times, the Soldier's load should not exceed 72 pounds. Sometimes, Soldiers must exceed the recommended weight. Leaders must realize how that excess weight impacts the unit's effectiveness. (FM 21-18 provides additional information on the Soldier's load.) This example, extracted from *The Soldier's Load* and *Mobility of a Nation*, by S. L. A. Marshall, details the dangers of excess loads.

Sustainment Operations

> **EXAMPLE**
> During the Normandy Invasion, many casualties were attributed to the excessive loads carried by US Soldiers as they tried to get ashore and across the beach. E Company, 16th Infantry suffered 105 casualties that day. Of these, 104 occurred on the beach, and most of them were due to their overloads. Many Soldiers fell prone at the water's edge and were drowned by the incoming tide. The Soldiers' packs were so heavy that they were able to walk only a few feet before falling to the sand. It took the company more than an hour to move 250 meters across the beach. The paratroopers that jumped into Normandy carried the following:
>
> | | | | |
> |---:|---|---:|---|
> | 1 | carbine or M1 rifle | 1 | steel helmet with liner |
> | 80 | rounds of ammunition | 1 | knit cap |
> | 2 | hand grenades | 1 | change of underwear |
> | 1 | mine | 2 | pairs of socks |
> | 6 | K-rations | 1 | gas mask |
> | 1 | impregnated jumpsuit | 1 | first-aid packet |
> | 1 | complete uniform | 1 | spoon |
> | 1 | entrenching tool | 2 | gas protective covers |
> | 1 | field bag | 1 | packet of sulfur |
> | 1 | escape kit | 1 | set toilet articles |
>
> Although they were required to jump heavy, once on the ground the individual Soldier discarded all unnecessary items and traveled light. They understood from their training that their success depended on mobility, stealth, and surprise. Even though these airborne units were without resupply for days, there is only one recorded incident where an airborne unit gave up ground due to ammunition shortages.

PLANS

11-10. The purpose of load plans is two-fold. First, it lets the Infantry company commander use the estimate of the situation to determine what ammunition, supplies, and equipment are essential. Second, it accounts for the potential impact of the Soldier-load problem and emphasizes the need to carry only what is necessary. The commander then arranges for the remainder of the load to be secured or transported. The company commander must consider METT-TC in determining the Soldier's load to be carried by the company. The company commander breaks down the company's equipment and supplies into one of the three echelons: combat load (approach march or fighting load), sustainment load, and contingency load (Figure 11-1, page 11-6).

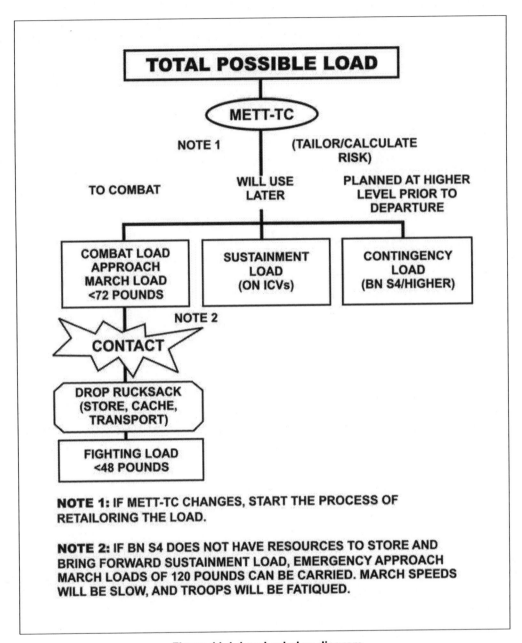

Figure 11-1. Load echelon diagram.

COMBAT LOAD

11-11. A combat load consists of the minimum mission-essential equipment, as determined by the mission commander. This includes only what is needed to fight and survive immediate combat operations. The two levels of combat load are fighting loads, which are carried on dynamic operations where contact with the enemy is expected, and approach march loads, which are carried when transportation cannot be provided for equipment over and above fighting loads.

Fighting Load

11-12. A fighting load is what the Soldier carries once contact has been made with the enemy. It consists only of essential items the Soldier needs to accomplish his task during the engagement. For close combat and operations requiring stealth, any load at all is a disadvantage. Cross loading of machine-gun ammunition, mortar rounds, antitank weapons, and radio equipment causes most combat loads to exceed 48 pounds. This is where risk analysis is critical. Excessive combat loads of assaulting troops must be configured so that the excess can be redistributed or shed (leaving only the fighting load) before or upon contact with the enemy.

Approach March Load

11-13. An approach march load is the load that the Soldier carries in addition to his fighting load. These items are dropped in an assault position, ORP, or other rally point before or upon contact with the enemy. On long dynamic operations, Soldiers must carry enough equipment and munitions to fight and exist until a planned resupply can take place. These loads vary and may exceed the goal of 72 pounds.

Sustainment Load

11-14. A sustainment load consists of the equipment required for sustained operations. This equipment is usually stored by the company supply section in the BSA and brought forward when needed. A sustaining load can include rucksacks, squad duffel bags, and sleeping bags. In combat, protective items for specific threats, such as joint-service lightweight integrated suit technology (JSLIST) might be stored in preconfigured unit loads. Commanders coordinate with the battalion S-4 to ensure that all sustainment load items are available.

Contingency Load

11-15. The contingency load includes all other items that are not necessary for ongoing operations, such as extra clothing, personal items, or even Javelins in a threat environment where the enemy lacks an armored capability. Contingency loads might be stored in duffel bags or palletized. Determining what goes in these loads and who is responsible for the storage and delivery of them, is a critical element for company commanders.

CALCULATION

11-16. The combat load for each Soldier consists of three components: common essential items carried (worn) by all Soldiers regardless of threat, environment, or mission; duty position load, consisting of the Soldier's assigned weapon (or components of the weapon system) plus ammunition; and variables, consisting of all other items carried, based on the commander's estimate of the situation. The latter are items that constitute the environmental, threat protection, and mission loads. When calculating load requirements, leaders should--

- Adjust combat loads so Soldiers carry less than 72 pounds.
- Divide combat loads into fighting loads and approach march loads.
- Have Soldiers pack rucksacks and assault packs accordingly.
- Place all other company equipment into the sustainment load.
- Once he decides what items Soldiers will carry on the mission, the leader decides how they will carry them. Soldiers need some items to be immediately available; other items can be carried in rucksacks.

MANAGEMENT

11-17. The key to load management is to carry only what is necessary to accomplish the mission. The following techniques assist the commander in load management.

- Make sure Soldiers distribute their loads evenly over the body, body armor, and load-carrying equipment (LCE), or load-bearing vest (LBV).
- Carry critical items within easy reach: carry water, ammunition, and a first aid pouch on the LCE, with other items in Army combat uniform (ACU) pockets. Ensure that placement of all items is standardized within the unit, and nothing that could prevent the Soldier from taking a well-aimed shot is allowed on the firing side of the LCE.
- Distribute loads throughout the unit. If bulk ammunition, rations, water, or demolitions must be manpacked, divide them into small loads consistent with METT-TC. This helps ensure they can be distributed on the battlefield where needed.
- Rotate heavy loads among several Soldiers. The unit can rotate radios, M240s, mortars, and Javelins if enemy contact is not imminent. Ensure that the assigned gunner stays near the weapons system components if they are rotated.

11-18. Upon contact with the enemy, drop rucksacks (if the tactical situation permits). The commander must consider how to recover the rucksacks and the effect not having the rucksacks might have on his unit) or leave them in an ORP, an assault position, or the assembly area. The leader can later request battalion transportation assets to bring them to his unit when possible (enemy situation dependent.) Soldiers mark their rucksacks by unit to facilitate quick recovery.

- Share or consolidate items; if the weather requires Soldiers to carry sleeping bags, carry only enough for those who will sleep at the same time; two or three Soldiers can share a sleeping bag. Soldiers can share the bags as they take turns rotating security duty.) In the same manner, two or three Soldiers can share a rucksack and take turns carrying it.
- Consider cutting rations to two or even one meal, ready to eat (MRE) per man per day for short periods. Remember that the MRE is a complete meal. It provides all the nutrition Soldiers need for sustained operations. Proper nutrition and fueling is critical for sustained Soldier performance.
- While carrying the rucksack, use water and rations carried in or on it first. If Soldiers must drop their rucksacks, what they carry in their ACUs and on the LCE or LBV remains available. Replace ammunition, water, and rations carried on LCE or LBV, or in ACU pockets, as soon as possible.
- When carrying radios in rucksacks, keep them attached to the backpack for access and use when rucksacks are dropped.
- Consider caches, supply linkups, captured stocks, and foraging to provide food, water, shelter, weapons, and equipment to reduce the need to manpack supplies.
- Avoid unnecessary movement and displacements. To conserve the Soldier's stamina, plan the mission as efficiently as possible. Do not move a platoon when moving a squad can do the job. If the leader becomes lost, he stops and determines his unit's location before moving and, if necessary, sends out someone to confirm the unit's location.

11-19. Supervise the Soldier's load closely. Not only must you check to ensure that the Soldier carries all items on the packing list, but also that he does not carry items not on the packing list. Do not let Soldiers carry unnecessary or 'comfort' items that will add weight and detract from their performance. Soldiers might be tempted to carry unnecessary items when they start on a mission and throw essential items away when they are tired. Packing lists for rucksack management and leader inspections before and during the mission ensure that only necessary items are carried. Rucksack management results in efficient use of a Soldier's energy and ensures that essential items are available when needed in combat.

- The company net does not always need the COMSEC equipment to function effectively. Ensure the threat warrants the extra weight on the radio operators.
- Consider distributing the approach march or sustainment loads to only two platoons. This allows the lead platoon to move with more stealth and alertness and to remain unburdened in case of contact. Platoons can then quickly swap rucksacks as they rotate the lead.

Section III. TRAINS

The logistical focal point is generally described as the trains. Sustainment personnel and equipment organic or attached to a force that provides support such as supply, evacuation, and maintenance services comprise the unit trains.

OVERVIEW

11-20. The company trains are the focal point for company sustainment operations. It is the most forward sustainment element, and provides essential medical treatment and critical resupply support. The size and composition of the Infantry company trains vary depending upon the tactical situation. The trains may consist of nothing more than preplanned locations on the ground (a control measure such as a checkpoint) during fast-paced offensive operations, or the trains may contain two to five tactical vehicles during resupply operations. The company trains are established to conduct evacuation (of WIAs, weapons, and equipment) and resupply as required. The company trains are located in a covered and concealed position, close enough to the company to provide responsive support, but out of enemy direct fire. The 1SG or XO will position the trains and supervise sustainment operations. Support to the company trains comes from the battalion combat trains.

SECURITY

11-21. Security of sustainment elements is critical to the success of the Infantry company and battalion missions. For this reason, the company trains must develop plans for continuous security operations. Company trains normally operate one terrain feature to the rear of the company. METT-TC factors dictate the actual distance. This location gives the company virtually immediate access to essential sustainment functions while allowing the trains to remain in a covered and concealed position behind the company combat elements. Where feasible, they may plan and execute a perimeter defense. The trains, however, may lack the personnel to conduct a major security effort. In such situations, they must plan and implement passive security measures to provide protection from enemy forces.

Section IV. SUPPLY AND TRANSPORTATION OPERATIONS

Each Infantry company normally deploys with 72 hours of supplies. The commander uses the unit basic load as the frame of reference for determining 72 hours worth of supplies. The Infantry company commander considers his situation to decide on the best means of resupplying his company. Resupply requests are classified as either routine or emergency. Cues and procedures for each method are specified in the company SOP and are rehearsed during company training exercises. The resupply method is typically either tailgate or service station depending on METT-TC. Infantry companies are supported by the Infantry battalion, which in turn is supported by the brigade through the forward support battalion (FSB).

CLASSES

11-22. Supplies are divided into 10 major categories, which are referred to as classes (Figure 11-2, page 11-11). The following paragraphs describe how these classes specifically relate to the Infantry company.

CLASS I

11-23. Subsistence supplies will be configured into unit-configured loads based on personnel strength reports. These loads are typically delivered by company supply sergeant LOGPACs to the battalion combat trains linking up with the company 1SG who makes further delivery to the companies. The company water trailer, supported by water from the BSB, supplies water to the unit. Units may also obtain water from within the theater of operations.

Class II

11-24. Limited stocks of Class II items (preventive medicine, field hygiene, weapons cleaning, and special tools) will be available at the BSB. This class also includes CTA-50 items, clothing, mission-oriented protective posture (MOPP) suits, tentage, tool sets, and administrative and housekeeping supplies and equipment. Class II supplies will be delivered with the LOGPACs.

Class III

11-25. Fuel support to the company will come from the battalion by organic fuel vehicles. Unit SOP determines how company vehicles are fueled. The two types of Class III are--

- Bulk--fuel.
- Package--coolant, oil, and lubricants, among other things.

Class IV

11-26. Company SOP specifies the use of Class IV items for the company. Requirements for class IV items such as concertina wire, sandbags, and pickets must be delivered as needed to the company area on battalion tactical vehicles.

Class V

11-27. The Infantry company deploys with a combat load of munitions. Ammunition resupply deliveries will be accomplished by the battalion ammunition vehicles and delivered to the battalion combat trains.

Class VI

11-28. The BSB does not stock Class VI supplies. After 30 days in theater, the supplement health and comfort pack (HCP) ration is usually issued with Class I rations.

Sustainment Operations

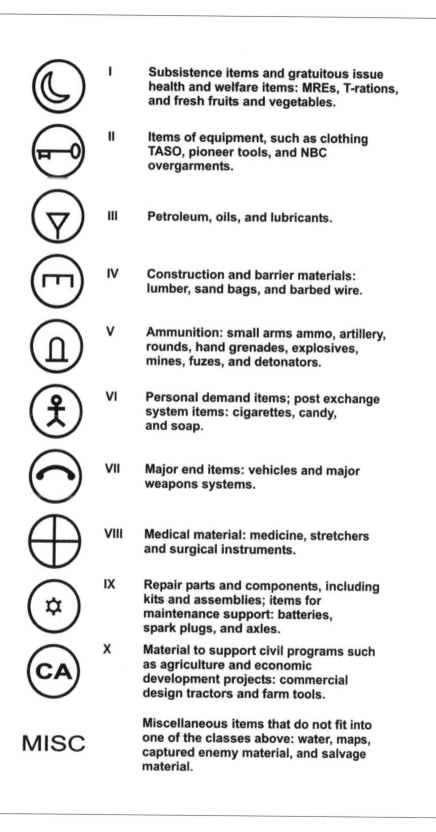

Figure 11-2. Classes of supply.

Chapter 11

CLASS VII

11-29. Class VII status is reported through command channels; it is intensively managed and command controlled. The BSB will receive replacement items as ready-to-fight systems. Ready-to-fight systems are sent forward with the LOGPAC.

CLASS VIII

11-30. Medical supplies, such as first aid dressings, refills for first aid kits, water purification tablets, and foot powder, are supplied to the battalion medical platoon by the BSB via LOGPAC, ambulance backhaul, or emergency delivery.

CLASS IX

11-31. The battalion stocks limited Class IX to perform organizational maintenance on small arms and communications equipment. The battalion either requests the appropriate repair parts in response to a specific request or repairs the piece of equipment by controlled exchange of serviceable parts. Rechargeable batteries for NVDs and man-portable radios may require one-for-one exchange. In combat situations, exchange and controlled substitution are the normal means of obtaining Class IX items.

ROUTINE RESUPPLY

11-32. Routine resupply operations cover items in Classes I, III, V, and IX, as well as mail and any other items requested by the company. Resupply operations normally occur once a day. Whenever possible, routine resupply should be conducted daily, ideally during periods of limited visibility.

LOGISTICS PACKAGE OPERATIONS

11-33. The company supply sergeant compiles and coordinates any unique supply request for the company and routes them through the battalion S-4. Based on the requests and the predetermined supply needs, he then organizes and assembles the LOGPAC in the battalion field trains. Supplies are usually configured to sustain the company for a 24-hour period or until the next scheduled LOGPAC. Other items to be included in the LOGPAC are coordinated by the appropriate staff officer and delivered to the field trains. These items may include, replacement personnel and Soldiers returning from medical treatment, vehicles returning to the company area from maintenance, and mail and personnel actions.

Movement of LOGPAC

11-34. Once the company LOGPAC has been formed in the field trains, it is ready to move forward under the control of the company supply sergeant. The forward support company XO or 1SG normally organizes a convoy of company LOGPACs to facilitate movement along a supply route to the LRP where the company's 1SGs take control. The 1SG controls distribution to the company using one of the various techniques discussed later. The convoy commander must also establish security measures for the LOGPAC along the MSR.

Actions at LRP

11-35. When the LOGPAC arrives at the LRP, the company 1SG assumes control of the company LOGPAC and continues tactical movement to the company resupply point. The LOGPAC stops at the LRP only when the tactical situation dictates or when ordered by the commander. Security is maintained at all times.

Resupply Procedures

11-36. The company can use the service station (Figure 11-3) or tailgate resupply (Figure 11-4) method. The time required for resupply is an important planning factor. Resupply must be conducted as quickly and efficiently as possible, both to ensure operational effectiveness and to allow the company LOGPAC vehicles to return to the LRP on time. Service station resupply of the company normally takes 60 to 90 minutes but may take longer. Tailgate resupply usually requires significantly more time than service station resupply. At times, leaders must use the in-position resupply method (Figure 11-5), but this takes much more time.

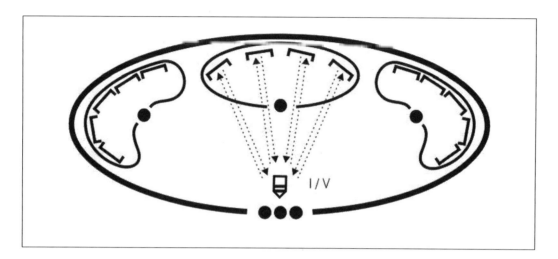

Figure 11-3. Service station resupply method.

Chapter 11

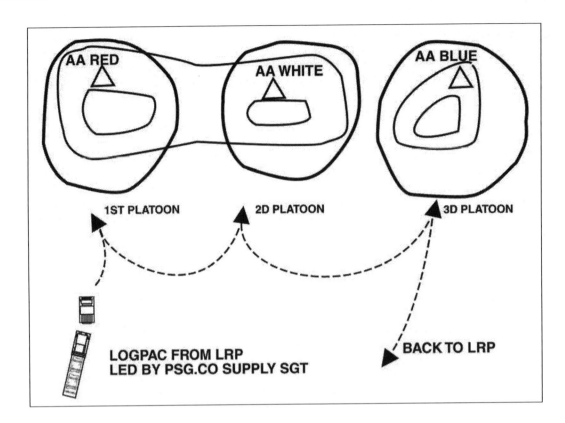

Figure 11-4. Tailgate resupply method.

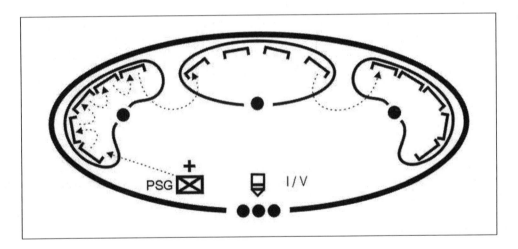

Figure 11-5. In-position method.

Return to LRP

11-37. Once resupply operations are complete, the LOGPAC vehicles are prepared for the return trip. Company vehicles requiring evacuation for maintenance are lined up and prepared for towing. Human remains and their personal effects are carried on cargo trucks, fuel trucks, or disabled vehicles. EPW ride in the cargo trucks. They are guarded by the walking wounded or by other company personnel. All supply requests and personnel action documents are consolidated for forwarding to the field trains, where the appropriate staff section processes them for the next LOGPAC. The supply sergeant leads the LOGPAC back to the LRP. The LOGPAC must keep moving through the LRP to avoid interdiction by enemy forces or artillery. Whenever possible, the reunited LOGPAC convoy returns to the BSA together. When METT-TC dictates or when the LOGPAC arrives too late to rejoin the larger convoy, the vehicles return to the BSA on their own.

RESUPPLY METHODS

11-38. As directed by the commander or XO, the 1SG establishes the company resupply point. Either he uses a resupply point (service station method) or he delivers resupply directly to the platoon (tailgate method). Occasionally the in-position resupply method might be required. He briefs each LOGPAC driver on which method to use. When he has the resupply point ready, the 1SG informs the commander. The company commander then directs each platoon or element to conduct resupply based on the tactical situation.

Service Station Resupply

11-39. The service station method allows platoons, squads, or Soldiers to move individually to a centrally located resupply point. This method requires the Soldiers to leave their fighting positions. Depending on the tactical situation, a squad or platoon moves out of its position, conducts resupply operations, and moves back into position. The squads or platoons rotate individually to eat, pick up mail and sundries, and replenish water. This process continues until the entire platoon has received its supplies. This technique is normally used in assembly areas and when contact is not likely. This technique also cuts down on the amount of vehicular movement such as the LOGPAC vehicles are stationary once they have established the resupply point.

Tailgate Resupply

11-40. When the platoons are dispersed or the tactical situation requires, the 1SG may use the tailgate method. The terrain must also permit the movement of multiple vehicles to each platoon position. The 1SG will bring LOGPAC to each platoon's area. Individual Soldiers or teams rotate back through the feeding area. While there, they pick up mail and sundries, and replenish other classes of supply. They centralize and guard any EPW. They take Soldiers KIA and their personal effects to the holding area, where the 1SG assumes responsibility for them.

In-Position Resupply

11-41. Occasionally, during operations when contact with the enemy is imminent, the in-position resupply method might be required to ensure adequate supplies are available to the squads. This method requires the company to bring forward supplies or equipment (or both) to individual fighting positions. The platoon will normally provide a guide to ensure the supplies (Class V) are distributed to the most critical position first. This method--

- Is used when an immediate need exits.
- Is used to resupply single classes of supply.
- Enables leaders to keep squad members in their fighting positions.

Note: If resupply contact is eminent or vehicles cannot move near platoon positions, platoon members may need to help the resupply personnel move supplies and equipment forward.

EMERGENCY RESUPPLY

11-42. Occasionally (normally during combat operations), the company may have such an urgent need for resupply that it cannot wait for a routine LOGPAC. Emergency resupply may involve Classes III, V, and VIII, as well as CBRN equipment and, on rare occasions, Class I. Emergency resupply can be conducted using either the service station or tailgate method, but more often the in-position method. The fastest appropriate means is normally used although procedures might have to be adjusted when the company is in contact with the enemy. In the service station method, individual squads may pull back during a lull in combat to conduct resupply and then return to the fight. With tailgate resupply, the company brings limited supplies forward to the closest concealed position behind each element.

PRESTOCKAGE OPERATIONS

11-43. Prestock resupply, which includes pre-positioning and caching, is most often required in defensive operations.

PRE-POSITIONED SUPPLIES

11-44. Normally only Class V items are pre-positioned. Prestock operations must be carefully planned and executed at every level. All leaders must know the exact locations of prestock sites, which they verify during reconnaissance or rehearsals. The company must take steps to ensure survivability of the prestock supplies. These measures include digging in prestock positions and selecting covered and concealed positions. The company commander must also have a plan to remove or destroy pre-positioned supplies to prevent the enemy from capturing them.

CACHES

11-45. A cache is a pre-positioned and concealed supply point. It can be used in any operation. Caches are an excellent tool for reducing the Soldier's load and can be set up for a specific mission or as a contingency measure. Cache sites have the same characteristics as an ORP or patrol base, with the supplies concealed above or below ground. An aboveground cache is easier to get to but is more likely to be discovered by the enemy, civilians, or animals. A security risk always exists when returning to a cache. A cache site should be observed for signs of enemy presence and secured before being used; it may have been booby-trapped and might be under enemy observation.

11-46. In the offense, advance elements may set up a cache along the intended route of advance to the objective. Caches may also be set up in-zone to support continuous operations without allowing the enemy to locate the company through air or ground resupply. Soldier's load considerations may limit the size of caches. Do not let the cache activities jeopardize the offensive mission. In some cases, special forces, allied forces, or partisans may set up caches.

11-47. In the defense, a defending unit may set up caches throughout the area of operations during the preparation phase. A cache should also be in each alternate or subsequent position throughout the depth of the defense sector. During stay-behind operations, or in an area defense on a fluid battlefield where the enemy is all around, caches might be the only source of supply for extended periods.

SECURITY

11-48. While these techniques are used in both offensive and defensive operations, the transfer of supplies to the company is usually conducted from a defensive posture. As such, the security considerations for a resupply operation are like those for a perimeter defense.

SUPPLY CONSIDERATIONS

11-49. The techniques described in the preceding paragraphs are the normal methods for resupply within the company. However, a basic understanding of nonstandard techniques, different modes of delivery, and specific supply issues is also required for the successful execution of the sustainment function.

FORAGING AND SCAVENGING

11-50. Foraging and scavenging are used infrequently and only under extreme conditions. Foraging is the gathering of supplies and equipment necessary to sustain basic needs, such as for food, water, and shelter, from within the area of operations. Scavenging is the gathering of supplies or equipment (friendly or enemy) from within the area of operations to help the user accomplish his military mission. Leaders must protect their Soldiers by determining whether the food or water is safe or whether the equipment is booby-trapped (FM 27-10).

AERIALLY RESUPPLYING

11-51. In using aerial resupply, the Infantry company commander must consider the threat's ability to locate his unit by observing the aircraft. Unless conducting the resupply in an area under friendly control and away from direct enemy observation (reverse slope of a defensive position with reconnaissance well forward), locate the drop zone or landing zone (DZ or LZ) away from the main unit in an area that can be defended for a short time. The delivered supplies are immediately transported away from the DZ or LZ.

CROSS-LEVELING

11-52. Cross-leveling is simply a redistribution of supplies throughout the unit. Usually done automatically between platoons and squads after every engagement, the company may cross-level supplies between platoons when resupply cannot occur. In some instances, supplies may not be evenly redistributed. For example, during preparation for an assault of an enemy trench system, the platoon with the task of support by fire might be required to give its hand grenades to the platoon with the task of clearing the trench.

BACKHAULING

11-53. Backhauling is a method used to make the most use of vehicular or manpack capabilities moving rearward. Backhauling returns supplies, equipment, or HAZMAT to the rear for disposition. Backhauling is also a means for nonstandard evacuation.

MANAGING CONSUMPTION OF WATER

11-54. Ensuring that Soldiers receive and drink enough water is a vital sustainment and leadership function at all levels in the company chain of command. Even in cold areas, everyone needs to drink at least two quarts of water a day to maintain efficiency. Soldiers must drink water at an increased rate in a combat environment.

11-55. Water is delivered to the unit under company or battalion control in 5-gallon cans, bottled water, water trailers, or collapsible containers. When a centralized feeding area is established, a water point is set up in the mess area and each Soldier fills his canteen as he goes through. When the company distributes rations, it can resupply water either by collecting and filling empty canteens or by distributing water cans to the platoons.

11-56. Water is habitually included in LOGPACs. The ability of the command to supply water is limited by the ability of the BSB's water section to purify, store, and distribute it. The logistics system may not always be able to meet unit needs, particularly during decentralized operations. In most environments, water is available from natural sources. Soldiers should be trained to find, treat (chemically or using field expedients), and use natural water sources. (See FM 3-05.70 for ways the unit can supply its own water, if needed.)

11-57. When water is not scarce, leaders must urge Soldiers to drink water even when they are not thirsty. The body's thirst mechanism does not keep pace with the loss of water through normal daily activity. The rate at which dehydration occurs depends on the weather conditions and the level of physical exertion.

11-58. If water is in short supply, be sparing in its use for hygiene purposes. Water used for coffee or tea might be counterproductive since both increase the flow of urine. Soups, however, are an efficient means of getting both water and nutrition when water is scarce. This is especially true in cold weather when heated food is desirable. When in short supply, water should not be used to heat MREs. A centralized heating point can be used to conserve water yet provide warmed MREs.

TRANSPORTATION

11-59. Movement of supplies, equipment, and personnel with the limited vehicle assets available requires careful planning and execution. Infantry companies have limited organic transportation for resupply operations. Vehicle assets from battalion or the forward supply company are provided for company resupply operations.

11-60. When extra vehicles are provided to the company, they must be employed to capitalize on their capability to execute the mission requirement, and they must be returned for follow-on company or parent-unit missions. Transportation assets are scarce, often resulting in trade-offs. For example, upload increased quantities of ammunition and less water, or carry unit rucksacks and be unavailable for resupply. The company commander must ensure that the asset is being employed to accomplish the most important mission. Time is critical and the company must reduce on-station time so that all company requirements can be met. Since most vehicles do not have radios, leaders must ensure that drivers know where they are going and how to get there. Land navigation training, marked routes, and strip maps referenced to landmarks are all ways to keep drivers from getting lost.

11-61. Because of the limited ground transportation, company personnel must know how to conduct aerial resupply (FM 90-4). An understanding of PZs/LZs selection, sling loading, bundle drops, and allowable cargo loads might be critical to company sustainment.

Section V. MAINTENANCE OPERATIONS

The maintenance of weapons and equipment is continuous. Every Soldier must know how to maintain his weapon and equipment in accordance with the related technical manual. The commander, XO, and 1SG must understand maintenance for every piece of equipment in the company.

MAINTENANCE REQUIREMENTS

11-62. Proper maintenance is the key to keeping vehicles, equipment, and other materials in serviceable condition. This continuous process starts when the operator of each piece of equipment or vehicle takes preventive measures and continues through repair and recovery of the equipment. Proper maintenance also includes inspecting, testing, servicing, repairing, requisitioning, recovering, and evacuating equipment.

11-63. Maintenance functions begin with PMCS, a daily responsibility for each piece of equipment to include inspection and maintenance forms (DA Form 2404 or DA Form 5988E) when required. These forms are the primary means through which the company obtains maintenance support or repair parts. The forms follow a pathway, described in the following paragraphs, from crew level to the BSA and back. Per unit SOP, the company XO or 1SG supervises the flow of these critical maintenance documents and parts. This flow normally occurs at LOGPAC.

- The unit SOP should detail when maintenance is performed (at least once a day in the field), to what standards, and who inspects it. The squad leader is often the one who inspects maintenance work whereas the platoon sergeant, platoon leader, 1SG, XO, and commander conduct spot-checks. A technique is for the 1SG or CO to spot check equipment at LOGPAC and ensure it is clean and 5988Es are complete before receiving CL I. Another technique is for each to spot-check a different platoon; whereas another is for each to check a single type of weapon or piece of equipment in all platoons daily. These instructions must be integrated into the SOP for patrol bases, assembly areas, defenses, and reorganization. They help ensure that Soldiers make a habit of maintenance and that they perform it without jeopardizing unit security.

- In addition to operator maintenance, selected Soldiers are trained to perform limited maintenance on damaged weapons and battle damage assessment and repair (BDAR).
- Inoperative equipment is fixed as far forward as possible. When a piece of equipment is damaged, it is inspected to see if it can be repaired on the spot. The company armorer keeps a small-arms repair kit in the company trains or on a company vehicle. If equipment cannot be repaired forward, it is evacuated immediately or returned with a LOGPAC. Even if the item cannot be evacuated at once, the sustainment system is alerted to prepare for repair or replacement. If a replacement is available from an evacuated Soldier or inoperative equipment, it is sent forward. If not, the leader works around it by prioritizing the use of remaining equipment, for example, he might use a squad radio for the company FM command net if the platoon radio is broken.
- Maintenance applies to all equipment. Items such as magazines, ammunition, and batteries are also maintained and inspected. While test firing in an assembly area, mark the magazines of weapons that have stoppages. If a magazine is marked more than twice, the magazine might be causing the stoppages. Inspect the ammunition belts for crew-served weapons along with the weapons. Dirty or corroded ammunition may also cause weapon malfunctions.

DESTRUCTION

11-64. When a vehicle or piece of equipment cannot be recovered or is damaged beyond repair, the platoon reports the situation to the company commander. The commander gives permission for destruction of the materiel if that is the only way to prevent enemy capture. Operators remove all salvageable equipment and parts and take all classified materials or paperwork that could be of intelligence value to the enemy. The platoon then destroys the vehicle or equipment IAW the company SOP.

Section VI. HEALTH SERVICE SUPPORT

Effective, timely medical care is an essential factor in sustaining the company's combat power during continuous operations. The company commander must ensure that the company's leaders and its medical personnel know how to keep Soldiers healthy, save their lives if they are wounded or injured, and make them well once injury or illness occurs.

HEALTH AND HYGIENE

11-65. The company commander and all leaders, in conjunction with the company senior trauma specialist and field sanitation team, emphasize and enforce high standards of health and hygiene at all times. (See Appendix A for more information.) This "*preventive medicine*" approach covers all aspects of the Soldier's health and well being, including--

- Daily shaving to ensure proper fit of the protective mask.
- Regular bathing and changing of clothes.
- Prevention of weather-related problems. These include cold injuries such as frostbite, trench foot, and immersion foot, and heat injuries like heat exhaustion and heat stroke. Soldiers must understand the effects of conditions such as sunburn and wind-chill.

- Prevention of diseases. Insect-borne diseases such as malaria and Lyme disease, and diarrhea diseases can be prevented with effective field sanitation measures, including unit waste control, water purification, rodent control, and use of insect repellents.
- Combat operational stress control, battle fatigue prevention, and strict implementation of the unit sleep plan.
- Prevent battle fatigue to include strict implementation of the unit sleep plan.

FIRST RESPONSE

11-66. First response is defined as the initial, essential stabilizing medical care rendered to wounded, injured, or ill Soldiers at the point of initial injury or illness. The first responder is the first individual to reach a casualty and provide first aid, enhanced first aid, or EMT. First aid can be performed by the casualty (self-aid) or another individual (buddy aid), while enhanced first aid is provided by the combat lifesaver (CLS). The individual who has medical MOS training is the combat medic (trauma specialist). He provides EMT for life-threatening trauma, stabilizes and prioritizes (triages) wounded for evacuation to the battalion aid station (BAS). At the BAS, wounded Soldiers receive advanced trauma medicine (ATM) by the treatment team composed of the surgeon, physician's assistant, and a senior trauma specialist.

COMBAT LIFESAVER

11-67. The CLS is a nonmedical Soldier trained to provide advanced first aid/lifesaving procedures beyond the level of self-aid or buddy aid. The CLS is not intended to take the place of medical personnel but to slow deterioration of a wounded Soldier's condition until treatment by medical personnel is possible. Each certified CLS is issued a CLS aid bag. Whenever possible, the company commander ensures there is at least one combat lifesaver in each fire team. An emerging "*first responder*" program expands CLS trauma treatment with increased emphasis on combat and away from training injuries.

11-68. Combat lifesavers are squad members trained in emergency medical techniques. They are the "*911*" medical assets for the squad until a medic or another more qualified medical person becomes available. Because combat lifesaving is an organic capability, the platoon and company should make it a training priority. The combat lifesaver ensures the squad CLS bag is packed, all IVs are present, and litters are properly packed, and identifies Class VIII shortages to the platoon medic. He participates in all casualty treatment and litter-carry drills. The combat lifesaver must know the location of the CCP and the SOP for establishing it. He has a laminated quick reference nine-line MEDEVAC card.

SENIOR TRAUMA SPECIALIST

11-69. The senior trauma specialist (company senior medic) is both the company's primary medical treatment practitioner and the supervisor of all battlefield medical operations. The latter role encompasses numerous responsibilities. The senior trauma specialist works closely with the company commander to ensure all members of the company understand what to do to provide and obtain medical treatment in combat situations. He oversees the training of combat lifesavers. Once combat begins, he will manage the company CCP, provide medical treatment, and prepare patients for MEDEVAC. He helps the 1SG arrange casualty evacuation. The senior trauma specialist is also responsible for monitoring the paperwork that is part of the medical treatment and evacuation process.

- He ensures that the casualty feeder report remains with each casualty until the Soldier reaches the battalion main aid station or field aid station.
- If a Soldier's remains cannot be recovered, the senior trauma specialist completes DA Form 1156 (*Casualty Feeder Card*), and gives it to the 1SG for processing as soon as possible.

Sustainment Operations

PLATOON MEDICAL OR TRAUMA SPECIALIST

11-70. The fact that platoon members commonly address their trauma specialist as "*doc*" or "*medic*" shows his critical role in providing them with competent, life-saving care. During combat planning and preparation he inspects platoon CLS bags, verifies IVs are placed in litters, and fills class VIII shortages. He will determine the location for the platoon CCPs and the SOP for establishing them. He rehearses casualty treatment and litter carries with all platoon members, not only aid and litter teams; and conducts CLS refresher training.

> *Note:* DA Form 1156 is collected at the aid station by designated medical personnel; it is forwarded to the S-1 section for further processing through administrative channels in the battalion field trains.

PLATOON SERGEANT

11-71. Although unit SOP dictates specific responsibilities, the platoon sergeant is typically responsible for ensuring that wounded or injured personnel receive immediate first aid and that the commander is informed of casualties. During critical operations, or when the platoon takes a lot of casualties, the platoon sergeant normally oversees the platoon casualty collection point (CCP). He coordinates with the 1SG and company senior trauma specialist for ground evacuation. He may ensure that the casualty feeder (DA Form 1156) form has been completed and routed to the proper channels. The platoon sergeant carries a laminated quick reference nine-line MEDEVAC card.

FIRST SERGEANT

11-72. The 1SG oversees the operation of the company CCP, particularly in critical operations or when casualties are high. He brings the full measure of his experience and authority to bear in the efficient treatment, collection, preparation, and transport of casualties. Successful casualty evacuation (CASEVAC) depends on his ability to anticipate, plan, and rehearse his CCP operation. METT-TC dictates the CCP site location. It must be accessible by both ground and air transport. The 1SG supervises and coordinates casualty operations, collects witness statements and submits them to the battalion S-1, and submits the battle loss report to the battalion TOC. These duties also relate to another important combat function of the 1SG: managing the company's personnel status. As needed, the 1SG cross-levels personnel to make up for shortages.

COMMANDER

11-73. The company commander has overall responsibility for medical services. His primary task is to position medical personnel at the proper point on the battlefield to treat casualties or to evacuate those casualties properly. The company commander designates the location for the company's CCP and ensures that the location is recorded on the appropriate overlays. He also develops and implements appropriate SOPs for casualty evacuation. Two key planning considerations follow.

11-74. The commander analyzes both fundamental categories of treatment and evacuation to determine if he must accept risk in one or the other and how he may mitigate identified risks. For example, where distances to available MTF are excessive and transportation assets stretched, the commander might request more medics during an operation.

11-75. Sites for casualty treatment and evacuation will vary widely on the noncontiguous battlefield and the commander tries to identify, disseminate, and coordinate with all available MTF accessible to his unit including those outside his organization.

CASUALTY EVACUATION

11-76. MEDEVAC is not the same as so-called "*casualty evacuation*" (CASEVAC).

- Casualty transport, commonly called CASEVAC, is the movement of casualties by nonmedical assets without specialized trauma care. For the purposes of this discussion, casualty transport or CASEVAC will mean that which is done when moving casualties from the point of injury (POI) to the platoon CCP or company CCP.
- MEDEVAC is the movement of casualties using medical assets while providing en route medical care. Ideally, casualties are transferred from a CCP to a MEDEVAC asset.

11-77. The two areas of medical support are treatment and evacuation. Effective CASEVAC has a positive impact on the morale of a unit. Casualties are cared for at the point of injury (or under nearby cover and concealment) and receive self- or buddy aid, advanced first aid from the combat lifesaver, or emergency medical treatment from the trauma specialist (company or platoon medic).

11-78. During the fight, casualties should remain under cover where they received initial treatment (self- or buddy aid). As soon as the situation allows, casualties are moved to the platoon CCP. From the platoon area, casualties are normally evacuated to the company CCP and then back to the BAS. The unit SOP addresses this activity, to include the marking of casualties in limited visibility operations. Small, standard, or IR chemical lights work well for this purpose. Once the casualties are collected, evaluated, and treated, they are prioritized for evacuation back to the company CCP. Once they arrive at the company CCP, the above process is repeated while awaiting their evacuation back to the BAS.

11-79. An effective technique, particularly during an attack, is to task-organize a logistics team under the 1SG. These Soldiers carry additional ammunition forward to the platoons and evacuate casualties to either the company or the battalion CCP. The leader determines the size of the team during his estimate.

11-80. When the company is widely dispersed, the casualties might be evacuated directly from the platoon CCP by vehicle or helicopter. Helicopter evacuation might be restricted due to the threat of enemy ground to air small arms, shoulder fired or other air defense weapons. In some cases, the casualties must be moved to the company CCP before evacuation. If the capacity of the battalion's organic ambulances is exceeded, unit leaders may rerole supply or other vehicles to backhaul or otherwise transport nonurgent casualties to the battalion aid station. In other cases, the platoon sergeant may direct platoon litter teams to carry the casualties to the rear.

11-81. Leaders minimize the number of Soldiers required to evacuate casualties. Casualties with minor wounds can walk or even assist with carrying the more seriously wounded. Soldiers can make field-expedient litters by cutting small trees and putting the poles through the sleeves of buttoned ACU blouses. A travois, or skid, might be used for casualty evacuation. Wounded are strapped on this type of litter, then one person can pull it. It can be made locally from durable, rollable plastic. Tie-down straps are fastened to it. In rough terrain, or on patrols, litter teams can evacuate casualties to the battalion aid station. Then, they are carried with the unit either until transportation can reach them or until they are left at a position for later pickup.

11-82. Unit SOPs and OPORDs address casualty treatment and evacuation in detail. They cover the duties and responsibilities of key personnel, the evacuation of chemically contaminated casualties (on separate routes from noncontaminated casualties), and the priority for operating key weapons and positions. They specify preferred and alternate methods of evacuation and make provisions for retrieving and safeguarding the weapons, ammunition, and equipment of casualties. Slightly wounded personnel are treated and returned to duty by the lowest echelon possible. Platoon aid men, evaluate sick Soldiers, and either treat or evacuate them as necessary. Casualty evacuation is rehearsed like any other critical part of an operation.

11-83. For procedures in the use of the casualty feeder report, DA Form 1156.

Note: Before casualties are evacuated to the CCP or beyond, leaders should remove all key operational or sensitive items and equipment, including COMSEC devices or SOIs, maps, position location devices. Every unit should establish an SOP for handling the weapons and ammunition of its WIAs. Protective masks must stay with the individual.

11-84. At the CCP, the senior trauma specialist conducts triage of all casualties, takes the necessary steps to stabilize their condition, and initiates the process of evacuating them to the rear for further treatment. He helps the 1SG arrange evacuation via ground or air ambulance, or by nonstandard means.

11-85. When possible, the HHC medical platoon ambulances provide evacuation and en route care from the Soldier's point of injury or the company's CCP to the BAS. The ambulance team supporting the company works in coordination with the senior trauma specialist supporting the platoons. In mass casualty situations, nonmedical vehicles might be used to assist in casualty evacuation as directed by the Infantry company commander. Plans for the use of nonmedical vehicles to perform casualty evacuation should be included in the unit SOP. Ground ambulances from the BSMC or supporting corps air ambulances evacuate patients from the BAS back to the BSMC medical treatment facility (MTF) located in the BSA.

Note: During entry operations, air ambulances might be unavailable for the first 96 hours.

SOLDIERS KILLED IN ACTION

11-86. The company commander designates a location for the collection of those KIA. Temporary remains holding areas should be established behind a natural barrier, such as a stand of trees, or shielded from the view of others by using either tents or tarpaulins. All personal effects remain with the body, but equipment and issue items become the responsibility of the squad leader until they can be turned over to the 1SG or supply sergeant. As a rule, human remains should not be transported on the same vehicle as wounded Soldiers. The commander sends a letter of condolence to the Soldier's next of kin, normally within 48 hours of the death.

Section VII. REORGANIZATION AND WEAPONS REPLACEMENT

To maintain effective, consistent combat power, the company must have specific plans and procedures that allow each element to integrate replacement personnel and equipment quickly. Unit SOP defines how Soldiers and equipment are prepared for combat, including areas such as uploading, load plans, PCIs, and in-briefings.

REPLACEMENTS AND CROSS-LEVELING OF PERSONNEL

11-87. Replacements for wounded, killed, or missing personnel are requested through the battalion S-1. Returning or replacement personnel arriving with the LOGPAC should have already been issued all TA-50 equipment, MOPP gear, and other items, including their personal weapons. Within the company, each platoon leader cross-levels personnel among his crews, with the 1SG controlling cross-leveling from platoon to platoon.

11-88. Integrating replacements into a company is important. A new arrival on the battlefield might be scared and disoriented as well as unfamiliar with local SOPs and the theater of operations. The following procedures help integrate new arrivals into a company.

- The company commander meets them and welcomes them to the unit. This is normally a brief interview. The company commander must have an SOP for reception and integration of newly assigned Soldiers.
- The platoon leader and platoon sergeant welcome them to the unit, inform them of unit standards, and introduce them to their squad leaders.
- The squad leader introduces them to the squad and briefs them on duty positions. He also ensures that each replacement has a serviceable, zeroed weapon, as well as ammunition, MOPP gear, and other essential equipment. The in-briefing should cover the squad and platoon's recent and planned activities.
- The new arrival is told about important SOPs and a paper copy should be given to the Soldiers on any special information concerning the area of operations. He might be given a form letter to

send to his next of kin. The letter should tell them where to mail letters and packages, tell them how to use the Red Cross in emergencies, and introduce them to the chain of command.

ENEMY PRISONERS OF WAR, DETAINEES, AND OTHER RETAINED PERSONS

11-89. All persons captured, detained, or retained by US Armed Forces during the course of military operations are considered "*detained*" persons until their status is determined by higher military and civilian authorities. The BCT has a military police platoon organic to the BSTB to take control of and evacuate detainees. However, as a practical matter, Infantry squads, platoons, companies and battalions capture and must provide the initial processing and holding for detainees.

11-90. All detained persons shall be given humanitarian care and treatment immediately. US Armed Forces will never torture, mistreat, or purposely place detained persons in positions of danger. No military necessity or exception exists that allows the violation of these principles. Tactical questioning of detainees is allowed relative to collection of CCIR. However, detainees must always be treated IAW the US Law of War Policy as set forth in the DOD Directive 5100.77, *DOD Law of War Program*.

11-91. In any tactical situation, the Infantry platoons and company have specific procedures and guidelines for handling prisoners and captured material. Unit SOPs determine exact EPW processing. Platoons usually have an EPW collection point and the company will have a separate collection point.

11-92. The five-S's and T method reminds Soldiers about the basic principles for handling EPW, which include tagging prisoners and all captured equipment and materiel (Table 11-1).

Action	Description
Search	Search each captive for weapons, items of intelligence value, and items that would make escape easier or compromise US security interests. Confiscate these items. Prepare a receipt when taking property. *Note:* When possible, conduct same gender searches. When not possible, perform mixed gender searches in a respectful manner. Leaders must carefully supervise Soldiers to prevent allegations of sexual misconduct. Captives may keep the following items found in a search. • Protective clothing and equipment that cannot be used as a weapon, such as helmets, protective masks, and clothing, for use during evacuation from the combat zone. • Retained property, such as ID cards or tags, personal property having no intelligence value and no potential value to others, such as photos and mementos, clothing, mess equipment (except knives and forks), badges of rank and nationality, decorations, religious literature, and jewelry. MI teams may take personal items, such as diaries, letters, and family pictures for review, but must return them to their owner(s). • Private rations of the detainee. Confiscate currency only on the order of a commissioned officer (AR 190-8) and provide a receipt and establish a chain of custody using DA Form 4137 (*Evidence/Property Custody Document*) or any other field expedient substitute.
Silence	Silence the detainees by directing them not to talk. Use gags if necessary, and check frequently to ensure that the detainee is still able to breathe.
Segregate	Segregate detainees based on perceived status and positions of authority. Segregate leaders from the remainder of the population. Segregate hostile elements such as religious, political, or ethnic groups hostile to one another. For their protection, normally segregate minor and female detainees from adult male detainees.

Table 11-1. Five S's and T method of detainee field processing.

Safeguard	Safeguard the detainees. Ensure detainees are provided adequate food, potable water, clothing, shelter, and medical attention. Ensure detainees are not exposed to unnecessary danger and are protected (afforded the same protective measures as the capturing force) while awaiting evacuation. Do not use coercion to obtain information from the captives. Provide medical care to wounded and sick detainees equal in quality to that provided to US forces. Report acts or allegations of abuse through command channels to the supporting judge advocate, and to the US Army Criminal Investigation Command.
Speed to a Safe Area/Rear	Evacuate detainees from the battlefield as quickly as possible, ideally to a collection point where military police take custody of the detainees. Transfer custody of all captured documents and other property to the US forces assuming responsibility for the detainees.
Tag	Use DD Form 2745 (Enemy Prisoner of War (EPW) Capture Tag) or a field expedient alternative, and include the following information. • Date and time of capture. • Location of capture (grid coordinates). • Capturing unit. • Circumstances of capture. Indicate specifically why the person has been detained. Use additional documentation when necessary and feasible to elaborate on the details of capture. • Documentation should answer the five Ws –who, what, where, why, and witnesses. • Use a form, such as DA Form 2823 (*Sworn Statement*) or an appropriate field expedient to document this information. • List all documents and items of significance found on the detainee. Attach Part A, DD Form 2745, or an appropriate field expedient capture card to the detainee's clothing with wire, string, or another type of durable material. Instruct the captive not to remove or alter the tag. Maintain a written record of the date, time, location, and personal data related to the detention. Attach a separate identification tag to confiscated property that clearly links the property with the detainee from whom it was seized.

Table 11-1. Five S's and T method of detainee field processing (continued).

11-93. In addition to initial processing, the capturing element provides guards and transportation to move prisoners to the designated EPW collection points. The capturing element normally carries prisoners on vehicles already heading toward the rear such as tactical vehicles returning from LOGPAC operations. The capturing element must also feed, provide medical treatment, and safeguard EPW until they reach the collection point.

11-94. Once the EPW arrive at the platoon collection point, the platoon sergeant assumes responsibility for them. He provides for security and transports them to the company EPW collection point where they are consolidated for evacuation to the rear. Normally the 1SG, often assisted by the supply section, moves the detainees to the vicinity of the combat trains for processing and subsequent interrogation by battalion or MI company personnel. Crews of vehicles undergoing repair, CS, or sustainment personnel can often be used as guards. They are then evacuated to the battalion EPW collection point, to the Brigade holding area or beyond.

11-95. Before an EPW detainee is evacuated, tag him with--

- Part A, DD Form 2745 (Figure 11-6).
- Part B, DD Form 2745 (Figure 11-7, the unit record copy).

Sustainment Operations

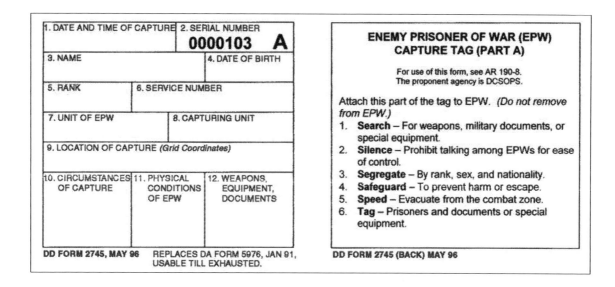

Figure 11-6. Enemy prisoner of war detainee tag.

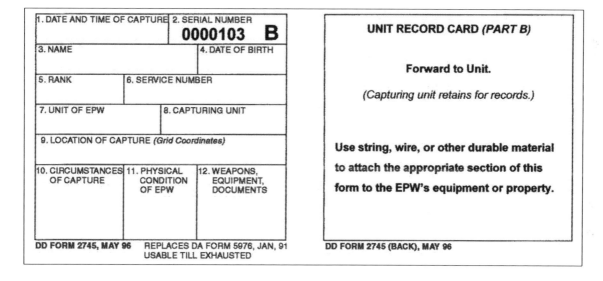

Figure 11-7. Unit record copy.

11-96. Before an EPW detainee is evacuated, tag his confiscated property with a DD Form 2745, C (Figure 11-8, page 11-28). Detainees are allowed to keep protective equipment such as protective masks. Other captured military equipment and detainee personal effects are inventoried on DA Form 4137. Soldiers then coordinate with the platoon or company headquarters to link up and turn the documents and prisoners over to designated individuals.

1. DATE AND TIME OF CAPTURE	2. SERIAL NUMBER **0000103 C**
3. NAME	4. DATE OF BIRTH
5. RANK	6. SERVICE NUMBER
7. UNIT OF EPW	8. CAPTURING UNIT
9. LOCATION OF CAPTURE *(Grid Coordinates)*	
10. DESCRIPTION OF WEAPONS, SPECIAL EQUIPMENT, DOCUMENTS	

DD FORM 2745, MAY 96 REPLACES DA FORM 5976, JAN 91 USABLE TILL EXHAUSTED.

DOCUMENT/SPECIAL EQUIPMENT WEAPONS CARD (PART C)

Attach this part of the tag to property taken. *(Do not remove from property.)*

As a minimum, the tag must include the following information:

Item 1. Date and time of capture *(YYYYMMDD).*
Item 8. Capturing unit.
Item 9. Place of capture *(grid coordinates).*
Item 10. Circumstances of *capture (how the EPW was captured).*

DD FORM 2745 (BACK), MAY 96

Figure 11-8. Enemy prisoner of war, document, and special equipment tag.

Chapter 12

Urban Operations

What is urban operations (UO)? It is all military actions, combat or not, that are conducted in any urban area.

All UO are full-spectrum operations. This means they can include peacetime military engagements, offensive, defensive, and stability operations, as well as support to civil authorities. The term replaces military operations on urbanized terrain (MOUT).

UO demand a high level of unit proficiency, coordination, training, and individual dedication. The most important lesson learned about UO is that powerful combined-arms teams work. Infantry must closely integrate with Armor, aviation, indirect fire support, reconnaissance, communications, civil affairs, PSYOP, and logistical elements. Leaders compose these teams after a detailed analysis of the OE and METT-TC (FM 3-06 and FM 3-06.11).

In this chapter, the term "*armored vehicles*" refers to both tracked and wheeled armored vehicles of all types, including tanks, armored fighting and reconnaissance vehicles, and Infantry carriers.

Section I. INTRODUCTION

Urban operations are not new. Throughout its history the US Army has fought many battles on urban terrain. What is new is that urban areas and urban populations have grown significantly and have begun to exert a much greater influence on military operations. The worldwide shift from rural to urban societies has created new challenges for leaders and Soldiers to not only win urban battles but also to establish stability, restore vital functions and to transition to a more secure environment. In many cases, the city's people and society have become even more of a factor affecting planning than the city's buildings and roads. Although companies and platoons will continue to operate as part of a battalion task force, they will often conduct mission with a degree of independence seldom seen previously. This section highlights the basic planning considerations for Infantry company commanders conducting full spectrum urban operations.

DEFINITIONS

12-1. To understand urban operations, you need to know more about urban areas and urban combat.

URBAN AREAS

12-2. An urban area is a concentration of structures, facilities, and people that form the economic and cultural focus for the surrounding area. An urban area consists of more than just buildings, roads, and bridges. It is a complex, interconnected system that includes both the physical structures and the people who work and live in them. Cities house many human endeavors: society, culture, religion, commerce, communication, manufacturing, and education, among others The population is ultimately the dominant feature in all urban operations, although the urban terrain does have significant tactical impact. Operations are affected by all five categories of urban areas. Cities, metropolises, and megalopolises with associated urban sprawl cover hundreds of square kilometers.

- Villages (population under 3,000).
- Towns (population between 3,000 and 100,000, but not part of a major urban complex).
- City (population over 100,000 to 1 million).
- Metropolis (population between 1 and 10 million).
- Megalopolis (population over 10 million).

URBAN COMBAT

12-3. Urban combat is a specific subset of urban operations. It collectively describes offensive and defensive operations conducted to defeat an enemy in an urban area both during high-intensity combat and during stability operations. Usually, that is mixed with noncombatants. Because of this intermingling, and the resulting necessity to limit collateral damage, the rules of engagement (ROE) and the constraints placed on the use of combat power may be more restrictive than under other combat conditions. This term replaces several older terms, including fighting in built-up areas (FIBUA) and combat in cities (CIC). Units may conduct operations in urban areas, including urban combat, when—.

- An assigned objective lies within an urban area and cannot be bypassed.
- The urban area and the population it contains is key or decisive in setting or shaping the conditions for current or future operations.
- An urban area is between two natural obstacles and cannot be bypassed.
- The urban area is in the path of a general advance and cannot be surrounded or bypassed.
- Political or humanitarian concerns require the control of an urban area or necessitate operations within it.
- Defending from urban areas supports a more effective overall defense or cannot be avoided.
- Occupation, seizure, and control of the urban area will deny the threat control of the urban area and the ability to impose its influence on both friendly military forces and the local civilian population. Therefore, friendly forces can retain the initiative and dictate the conditions for future operations.

CONDITIONS

12-4. Due to political, technological, and societal changes, and to the Army's growing role in maintaining regional stability, urban operations are conducted across the full spectrum of offense, defense, stability, and support. This spectrum affects how units plan and execute assigned missions. The enemy's actions significantly affect the conditions of UO, which may transition from one condition to another rapidly. Units might be conducting operations under different conditions at two locations at the same time. The following definitions of the three general conditions of UO provide clarity, focus, and a mental framework for commanders and leaders conducting tactical planning for urban operations.

SURGICAL CONDITIONS

12-5. Operations conducted under surgical conditions include special-purpose raids, small precision strikes, or small-scale personnel seizures or arrests, focused psychological or civil affairs operations, or recovery operations. Many operations under this condition are planned and conducted by Special Operations Forces. This condition is the most tightly focused and normally the least physically destructive of all the conditions of UO. Operations under this condition may involve cooperation between US forces and host nation police or paramilitary security forces. In many cases, conventional Infantry units are not directly involved in operations under surgical conditions. They may support SOF forces it by isolating the area or providing security or crowd control. Very often, Infantry forces are deeply involved in follow-up operations executed due to SOF actions.

PRECISION CONDITIONS

12-6. Operations under precision conditions include full-spectrum urban operations in areas. The enemy might be mixed with noncombatants, or strategic or political considerations might restrict the use of combat power. Infantry units must routinely expect to operate under precision conditions.

12-7. Urban operations under precision conditions normally involve combat action, and usually close combat. Some of this combat can be quite violent for short periods. It is marked, however, by the conscious acceptance by US forces of the need to focus and sometimes restrain the combat power used. The commander always attempts to bring overwhelming force to bear, but only on specific portions of the urban area occupied by the threat. He may choose different tactics or techniques in order to remain within the bounds of more restrictive ROE. He may make use of non-lethal munitions and equipment or use the unit's weapons in less lethal ways in order to reduce collateral damage or injury of non-combatants.

12-8. When preparing for urban operations under precision conditions, commanders and leaders must realize that, not only could the ROE change, but also the tactics and techniques used. These changes require that Soldiers be given time to prepare, to adjust small-unit SOPs.

HIGH INTENSITY CONDITIONS

12-9. Operations under high intensity conditions include combat actions against a determined enemy occupying prepared positions or conducting planned attacks. Urban operations under these type conditions require the coordinated application of the full combat power of the joint combined arms team. Infantry units must be prepared at all times to conduct violent combat under conditions of high-intensity UO.

12-10. An Infantry unit's mission during UO under high intensity conditions is normally to seize, clear, or defend urban terrain, engaging and defeating the enemy by using appropriate levels of force. Although the changing world situation may have made high-intensity UO less likely, it represents the high end of the combat spectrum, and units must be trained for it.

12-11. Urban combat under high-intensity conditions is the most stressful of all operations in urban areas and can be casualty-intensive for both sides. Even though the fully integrated firepower of the joint combined arms team is being used, commanders must still strive to prevent unnecessary collateral damage and casualties among noncombatants.

Section II. URBAN BATTLESPACE

Urban areas mainly consist of fabricated features such as buildings that provide cover and concealment, limit fields of observation and fire, and block movement of forces (especially heavy forces). Thick-walled buildings provide ready-made fortified positions. Thin-walled buildings may have fields of observation and fire that may prove important. Another important aspect is that urban areas complicate, confuse, and degrade the commander's ability to identify and control his forces. All these factors will influence the urban battle space. Commanders and leaders can enhance situational understanding by maintaining a clear understanding of their urban battle space.

TYPES

12-12. Urban battle space includes--

URBAN AIRSPACE

12-13. Airspace provides a rapid avenue of approach into an urban area. While aviation assets are unaffected by obstacles such as rubble, vehicles, or constructed barriers, they must consider power lines, towers, sign poles, and billboards when flying. Task force reconnaissance elements can locate, identify, and report these obstacles to allow for improved flight planning. The proliferation of UASs has also complicated the company commander's task.

SUPERSURFACES (TOPS OF BUILDINGS)

12-14. The term "*supersurface*" refers only to the top, roof, or apex of a structure. These areas can provide cover and concealment, limit or enhance observation and fields of fire, and, depending on the situation, enhance, restrict, canalize, or block movement. Supersurface areas can also provide concealed positions for snipers, automatic weapons, light and medium antitank weapons, and man-portable air defense systems. In many cases, they enable top-down attacks against the weakest points of armored vehicles and unsuspecting aircraft.

INTRASURFACES (INTERIOR OF BUILDINGS)

12-15. The intrasurface refers to the floors within the structural framework—the area from the surface level (ground) up to, but not including, the structure's permanent roof, or apex. Intense combat engagements often occur in this intrasurface area, which is also known for its widely diverse and complex nature. The intrasurface of a building greatly limits what can be accomplished by reconnaissance and surveillance systems, but at the same time, enhances cover and concealment. Also, the intrasurface areas provide mobility corridors within and between structures at upper levels for both friendly and enemy forces. Intrasurface areas may also provide concealed locations for snipers, automatic weapons, light and medium antitank weapons, and man-portable air defense systems. In many cases, they enable top-down attacks against the weakest points of armored vehicles and unsuspecting aircraft.

SURFACES (GROUND, STREET, AND WATER LEVEL)

12-16. Streets are usually avenues of approach. Streets and open areas provide a rapid approach for ground movement in urban terrain. Units moving along streets can be canalized by buildings and have little space for maneuver, while approaching across large open areas such as parks, athletic fields and parking areas. Streets also expose forces to observation and engagement by enemy elements. Using obstacles on city streets usually works better than it does in open terrain, because in cities they are harder to bypass.

SUBSURFACES (UNDERWATER AND SUBTERRANEAN)

12-17. Common subsurface areas, which include subways, sewers, public utility systems, and cellars, can be used as avenues of movement for dismounted elements. Both attacker and defender can use subterranean routes to outflank or turn the opposition, or to conduct infiltration, ambushes, counterattacks, and sustaining operations. Subsurface systems in some urban areas are easily overlooked but can be important to the outcome of operations.

12-18. Commanders and leaders must be able to identify building types, construction materials, and building design and must understand the effectiveness and limitations of weapons against these factors. They must also understand that urban combat will require them to visualize a three-dimensional battle space. Commanders and leaders must be aware of how their urban battle space is changing as friendly and enemy forces and civilians move and as weather and environmental conditions change. They can react to changes within their battle space with the timely movement of assault, support, and breaching elements in the offense; repositioning of platoons and squads in the defense; and synchronization of CS and sustainment assets. Other factors that impact battle space include--

- CASEVAC and resupply procedures.
- Procedures for handling EPW and noncombatants.
- Rules of engagement.
- Battlefield obscuration.
- Communications.
- Movement of vehicles, that is, how the battlespace will affect movement and target engagement.

ZONES

12-19. The commander subdivides the area of operations (AO) and the area of interest (AI) into appropriate types of zones, as follows (see FM 34-130 for more information).

CITY CORE

12-20. The city core is the heart of the urban area—the downtown or central business district. It may be relatively small and compact, but it often contains a larger percentage of the urban area's shops, offices, and public institutions. It normally contains the highest density of multistory buildings and subterranean areas. In most cities, the core has undergone more recent development than the core periphery. As a result, the two regions are often quite different. Typical city cores of today are made up of buildings that vary greatly in height.

CORE PERIPHERY

12-21. The core periphery is located at the edges of the city core. The core periphery consists of streets 12 to 20 meters wide with continuous fronts of brick or concrete buildings. The building heights are uniform—two or three stories in small towns, five to ten stories in large cities. Dense random and close orderly block are two common construction patterns that can be found within the city core and core periphery zones.

Dense Random Construction

12-22. This construction is a typical old inner city pattern with narrow winding streets radiating from a central area in an irregular manner. Buildings are closely located and frequently close to the edge of a roadway.

Close Orderly Block Construction

12-23. Wider streets generally form rectangular patterns in this area. Buildings frequently form a continuous front along the blocks. Inner-block courtyards are common.

DISPERSED RESIDENTIAL AREA

12-24. This type area is normally contiguous to close-orderly block areas in Europe. The pattern consists of row houses or single-family dwellings with yards, gardens, trees, and fences. Street patterns are normally rectangular or curving.

HIGH-RISE AREA

12-25. Typical of modern construction in larger cities and towns, this area consists of multistoried apartments, separated open areas, and single-story buildings. Wide streets are laid out in rectangular patterns. These areas are often contiguous to industrial or transportation areas or interspersed with close-orderly block areas.

INDUSTRIAL-TRANSPORTATION AREA

12-26. Industrial-transportation areas are generally located on or along major rail and highway routes in urban complexes. Older complexes might be located within dense, random construction or close-orderly block areas. New construction normally consists of low, flat-roofed factory and warehouse buildings. High-rise areas providing worker housing is normally located adjacent to these areas throughout the Orient. Identification of transportation facilities within these areas is critical because these facilities, especially rail facilities, pose significant obstacles to military movement.

PERMANENT OR FIXED FORTIFICATIONS

12-27. These include any of several different types, and might be considered isolated forts. While most of these fortifications are found in Western Europe, many can be found in the Balkans, Middle East, Asia, Africa, and South America. Those in the United States are mostly of the coast defense type. Permanent fortifications can be made of earth, wood, rock, brick, concrete, steel-reinforced concrete, or any combination of the above. Some of the latest variants are built underground and employ heavy tank or warship armor, major caliber and other weapons, internal communications, service facilities, and CBRN overpressure systems.

SHANTYTOWNS

12-28. Shantytowns do not necessarily follow any of the above patterns and might be found in many different zones within urban areas. Many underdeveloped countries are composed of small towns and villages and very few large cities. Most of the structures in the small towns and villages might be constructed from materials ranging from cardboard to concrete block. Some countries in arid regions depend on adobe for construction. Even the larger cities can have shantytowns at the edge that consist of cardboard or tin shacks.

BUILDING ANALYSIS

12-29. This part of the analysis is very important for commanders, leaders, and staffs. Commanders and leaders must be capable of identifying the types of buildings that are in their company sectors, objective areas, and areas of influence. They must also understand the effects of weapons that are used against those buildings. The capability of identifying building types and understanding weapons effects enables commanders to give clear instructions to their subordinates concerning mission execution. It also helps the platoon and the squad leaders to choose the appropriate weapons or explosives to accomplish their respective missions.

TYPES OF MASS-CONSTRUCTION BUILDINGS

12-30. Mass-construction buildings are those in which the outside walls support the weight of the building and its contents. Additional support, especially in wide buildings, comes from using load-bearing interior walls, strongpoints (called pilasters) on the exterior walls, cast-iron interior columns, and arches or braces over the windows and doors.

Modern Mass-Construction Buildings

12-31. Modern types of mass construction buildings are wall and slab structures such as many modern apartments and hotels as well as tilt-up structures commonly used for industry or storage. Mass-construction buildings are constructed in many ways.

- The walls can be built in place using brick, block, or poured-in-place concrete.
- The walls can be prefabricated and "*tilt-up*" or reinforced-concrete panels.
- The walls can be prefabricated and assembled like boxes.

Brick Buildings

12-32. Brick buildings are the most common mass-construction buildings. In Europe, brick buildings are commonly covered with a stucco veneer so that bricks do not show. One of the most common uses of brick buildings is the small store. These buildings are found in all urban areas but are most common in the dense random construction and close-orderly block areas.

Warehouses

12-33. Another common mass-construction building in the industrial transportation zone is the warehouse. It is built of poured-in-place concrete reinforced with steel bars or of prefabricated "*tilt-up*" walls. The walls of warehouses provide good cover, although the roof is vulnerable. The warehouses' large open bays permit ATGM firing and, because they are normally found in outlying areas, often afford adequate fields of fire. These buildings are built on slabs, which can normally support the weight of vehicles and can provide excellent cover and concealment for tanks.

Box-Wall Types

12-34. Another mass-construction building is the box-wall principle type. It is made from prefabricated concrete panels, which are made of 6- to 8-inch-thick reinforced concrete. The outside wall is often glass. The box-wall principle building provides good cover, except at the glass wall. The rooms are normally too small for ATGMs to be fired. A good circulation pattern exists from room to room and from floor to floor. These buildings are commonly used as hotels or apartments and are located in high-rise areas.

Public Gathering Places

12-35. Public gathering places (churches, theaters) are mass construction buildings with large, open interiors. The walls provide good cover, but the roof does not. The interior walls are not load bearing and are normally easy to breach or remove. These buildings have adequate interior space for firing ATGMs. They are often located next to parks or other open areas and, therefore, have fields of fire long enough for ATGMs. Public gathering places are most common in the dispersed residential and high-rise areas.

TYPES OF FRAMED BUILDINGS

12-36. Framed buildings are supported by a skeleton of columns and beams and are usually taller than frameless buildings. The exterior walls are not load bearing, and are referred to as either heavy clad or light clad. Another type of framed building often found in cities is the garage, which has no cladding.

Heavy-Clad Framed Buildings

12-37. Heavy-clad buildings were common when framed buildings were first introduced. Their walls are made of brick and block that are sometimes almost as thick as frameless brick walls, although not as protective. Heavy-clad framed buildings are found in the city core or core periphery. They can be recognized by a classic style or architecture in which each building is designed with three sections: the pediment, shaft, and capital. Unlike the brick building, the walls are the same thickness on all floors, and the windows are set at the same depth throughout. Often the frame members (the columns) can be seen, especially at the ground floor. The cover provided from the cladding, consisting of layers of terra cotta blocks, brick, and stone veneer, is not as good as cover from the walls of brick buildings. It protects against small arms fire and light shrapnel but does not provide much cover against heavy weapons.

Light-Clad Buildings

12-38. Light-clad buildings are more modern and might be constructed mostly of glass. Most framed buildings built since World War II are light-clad buildings. They are found in both core and outlying high-rise regions. Their walls consist of a thin layer of brick, lightweight concrete, or glass. Such materials provide minimal protection against any weapon. However, the floors of the buildings are much heavier and provide moderate overhead cover. The rooms in light-clad framed buildings are bigger than those in heavy-clad buildings. This feature, along with the fact that the buildings usually stand detached from other buildings, favors the employment of ATGMs. The interior partitions are thin, light, and easy to breach.

Chapter 12

Parking Garages

12-39. The parking garage is one of the few buildings in an urban area in which all floors support vehicles. They provide the means to elevate vehicle-mounted TOWs. Their open interiors permit firing of ATGMs. Garages are normally high enough to provide an all-round field of fire for antiaircraft weapons. For example, a Soldier equipped with a Stinger could hide under the top floor of the garage, come out to engage an aircraft, and then take cover back inside.

Section III. CHARACTERISTICS

Many characteristics separate urban operations from operations in other environments. One of the most significant is the intimate interaction between US forces and the populace. These affect both combat and non-combat actions, especially at the small unit tactical level. Language and cultural misunderstandings have the potential to create problems and hinder mission accomplishment. Urban combat places friendly and enemy forces in close proximity while at the same time limiting lines of sight and fields of fire. US technological advantages are often reduced during operations in dense urban areas. Air power may not be as useful to an Infantry force fighting in a building complex. An adept enemy may try to use the technique of "*hugging*" American forces to deny them use of their overwhelming firepower. Urban combat is primarily a small unit fight, requiring significant numbers of Infantry to accomplish the mission; however, combined arms must support the Infantry. Urban combat is characterized by moment-to-moment decisions by individuals and small-unit leaders. This demonstrates the importance of every Soldier understanding the ROE. Commanders and leaders must anticipate what subordinates will need to accomplish their mission. The greatest threats are normally small arms fire, snipers, booby traps, and rocket-propelled grenades (RPGs). Soldiers can face booby traps on doorways and windows and on entrances to underground passageways. Our own increased emphasis on sniper training and experience also serves to counter the sniper threat and enhances friendly mission success. (See Appendix E for more information.)

CHANGING CONDITIONS

12-40. Platoons and squads may find themselves executing missions in changing conditions during UO. The change from stability operations to combat operations and vice-versa will often change conditions from high-intensity to precision or the opposite. METT-TC factors and the ROE determine this change. ROE changes are normally made at echelons much higher than company and battalion, but they normally require that units modify the way they fight in urban areas. Squads and platoons will be required to select different tactics and techniques based on the conditions they face. The ROE will ultimately determine these conditions for the Infantry platoon and squad.

SMALL-UNIT BATTLES

12-41. Units fighting in urban areas often become or feel isolated, making combat a series of small-unit battles. Soldiers and squad or team leaders must have the initiative, skill, and courage to accomplish their missions while isolated from their parent units. A skilled, well-trained defender has tactical advantages over the attacker in this type of combat. The defender may occupy strong covered and concealed static positions and conduct three-tier ambushes, whereas the attacker must be exposed in order to advance. Greatly reduced line-of-sight ranges, built-in obstacles, and compartmented terrain may require the commitment of more troops for a given frontage. While the defense of an urban area can be conducted effectively with relatively small numbers of troops, the troop density required for an attack in urban areas might be greater than for an attack in open terrain. Individual Soldiers must be trained and psychologically ready for this type of operation.

COMMUNICATIONS

12-42. Urban operations require centralized planning and decentralized execution. Therefore, effective vertical and horizontal communications are critical. Leaders must trust their subordinates' initiative and skill, which can only occur through training. The state of a unit's training and cohesion are vital, decisive factors in the execution of operations in urban areas.

12-43. Structures and a high concentration of electrical power lines normally degrade radio and FBCB2 communications in urban areas. Many buildings are constructed so that radio waves will not pass through them. Frequently, units may not have enough radios to communicate with subordinate elements as they enter buildings and move through urban canyons and defiles.

12-44. Visual signals may also be used, but are often ineffective due to the screening effects of buildings, walls, and so forth. Signals must be planned, widely disseminated, and understood by all assigned, attached, or OPCON units. Increased noise makes the effective use of sound signals difficult. Verbal signals may also reveal the unit's location and intent to the enemy.

12-45. Messengers and wire can be used as other means of communication. Messengers are slow and susceptible to enemy fire when moving between buildings or crossing streets. Wire is the primary means of communication for controlling the defense of an urban area. It should be considered as an alternate means of communications during offensive operations, if assets are available. However, wire communications can often be cut by falling debris, exploding munitions, and moving vehicles.

NONCOMBATANTS

12-46. Because urban areas are population centers, they have noncombatants, who will affect both friendly and threat courses of action across the spectrum of UO. Besides the local inhabitants, urban areas are likely to contain refugees, governmental and nongovernmental officials, and members of the international media. For example, during fighting in Grozny, 150,000 refugees expanded the pre-battle population of 450,000 to 600,000, and 50,000 civilians were killed or wounded in the fighting. Units must prepare to deal with all categories of noncombatants.

AMMUNITION

12-47. Units conducting urban combat must expect to use large quantities of ammunition because of short ranges, limited visibility, briefly exposed targets, constant engagements, and requirements for suppression. Shoulder-launched munitions, rifle and machine gun ammunition, 40-mm grenades, hand grenades, and explosives are high-usage items in this type of fighting. When possible, those items should be either stockpiled or brought forward on-call, so that they are easily available.

CASUALTIES

12-48. Additional, incidental or accidental casualties are caused by shattering glass, falling debris, rubble, ricocheting rounds, urban fires, and falls from heights. Difficulty in maintaining situational awareness increases the risks of fratricide and thus contributes to the casualties. Stress-related casualties and nonbattle injuries resulting from illnesses or environmental hazards, such as contaminated water, toxic industrial materials (TIM), also increase the number of casualties, even during non-combat urban operations. The dispersed nature of urban operations often means that a trained medic may not be immediately available to treat casualties during the "*Golden Hour*," that fleeting period when casualties have their best chance of survival. This puts a premium on planning for casualty evacuation, training and certifying combat lifesavers, and on the (junior leaders') ability to call for and coordinate MEDEVAC.

MANEUVER SPACE

12-49. Buildings, street width, rubble, debris, and noncombatants all contribute to limited mounted maneuver space inside urban areas. Armored vehicles will rarely be able to operate inside an urban area without Infantry support. This does not mean that armored vehicles cannot operate or fight in urban areas. Modern armored fighting vehicles are incredibly powerful and survivable, even in high-intensity urban combat. Fighting together, however, makes both mounted and dismounted forces more effective.

THREE-DIMENSIONAL TERRAIN

12-50. Friendly and threat forces will conduct operations in a three-dimensional battle space. Engagements can occur on the surface, above the surface, or below the surface of the urban area. Also, engagements can occur inside and outside of buildings. Multistory buildings will present the additional possibility of different floors within the same structure being controlled by either friendly or threat forces.

COLLATERAL DAMAGE

12-51. Depending on the nature of the operation and METT-TC factors, significant collateral damage may occur, especially under conditions of high-intensity urban combat. Commanders and leaders must ensure that ROE are disseminated and enforced.

HUMAN INTELLIGENCE

12-52. Until technological provides better ways to gather information, an increased need exists for human intelligence (HUMINT). Reconnaissance efforts of battalion and brigade assets can assist as well as the shaping operations of division or joint task force assets. Companies and below normally have to continue to rely on information provided to them from human sources.

COMBINED ARMS

12-53. While urban operations historically have generated the need to accomplish many Infantry-specific tasks, urban combat conducted purely by Infantry units have often proven to be unsound. Appropriately tasked-organized, combined arms teams, consisting mostly of Infantry, Engineers, and Armor supported by other combat, combat support, and logistical assets, have been more successful in offense and defense. The same concept is true for stability operations, when the decisive operation may not necessarily be performed by combat units.

CRITICAL POINTS

12-54. During offensive operations, companies, platoons, and squads will be assaulting buildings and clearing rooms. More often, assets will not exist to isolate large portions of the urban area. Therefore, skillful use of direct and indirect fires, obscurants, and maneuver must occur to isolate key buildings or portions of buildings in order to secure footholds and clear.

SNIPERS

12-55. Historically, snipers have been especially useful in urban areas. They can provide long- and short-range precision fires and can help with company- and platoon-level isolation efforts. Snipers also provide valuable precision fires during stability operations. Along with engaging assigned targets, snipers are a valuable asset to the commander for providing observation along movement routes and suppressive fires during an assault.

SUPPORT-BY-FIRE POSITIONS

12-56. Buildings, street width, rubble, debris, and noncombatants all dictate the positioning and fields of fire for crew-served and key weapons in urban areas.

Section IV. WEAPONS AND DEMOLITIONS

The characteristics and nature of combat in urban areas affect the employment of weapons and the results they can achieve. Leaders at all levels must consider the following factors in various combinations.

SURFACES

12-57. Hard, smooth, flat surfaces are characteristic of urban targets. Rarely do rounds impact perpendicular to these flat surfaces; rather, they impact at some angle of obliquity, which reduces the effect of a round and increases the threat of ricochets.

ENGAGEMENT RANGES

12-58. Engagement ranges are close. Studies and historical analyses have shown that only 5 percent of all targets are more than 100 meters away. About 90 percent of all targets are located 50 meters or less from the identifying soldier. Few personnel targets will be visible beyond 50 meters and engagements usually occur at 35 meters or less. Minimum arming ranges and troop safety from backblast or fragmentation effects must be considered.

ENGAGEMENT TIMES

12-59. Engagement times are short. Enemy personnel present only fleeting targets. Enemy-held buildings or structures are normally covered by fire and often cannot be engaged with deliberate, well-aimed shots.

DEPRESSION AND ELEVATION

12-60. Depression and elevation limits for some weapons create dead space. Tall buildings form deep canyons at street level that are often safe from indirect fires. Target engagement from oblique angles, both horizontal and vertical, demands superior marksmanship skills.

REDUCED VISIBILITY AND INCREASED NOISE

12-61. Smoke from burning buildings, dust from explosions, shadows from tall buildings, and the lack of light penetrating inner rooms all combine to reduce visibility and to increase a sense of isolation. Added to this is the masking of fires caused by rubble and man-made structures. Targets, even those at close range, tend to be indistinct. Urban combat creates intense noise that may totally prevent Soldiers, even those nearby, from hearing voice commands.

FRIENDLY FIRE

12-62. Urban fighting often becomes confused melees with several small units attacking on converging routes. The risks from friendly fires, ricochets, and fratricide must be considered during planning. Control measures must be continually adjusted to lower the risks. Soldiers and leaders must maintain a sense of situational awareness and clearly mark their progress IAW unit SOP to avoid fratricide.

Chapter 12

CLOSE COMBAT

12-63. Both the shooter and target may be inside or outside buildings and they may both be inside the same or separate buildings. The enclosed nature of combat in urban areas means the weapon's effects, such as muzzle blast and back blast, must be considered as well as the round's impact on the target.

MAN-MADE STRUCTURES

12-64. Often man-made structures must be attacked before enemy personnel inside are attacked. Weapons and demolitions often must be chosen for employment based on their effects against masonry and concrete rather than against enemy personnel.

MODERN BUILDINGS

12-65. Modern engineering and design improvements mean that most large buildings constructed since World War II are resilient to the blast effects of bomb and artillery attack. They may burn easily, but usually retain their structural integrity and remain standing. Once high-rise buildings burn out, they are still useful to the military and are almost impossible to damage further. A large structure can take 24 to 48 hours to burn out and become cool enough for soldiers to enter.

Section V. FUNDAMENTALS

The fundamentals described in this paragraph apply to UO regardless of the mission or geographical location. Some fundamentals may also apply to operations not conducted in an urban environment, but are particularly relevant in an environment dominated by manmade structures and a dense noncombatant population. Commanders should use these fundamentals when planning UO.

PERFORM FOCUSED INFORMATION OPERATIONS AND AGGRESSIVE INTELLIGENCE, SURVEILLANCE, AND RECONNAISSANCE

12-66. During OIF and OEF, there were numerous cases of valuable intelligence collection and targeting efforts through the employment of UAS and SUAS. In the near future, more advanced models of unmanned aerial and ground sensors will enter into service. The company commander must exploit such systems in UO for the purposes of intelligence, surveillance, and reconnaissance to destroy the enemy, protect his own force, and to accomplish both combat and non-combat missions. Information superiority efforts aimed at influencing non-Army sources of information are critical in UO. Because of the density of noncombatants and information sources, the media, the public, allies, coalition partners, neutral nations, and strategic leadership will likely scrutinize how units participate in UO. The proliferation of cell phones, Internet capability, and media outlets ensure close observation of unit activities. With information sources rapidly expanding, information about Army operations reaches the public faster than the internal military information system (INFOSYS) can process it. Units can aggressively integrate information operations into every facet and at all levels of the operation to prevent negative impacts. Under media scrutiny, the actions of a single Soldier may have significant strategic implications. The goal of information operations is to ensure that the information available to all interested parties, the public, the media, and other agencies, is accurate and placed in the proper context of the Army's mission. While many information operations will be planned at levels above the brigade, tactical units conducting UO may often be involved in the execution of information operations such as military deception, operations security (OPSEC), physical security, and psychological operations.

CONDUCT CLOSE COMBAT

12-67. Close combat is required in offensive and defensive urban operations. The capability must be present and visible in stability operations and might even be required, by exception, during urban operations in support of civil authorities. Close combat in any urban operation requires a lot of resources and properly trained and equipped forces. It also has the potential for high casualties. Close combat can, of course, achieve decisive results when properly conducted. Infantry units must always be prepared to conduct close urban combat as part of a combined arms team.

AVOID ATTRITION APPROACH

12-68. Previous doctrine was inclined towards a systematic linear approach to urban combat. This approach placed an emphasis on standoff weapons and firepower. It can result in significant collateral damage, a lengthy operation, and be inconsistent with the political situation and strategic objectives. Enemy forces that defend urban areas often want units to adopt this approach because of the likely costs in resources. Commanders should only consider this tactical approach to urban combat only when the factors of METT-TC warrant its use.

CONTROL ESSENTIALS

12-69. Many modern urban areas are too large to be completely occupied or even effectively controlled. Therefore, units must focus their efforts on controlling only the factors essential to mission accomplishment. At a minimum, this requires control of key terrain. The definition of key terrain remains standard: terrain whose possession or control provides a marked advantage to one side or another. In the urban environment, functional, political, or social significance might be what makes terrain key. For example, a power station or a major government building might be key terrain. Units focus on control of the essential so they can concentrate combat power where it is needed and conserve it for use elsewhere. This implies risk in those areas where units choose not to exercise control in order to be able to mass overwhelming power where it is needed.

MINIMIZE COLLATERAL DAMAGE

12-70. Units should use precision standoff fires, information operations, and nonlethal tactical systems to the greatest extent possible consistent with mission accomplishment and the battalion commander's intent. Operational commanders may develop unique ROE for each urban area and mandate firepower restrictions. Information operations and nonlethal systems may compensate for some of these required restrictions. Moreover, commanders must always consider the short and long-term effects of firepower on the population, the infrastructure, and on subsequent unit missions.

SEPARATE COMBATANTS FROM NONCOMBATANTS

12-71. Promptly separating noncombatants from combatants may make the operation more efficient and diminish some of the enemy's advantages. Separation of noncombatants may also reduce some of the restrictions on the use of firepower and enhance force protection. This important task becomes more difficult when the adversary is an unconventional force and can mix easily with the civil population. Efforts should be made at all levels to encourage the populace to separate from enemy combatants.

RESTORE ESSENTIAL SERVICES

12-72. Tactical units might have to plan for the restoration of essential services that may fail to function upon their arrival or cease to function during an operation. Essential services include shelter, power generation and distribution, food and water delivery, sewage removal and treatment, medical services, and civil order. The use of nonlethal and less destructive munitions and capabilities can help ensure that potentially vital infrastructure remains intact. Initially, Army forces might be the only force able to restore or provide essential services. However, units must transfer responsibility for providing essential services to other agencies, nongovernment organizations (NGOs), or the local government as quickly as possible.

PRESERVE CRITICAL INFRASTRUCTURE

12-73. Attempts to preserve the critical elements for post-combat sustainment operations, stability operations, civil support operations, or the health and well-being of the indigenous population often will be required. This requirement differs from simply avoiding collateral damage. Units might have to initiate actions to prevent the removal, damage, or destruction of infrastructure that will be required in the future. In some cases, preserving critical infrastructure might be the assigned objective of the operation itself.

UNDERSTAND HUMAN DIMENSION

12-74. Commanders must carefully consider and manage the allegiance and morale of the civilian population that may decisively affect operations. The assessment of the urban environment must identify clearly and accurately the attitudes of the urban population toward units. Guidance to subordinates covering numerous subjects including ROE, force protection, logistics operations, and fraternization, is one of the many outcomes of this assessment. Commanders may also be required to consider the demographic variance in the attitudes of an urban population. Western cultural norms may not be appropriate if applied to a nonwestern urban population. Commanders must make their assessments based on a thorough understanding and appreciation of the local social and cultural norms of the population. Sound policies, discipline, and consideration will positively affect the attitudes of the population toward Army forces. Also, well-conceived information operations can also enhance the position of units relative to the urban population. Even during combat operations against a conventional enemy force, the sensitivity and awareness of units toward the civilian population will affect the post combat situation. The human dimension of the urban environment often has the most significance and greatest potential for affecting the outcome of an urban operation.

CONTROL TRANSITION

12-75. Urban operations of all types are resource intensive and thus commanders must plan to conclude UO expediently, yet consistent with successful mission accomplishment. The end state of all UO ultimately transfers control of the urban area to another agency or returns it to civilian control. This requires the successful completion of the Army force mission and a thorough transition plan. The transition plan may include returning control of the urban area to another agency a portion at a time as conditions permit. For brigades and below, transition may also include changing missions from combat operations to stability operations or vice versa.

Section VI. ARMOR

This section discusses employment considerations for company-size combined arms teams; limitations, strengths; and employment of Infantry and Armored vehicles; task organization with tanks at company team level and with Bradleys at company team level; armored vehicle positions; transportation of Infantry; armored vehicular, weapon, and munitions considerations.

EMPLOYMENT CONSIDERATIONS FOR COMPANY-SIZE COMBINED-ARMS TEAMS

12-76. Urban combat by units composed entirely of Infantrymen is a historical anomaly. Across the spectrum of combat action in urban areas, powerful combined-arms teams produce the best results. Infantry units operating alone suffer from critical shortcomings that can only be compensated for by appropriate

Urban Operations

task organization with heavy Infantry, armor, and engineers. These teams must be supported by closely integrated aviation, fire support, communications, and logistical elements. This section discusses employment of Infantry and armored vehicles during the execution of UO (see also Appendix C).

STRENGTHS AND LIMITATIONS OF INFANTRY AND ARMORED VEHICLES

12-77. Due to the decentralized nature of urban combat and the need for a high number of troops to conduct operations in dense, compact terrain, Infantrymen will always represent the bulk of forces. At the small-unit tactical level, Infantry forces have disadvantages that can be compensated for by heavy Infantry or armor units. Conversely, tanks and heavy Infantry face problems in the confines of urban areas that place them at a severe disadvantage when operating alone. Only together can these forces decisively accomplish their mission.

INFANTRY STRENGTHS

12-78. Infantry strengths include--

- Infantry small-arms fire within a building can destroy resistance without seriously damaging the structure.
- Infantrymen can infiltrate into position without alerting the enemy. Infantrymen can move over or around most urban terrain, regardless of the amount of damage to buildings.
- Infantrymen have excellent all-round vision and can engage targets with small arms fire under almost all conditions.
- Only Infantrymen are capable of providing the close in flank, rear, and overhead security necessary by armored forces operating in urban terrain.

INFANTRY LIMITATIONS

12-79. Infantry limitations include--

- Infantry forces lack heavy supporting firepower, protection, and long-range mobility and speed.
- Exposed Infantry forces are subject to taking a high number of casualties between buildings and crossing roads when operating close to armored vehicles if, and when, anti armor or heavy weapons are directed against the armor forces.
- Infantry forces are more subject to fratricide-related casualties from friendly direct and indirect fire.

ARMORED VEHICLE STRENGTHS

12-80. Armored vehicle strengths include--

- The thermal sights on armored vehicles can detect enemy activity through darkness and smoke.
- Armored forces deliver devastating fires; are fully protected against antipersonnel mines, fragments and small arms; and have excellent mobility along unblocked routes.
- Armored vehicles project a psychological presence, and an aura of invulnerability that aids the friendly forces in deterring violence. Mounted patrols by armored vehicles can monitor large areas of a city while making their presence known to the entire populace, both friendly and unfriendly.
- Armored vehicles can move mounted Infantrymen rapidly to a different area. With their long-range sights and weapons, armored vehicles can dominate large expanses of open area and thus free Infantry to operate in more restrictive terrain and visual dead space.

- The mobile protected firepower of armored vehicles can be used to add security, resupply convoys, and extract wounded personnel under fire. The armored vehicle's smoke-generation capability can aid this and other small-unit actions.

ARMORED VEHICLE LIMITATIONS

12-81. Armored vehicle limitations include--

- Crewmembers in armored vehicles have poor all-round vision through their vision blocks; they are easily blinded by smoke or dust. Tanks cannot elevate or depress their main guns enough to engage targets very close to the vehicle or those high up in tall buildings. Armored vehicle thermal sights are not able to detect infrared or laser-type marking signals used by most Infantry forces.
- If isolated or unsupported by Infantry, armored vehicles are vulnerable to enemy hunter or killer teams firing light and medium antiarmor weapons. Because of the abundance of cover and concealment in urban terrain, armored vehicle gunners may not be able to identify enemy targets easily unless the commander exposes himself to fire by opening his hatch or Infantrymen directing the gunner to the target.
- Armored vehicles are noisy. Therefore, they are unlikely to arrive anywhere undetected. Improvised barricades, narrow streets and alleyways, or large amounts of rubble can block armored vehicles.
- Due to the length of the tank main gun, the turret will not rotate if it hits a solid object. Heavy fires from armored vehicles can cause unwanted collateral damage or destabilize basic structures.
- The main gun of an M1A2 can only elevate +20 degrees and depress -9 degrees. Examples of minimum distances from buildings where a HEAT round is used are--
    - 1st (ground) floor—2.5 meters from the target.
    - 3d floor—23 meters from the target.
    - 18th floor—132 meters from the target.

EMPLOYMENT OF INFANTRY AND ARMORED VEHICLES

12-82. Heavy Infantry or armored units (operating in platoon, company team, and battalion task force strength) combine mobility, protection, and firepower to seize the initiative from the enemy and greatly aid friendly success. Caution must be exercised when working with armored vehicles. Keeping up with the locations of Infantry Soldiers is difficult for vehicle commanders. Sudden movement or displacement of armored vehicles must be anticipated. This is particularly critical in limited visibility, incoming indirect fires, or sudden contact with enemy forces. Caution around armored vehicles is applicable in all situations and is not unique to UO. The Infantryman is responsible for maintaining situational awareness of armored vehicle location, while staying out of the way of armored vehicles operating in urban terrain. Armored vehicles can support Infantry during urban combat operations by (Figure 12-1).

- Providing shock action and firepower.
- Isolating objectives with direct fire to prevent enemy withdrawal, reinforcement, or counterattack.
- Providing thermal observation of urban areas through limited visibility or obscuration outside small arms range.
- Neutralizing or suppressing enemy positions with smoke, HE, and automatic weapons fire as Infantry closes with and destroys the enemy.
- Assisting opposed entry of Infantry into buildings when doorways are blocked by debris, obstacles, or enemy fire.
- Smashing through street barricades or reducing barricades by fire.
- Obscuring enemy observation using smoke generators or grenade launchers.

Urban Operations

- Securing cleared portions of the objective by covering avenues of approach.
- Attacking by fire any other targets designated by the Infantry.
- Establishing roadblocks or checkpoints.
- Suppressing identified sniper positions.

> **CAUTION**
> When operating close to Infantry during combined-arms urban combat, tanks should employ heat shields, normally used for towing, to deflect the intense heat caused by the exhaust.

Figure 12-1. Tank in direct fire, supported by Infantry.

TASK ORGANIZATION WITH TANKS AT COMPANY TEAM LEVEL

12-83. An attached or OPCON Bradley fighting vehicle (BFV) platoon will have Infantry squads that can be employed in the scheme of maneuver. Therefore, platoon integrity with a BFV platoon should be maintained in urban combat and the BFV platoon should be used as a maneuver element. The information in this paragraph refers to tank platoons. Normally, a tank platoon would be OPCON to an Infantry company during combined-arms operations at the company team level. Armored vehicles do not usually operate in elements smaller than sections. Their tactics, training, and communication systems are designed to work at section level and higher. The four basic techniques of task organizing the tank platoon into the Infantry company for urban combat follow.

TANK PLATOON AS A MANEUVER ELEMENT

12-84. In this technique, the tank platoon leader is responsible for maneuvering the tanks IAW the company team commander's intent. With this task organization, likely missions for the tanks are to support by fire or to overwatch the movement of the Infantry. This task organization is the most difficult to maneuver tanks with the Infantry. However, in order to execute the mission, the tank platoon leader can choose to maneuver the platoon by sections. This provides greater flexibility in supporting the Infantry during the close fight. This technique will allow the company team commander to employ the tank platoon as a reserve or counter, attack force, and capitalize on mobility, and shock effect of the tank platoon.

TANK SECTIONS UNDER INFANTRY PLATOON CONTROL

12-85. In this technique, tanks are organized into two sections and each section is placed under the OPCON of an Infantry platoon, and employed IAW the platoon leader's scheme of maneuver. The company team commander relinquishes direct control of the tank maneuver to the Infantry platoon leaders. This technique is very effective in maintaining the same rate of progress between the tanks and the Infantry. However, Infantry platoon leaders are burdened with the additional responsibility of employing tanks. The general lack of experience with tanks and the overall battlefield focus of the Infantry platoon leader can also affect this technique. This technique is best suited when contact with the enemy is expected and close continuous support is required for movement or clearing buildings.

TANK SECTIONS UNDER COMPANY AND PLATOON CONTROL

12-86. In this technique, the tank platoon is organized into two sections: one under company control and the other under platoon control. The selected Infantry platoon would have a tank section available to support the close fight. With this technique, the company team commander has a tank section to deploy at the critical place and time of his choosing. This task organization still allows support to the Infantry close fight while keeping additional support options in reserve for the commander to employ. The disadvantages to this technique are that an Infantry platoon leader is employing tanks, instead of the tank platoon leader, and the tanks directly available to the company team commander are cut in half. This technique requires detailed planning, coordination, and rehearsals between the Infantry platoons and tank sections.

INFANTRY SQUADS UNDER TANK PLATOON CONTROL

12-87. In this technique, the company team commander has the option of placing one or more Infantry squads OPCON to the tank platoon leader. He may also retain all tanks under the control of the tank platoon leader or place a tank section OPCON to an Infantry platoon leader. This technique will give the company team commander a fourth maneuver platoon, and involves the tank platoon leader in the fight. It can work well in a situation where a mobile reserve that needs Infantry protection is required. This technique requires detailed planning, coordination, and rehearsals between the Infantry squads and tank platoon or sections. Major disadvantages to this technique are the transportation of the Infantry squads and the ability of the Infantry squads to communicate with the tank platoon.

GUIDELINES

12-88. None of the techniques described above are inherently better than the other one. The task organization is tailored to accomplish the mission. Regardless of the technique selected, the guidelines below should be followed.

- Tanks should be used as sections. Single tanks may operate in support of Infantry; however, tanks should operate as sections. If using tanks to shield squads and teams from building to building as part of the maneuver plan, the leader of the forward element needs to control the tanks.
- If the company commander is controlling the tanks, he needs to move forward to a position where he can effectively employ the tanks in support of the Infantry.
- The task organization should support the span of control. If the company commander is going to control the tanks, then there is no reason to task-organize the tanks by section under Infantry platoons.

12-89. Tanks need Infantry support when the two elements are working together. Do not leave tanks alone because they are not prepared to provide local security during the operation. Tanks are extremely vulnerable to dismounted attack when operating on urban terrain. Tanks are most vulnerable and need local flank, rear, and overhead security when Infantry are in the process of clearing buildings. Tanks must remain relatively stationary for prolonged periods while Infantry clear the building allowing threat antitank (AT) teams to maneuver to a position of advantage.

MUTUAL SUPPORT

12-90. Infantry or tank teams work together to bring the maximum combat power to bear on the enemy. The Infantry provides the eyes and ears of the team. The Infantry locates and identifies targets for the tank to engage. It maneuvers along covered and concealed routes to assault enemy elements fixed or suppressed by tank fire. It provides protection for the tank against attack by enemy Infantry. Meanwhile, the tank provides heavy, continuous supporting fires against enemy positions.

MOVEMENT

12-91. The Infantry normally leads movement through urban areas. The tanks follow and provide overwatch. However, the Infantry must still ensure the tanks are provided flank and rear security. If the Infantry discovers an enemy position or encounters resistance, the tanks immediately maneuver to a position where they can respond with supporting fire to fix the enemy or suppress him and allow the Infantry to develop the situation. The Infantry leader directs the tank to move, if necessary, and identifies specific targets for the tank to engage.

COORDINATION

12-92. Coordination between tank and Infantry leaders is close and continuous. The tank commander or driver, accompanied by the Infantry leader, may need to dismount and move to a position where the route or target is more visible. Signals for initiating, shifting, or ceasing fires must be understood by all. One of the greatest barriers to coordination, command, and control in urban combat is the intense noise. Verbal commands should be backed up by simple, nonverbal signals.

COMMUNICATIONS

12-93. The tank platoon leader and platoon sergeant maintain communications with the company team commander via conventional FM radio. Individual tanks and Infantrymen communicate with each other using one or more of these techniques.

Visual Signals

12-94. Visual signals, either prescribed by TSOP or coordinated during linkup, facilitate simple communications.

Wire

12-95. M1-series tank crewmembers can route WD-1 wire from the AM-1780 through the loader's hatch or vision block and attach it to a field phone on the back of the tank. WD-1 wire can also be run on a more permanent basis starting through the engine compartment, through the hull or subturret floor, and attached to the turret's intercom system via the driver's communications box. These techniques work better in a defensive situation rather than in an attack, where wire might hinder tank movement or reaction time.

FM Radios

12-96. FM radios or other short-range hand-held radios can be distributed during the linkup to provide a reliable means of communications between Infantry and supporting tank commanders (TCs). These radios allow the Infantry to use terrain more effectively in providing close in protection for the tank; Infantrymen can watch for enemy elements while limiting exposure to enemy fires directed against the tank. Information in SOIs is used by the tank platoon or sections and the company team headquarters or the

Infantry platoons. This SOI information is a fast reliable method of communications that does not require additional assets.

SMOKE

12-97. The tank's smoke grenade launchers or smoke generator might be used to protect the tank from enemy fire and to provide concealment for the Infantry forces as they move either across open areas or recover wounded. The use of smoke must be carefully coordinated. Although the tank's sights can see through most smoke, Infantrymen are at a significant disadvantage when enveloped in dense smoke clouds. The smoke grenade launchers on the tank provide excellent, rapidly developed local smoke clouds, but the grenades produce burning fragments that are hazardous to Infantrymen near the tank and can ignite fires in urban areas.

HEAVY DIRECT-FIRE SUPPORT

12-98. Tanks are valuable tools for assisting the assaulting forces during actions in the objective area. When possible, tanks are positioned where their fires can be used to prevent enemy reinforcement and engage enemy forces withdrawing from the objective. Due to the nonlinear nature of urban engagements, enemy forces may move to the rear or flanks of the tanks and destroy them. If a small element of Infantry cannot be spared to support the tanks, both vehicles in the section should move to positions of cover and mutual support. Loaders and vehicle commanders should be alert, especially for enemy Infantry approaching from above, the rear, or from the flanks.

OTHER CONSIDERATIONS

12-99. Other considerations for employing tanks at company team level are--

- In planning, pay close attention to available terrain that supports tank cross-country movement. While the pace might be slower, security might be significantly enhanced.
- Involve tank platoon leaders and sergeants in the planning process. Their expertise will hasten the understanding of what tanks can and cannot do and aid the Infantry company commander in making the best employment decision.
- Tanks can be used to carry ammunition, water, and other supplies to support the urban fight.
- To keep tanks mission capable requires planning for refueling and rearming. There may also be a requirement to recover disabled vehicles. The company XO coordinates with the battalion S-4 to ensure that the proper logistical support is provided for the tanks. Infantry companies and battalions do not have resources to support a tank platoon with CL III, V, or recovery and need sustainment augmentation. Normally the attached tank platoon or company comes with their sustainment slice elements.
- Infantry company commanders allocate time in the planning process for precombat inspections (PCIs) for the tanks.

12-100. Conduct a combined-arms rehearsal at the level that the tanks are task-organized. Try to replicate conditions for mission execution during rehearsals, for example, in daylight and in limited visibility, civilians on the battlefield, host nation support, and ROE. Also include the following.

- Communications.
- Direct fire plans.
- Breach drills.
- Procedures for Infantry riding on tanks. (Tanks can move a maximum of nine personnel.)
- Techniques for using tanks as Infantry shields.

12-101. To minimize casualties when moving outside or between buildings--

- Cover all possible threat locations with observation and fire.

Chapter 12

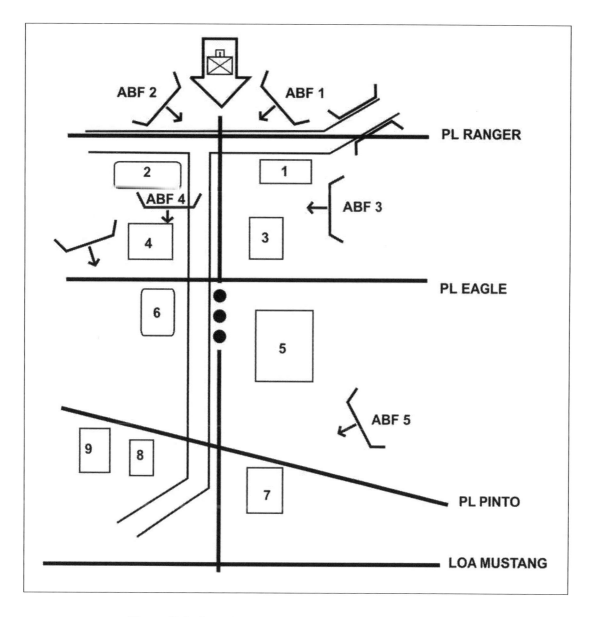

Figure 12-2. Graphic control measures for Infantry and heavy.

12-102. Rely on the radio to help control the battle. Platoon leaders and radio operators must be well trained in sending reports. Constant reporting from the subordinate elements to the commander is critical for mission success.

ARMORED VEHICLE POSITIONS

12-103. Fighting positions for tanks are essential to a complete and effective defensive plan in urban areas. Armored vehicle positions are selected and developed to obtain the best cover, concealment, observation, and fields of fire while retaining the vehicle's ability to move.

HULL DOWN

12-104. If fields of fire are restricted to streets, hull-down positions should be used to gain cover and fire directly down streets (Figure 12-3, page 12-20). From those positions, tanks are protected and can move to

Urban Operations

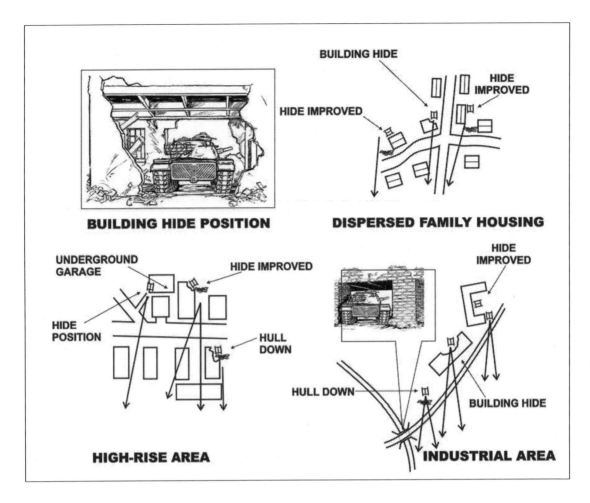

Figure 12-5. Building hide position.

TRANSPORTATION OF INFANTRY

12-107. At times, the tank platoon might be required to transport Infantrymen. This is a hazardous undertaking and must not be done if the unit is expecting contact. The proliferation of remotely detonated mines and IEDs has made this technique even more hazardous. It should be used only after careful consideration or in extremis. If the tank platoon is moving as part of a larger force and is tasked to provide security for the move, the lead section or element should not carry Infantry (Figure 12-6, page 12-26).

Chapter 12

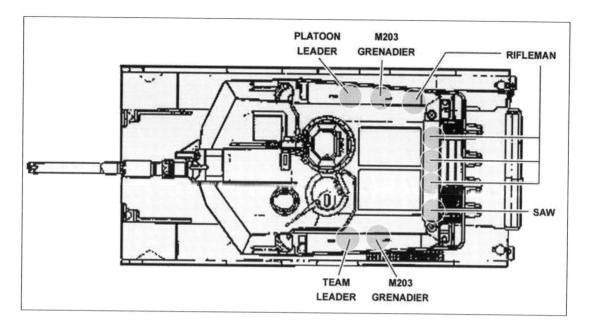

Figure 12-6. Example positions for Infantry riding on a tank.

PROCEDURES, PRECAUTIONS, AND CONSIDERATIONS

12-108. Infantry and armor leaders must observe the following procedures, precautions, and considerations when Infantrymen ride on tanks.

- Infantrymen should thoroughly practice mounting and dismounting procedures and actions on contact.
- Infantrymen must always alert the TC before mounting or dismounting. They must follow the commands of the TC.
- Infantry platoons should be broken down by squads, similar to air assault chalks, with the Infantry platoon leader on the armor platoon leader's vehicle and the Infantry platoon sergeant on the armor platoon sergeant's vehicle.
- Platoon leaders, platoon sergeants, and team leaders should position themselves near the TC's hatch, using the external phone (if available) to talk to the TC and relay signals to the unit.
- If possible, the lead vehicle should not carry Infantrymen. Riders restrict turret movement and are more likely to be injured or killed on initial contact.
- Whenever possible, Infantrymen should mount and dismount to the front of a tank, while ensuring the driver and, or, the TC have made eye contact and granted permission to climb on to, or off, the tank to Infantrymen.
- Infantrymen must always have three points of contact with the vehicle, and watch for low-hanging objects such as tree branches.
- Infantrymen should wear hearing protection.
- Infantrymen should not ride with anything more than their battle gear. Rucksacks should be transported by other means.
- Infantrymen should scan in all directions while riding. They might be able to spot a target the vehicle crew does not see.

ACTIONS ON CONTACT

12-109. Infantrymen should be prepared to take the following actions on contact.

- Wait for the vehicle to stop.
- At the TC's command, dismount immediately (one fire team on each side). Do not move forward of the turret. Do not dismount a vehicle unless ordered or given permission to do so.
- Move at least 5 meters to the either side of the vehicle. NEVER move behind or forward of the vehicle.
- NEVER move in front of vehicles unless ordered to do so. Overpressure from firing a main gun can inflict serious injury to Infantrymen dismounted forward (Figure 12-7, page 12-28).
- NEVER dangle arms or legs, equipment, or anything else off the side of a vehicle; they could catch in the tracks, causing damage to the equipment or vehicle and death or injury to the Soldier.
- NEVER place too many riders on the vehicle.
- NEVER fall asleep when riding. The warm engine may induce drowsiness; a fall could be fatal.
- NEVER smoke when mounted on a vehicle.
- NEVER stand near a moving or turning vehicle at any time. Tanks have a deceptively short turning radius.

WARNING

The overpressure from the tank's 120-mm cannon can seriously injure dismounted Infantry within a 90-degree arc extending from the muzzle of the gun tube out to 200 meters.

DANGER

FROM 200 TO 1,000 METERS ALONG THE LINE OF FIRE, ON A FRONTAGE OF ABOUT 400 METERS, DISMOUNTED INFANTRY MUST BE AWARE OF THE DANGER FROM DISCARDING SABOT PETALS, WHICH CAN CAUSE SERIOUS INJURIES OR DEATH TO NEARBY PERSONNEL.

Chapter 12

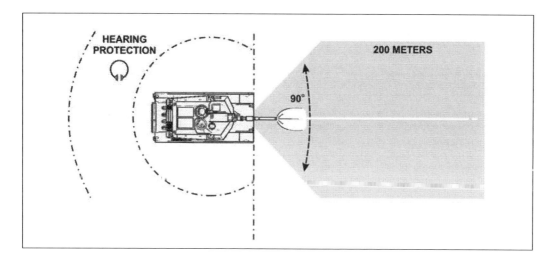

Figure 12-7. Danger areas around a tank firing a 120-mm main gun.

ADDITIONAL CONSIDERATIONS

12-110. Additional considerations and preparations for transporting Infantrymen include--

Armor

12-111. Uses main-gun fire to reduce obstacles or entrenched positions for the Infantry.

12-112. Takes directions from the Infantry ground commander (platoon leader or platoon sergeant or squad leader) to support their fire and maneuver.

12-113. Provides reconnaissance by fire for the Infantry.

12-114. Should know and understand how the Infantry clears buildings, and how they mark cleared buildings, the casualty evacuation plan, signal methods, engagement criteria for tank main gun, front line trace reporting, and ground communication from the tank with the dismounted personnel.

12-115. Uses its night vision capability to augment and supplement the Infantry's night vision capabilities.

Infantry

12-116. Provides information for the tank crewmembers to help them overcome tank noise and the lack of ground SA. Provides reconnaissance and fire direction of enemy positions for main gun attack. Considerations for dismounted tank security include--

- Tank crewmembers should rehearse the mounting and dismounting of Infantrymen from their vehicle, briefing the Infantrymen on safety procedures for the vehicle and weapon systems.
- Tank commanders need to rehearse communicating with dismounted Soldiers via TA-1 and DR-8 in the bustle rack.

Vehicle Preparation

12-117. Vehicle preparation for combat in urban terrain should include at least the following actions.

- Keep at least one ballistic shield to the integrated sight unit (ISU) closed (most engagements will be under boresight range and the battlesight technique will suffice).
- Place sandbags around antenna connections and electrical wiring on the turret top.

- Place extra coax ammunition inside the turret.
- Remove all highly flammable products from the outside of the vehicle and from the sponson boxes in order to prevent vehicle fires due to small arms or RPG detonations.

CONSIDERATIONS FOR ARMORED VEHICLES, WEAPONS, AND MUNITIONS

12-118. Numerous factors related to tanks and their organic weapons and munitions affect the tank platoon's UO planning and execution, including--

- The preferred main gun rounds in the urban environment are HEAT, multipurpose antitank (MPAT) (ground mode), and MPAT-OR (XM908). These all perform much better than sabot rounds against bunkers and buildings.
- HEAT ammunition will open a larger hole in reinforced concrete or masonry structures than MPAT or MPAT-OR (XM908). Both MPAT and MPAT-OR, however, offer greater incapacitation capability inside the structure.
- HEAT ammunition arms about 60 feet from the gun muzzle. It loses most of its effectiveness against urban targets at ranges of less than 60 feet.
- MPAT and MPAT-OR rounds arm about 100 feet from the muzzle of the gun. Because of the shape and metal components of the projectiles, however, this ammunition remains effective at ranges of less than 100 feet.
- Sabot petals, including those on MPAT and MPAT-OR, endanger accompanying Infantry elements. They create a hazard area extending 70 meters on either side of the gun-target line out to a range of 1 kilometer.
- The tank's main gun can depress only to -10 degrees and can elevate only to +20 degrees, which creates considerable dead space for the crew at the close ranges that are typical in the urban environment.
- The external M2 HB machine gun can deliver a heavy volume of suppressive fire and penetrate light construction, buildings, and most barricades. The M2 HB machine gun can elevate to +36 degrees. However, the TC must be unbuttoned to fire the M2 on the M1A2 or M1A2 SEP.
- The M240 coax machine gun can effectively deliver suppressive fires against enemy personnel and against enemy positions that are behind light cover.
- The loader's M240 machine gun can effectively deliver suppressive fire against enemy personnel and against enemy positions that are behind light cover; however, the loader must be unbuttoned to operate it.
- When buttoned up, the tank crew has limited visibility to the sides and rear and no visibility to the top.

TASK ORGANIZATION WITH BRADLEYS AT COMPANY TEAM LEVEL

12-119. The heavy platoon provides a very flexible heavy direct fire support asset to Infantry companies conducting operations on urban terrain. The 25-mm cannon and 7.62-mm coax machine gun, combined with the additional Infantry, Javelin, and TOW ATGMs, provide the company team commander powerful combat multipliers during urban combat.

Target Engagement

12-120. Streets and alleys are natural firing lanes and killing zones. Because of this, all vehicular traffic is greatly restricted and canalized, and subject to ambush and short-range attack. Tanks are at a disadvantage because their main guns cannot be elevated enough to engage targets on the upper floors of tall buildings. The BFV, with +60 to -10 degrees elevation of the 25-mm gun and 7.62-mm coax machine gun, has a much greater ability to engage targets in urban terrain.

Chapter 12

GENERAL CONSIDERATIONS WHEN USING BRADLEY FIGHTING VEHICLES

12-121. Infantry companies might be task-organized with heavy platoons when conducting operations in urban terrain. A BFV platoon can provide its own Infantry support. Generally, BFVs should not be separated from their Infantry. Working as a team, Infantrymen (the rifle squads) provide security for the vehicles; the BFVs provide critical fire support for the Infantry company team.

Movement

12-122. When moving, if the street is large enough, BFVs should stay close to a building on either side of the street. This allows each BFV to cover the opposite side of the street. BFVs can button up for protection, but the BFV crew must remain alert for signals from Infantry. Coordination between mounted and dismounted elements is critical in urban terrain.

Close Combat Missiles

12-123. The BFV lacks adequate armor protection to withstand medium to heavy ATGM fire. It is normally employed after the area has been cleared of ATGM positions or on terrain dominating the city to provide long-range antiarmor support or fire suppression. Shoulder-launched munitions or Javelins provide a significant amount of the BFV platoon's short-range antiarmor fires in urban areas; the TOWs provide long range antiarmor fires. The BFV's 25-mm gun and machine gun are employed while providing direct fire support.

ORGANIZATION AND TASKS

12-124. The BFV platoon is comprised of mounted and dismounted elements. Based on the company commander's guidance and the METT-TC factors, the BFV platoon leader will normally determine how his elements will be deployed.

Section VII. OFFENSE

Offensive operations in urban areas are based on offensive doctrine applied to urban terrain. Urban combat imposes a number of demands that are different from ordinary field conditions such as problems with troop requirements, maneuver, and use of equipment. As with all offensive operations, the company commander must retain his ability to maneuver against enemy positions.

PLANNING CONSIDERATIONS

12-125. Combat operations in a built-up area have a slower pace and tempo than operations in open terrain. However, commanders still maintain a momentum that is relentless in order to keep the enemy off balance. Due to the close environment (and the restricted ability to use all available weapons systems), synchronization of maneuver and combat support assets is one of the Infantry company commander's main challenges. Missions in UO are more methodical. Normally, the Infantry company conducts missions as part of a battalion operation, but must be prepared to operate independently. The company must also be prepared to conduct different but mutually supporting missions simultaneously such as establish a checkpoint and clear a block at the same time.

TROOP REQUIREMENTS

12-126. Due to the nature of combat in built-up areas, more units are required to conduct operations in an area of like size of rural terrain. The physical size of an AO is reduced. This is due to the nature of the terrain. Urban terrain is more compressed, has three dimensions in which maneuver forces operate, and operations are characterized by small units operating independently.

12-127. Commanders consider Soldier fatigue. Room-clearing techniques are physically demanding, so they quickly tire a force. Commanders plan for the relief or rotation of their forces before they reach the point of exhaustion.

12-128. Additional forces might be needed to control civilians in the built-up area. These forces protect civilians, provide first aid, and prevent them from interfering with the tactical plan.

12-129. Fighting in a built-up area normally results in a greater number of friendly casualties. The ability to see the enemy is fleeting and confined to very short ranges compared to combat in open terrain. Fratricide prevention is a challenge and must be addressed in detail by the commander. The command is likely to operate under some form of a restrictive ROE based on noncombatants on the battlefield. Plan for CASEVAC and instruct subordinate units to conduct this task.

MANEUVER

12-130. The complex nature of the urban environment makes it difficult for commanders to maneuver their Infantry company and its attachments quickly. The presence of large numbers of civilians, and clearing buildings and looking for antiarmor ambushes, snipers, and booby traps degrade the company's ability to maneuver platoons and squads. Due to the dense environment and its effects on weapons systems, the synchronization of combat power is one of the commander's main challenges. Offensive operations are planned in detail, with subordinate elements given specific instructions and on-order missions.

LIMITATIONS

12-131. Infantry company commanders attacking a built-up area must recognize some important limitations in the use of available assets.

- The presence of large numbers of civilians or non combatants in urban areas must be factored into all of the commander's analysis. The presence, flow, and actions of civilians on the battlefield will have a huge effect on the actions of the Infantry company. Normally, the use of indirect fires, especially field artillery, is much more restricted in built-up areas than in open terrain. Leaders must consider the effects of indirect fire on the urban area and civilians, especially when restrictive ROE are in effect. Indirect fires must be fired in greater mass to achieve the desired effect.
- The rubble caused by massive indirect fires adversely affects a unit's mobility during the attack.
- The commander and leaders consider the effect that city lights, fires, and background illumination have on NVDs. These elements may limit the effectiveness of NVDs and make thermal imagery identification difficult.
- Communications equipment may not function to its maximum effectiveness because of the density in building construction. Therefore, execution of the operation relies on all Soldiers understanding the plan and actions being event driven, rather than waiting for instructions.

METT-TC FACTORS

12-132. The Infantry company commander's analysis of the METT-TC factors is critical for successful planning and execution during UO.

MISSION

12-133. The Infantry company commander must receive, analyze, and understand the mission before he begins planning. He and his troops must clearly understand the conditions of the operation and the ROE. The company commander must fully analyze, understand, and apply the effects of civilians and non combatants on the battlefield. The company commander might be required to conduct different missions simultaneously.

Common Missions

12-134. Infantry companies should expect to receive similar types of offensive missions in urban terrain that they receive in other terrain. The following are common company missions in urban combat.

Isolate an Urban Objective

12-135. The Infantry company normally conducts this mission as part of a battalion. The Infantry company deploys its platoons to secure the area around or near a building, block, or village in order to kill or capture any withdrawing enemy forces and prevent reinforcement of (or a counterattack against) the objective. Engineers or other CS and sustainment assets may reinforce the company based on the ROE and METT TC factors. In view of the fact that many casualties might be inflicted on friendly units moving between buildings or down streets, this mission takes on significant importance.

Assault a Building

12-136. Infantry companies normally conduct this mission as part of a battalion operation when the building is too large for a platoon to assault and clear, and the enemy defending the building employs a force larger than a platoon does. The company isolates the building, gains a foothold, and clears the building. The battalion commander normally helps by directing another Infantry company (or other companies) to isolate the building. Engineers or other CS and sustainment assets usually reinforce the Infantry company consistent with the ROE and the METT-TC factors.

Attack a Block or Group of Buildings

12-137. Infantry companies may attack a block or group of buildings. Again, the company normally conducts this mission as part of a battalion operation. If the company attacks a block as part of a battalion operation, it might be the battalion's main or shaping operation. Another friendly unit may isolate the objective, or the company may find that it must isolate the objective area in whole, or in part. If an Infantry company receives the mission to assault a block independently, then the company must isolate the objective area by itself.

Move to Contact

12-138. Infantry companies in UO may move through urban terrain in order to gain and maintain contact with the enemy. This mission typically includes movement (often rapid) through an urban area to develop the situation by seizing or clearing blocks and buildings.

Conduct Hasty Attack of a Village

12-139. Infantry companies may conduct a hasty attack of a village either independently or as part of an Infantry battalion operation. Normally, the purpose of this mission is to reduce enemy control of a position and facilitate movement for other operations.

Seize Key Urban Terrain

12-140. Infantry companies can seize key terrain to give friendly forces an advantage. Key terrain can include, among other things, overpasses, building complexes, traffic circles, surrounding natural terrain, or bridges. The Infantry company usually seizes key urban terrain independently either to facilitate its own movement or to support other operations.

Raid

12-141. Infantry companies plan urban raids much as they do other raids. Objectives might be in built-up areas, which would require the company to move through urban and other terrain in order to arrive at the objective. Although the company may conduct a raid independently, it could also do so in support of a battalion area raid.

Analysis of Mission

12-142. When conducting his analysis, the Infantry company commander considers his battalion commander's intent and the end state of the operation. For example, the company commander must determine if clearance means every building, block by block (systematic clearance), or if the seizure of key terrain requires clearing only along the axis of advance (selective clearance). The company commander must also consider how and where the company must be postured in order to conduct follow-on missions and to facilitate the battalion and brigade missions. This influences the missions he assigns to his platoon and attached element leaders.

12-143. When the company is involved in clearing operations, bypassing buildings increases the risk of attack from the rear or flank unless planned support isolates and suppresses those buildings. Normally, the clearing platoons must not only clear each building in the company's zone but also leave security behind to prevent enemy reoccupation of buildings. This may not be feasible due to the nature of the mission, but if it is part of the plan, it should be made clear to the platoon leaders when orders are issued.

12-144. The engagement can transition quickly from precision to high intensity conditions, a transition that might be caused by enemy actions. An assault against a deliberate, prepared defense with obstacles becomes high intensity. Indications of an enemy-forced change of engagement criteria (and a change from precision conditions to high intensity) include--

- The requirement to breach multiple obstacles.
- The use of booby traps by the enemy.
- The requirement to use repetitive explosive breaching to enter a building.
- Rooms that are so well prepared or barricaded that normal movement and clearing techniques cannot be employed.

Movement

12-145. Moving from building to building or between buildings presents a challenge to platoons. Historical examples have shown that many casualties occur during movement from building to building and down streets. Therefore, Infantry company commanders plan operations in a manner that allows subordinate elements to take maximum advantage of covered and concealed routes within the urban area. Also, company commanders must carefully analyze which buildings must be isolated, suppressed, and obscured, consistent with the ROE. If working with heavy units, they may use any available tanks and BFVs as shields for maneuvering platoons.

Coordination of Fire Support

12-146. Most fire support coordination occurs at battalion level to take into account the ROE. Prior coordination determines the techniques and procedures to use for communicating, identifying targets, and shifting fires. The company FSO is extensively involved in this portion of the planning process. The company plans fires consistent with the ROE, considering civilians, houses of worship, medical centers, schools, public services, and historical monuments.

ENEMY

12-147. Key factors that affect the Infantry company commander's analysis are what phase of the OE (regional, transitional, or adaptive) are enemy forces operating in, the type of enemy force that is expected in the urban area, the enemy's probable courses of action, and the ROE. More restrictive ROE support the defender; less restrictive ROE support the attacker. The type of threat is one factor used to determine how the company should be task-organized and how combat power should be synchronized to accomplish the mission. For example, if the company has the mission to secure a water treatment facility that is determined to be key terrain, the commander needs to consider possible threats to the facility that may not be direct force-on-force actions.

Conventional Forces and Regional or Transitional Operations

12-148. Many third world countries have adopted techniques of urban combat from either the United States or the Commonwealth of Independent States. Therefore, a future threat might consider the motorized or mechanized rifle battalion the most effective unit for urban combat due to its mobility, armor protection, and ability to adapt buildings and other structures for defense quickly. In countries whose forces are equipped and trained as in the former Warsaw Pact, standard urban defenses include--

- Threat defenses that are organized into two echelons to provide greater depth and reserves.
- Company strongpoints are prepared and form the basis for the battalion defensive position.
- The reserve is located in a separate strongpoint.
- Ambush locations are established in the gaps of the strongpoints, and dummy strongpoints are constructed to deceive the attacker.
- Positions for securing and defending the entrances to and exits from underground structures and routes are established.
- Security positions are prepared forward of first echelon defensive positions.
- A motorized or mechanized rifle company may defend several buildings or a single large building with mutually supporting fires.
- Each platoon defends one or two buildings, or one or two floors of a single building.

12-149. In many third world countries, the forces are predominantly light with some outdated armored vehicles. Some countries may not have actual armed forces but have some form of armed militia(s). These forces normally do not fight a defense in the former Warsaw Pact style, but rather offer uncoordinated resistance, often extremely intense, as experienced in Somalia.

Unconventional Forces and Adaptive Operations

12-150. Enemy analysis of unconventional forces during adaptive operations is similar to that for low intensity conflict during urban counterinsurgency, counterguerrilla, and counterterrorist operations. (See Chapter 2, FM 3-06.11, for more on conventional and unconventional threat analysis in UOs.)

TERRAIN

12-151. Offensive operations are tailored to the urban environment based on an analysis of each urban terrain setting, its types of built-up areas, and existing structural forms. Commanders and subordinate leaders incorporate the following special planning considerations for an urban environment when conducting an offensive operation.

- Military maps may not provide enough detail for urban terrain analysis nor reflect the underground sewer system, subways, underground water system, mass transit routes, and utility facilities. When available, the commander uses building or city plans, engineering prints, aerial photographs, tourist maps, or other aids that may assist him in his analysis of the terrain.
- Natural terrain surrounding the built-up area.
- Key and decisive terrain such as intersections, building that dominate the AOs, and entrances to subsurface avenues that allow covered repositioning of forces.
- Construction and structural composition of buildings.
- Confined spaces that limit observation, fields of fire, maneuver, or prevent the concentration of fires at critical points.
- Covered and concealed routes to and within the built-up area.
- Limited ability to employ maximum fire power due to the need to minimize damage and rubbling effects (based on ROE).

- Challenges with conducting effective reconnaissance during conventional regional and transitional operations. During operations in the adaptive phase of a conflict or operations under restrictive ROE, the opposite is true. Reconnaissance and security are more easily accomplished by both sides and are more difficult to prevent.
- ROE that limits the use of firepower.
- Significant numbers of civilians who might have to be evacuated, some forcibly. Civilians may hinder operations on purpose or merely operations by their presence.

TROOPS AVAILABLE

12-152. An Infantry company normally participates in an attack as part of an attacking battalion. Consideration is given to the mission and the assets available to accomplish the mission. The commander makes a determination of whether he has enough of the proper assets to be successful. If available, towed 105-mm howitzers can use direct fire to destroy bunkers, heavy fortifications, or enemy positions in reinforced concrete buildings (Figure 12-8). The towed 105-mm howitzer may also clear or create an avenue of approach. Whenever artillery is used in the direct-fire role, it must be close to the Infantry providing security against enemy ground attack. Prior coordination is necessary so the bulk of the field artillery unit's shells are HE.

Figure 12-8. Artillery in direct-fire role.

TIME

12-153. Offensive operations in built-up areas have a slower pace and operational tempo. Consider the following issues when analyzing time available for an attack in urban terrain.

- Time to prepare.
- Movement rates.

Chapter 12

- Time needed to clear a building.
- Impact of slower movement rates and time to accomplish tasks on the synchronization of supporting assets.

12-154. Plan additional time to recover from fatigue. Troops tire more quickly because of stress and the additional physical exertion related to clearing urban terrain.

12-155. Allow additional time for thorough reconnaissance and rehearsals in order to prevent excessive casualties and fratricide.

CIVIL CONSIDERATIONS

12-156. Civil considerations have been discussed in detail in chapter 2. They include the influences of man-made infrastructure, civilian institutions, and the attitudes and activities of civilian leaders, populations, and organizations within an AO, with regard to the conduct of military operations. These considerations, regardless of the environment should be analyzed and planned for, especially in the urban environment where the civilians are such a critical component of how an urban area works. Civil considerations of the urban environment can either help or hinder friendly or enemy forces. Leaders must understand the impact of their actions--as well as their subordinate's actions--on the civilian population, and the effect they will have on current and future operations.

COMMAND AND CONTROL

12-157. Units in built-up areas frequently fight separated and isolated from one another. Planning is centralized, but execution is decentralized. Therefore, the commander must clearly describe his vision of the terrain and the enemy to his platoon and squad leaders. In all situations, leaders position themselves where they can control the action and assist subordinate leaders. This is difficult in urban terrain due to obstacles, poor visibility, difficulty in communications, and the intensity of urban combat. Infantry commanders must demand timely, accurate, and complete reporting.

COMMAND

12-158. Infantry commanders issue orders and develop control measures to facilitate decentralized execution. Increased difficulties in command, control, and communications from higher headquarters demand increased responsibility and initiative from subordinate leaders. Graphic control measures common to other tactical environments are also used in combat in built-up areas. These and other control measures ensure coordination throughout the chain of command.

CONTROL

12-159. Thorough rehearsals and briefbacks enhance control. Subordinate leaders must clearly understand the commander's intent (two levels up) and desired mission end state in order to facilitate control. Infantry company commanders should consider using subordinate leaders to control certain portions of the fight when the commander's attention needs to be focused elsewhere, for example, using the XO to control the support element while the commander controls the assault elements.

Communications

12-160. In built-up areas, radio and FBCB2 communications are often less effective than field telephones and messengers. Communications equipment may not function properly because of the materials used in the construction of buildings and the environment. Wire laid at street level is easily damaged by rubble and vehicle traffic. Pyrotechnic signals are hard to see because of buildings and smoke. The high noise level of engagements in and around buildings makes sound signals and voice alerts difficult to hear. Besides, voice communications can signal the unit's intent and location to the enemy. Line-of-sight (LOS) limitations affect both visual and radio communications. Therefore, the time needed to establish an effective communications system might be greater in an urban environment. Leaders consider these effects when they allocate time to establish communications. Since the effectiveness of normally dependable communications might be uncertain in UO, units may fight without continuous communications.

Graphic Control Measures

12-161. Graphic control measures are used to convey intent and control the operation. This paragraph describes some considerations for using graphic control measures in UO.

12-162. Use numbers to identify buildings. When attacking to seize a foothold, the Infantry company normally assigns a building or a few small buildings as a platoon's objective. When an objective extends to a street, only the near side of the street is included in the objective area. Key buildings or groups of buildings may also be assigned as intermediate objectives. To simplify assigning objectives and reporting, buildings along the direction of attack should be identified. An example using numbers is shown in Figure 12-9. Floors can be lettered in order to identify them. A possible method is to letter the ground floor A (Alpha), second floor B (Bravo), and so on. Then, referencing a particular location is easier. For instance, the third floor in Building 5 would be referred to as 5C.

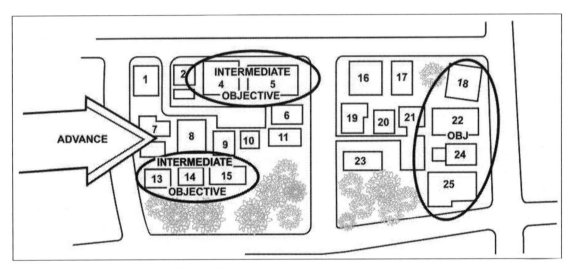

Figure 12-9. Example numbering system.

12-163. Phase lines can be used to report progress or to control the advance of attacking units (Figure 12-10). Principal streets, rivers, and railroad lines are easily identifiable and are suitable phase lines. Phase lines are shown on the near side of the street or open area. In systematic clearing, a company may have the mission to clear its zone up to a phase line. In that case, the company commander chooses his own objectives when assigning missions to his subordinate units.

Chapter 12

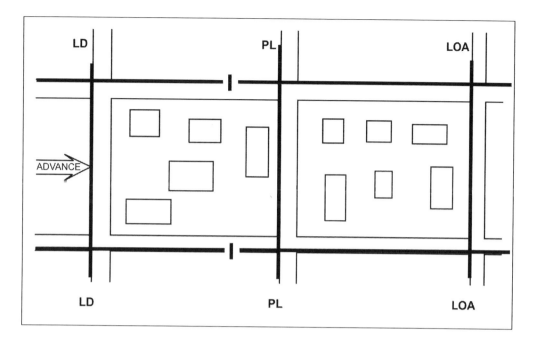

Figure 12-10. Zones, boundaries, and phase lines.

12-164. Set company boundaries within blocks, so that it is clear whether streets are included in a company zone. Ensure that both units understand which has responsibility for the street itself. To facilitate command and control, try to place boundaries to ensure both sides of a street are in the zone of one unit.

12-165. Plan checkpoints and contact points at street corners, buildings, railway crossings, bridges, or any other easily identifiable urban feature.

12-166. Forward units may occupy an attack position for last-minute preparation and coordination. The attack position is often behind or inside the last covered and concealed position, such as a large building, before crossing the LD. The LD should be the near side of a street or rail line.

12-167. A unit's assigned frontage for the attack of a built-up area depends on the size of buildings and the resistance anticipated. Based on city blocks averaging 175 meters wide, a company normally attacks on a one- to two-block front. A battalion attacks on a two-to four-block front.

TASK ORGANIZATION INTO THREE ELEMENTS

12-168. The Infantry company commander normally task-organizes his company into three elements: an assault force, a support force, and a reserve. In UOs, all elements must be prepared to breach. The support force might be given a number of tasks to conduct sequentially or simultaneously. Specifically, the support force may support by fire, isolate the objective, or breach. The tactical situation dictates whether separate elements need to be task-organized in order to conduct these support missions. If available, engineers usually support the force performing a breach. The size and composition of the force are determined by METT-TC. If the company is part of a battalion operation, the company may have the mission to conduct one or more of the tasks mentioned above. If conducting an urban attack independently, the Infantry company must perform both assault and support force tasks.

ASSAULT FORCE

12-169. The purpose of the assault force is to destroy the enemy, capture the enemy, or force the withdrawal of the enemy from any urban objective. The assault force of an Infantry company may consist of two or more Infantry platoons usually reinforced with engineers. Building clearing and room clearing are normally conducted at platoon and squad levels. The assault force may also conduct a breach.

SUPPORT FORCE

12-170. The purpose of the support force is to provide any support that might be required by the assault force. The support force at the Infantry company level normally consists of the company's organic assets, such as Infantry platoons, mortars, and weapons squads; its attachments; and units that are under the OPCON of the company commander. The support force must be prepared to breach. This assistance includes, but is not limited to--

- Suppressing or obscuring the enemy within the objective building(s) and adjacent structures.
- Isolating the objective building(s) to prevent enemy withdrawal, reinforcement, or counterattack.
- Breaching walls en route to and in the objective structure.
- Destroying or suppressing enemy positions with direct fire weapons.
- Securing cleared portions of the objective.
- Providing resupply of ammunition, explosives, and personnel.
- Evacuating casualties, EPW, and civilians.

RESERVES

12-171. Infantry companies fighting in urban terrain should designate a reserve when feasible. The company reserve should be mobile and prepared for commitment. Because of the available cover in built-up areas, the reserve can stay close to forward units. The reserve follows within the same block so that it can immediately influence the attack. Platoons might be detached from the company to form a battalion reserve. The reserve force should be prepared to breach. A unit with a reserve mission might be called upon to perform one or more of the following tasks.

- Assume the mission of the assault force.
- Attack from another direction.
- Exploit an enemy weakness or friendly success.
- Clear bypassed enemy positions.
- Secure the rear or a flank.
- Maintain contact with adjacent units.
- Support or counterattack by fire.

BREACHING ELEMENT

12-172. At the Infantry company level, the assault, support, or reserve force may conduct breaching. However, a separate breaching force might be created, or platoons might be given this task and organized accordingly. The purpose of breaching is to provide the assault force with access to an urban objective, using explosive, ballistic, or mechanical methods. Explosive breaching includes using nonelectrical demolition systems; ballistic breaching includes using direct fire weapons; and mechanical breaching includes using crowbars, axes, saws, hooligan's tools, and sledgehammers. Attached engineers or members of the company who have additional training in explosive, ballistic, and mechanical breaching techniques may conduct the breach.

MOVEMENT

12-173. When moving in built-up areas (BUAs), an Infantry company follows the same fundamentals and principles and uses the same movement techniques as in other areas. Enemy actions against the company might consist of ambushes on the street, enfilade fire down the streets, sniper fire, fire from rooftops and from within buildings, or artillery or mortar fire. The company can minimize the effects of enemy defensive fires during movement by--

- Using covered routes (moving through buildings).
- Moving only after defensive fires have been suppressed or obscured.
- Moving at night or during other periods of limited visibility.
- Selecting routes that will not mask friendly suppressive fires.
- Crossing open areas (streets and spaces between buildings) quickly under the concealment of smoke with suppression provided by support forces.
- Moving on rooftops that are not covered by enemy direct fires.
- Using the concealment provided by shaded areas.
- Using cover provided by attached armored vehicles.
- Creating deceptions.
- Using suppressive fires on known or suspected enemy positions, as allowed by ROE.

Movement Down Streets

12-174. Should the situation allow or require movement down a street, platoons move in file along one or both sides of the street with overwatching fires from supporting weapons. Individual Soldiers disperse, move quickly, and observe and cover certain areas as detailed, for example, second-floor windows on the opposite side of the street. As in all urban situations, platoons must search for defenders all-round and in all three dimensions (front, flanks, rear, upper stories, basements, and rooftops).

Speed of Movement

12-175. The speed of movement depends on the type of operation, terrain, and degree of enemy resistance. As in any other terrain, speed and security are trade-offs: the faster the unit moves, the less secure the movement. Slowing down allows for improved security.

12-176. In lightly defended areas, the mission or the "*need for speed*" can mean moving through the streets and alleys without clearing all buildings in order to reach and secure key terrain. More importantly, the company commander must establish and enforce the tempo of the operation.

Danger Areas

12-177. As in any other type of terrain, the company avoids danger areas if possible. Unlike in other terrain, almost everything is a potential danger area in urban terrain. Types of urban danger areas include, among others--

- Open areas.
- Parking lots and garages.
- Intersections.
- Streets, alleys, and roadways.
- Traffic circles and cul-de-sacs.
- Bridges, overpasses, and underpasses.
- Subterranean areas.
- Rooftops.
- Areas where large numbers of civilians gather.

DELIBERATE ATTACK

12-178. At the company level, a deliberate attack of an urban area usually involves the sequential execution of the following.

RECONNOITER OBJECTIVE

12-179. This involves making a physical reconnaissance of the objective with company assets and those of higher headquarters, as the tactical situation permits. It also involves a map reconnaissance of the objective and all the terrain that affects the mission, to include the analysis of aerial imagery, photographs, or any other detailed information about the building or other urban terrain, for which the company is responsible. Also, any human intelligence (HUMINT) collected by reconnaissance and surveillance units, such as the battalion reconnaissance platoon, snipers, is considered during the planning process.

MOVE TO OBJECTIVE

12-180. Movement is made rapidly without sacrificing security. Movement is along covered and concealed routes and can involve moving through buildings, down streets, subsurface areas, or a combination of all three. Urban movement must take into account the three-dimensional aspect of the urban area.

ISOLATE OBJECTIVE

12-181. Isolating the objective involves seizing terrain that dominates the area so that the enemy cannot supply, reinforce, or withdraw its defenders. Companies might be required to isolate an objective as part of a battalion operation or might be required to do so independently.

SECURE A FOOTHOLD

12-182. Securing a foothold involves seizing an intermediate objective that provides cover from enemy fire and a location for attacking troops to enter the urban area. The size of the foothold is METT-TC dependent and is usually a company intermediate objective. In some cases, a large building might be assigned as a company intermediate objective (foothold). As the company attacks to gain a foothold, it should be supported by suppressive fire and smoke.

CLEAR AN URBAN AREA

12-183. Before determining how much to clear the urban area, leaders must consider the factors of METT-TC. The ROE influence the tactics, techniques, and procedures (TTP) that platoons and squads select as they move through the urban area and clear individual buildings and rooms. The two categories of clearing urban areas are selective and systematic.

Selective Clearing

12-184. The commander may decide to clear only those parts necessary for the success of his mission if--

- An objective must be seized quickly.
- Enemy resistance is light or fragmented.
- The buildings in the area have large open areas between them. In this case, the commander would clear only those buildings along the approach to his objective or only those buildings necessary for security (Figure 12-11, page 12-43).

Systematic Clearing

12-185. An Infantry company may have a mission to clear an area of all enemy forces systematically. Through detailed analysis, the commander may anticipate that he will be opposed by a strong, organized resistance or will be in areas having strongly constructed buildings that are close together. Therefore, one or two platoons may attack on a narrow front against the enemy's weakest sector. They move slowly through the area, clearing systematically from room to room and building to building. The other platoon supports the clearing units and is prepared to assume their mission.

CONSOLIDATE, REORGANIZE, AND PREPARE FOR FUTURE MISSIONS

12-186. Consolidation occurs immediately after each action. Consolidation is security and allows the company to prepare for counterattack and to facilitate reorganization. In an urban environment, units must consolidate and reorganize rapidly after each engagement. The assault force in a cleared building must be quick to consolidate in order to repel enemy counterattacks and to prevent the enemy from infiltrating back into the cleared building. After securing a floor, selected members of the assault force are assigned to cover potential enemy counterattack routes to the building. Priority must be given to securing the direction of attack first. Those Soldiers alert the assault force and place a heavy volume of fire on enemy forces approaching the building. Reorganization occurs after consolidation. Reorganization actions prepare the unit to continue the mission; many actions occur at the same time.

Consolidate

12-187. Platoons assume hasty defensive positions after the objective has been seized or cleared. Based upon their specified and implied tasks, assaulting platoons should be prepared to assume an overwatch mission and support an assault on another building, or another assault within the building. Commanders must ensure that platoons guard enemy mouseholes between adjacent buildings, covered routes to the building, underground routes into the basement, and approaches over adjoining roofs.

Reorganize

12-188. After consolidation--

- Resupply and redistribute ammunition, equipment, and other necessary items.
- Mark the building to indicate to friendly forces that the building has been cleared.
- Move support or reserve elements into the objective if tactically sound.
- Redistribute personnel and equipment on adjacent structures.
- Treat and evacuate wounded personnel.
- Treat and evacuate wounded EPW and process remainder of EPW.
- Segregate and safeguard civilians.
- Reestablish the chain of command.
- Redistribute personnel on the objective to support the next phase or mission.

Prepare for Future Missions

12-189. The company commander anticipates future missions and prepares the company chain of command for transition.

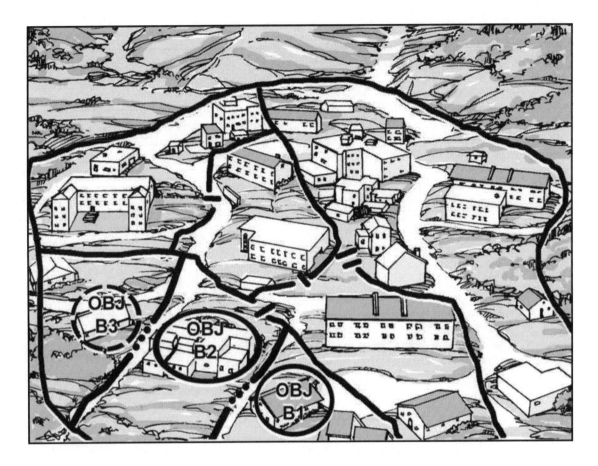

Figure 12-11. Clearing of selected buildings within sector.

ISOLATION OF URBAN OBJECTIVE

12-190. Infantry companies isolate urban objectives in order to prevent reinforcement of (or counterattack against) the objective and to kill or capture any withdrawing enemy forces. When planning the isolation, commanders consider three-dimensional and in-depth isolation of the objective (front, flanks, rear, upper stories, basements, and rooftops). They employ all available direct and indirect fire weapons consistent with the ROE. Isolating the objective is a key factor in facilitating the assault and preventing casualties. The company may perform this mission as the support element for a battalion operation, or it may assign the task to its own internal support element for a company attack. In certain situations, Infantry companies may isolate an objective or an area for special operations forces. When possible, the company should isolate the objective using stealth and rapid movement in order to surprise the enemy. Depending on the tactical situation, Infantry companies may use infiltration in order to isolate the objective. Likely tasks include, among others, the following.

BATTALION ATTACK

12-191. An Infantry company may isolate the objective as the support element for a battalion operation. When an Infantry company has this mission, the objective is normally a larger structure, block, or group of buildings. The company commander task-organizes his platoons and assigns them sectors of fire based on the METT-TC factors. In addition to isolating the objective, the company (support element) may have additional tasks to conduct on order or simultaneously. Examples of these additional tasks include providing the battalion reserve, assuming assault element missions, handling civilians and EPW, and performing CASEVAC.

Company Attack

12-192. When an Infantry company conducts an attack, the task organization and tasks given to the company support element are determined by the METT-TC factors. If the company conducts a company attack, the objective can be a building(s), block, traffic circle, or village (Figure 12-12). Figure 12-13, page 12-45, shows how to control direct fires during the assault.

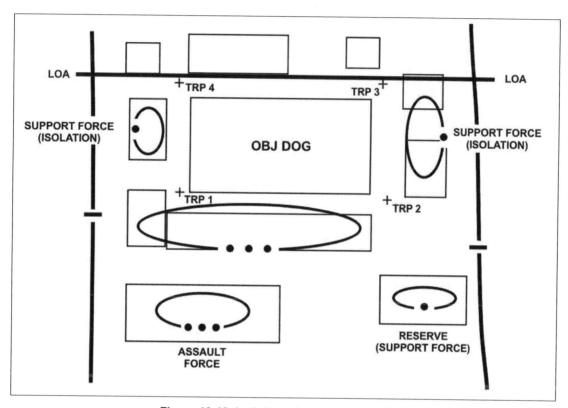

Figure 12-12. Isolation of an urban objective.

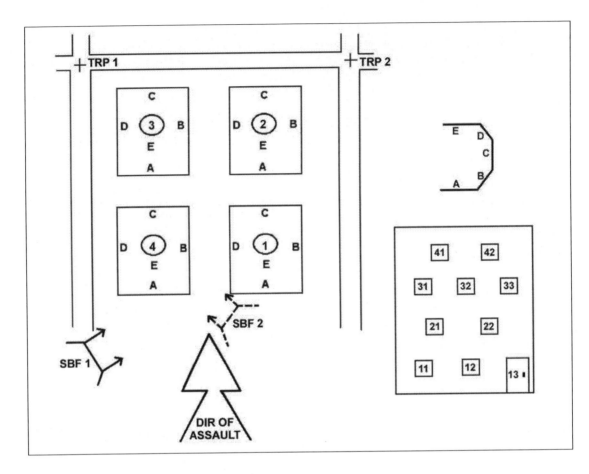

Figure 12-13. Direction-of-attack technique for direct-fire planning and control.

ASSAULT OF A BUILDING

12-193. The Infantry company conducts this task independently or as part of the assault element of a battalion. (Independently is defined here as an Infantry company having to provide its own support element, as opposed to conducting an operation without flank and rear support such as a raid or ambush.) If the company is the assault force of a battalion, it most frequently conducts the assault against a large building defended by a strong enemy force, for example, a reinforced platoon. The company commander must clearly understand the specified and implied tasks required to accomplish the mission, as well as the battalion commander's intent and the concept of the operation. This allows the company commander to task-organize and issue specific missions to his subordinate elements concerning which floors and rooms to clear, seize, or bypass. As an example, Figure 12-14 shows an Infantry battalion assigned the task of clearing the objectives in its zone (DOG and CAT). Company B has the task of seizing OBJ CAT. The company commander has decided to assign an intermediate objective (MOUSE) to 1st Platoon. 3d Platoon is a supporting element with the task of isolating MOUSE (1st and 2d squads to occupy the positions indicated) and providing one squad to act as the company reserve (3d squad). 2d Platoon (+) will pass through 1st Platoon, which will mark a passage lane and seize CAT.

Chapter 12

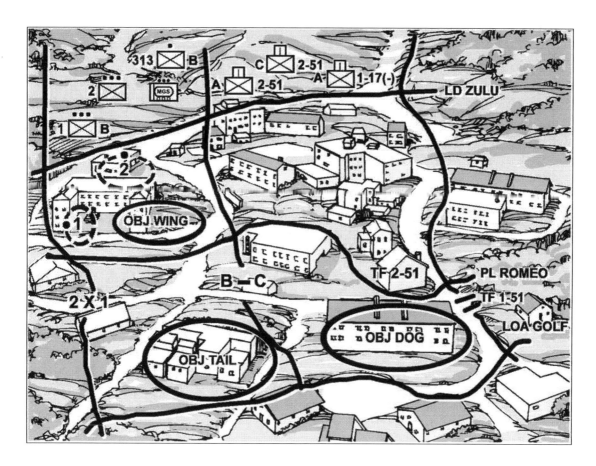

Figure 12-14. Assault of a building.

EXECUTION

12-194. Platoons may trade off with each other to clear alternate floors when clearing a multistory building. This permits troops to rest after a floor has been cleared. Platoons will most likely need to leave security on cleared floors and in cleared rooms. This facilitates the passage of another platoon, allowing the continuation of the assault. The assault element must quickly and violently execute its assault and subsequent clearing operations. Once it gains the momentum, the assault force must maintain this momentum to prevent the enemy from organizing a more determined resistance on other floors or in other rooms. If platoons find rooms, hallways, or stairwells that are barricaded with furniture or where obstacles have been placed, they should first try to bypass the barricade or obstacle and maintain the momentum of the attack. If they cannot bypass the barricade or obstacle, they should place security on it, check it for booby traps, and then reduce it. Subordinate leaders should continue the momentum of the assault.

AMMUNITION AND EQUIPMENT

12-195. METT-TC factors and the ROE determine how the assault element is equipped and armed. Commanders carefully manage the Soldier's load during the assault. Ammunition, water, special assault weapons and equipment, and medical supplies are normally the only items carried in the assault. The assault force carries only a fighting load of equipment and as much ammunition as is practical. (See Chapter 11, Section II for more on load management.) The battalion and Infantry company trains maintain control of additional ammunition and equipment not immediately needed by the assault force. An often-overlooked munition in an urban battle is the light antitank weapon. Soldiers can use these for a variety of purposes such as suppressing a manned position or supporting a breach into a structure. Resupply should be pushed to the assault element by the support element.

LOCATIONS

12-196. The assault may begin from the top or bottom of the building.

Top Entry

12-197. Entering at the top and fighting downward is the preferred method of clearing a building. This method is only feasible, however, when the company can gain access to an upper floor or rooftop by ladder or from the windows or roofs of adjoining, secured buildings, or by helicopter if enemy air defense weapons can be suppressed. The company can also gain access to the roof by entering at ground level and fighting up a stairwell or elevator shaft. They then clear the remainder of the building from the top to bottom. This will afford the Soldiers a covered and concealed route to the upper floors of the building. Rooftops are danger areas when surrounding buildings are higher and forces can be exposed to fire from those buildings. Helicopters should land only on those buildings that have special heliports on the roofs or on parking garages, but Soldiers can rappel or fast rope onto the roof or dismount as the helicopter hovers a few feet above the roof. Troops can then breach the roof or common walls. They may use ropes or other means to enter the lower floors through the holes created. The use of ladders to assault an upper level should be a last resort.

Bottom Entry

12-198. Entry at the bottom is common and might be the only option available. When entering from the bottom, breaching a wall (ROE dependant) is the preferred method because doors and windows might be booby-trapped and covered by fire from inside the structure. If the assault element must enter through a door or window, it should enter from a rear or flank position. Prior to entering the building, the commander must ensure the platoons have the capability to create entry points from covered and concealed positions.

SUPPRESSIVE FIRES

12-199. The support force provides suppressive fire while the assault force systematically clears the building. It also provides suppressive fire on adjacent buildings to prevent enemy reinforcements or withdrawal. Suppressive fires can consist of firing at known and suspected enemy locations or, depending on the ROE, only of firing at identified targets, or of returning fire. The support force destroys or captures any enemy personnel trying to exit the building. The support force must also deal with civilians displaced by the assault.

CLEARING OF ROOMS

12-200. Clearing platoons carry enough room marking equipment and plainly mark cleared rooms from the friendly side IAW unit TSOP (Figure 12-15). Markings must be visible to friendly units even if the operation occurs in limited visibility. The support force must understand which markings will be used and ensure that suppressive fires do not engage cleared rooms and floors. The commander must know where the assault teams are and which rooms and floors have been cleared. It is a key command and control function for the company commander.

Chapter 12

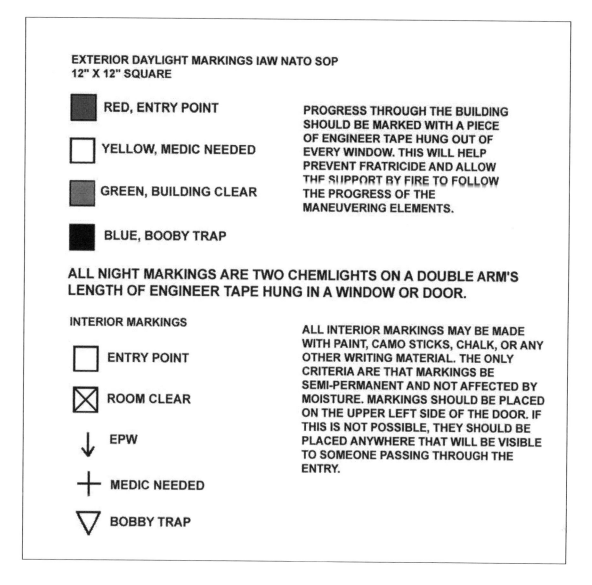

Figure 12-15. Example marking SOP.

ATTACK OF BLOCK OR GROUP OF BUILDINGS

12-201. An Infantry company normally attacks a block or group of buildings as part of a battalion attack.

EXECUTION

12-202. Platoon attacks supported by both direct and indirect fires characterize the execution of this mission. Success depends on isolating enemy positions (which often become platoon objectives), suppressing enemy weapons, seizing a foothold in the block, and clearing the block's buildings room by room.

DIRECT-FIRE WEAPONS

12-203. Machine guns and other direct fire support weapons fire on the objective from covered positions, consistent with the ROE. These weapons should not be fired for prolonged periods from one position. The gunners should use a series of positions and displace from one to another to gain better fields of fire and to avoid being targeted by the enemy. Direct fire support tasks can be assigned as follows.

- Machine guns fire along streets and into windows, doors, mouseholes, and other probable enemy positions. ROE may restrict firing only to known enemy locations.
- M203 suppress enemy positions.
- Riflemen engage targets of opportunity.

OBSCURATION AND ASSAULT

12-204. Before an assault, the Infantry company commander may employ smoke to conceal the assaulting platoons. He secures their flanks with direct fire weapons and employment of the reserve, if necessary. Concealed by smoke and supported by direct fire weapons, an assaulting platoon attacks the first isolated building. The assault force uses the cover of suppressive fires to gain a foothold. The company commander must closely coordinate the assault with its supporting fire so that the fire is shifted at the last possible moment. After seizing the block, the company consolidates and reorganizes to repel a counterattack or to continue the attack.

CONSOLIDATION AND REORGANIZATION

12-205. Consolidation occurs immediately after each action. Consolidation provides security and allows a unit to reorganize and prepare for counterattack. In an urban environment, units must consolidate rapidly after each engagement. The assault force in a cleared building must be quick to consolidate in order to repel enemy counterattacks and to prevent the enemy from infiltrating back into the cleared building. Many actions occur simultaneously. After securing a floor, selected members of the assault force cover potential enemy counterattack routes to the building. Priority must be given to securing the direction of attack first. Those Soldiers alert the rest of the assault force and place a heavy volume of fire on enemy forces approaching the building. Reorganization actions prepare the unit to continue the mission.

CONSOLIDATION

12-206. Platoons assume hasty defensive positions once the objective has been seized or cleared. Based upon their mission, assaulting platoons should be prepared to assume a support by fire position in order to support an assault on another building or continue the attack with an assault within the building. Commanders must ensure that platoons guard--

- Enemy mouseholes between adjacent buildings.
- Covered routes to the building.
- Underground routes into the basement.
- Approaches over adjoining roofs.

REORGANIZATION

12-207. Reorganization actions include--

- Resupply and redistribute ammunition.
- Mark the building to indicate to friendly forces that it has been cleared.
- Move support or reserve elements into the objective (if tactically sound).
- Redistribute personnel and equipment on adjacent structures.
- Treat and evacuate wounded personnel.
- Treat and process EPW.
- Segregate and safeguard civilians.
- Reestablish the chain of command.
- Redistribute personnel on the objective to support the next phase or mission.

Chapter 12

> *Note:* During evacuation of casualties, the commander must ensure that he does not allow the evacuation to interfere with his on-going operation. He ensures adequate forces are maintained to prevent the enemy from successfully counterattacking and reoccupying the building(s) that the company seized and cleared.

Section VIII. DEFENSE

In a built-up area, the defender takes advantage of inherent cover and concealment afforded by urban terrain. He also considers restrictions to the attacker's ability to maneuver and observe. By using the terrain and fighting from well-prepared and mutually supporting positions, a defending force can delay, block, fix, or destroy a much larger attacking force. The defense of a built-up area is organized around key terrain features, buildings, and areas that preserve the integrity of the defense and provide the defender ease of movement. The defender organizes and plans his defense by considering obstacles, AA, key terrain, observation and fields of fire, cover and concealment, fire hazards, and communications restrictions.

METT-TC FACTORS

12-208. Procedures and principles for planning and organizing the defense of a built-up area are the same as for other defensive operations. In developing a defensive plan, the defender considers the METT-TC factors. Planning for the defense of a built-up area is detailed and centralized. Execution is decentralized. Therefore, the company commander and his subordinate leaders must understand the mission, end state, and the commanders' intent two levels up.

Mission

12-209. The Infantry commander may receive the mission as a FRAGO or OPORD. The Infantry company may defend independently or as part of a larger force. Mission planning is essentially the same for all defensive operations.

Enemy

12-210. The commander analyzes the type of enemy force he may encounter. If the attacker is mostly dismounted Infantry, the greatest danger is allowing him to gain a foothold. If the attacker is mostly armored or mounted motorized Infantry, the greatest danger is that he will mass direct fire and destroy the defender's positions.

Terrain

12-211. Terrain in built-up areas is three-dimensional: ground level (streets and parks), above ground (buildings), and below ground (subways and sewers). Analyze all manmade and natural terrain features when planning to defend on built-up terrain. The type of built-up area in which he will operate affects the commander's defensive plan. Commanders emphasize obtaining and using all information. The items of information peculiar to combat in built-up areas include--

- Street, water, and sewer plans.
- Key installations and facilities.
- Key civilians.
- Civilian police and paramilitary forces.
- Sources of food.
- Communications facilities and plans.
- Power stations.

Villages

12-212. Villages are often on choke points in valleys, dominating the only high-speed AA through the terrain. If the buildings in such a village are well constructed and provide good protection against both direct and indirect fires, a formidable defense can be mounted by placing a company in the town while controlling close and dominating terrain with other battalion elements.

12-213. If the terrain allows easy bypass and no other villages are located on defensible terrain within a mutually supporting distance, defending the village is unwise. Doing so would allow friendly forces to be easily bypassed or isolated.

12-214. Commanders may use villages on approaches to large towns or cities to add depth to the defense or to secure the flanks. These villages are often characterized by clusters of houses and buildings (stone, brick, or concrete). Company-size BPs in these small villages can block approaches into the main defensive positions.

Strip Areas

12-215. Strip areas consist of houses, stores, and factories and are built along roads or down valleys between towns and villages. They afford the defender the same advantages as villages. If visibility is good and fields of fire are available, a unit acting as a security force need occupy only a few strong positions spread out along the strip. When engaged at long ranges, this will deceive the enemy into thinking the strip is an extensive defensive line. Strip areas often afford covered avenues of withdrawal to the flanks once the attacking force is deployed and before the security force becomes decisively engaged.

Towns and Cities

12-216. Sometimes a small force must face a mostly armored enemy in a small urban area that forms a choke point. When this happens, the force can gain the advantage by dominating critical approaches. To deny the enemy the ability to bypass the town or city, the defending force must control key terrain and coordinate with adjacent forces. Reserve forces are positioned to reinforce critical areas quickly. Obstacles and minefields help slow and canalize the attacker. In urban areas, finding positions that offer both good fields of fire and cover is often difficult. The forward edge of a town usually offers the best fields of fire, but are easily targeted by enemy overwatch and supporting fire. These areas often contain residential buildings constructed of light materials. Factories, civic buildings, and other heavy structures, which provide adequate cover and are more suitable for a defense, are usually deeper in the town, but offer limited fields of fire on likely AAs. Of course, since the forward edge of a town is the obvious position for a defender, it should usually be avoided. However, if the terrain limits the enemy's ability to engage, or if the town has strongly constructed buildings that offer adequate protection, then the defender can set up his position there. At first, a force might be assigned to BPs on the forward edge of the town. This allows it to provide early warning of the enemy's advance. The force engages the enemy at long range, and deceives the enemy as to the true location of the defense. Then, the force withdraws in time to avoid decisive engagement. If the forward edge offers limited observation, from the forward edge, then the force should be positioned on better terrain forward or to the flanks of the town. This will improve observation and allow engagement of the enemy at long ranges. To prevent airmobile or airborne landings within the city or town, the commander must emplace obstacles on probable LZs and DZs, to include parks, stadiums, and large rooftops and heliports. Direct and indirect fires should also cover these areas.

Large Built-Up Areas

12-217. In large built-up areas, tall buildings are normally close together. This requires a higher density of troops and smaller defensive sectors than in other urban terrain. The density of buildings, rubble, and street patterns dictate the depth and frontage of the unit (Table 12-1).

Unit	Frontages	Depths
Infantry battalion	4 to 8 blocks	3 to 6 blocks
Infantry company	2 to 4 blocks	2 to 3 blocks
Infantry platoon	1 to 2 blocks	1 block

Note: An average city block has a frontage of about 175 meters. These minimum figures apply in areas of dense block-type construction, multistory buildings, and underground passages.

Table 12-1. Approximate frontages and depths in large built-up areas.

12-218. In a large built-up area, an Infantry company has a sector, BP(s), or a strongpoint to defend. Although mutual support between positions should be maintained, built-up terrain often allows for infiltration routes that the enemy may use to pass between positions. Therefore, the defender must identify the following.

- Positions that enable him to place effective direct fires on the infiltrating enemy.
- Covered and concealed routes for friendly elements to move between positions (subways and sewers).
- Structures that dominate large areas.
- Areas where antiarmor weapons have effective fields of fire such as parks, boulevards, rivers, highways, and railroads.
- Firing positions for mortars.
- Command and control locations that offer cover, concealment, and ease of communications.
- Protected storage areas for supplies.

12-219. Leaders choose buildings that add most to the general plan of defense for occupation. Mutual support between these positions is vital to prevent the attacker from maneuvering and outflanking the defensive position, making it untenable. Buildings chosen for occupation as defensive positions should have the following characteristics.

- Good protection.
- Strong floors to keep the structure from collapsing under the weight of debris.
- Thick walls.
- Construction consisting of nonflammable materials (avoid wood).
- Strategic locations (corner buildings or prominent structures).
- Adjacent to streets, alleys, vacant lots, and parks. (These buildings usually provide better fields of fire and are more easily tied in with other buildings.)
- Covered by friendly fire and offering good escape routes.

Obstacles

12-220. A built-up area is itself an obstacle since it canalizes and impedes an attack. Likely AA should be blocked by obstacles and covered by fire (Figure 12-16, page 12-48).

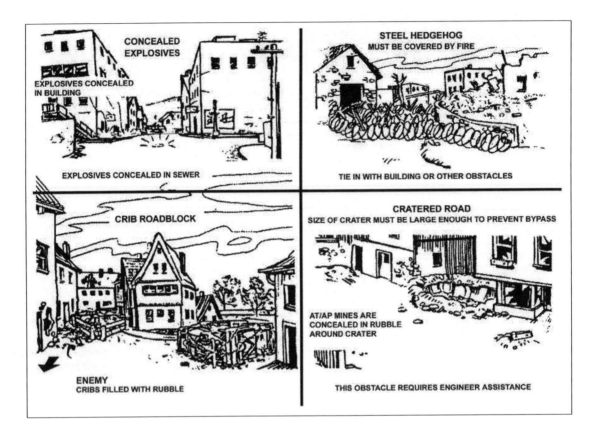

Figure 12-16. Example of urban obstacles.

Avenues of Approach

12-221. The defender considers conventional AA into and out of the city, and the avenues above and below ground level within built-up areas. Normally the defender initially has the advantage. He knows the built-up area and can move rapidly from position to position through buildings and underground passages. Control of these above-and below-ground AA becomes more critical when the defense of key terrain must be oriented against terrorism and sabotage. All AA (three-dimensionally) must be denied. Infantry company commanders exploit the use of field-expedient obstacles, such as cars and light poles, or the emplacement of command detonated antipersonnel mines and antitank mines. Commanders must clearly understand the ROE and what they are permitted to emplace. When necessary, obstacles can be emplaced without mines and covered by fire within the parameters of the ROE.

Key Terrain

12-222. Key terrain is any place where seizure, retention, or control affords a marked advantage to either combatant. Examples of key terrain during UO are bridges over canals or rivers, building complexes, public utilities or services, and parks. The population of a built-up area may also be considered key terrain. The identification of key terrain allows the defender to select his defensive positions and helps in determining the enemy's objectives.

Observation and Fields of Fire

12-223. The defender positions weapons to obtain maximum effect and mutual supporting fire. The defender strives for long-range engagements out to the maximum effective ranges. Observers should be well above street level to adjust fires on the enemy at maximum range. Fires and FPFs are preplanned (if possible and ROE permitting) and preregistered on the most likely approaches to allow for their rapid shifting to threatened areas.

Cover and Concealment

12-224. The defender prepares positions using the protective cover of walls, floors, and ceilings. Soldiers improve positions using materials on hand. When the defender must move, he can reduce his exposure by--

- Using prepared breaches through buildings.
- Moving through reconnoitered and marked subterranean systems.
- Using trenches.
- Using the concealment offered by smoke and darkness to cross open areas.

Troops Available

12-225. UOs are often Soldier intensive, particularly for units conducting offensive operations. Due to the fortified nature of structures available to the enemy, the attacker must carry large quantities of bulky and heavy systems, equipment, ammunition, and explosives to remove obstacles or penetrate walls. The result is soldiers suffering from fatigue much sooner than in a non-urban environment. This also slows the attacker and increases his exposure to hostile fire. This results in higher casualties and may make casualty collection and evacuation extremely difficult. Maintaining security is another significant challenge. Encountering large numbers of buildings and rooms makes it nearly impossible to secure much of the ground taken during an attack. The enemy will always attempt to discover and reoccupy unsecured areas to attempt to counter-attack. To the maximum extent possible, commanders must use as large a force as possible, and assure they are physically fit and well rested for urban operations.

Employment of Platoons and Organic Assets

12-226. The commander decides where to engage the enemy. He then selects platoon BPs or assigns sectors where platoons can implement his fire plan. The frontage for a platoon is about one to two city blocks long. Platoons can occupy about three small structures or a larger two- to three-story building (Figure 12-17). Along with platoon primary, alternate, and subsequent positions the commander may also direct supplementary position(s) to reorient the defense to meet enemy threats from another direction.

Urban Operations

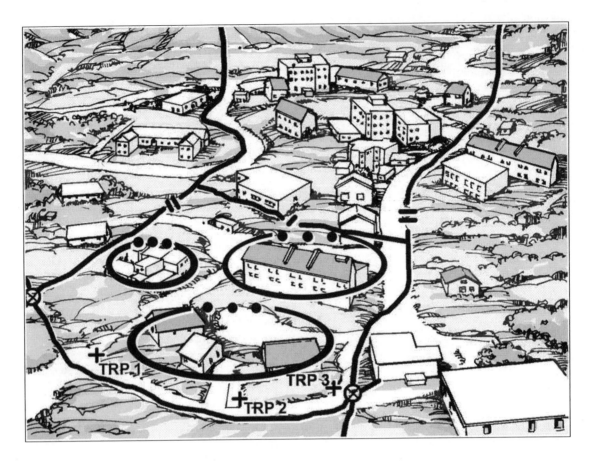

Figure 12-17. Platoon battle positions in a company sector.

Mortars

12-227. Company mortars are employed to maximize the effect of their high-angle fires. Due to UO considerations of distance, obstacles, and suppression effects on the target, the best use of 60-mm mortars might be in the direct lay mode and direct alignment. For indirect fire assets in the indirect fire mode, employ mortars and artillery one-half and five times the height of the building to be cleared for minimum planning distances in support of Infantry assaults. Mortars are used to engage--

- Enemy overwatch positions.
- Enemy Infantry before they seize a foothold.
- Targets on rooftops.
- Enemy reinforcements within range.

Javelins

12-228. Based on the Javelin's capabilities and limitations, commanders give the platoons missions that can use antiarmor systems to support the defensive scheme of maneuver.

Employment of Fire Support

12-229. Fire planning must be comprehensive due to the proximity of buildings to targets, minimum range restrictions, repositioning requirements, and ROE. Mortar and artillery fires are planned on likely enemy support positions, AA and on top of, and immediately around defensive positions for close support.

Artillery

12-230. Artillery fire should be used to--

- Suppress or obscure enemy overwatch elements.
- Disrupt or destroy an assault.
- Provide counterbattery fire.
- Support counterattacks.
- Provide direct fire when necessary.

Final Protective Fires

12-231. FPFs are planned to disrupt dismounted assaults threatening to overwhelm the defensive positions. Fires within the city are planned along likely routes of advance to destroy the enemy as he tries to deepen a penetration.

Priorities

12-232. The Infantry company commander establishes priorities of fire and priority targets. Priority targets are based on enemy AA and enemy systems that present the greatest danger to the defense. Mortar and artillery fires should suppress enemy ATGMs and overwatch positions or elements. If enemy formations secure a foothold, priority shifts to the destruction of enemy forces within the penetration.

Support of Counterattacks

12-233. When initiated, counterattacks are given priority of supporting fires. When artillery executes the missions mentioned above, it must remain mobile and be prepared to displace to preplanned positions to avoid enemy counterbattery fire.

Indirect Fire Planning

12-234. At company and platoon level, indirect fire plans include fires of organic, attached, and supporting weapons. The Infantry company commander also plans his own mortar and artillery fires on and immediately around his BPs for close support.

Employment of Air Defense Assets

12-235. Air defense assets available to the commander, such as Stinger and Avenger, are normally employed to ensure all-round air defense. These assets are normally controlled at battalion level, but they might be placed under the Infantry company commander's control when METT-TC factors warrant. The lack of good firing positions (for long-range, air defense missile systems in the built-up area) might limit the number of deployed weapons. In the defense, weapons systems might have to be winched or airlifted into positions. Rooftops and parking garages are good firing positions because they normally offer a better line of sight. Stingers and Avengers might be assigned to protect specific positions or function in general support of the battalion.

Employment of Engineers

12-236. Engineers are employed under battalion control or attached to Infantry companies. Infantry company commanders may receive an engineer squad to assist them in preparing the defense. Commanders must consider engineer tasks that enhance survivability, mobility, and countermobility. Tasks that engineers can accomplish in the defense of a built-up area include--

- Constructing obstacles and rubbling.
- Clearing fields of fire.

- Laying mines.
- Preparing mobility routes between positions.
- Preparing fighting positions.
- Fight as Infantry.

Employment of the Reserve

12-237. The commander's defensive plan always considers the employment of a reserve. The reserve force should be prepared to counterattack to complete the destruction of the enemy force, counterattack to regain key positions, to block enemy penetrations, to protect the flanks, or to assist by fire in the disengagement and withdrawal of positions. For combat in a built-up area, a reserve force--

- Normally consists of Infantry.
- Must be as mobile as possible.

Time Available

12-238. The commander establishes and monitors the progress of the priorities of work. Many tasks are accomplished simultaneously, but priorities for preparation should be IAW the commander's order. A sample priority of work sequence follows.

Establish Security

12-239. The unit establishes all-round security by placing forces on likely AAs. METT-TC factors determine the level of security, for example, 30 or 50 percent. The reconnaissance and counterreconnaissance plan is initiated.

Assign Areas of Responsibility

12-240. Boundaries define areas of responsibility. They include areas where units may fire and maneuver without interference or coordination with other units. Responsibility for the primary AA is never split. In areas of semidetached construction, where observation and movement are less restricted, boundaries are established along alleys or streets to include both sides of a street in a single sector. Where buildings present a solid front along streets, boundaries might have to extend to one side of the street. Responsibilities are assigned for EAs and BPs designated to cover the EA. BPs should be specifically assigned to platoons, as required by METT-TC. The Infantry company commander should specify which building(s) comprises the platoon BP or strongpoint. Positions should be clearly designated so that no doubt remains as to which platoon has responsibility for occupation or control.

Clear Fields of Fire

12-241. In built-up areas, commanders may need to rubble certain buildings and structures to provide greater protection and fields of fire to the defender. If the ceiling of a lower-story room can support the weight of the rubble, collapsing the top floor of a building before the engagement starts may afford better protection against indirect fires. Rubbling an entire building can increase the fields of fire and create an obstacle to enemy movement. However, defenders must be careful. Rubbling buildings too soon or rubbling too many may give away exact locations and destroy cover from direct fire. Planning must be extensive so that rubbled buildings will not interfere with planned routes of withdrawal or counterattack. Vehicles may also have to be moved to clear fields of fire.

Select and Prepare Initial Fighting Positions

12-242. The Infantry company commander selects positions in depth. The unit prepares positions as soon as troops arrive and continue preparing as long as positions are occupied. Enemy infiltration or movement sometimes occurs between and behind friendly positions. Therefore, each position is organized for all-round defense. The defender--

- Makes minimum changes to the outside appearance of buildings where positions are located.
- Screens or blocks windows and other openings to keep the enemy from seeing in or tossing in hand grenades. Include all windows so that the enemy cannot tell which openings the defenders are covering.
- Removes combustible material to limit the danger of fire. Fires are dangerous to defenders and create smoke that could conceal attacking troops. For these reasons, defenders should remove all flammable materials and stockpile firefighting equipment such as water and sand. The danger of fire also influences the type of ammunition used in the defense. Do not use tracers or incendiary rounds extensively if threat of fire exists.
- Turns off electricity and gas at the facility that serves the urban area. Both propane and natural gas are explosive. Natural gas is also poisonous, displaces oxygen, and is not filtered by a protective mask. Propane gas, although not poisonous, is heavier than air. If it leaks into an enclosed area, it displaces the oxygen and causes suffocation.
- Locates positions so as not to establish a pattern. The unit avoids obvious firing locations.
- Camouflages positions.
- Reinforces positions with all materials available such as mattresses and furniture. Use caution because mattresses and fabric furniture are flammable. Fill drawers and cabinets with earth or sand to provide cover. Consider placing vehicles, such as trucks or buses, over positions outside buildings. Drain flammable fluids from vehicles and remove other flammables such as seats. Fill gas tanks with water.
- Blocks stairwells and doorways with wire or other material to prevent enemy movement. Create holes between floors and rooms to allow covered and concealed movement within a building.
- Prepares range cards, fire plans, and sector sketches.
- Considers how to use basements. If grazing fire can be achieved from basement windows, emplace machine guns in basements. When not using basements, seal them to prevent enemy entry.
- Caches resupply of ammunition, water, and medical supplies.

Establish Communications

12-243. When allocating time to establish communications, commanders consider the effects of built-up areas. LOS limitations affect both visual and radio communications. Wire laid at street level is easily damaged by rubble and vehicle traffic. The noise of built-up area combat is much louder than in other areas, making sound signals difficult to hear. Therefore, the time needed to establish an effective communications system in urban terrain might be greater than in other terrain. Infantry company commanders consider the following techniques when planning for communications.

- Emplace LOS radios and retransmission sites on the upper floors of buildings.
- Use existing telephone systems. However, telephones are not secure even though many telephone cables are underground.
- Use messengers at all levels since they are the most secure means of communications.
- If assets are available, lay wire through buildings for maximum protection.

Emplace Obstacles and Mines

12-244. To save time and resources in preparing the defense, commanders emphasize using all available materials (to include automobiles, railcars, and rubble) to create obstacles. Civilian construction equipment and materials must be located and inventoried. This equipment can be used with engineer assets or in place of damaged equipment. Coordination must be made with proper civilian officials before use, which is normally a brigade or battalion staff responsibility. Engineers can provide advice and resources as to the employment of obstacles and mines.

12-245. The principles for employing mines and obstacles do not change in the defense of a built-up area, but techniques do change. For example, concrete and asphalt make burying and concealing mines in streets difficult. Consider placing mines in sandbags so they are not visible and using fake mines placed in sandbags in order to deceive the enemy. Mines and obstacles are emplaced consistent with the ROE. Any antipersonnel mines must be command detonated.

12-246. Obstacles are tied to buildings and rubbled areas to increase effectiveness and to canalize the enemy. Family of scatterable mines (FASCAM) might be effective on the outskirts of an urban area or in parks, but in a city core, areas might be too restrictive.

12-247. Riot control agents might be employed to control noncombatant access into defensive areas.

Improve Fighting Positions

12-248. When time permits, all positions, to include supplementary and alternate positions, are reinforced with sandbags and provided overhead cover. Attached engineers can help in this effort by providing advice and assisting with construction.

Establish and Mark Routes Between Positions

12-249. Reconnaissance by all defending elements aids in route selection for use by defenders moving between positions. Movement is crucial in fighting in built-up areas. Early selection and marking of routes adds to the defender's advantages.

Civil Considerations

12-250. International law and moral imperatives require the Infantry company commander to consider the effects of operations on the civilian population. The company commander also considers cultural, economical, and political boundaries as they may have a direct impact on the range of tactical options available to him.

12-251. Commanders might be precluded from countermobility operations directed at economically important roads, railways, and bridges. They consider civilian movement when emplacing minefields. Commanders implement restrictive fire control measures consistent with ROE.

12-252. Units with large civilian populations in their AO often conduct support operations while preparing a defense. When Army forces must damage areas that are important to civilians, they ensure that civilian leaders and populations understand why these actions are necessary.

Fire Hazards

12-253. The defender's detailed knowledge of the terrain permits him to avoid areas that are likely to be fire hazards. All urban areas are vulnerable to fire, especially those with many wooden buildings. The defender can deliberately set fires.

- To disrupt and disorganize the attackers.
- To canalize the attackers into more favorable EAs.
- To obscure the attacker's observation.

12-254. Likewise, the enemy may cause fires to confuse, disrupt, or constrain friendly forces and efforts. Company commanders anticipate this possibility and ensure that fire-fighting equipment is on hand when conducting this type of operations.

Chapter 12

COMMAND AND CONTROL

12-255. In all defensive situations, the Infantry company commander positions himself where he can control the action. The leader must see and feel the battlefield. In urban terrain, this is a greater challenge due to obstacles, poor visibility, difficulty in communication, and intense fighting.

GRAPHIC CONTROL MEASURES

12-256. Phase lines are used to monitor and control friendly units. Principal streets, rivers, and railroad lines are suitable phase lines. They should be clearly and uniformly marked on the near or far side of the street or open area. Checkpoints aid units in reporting locations and controlling movement. Contact points designate specific points where units make physical contact. TRPs facilitate fire control. These and other control measures ensure coordination throughout the chain of command.

COMMAND POSTS

12-257. Command posts should be located underground, if possible. Their vulnerability requires all-round security. Since each CP might have to secure itself, it should be near the reserve unit for added security. When collocated with another unit, however, CPs might not need to provide their own security. A simplified organization for command posts is required for ease of movement. Alternate CP locations and routes must also be identified.

ACTIONS ON CONTACT

12-258. When enemy forces attack to seize initial objectives, the defender employs all available fires to destroy and suppress the direct fire weapons that support the ground attack. Enemy tanks and APCs are engaged as soon as they come within the effective range of antiarmor weapons. As the enemy attack develops, the actions of small-unit leaders assume increased importance. Squad and platoon leaders are often responsible for fighting independent engagements. Thus, all leaders must understand their commander's concept of the defense (two levels up). Where the enemy's efforts are likely to result in his gaining a foothold, violent counterattacks must deny him access into the MBA.

REAR AREA

12-259. Infantry companies do not normally deploy maneuver elements in the rear area; however, squads and platoons might be detached in order to protect sustainment elements. In certain cases, the company trains may collocate with the battalion combat trains.

COUNTERATTACKS

12-260. Reserves should be prepared to counterattack to complete the destruction of the enemy, regain key positions, block enemy penetrations, provide flank protection, and assist by fire the disengagement and withdrawal of endangered positions. Enemy footholds must be repelled violently. When the reserves are committed to counterattack to reinforce a unit, they might be attached to the unit in whose sector the counterattack is taking place. Otherwise, the counterattack becomes the main effort. This makes coordination easier, especially if the counterattack goes through the unit's positions.

DEFENSE DURING LIMITED VISIBILITY

12-261. Infantry company commanders can expect the attacker to use limited visibility conditions to conduct necessary operations to sustain or gain daylight momentum. Commanders employ the following measures to defend against attacks in limited visibility.

- Shift defensive positions and crew-served weapons to an alternate position or a hasty security position just before dark to deceive the enemy as to the exact location of the primary position.

- During limited visibility, consider the need to occupy, block, or patrol unoccupied areas between units, which can be covered by observed fire during daylight. Install early warning devices.
- Emplace radar, remote sensors, and night observation devices to cover streets and open areas.
- Position nuisance mines, noise-making devices, tanglefoot tactical wire, and OPs on all AA for early warning and to detect infiltration.
- Plan for artificial illumination, for example, street lamps, stadium lights, or pyrotechnics.
- Use indirect fire, grenade launchers, and hand grenades when probing defenses. This helps prevent disclosure of defensive positions.
- Plan a signal to initiate FPFs. Crew-served weapons, armored vehicle mounted weapons (if available), and individual riflemen fire within their assigned sectors. Grenades and command-detonated mines should supplement other fires as the enemy approaches the positions.
- Move to daylight positions before BMNT. To facilitate movement, mark buildings from the friendly side IAW unit SOP.

COMMUNICATIONS RESTRICTIONS

12-262. Radio communications are at first the best way to control the defense of a built-up area and to enforce security. Structures and a high concentration of electrical power lines may degrade radio communication in built-up areas. Wire should be emplaced and used for communications as time permits. However, wire can be compromised if interdicted by the enemy. Messengers can be used as another means of communication. Visual signals may also be used but are less effective because of the screening effects of buildings and walls. Signals must be planned, widely disseminated, and understood by all assigned and attached units. Increased battle noise makes the effective use of sound signals difficult.

HASTY DEFENSE

12-263. A likely defensive mission for the Infantry company in urban terrain is to conduct a hasty defense, which is characterized by reduced preparation time. All the TLPs are the same. The priorities of work are about the same, but many occur at the same time. Sectors of fire are determined, units are deployed, weapons emplaced, and positions prepared IAW the amount of time the company commander has available.

OCCUPATION AND PREPARATION OF POSITIONS

12-264. Preparations for the hasty defense vary with the time available. Security is the first priority. All other preparations revolve around the ability to execute the direct fire plan. Camouflage and conceal the presence of the hasty fighting positions and provide as much protection as possible for the Soldiers operating them. Construct positions using appliances, furniture, and other convenient items and materials. Locate positions back from the windows in the shadows of the room.

Position Crew-Served and Special Weapons

12-265. Generally, position crew-served and special weapons inside buildings unless an outside position is preferable and can be protected and camouflaged.

Emplace Barriers and Obstacles

12-266. The company establishes two belts of barriers and obstacles that are not as extensive as in a defense that permits more time. The company covers all obstacles with observation and fires.

Prepare Positions

12-267. Consider the following work sequence.

- Gather available materials, such as tables, dressers, and appliances, to construct positions.
- Construct stable firing platforms for the weapons.
- Use the material gathered to build frontal and side protection. Fill cabinets, dressers, end tables, and other furnishings with materials to stop small arms fire.
- Do not disturb firing windows. Curtains and other aspects of the original setting are components of camouflage.
- Construct alternate firing positions similar to the primary positions.
- Emplace rear and overhead cover on the primary positions (after constructing alternate positions).
- Remove fire hazards. Pre-position firefighting equipment.
- Construct dummy positions in rooms above, below, and next to primary and alternate positions in order to draw enemy suppressive fire away from primary positions.
- Walk the positions from the enemy side.

Rehearse

12-268. Conduct rehearsals with leaders and Soldiers on orienting the defense, selecting unit positions and crew-served weapons positions, planning a counterattack, and a withdrawing plan, for example.

Enhance Movement

12-269. Little time is available to improve movement within the defense. Units should plan to use tunnels, underground routes, and routes through buildings. The movement enhancement priority is to remove obstructions to alternate positions and the counterattack route.

Communicate

12-270. Check communications. Communications are primarily radio. Plan and improve routes for messengers. If time is available, emplace wire as an improvement to the defense.

IMPROVEMENT OF DEFENSE

12-271. As time permits, consider the following areas and prioritize them IAW the METT-TC factors.

- Barrier and obstacle improvement.
- Improvement of primary, alternate, and subsequent positions.
- Preparation of supplementary positions.
- Additional movement enhancement efforts.
- Initiation of patrols.
- Improvement of camouflage.
- Continued rehearsals for counterattack and withdrawal.
- Rest plan.

COMPANY DEFENSE OF A VILLAGE

12-272. A village is characterized by a built-up area surrounded by other types of terrain. Normally, an Infantry company defends a village as part of a battalion defense, establishing BPs and strongpoints with other Infantry companies defending from key or decisive terrain. Once the Infantry company commander has completed his reconnaissance of the village, he reconnoiters the surrounding terrain and, with the information assembled, develops his plan for the defense. One of his first decisions is whether to defend on the leading edge of the village or farther back within the confines of the village. Normally, defending on the leading edge, where the defending company can take advantage of longer-range observation and fields of fire, is more effective against an armor-heavy force. Defending in depth within the village to deny the enemy a foothold is more effective against a force that is primarily Infantry. This decision is based on the METT-TC factors. The company may need to coordinate with adjacent units to plan for the defense or control of the open terrain that typically surrounds a village.

FACTORS

12-273. Several factors influence the commander's decision. First, he must know the type of enemy. If the threat is mainly Infantry, the greatest danger is allowing them to gain a foothold in the town. If the threat is armored or motorized Infantry, the greatest danger is that massive direct fire will destroy the Infantry company's defensive positions. The Infantry company commander must also consider the terrain forward and to the flanks of the village where the enemy can direct fires against his positions.

PLATOON BATTLE POSITIONS

12-274. Platoons are given a small group of buildings in which to prepare their defense, permitting the platoon leader to establish mutually supporting squad-size positions. This increases the area that the platoon can control and hampers the enemy's ability to isolate or bypass a platoon. A platoon might be responsible for the road through the village. The rest of the Infantry company is then positioned to provide all-round security and defense in depth.

COMPANY MORTARS AND JAVELINS

12-275. The positioning of the Infantry company's mortars must protect the mortars from direct fire and allow for overhead clearance. Javelin positions must allow them to engage targets at maximum ranges with alternate firing points.

DEFENSE OF A BLOCK OR GROUP OF BUILDINGS

12-276. An Infantry company normally conducts a defense of a city block or group of buildings as part of a battalion conducting a sector defense in a built-up area. Company commanders may assign their platoons strongpoints, BPs, sectors, or any combination of these. An Infantry company operating in urban terrain might have to defend a city block or group of buildings in a core periphery or residential area. The company conducts this operation IAW the battalion's defensive scheme of maneuver. The operation is coordinated with the action of security forces that are charged with delaying to the front of the company's position. The defense should take advantage of the protection of buildings that dominate the AA into the MBA. This mission differs from defense of a village in that it is more likely to be conducted completely on urban terrain, without the surrounding open terrain that characterizes the defense of a village. An Infantry company is particularly well suited for this type of mission since the fighting requires the enemy to move Infantry into the built-up area to seize and control key terrain.

RECONNAISSANCE AND SECURITY

12-277. Reconnoiter the terrain and prepare obstacles and fire lanes. Patrols supplement the OPs, mainly during periods of limited visibility. The company should use wire communications. Platoons should have the mission to provide one OP in order to provide spot reports concerning the size, location, direction, and rate of movement; and the type of enemy assaulting the company sector or BP.

TASK ORGANIZATION

12-278. METT-TC factors determine how the company commander task-organizes the company to accomplish the mission.

Chapter 12

EXECUTION

12-279. The defensive forces engage the enemy with direct and indirect fire on the AA, cover the obstacles by fire, and prepare a strong defense inside the buildings. Reserve forces should be near the front of the company sector in covered and concealed positions with a number of planning priorities. Counterattack forces should have specific instructions as to what their actions will be after the enemy assault has been repelled, for example, to stay in sector or to revert back to reserve status. The company conducts rehearsals both day and night.

DEFENSE OF KEY TERRAIN

12-280. An Infantry company defends key terrain independently or as part of a battalion. It may form a perimeter defense around key terrain such as a public utility (a gas, electric, or water plant), a communications center (a radio or television station), a government center, a command and control facility, or a traffic circle that enhances movement, for example. The Infantry company can occupy and defend buildings and other dominant terrain, or it can establish and operate checkpoints and roadblocks in conjunction with this defense.

12-281. An Infantry company can defend a traffic circle or similar terrain to prevent the enemy from seizing it. This is characterized by the occupation and defense of the buildings around the traffic circle that control the AA into and out of the objective area. This defense might be part of conventional operations or might be an adjunct to a mission of stability operations or support operations. In many cases, an unclear enemy situation and extremely restrictive ROE characterize this mission. The METT-TC factors determine how to defend the objective.

TASK ORGANIZATION

12-282. The METT-TC factors determine the task organization of the Infantry company.

TASKS

12-283. Some of the following tasks might be necessary.

- Provide inner and outer security patrols.
- Establish OPs.
- Establish checkpoints and roadblocks.
- Conduct civilian control and evacuation.
- Conduct coordination with local authorities.
- Prevent collateral damage.
- Supervise specific functions associated with operation of the facility such as water purification tests and site inspections.

EXECUTION

The Infantry company commander does not have to occupy the key terrain. He deploys his units in such a manner to prevent the enemy from controlling the key terrain. The company emplaces machine guns and antitank weapons to cover the dismounted and mounted AA respectively. It normally uses wire obstacles to deny entry into the area and uses antitank and command-detonated mines consistent with the ROE. Obstacles are covered by fire and rigged with detection devices and trip flares. The company is prepared to defend against a direct attack such as a raid or sabotage. The company commander positions the 60-mm mortar section to provide all-round fire support, and positions the AT section to engage vehicular targets. If the threat does not require the employment of mortars or AT weapons, the commander can give these sections other tasks.

OTHER CONSIDERATIONS

12-284. Depending on the mission requirements and threat, the Infantry company commander might have to consider the need for the following.

- Artillery and attack helicopter support.
- Air defense artillery assets to defend against air attack.
- Engineer assets to construct obstacles.
- Interpreters to assist in the functioning of a facility and operation of the equipment.
- Military police (MP), civil affairs, and PSYOP assets for civilian control and liaison.
- Coordination with local police and authorities.

FORCE PROTECTION

12-285. The Infantry company might be required to conduct a perimeter defense as part of a force protection mission such as defending a friendly base camp on urban terrain. The same techniques of establishing a perimeter defense described above are used. The company maintains the appropriate level of security, for example, 100 percent, 50 percent, or 30 percent, consistent with the commander's plan and the enemy situation. Additional tasks may include--

- Set up roadblocks and checkpoints.
- Search individuals and vehicles before they enter the camp.
- Maintain a presence as a show of force to the population outside the base camp.
- Conduct inner and outer security patrols.
- Clear urban terrain of any enemy that overwatches the base camp.
- Conduct ambushes to interdict any enemy forces moving toward the base camp.
- Restrict access to locations within the base camp and conduct surveillance of these locations from (or from within) adjacent structures or positions.
- Conduct reaction force duties inside and outside the perimeter of the camp.

DEFENSE OF AN URBAN STRONGPOINT

12-286. A company might be directed to construct a strongpoint as part of a battalion defense (Figure 12-18). In order to do so, it must be augmented with engineer support, more weapons, and sustainment resources. A strong point is defended until the unit is formally ordered out of it by the commander directing the defense. Urban areas are easily converted to strongpoints. Stone, brick, or steel buildings provide cover and concealment. Buildings, sewers, and some streets provide covered and concealed routes and can be rubbled to provide obstacles. Telephone systems can provide communications.

12-287. The specific positioning of units in the strongpoint depends on the commander's mission analysis and estimate of the situation. The same considerations for a perimeter defense apply in addition to the following.

- Reinforce each individual fighting position (to include alternate, subsequent and supplementary positions) to withstand small-arms fire, mortar fire, and artillery fragmentation. Stockpile food, water, ammunition, pioneer tools, and medical supplies in each fighting position.
- Support each individual fighting position with several others. Plan and construct covered and concealed routes between positions and along routes of supply and communication. Use these to support counterattack and maneuver within the strongpoint.
- Divide the strongpoint into several independent, but mutually supporting, positions or sectors. If one of the positions or sectors must be evacuated or is overrun, limit the enemy penetration with obstacles and fires and support a counterattack.

Chapter 12

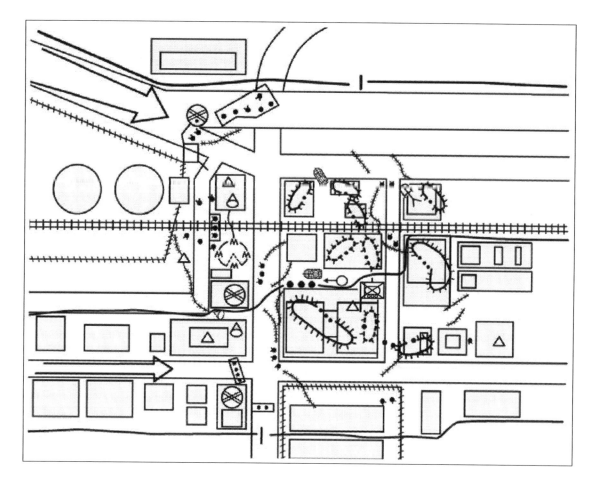

Figure 12-18. Urban strongpoint.

12-288. Construct obstacles and minefields to disrupt or canalize enemy formations, to reinforce fires, and to protect the strongpoint from the assault. Place the obstacles and mines out as far as friendly units can observe them, within the strongpoint, and at points in between where they will be useful.

12-289. Prepare range cards for each position and confirm them by fires. Plan indirect fires in detail and register them. Indirect fires should also be planned for firing directly on the strongpoint using proximity fuzes.

12-290. Plan and test several means of communication within the strongpoint and to higher headquarters to include radio, wire, messenger, pyrotechnics, and other signals.

12-291. Improve or repair the strongpoint until the unit is relieved or withdrawn. More positions can be built, routes to other positions marked, existing positions improved or repaired, and barriers built or fixed.

12-292. A strong point might be part of any defensive plan. It might be built to protect vital units or installations, as an anchor around which more mobile units maneuver, or as part of a trap designed to destroy enemy forces that attack it.

DELAY

12-293. The intent of a delay is to trade space for time: slow the enemy, cause him casualties, and stop him, if you can, without becoming decisively engaged. This procedure is done by defending, disengaging, moving, and defending again. A company delay is normally conducted as part of the battalion task force's plan. The delay destroys enemy reconnaissance elements forward of the outskirts of the urban area, prevents the penetration of the urban area, and gains and maintains contact with the enemy to determine the strength and location of the main attack. Infantry companies are well suited for this operation, because they can take advantage of the cover and concealment provided by urban terrain and inflict casualties on the enemy at close range. Delays are planned by assigning platoon BPs, platoon sectors, or both. Figure 12-19 shows a company delay in urban terrain with the company commander assigning platoon BPs. Routes are planned to each BP or within the sector. Routes are also planned to take advantage of the inherent cover and concealment afforded by urban terrain such as going through and hugging buildings, using shadows, and subsurface areas.

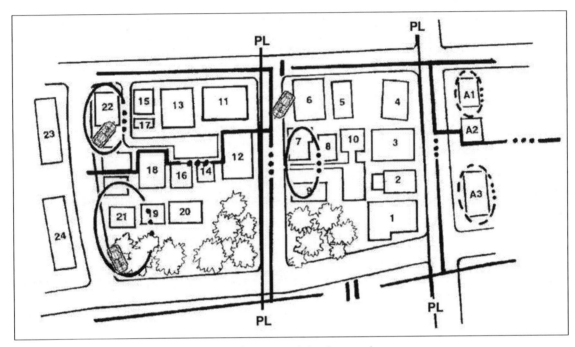

Figure 12-19. Company delay in an urban area.

12-294. The company's sector should be prepared with obstacles to increase the effect of the delay. Engineers prepare obstacles on main routes but avoid some covered and concealed routes that are used by the friendly troops for reinforcement, displacement, and resupply. These routes are destroyed and obstacles are executed when no longer needed.

12-295. Antiarmor weapon systems and combat vehicles (tanks, BFVs if available) should be positioned on the outskirts of the urban area to destroy the enemy at maximum range. They are located in defilade positions or in prepared shelters. They fire at visible targets and then displace to alternate positions. If available, platoons are reinforced with sensors or ground surveillance radars (GSRs), which can be emplaced on the outskirts or on higher ground to attain the maximum range in the assigned sector. Platoons delay by detecting the enemy early and inflicting casualties on him using patrols, OPs, and ambushes, and by taking advantage of all obstacles. Each action is followed by a disengagement and displacement. Displacement occurs on covered and concealed routes through buildings or underground. By day, the defense is dispersed; at night, it is more concentrated. Close coordination and maintaining awareness of the current friendly and enemy situation are critical aspects of this operation.

Appendix A
Risk Management, Fratricide Avoidance, and the Effects of Continuous Operations

Risk is the possibility of loss or injury (damage) to personnel or equipment. Levels of risk vary, but risk itself is constant across the full spectrum of operations. Commanders at every level manage risk at all times, during every phase of every operation. Risk is an integral part of planning. Accidents cost valuable human and other resources. Every Leader and Soldier must know how to manage risk and avoid fratricide to ensure the company executes the mission as safely as mission constraints allow. Managing risk, avoiding fratricide, and understanding the effects of continuous operations helps commanders avoid loss of combat power.

This edition adds discussions of risk management, fratricide avoidance, and continuous operations as quick references, along with key safety and force protection information.

Section I. RISK MANAGEMENT

Risk is the chance of injury or death to individuals and of damage to or loss of vehicles and equipment. Risk is always present in every combat and training situation that Infantry companies might face. Risk management must take place at all levels of the chain of command during each phase of every operation; it is integral to tactical planning. All company leadership must know how to apply the risk-management process and supervise procedures to mitigate risk. Soldiers must follow risk management guidelines and instructions. This, along with fratricide reduction measures, helps ensure that the mission is executed in the safest possible environment, within mission constraints.

TYPES OF RISK

A-1. The two types of risk are tactical and accidental.

TACTICAL RISK

A-2. Tactical risk is a risk concerned with hazards that exists because of the presence of either the enemy or an adversary. Intelligence plays a critical role in identifying hazards associated with tactical risk. IPB is a dynamic leader process that continually integrates new information and intelligence that ultimately becomes input to the leaders risk assessment process. Intelligence helps identify hazards during operations by—.

- Identifying the opportunities and risks (constraints) of the battlefield environment to both friendly and threat forces.
- Thoroughly portraying threat capabilities and vulnerabilities.
- Collecting information on populations, governments, and infrastructures.

ACCIDENT RISK

A-3. Accident risk includes all operational risk other than tactical risk. It includes the risks posed by an operation to the friendly force and to civilians. The risk assessment enhances awareness of the situation. The awareness builds confidence, and allows Soldiers and units to take timely, efficient, and effective protective measures.

Appendix A

STEPS

A-4. The five steps commanders take to manage risk follow.

STEP 1, IDENTIFY HAZARDS

A-5. A hazard is a source of danger. It is any existing or potential condition that could cause injury, illness, or death to personnel; damage to or loss of equipment and property; or other conditions that can degrade the mission. Tactical and training operations pose many types of hazards. The commander must identify the hazards associated with all aspects and phases of his unit's mission, paying particular attention to the METT-TC factors. Risk management must never be an afterthought; leaders must begin the process during their TLP and continue it throughout the operation. Table A-1 lists possible sources of battlefield hazards. The list is organized according to the METT-TC factors.

Mission	• Duration of the operation. • Complexity and clarity of the plan. (Is the plan well-developed and easily understood?) • Proximity and number of maneuvering units.
Enemy	• Knowledge of the enemy situation. • Enemy capabilities. • Availability of time and resources to conduct reconnaissance.
Terrain and Weather	• Visibility conditions including light, dust, fog, and smoke. • Precipitation and its effect on mobility. • Extreme heat or cold. • Additional natural hazards such as broken ground, steep inclines, or water obstacles.
Troops and Equipment	• Equipment status. • Experience the units conducting the operation have working together. • Danger areas associated with the platoon's weapons systems. • Soldier and leader proficiency. • Status of Soldier and leader rest. • Acclimatization. • Impact of new leaders or crewmembers. • Other aspects of friendly unit situation. • NATO or multinational military actions combined with US forces.
Time Available	• Time available for TLP and rehearsals by subordinates. • Time available for PCCs and PCIs.
Civil Considerations	• Applicable ROE or ROI. • Stability and reconstruction as well as homeland security operations (both of which significantly increase contact with civilians). • Potential for media contact and inquiries. • Interaction with host nation or other participating nation support.

Table A-1. Examples of potential hazards.

STEP 2, ASSESS HAZARDS TO DETERMINE RISKS

A-6. Hazard assessment is the process of determining the direct impact of each hazard on an operation. Use the following steps.

- Determine hazards that can be eliminated or avoided.
- Assess each hazard that cannot be eliminated or avoided to determine the probability that the hazard will occur.
- Assess the severity of hazards that cannot be eliminated or avoided. Severity, defined as the result or outcome of a hazardous incident, is expressed by the degree of injury or illness

(including death), loss of or damage to equipment or property, environmental damage, or other mission-impairing factors such as unfavorable publicity or loss of combat power.
- Taking into account both the probability and severity of a hazard, determine the associated risk level (extremely high, high, moderate, and low). Table A-2 summarizes the four risk levels.
- Based on the factors of hazard assessment (probability, severity, and risk level, as well as the operational factors unique to the situation), complete the risk management worksheet. Figure A-1 shows an example completed risk-management worksheet.

Risk Level	Mission Effects
Extremely High (E)	Mission failure if hazardous incidents occur in execution.
High (H)	Significantly degraded mission capabilities in terms of required mission standards. Not accomplishing all parts of the mission or not completing the mission to standard (if hazards occur during mission).
Moderate (M)	Expected degraded mission capabilities in terms of required mission standards. Reduced mission capability (if hazards occur during the mission).
Low (L)	Expected losses have little or no impact on mission success.

Table A-2. Risk levels and impact on mission execution.

A. Mission or Task: Conduct a deliberate attack	B. Date/Time Group Begin: 010035R May XX End: 010600R May XX	C: Date Prepared: 29 April XX
D. Prepared By: (Rank, Last Name, Duty Position) CPT Smith, Cdr			

E. Task	F. Identify Hazard	G. Assess Hazard	H. Develop Controls	I. Determine Residual Risk	J. Implement Controls (How To)
Conduct obstacle breaching operations	Obstacles	High (H)	Develop and use obstacle reduction plan	Low (L)	Unit TSOP, OPORD, training handbook
	Inexperienced soldiers	High (H)	Additional training and supervision	Moderate (M)	Rehearsals, additional training
	Operating under limited visibility	Moderate (M)	Use NVDs, use IR markers on vehicles	Low (L)	Unit TSOP, OPORD
	Steep Cliffs	High (H)	Rehearse using climbing ropes	Moderate (M)	FM 3-97.6, Mountain Operations; TC 90-6-1, Mountaineering
	Insufficient planning time	High (H)	Plan and prepare concurrently	Moderate (M)	OPORD, Troop-leading procedures

K. Determine overall mission/task risk level after controls are implemented (circle one)

LOW (L)   **MODERATE (M)**   HIGH (H)   EXTREMELY HIGH (E)

Figure A-1. Example completed risk management worksheet.

Appendix A

STEP 3, DEVELOP CONTROLS AND MAKE RISK DECISIONS

A-7. This step actually consists of two substeps: develop controls and make risk decisions. These substeps are accomplished during the "*make a tentative plan*" step of the TLP.

Develop Controls

A-8. After assessing each hazard, develop one or more controls that will either eliminate the hazard or reduce the risk (probability, severity, or both) of potential hazardous incidents. When developing controls, consider the reason for the hazard, not just the hazard itself. For example, driving can be a hazard; driving in inclement weather or with limited sleep cause driving to be hazardous.

Make Risk Decisions

A-9. A key element in the process of making a risk decision is determining whether accepting the risk is justified or, conversely, unnecessary. The decision-maker must compare and balance the risk against mission expectations. He alone decides if the controls are sufficient and acceptable and whether to accept the resulting residual risk. If he determines the risk is unnecessary, he directs the development of additional controls or alternative controls; as another option, he can modify, change, or reject the selected COA for the operation.

STEP 4, IMPLEMENT CONTROLS

A-10. Controls are the procedures and considerations the unit uses to eliminate hazards or reduce their risk. Implementing controls is the most important part of the risk management process; it is the chain of command's contribution to the safety of the unit. Implementing controls includes coordination and communication with appropriate superior, adjacent, and subordinate units and with individuals executing the mission. The commander must ensure that specific controls are integrated into OPLANs, OPORDs, SOPs, and rehearsals. The critical check for this step is to ensure that controls are converted into clear, simple execution orders understood by all levels. If the leaders have conducted a thoughtful risk assessment, the controls will be easy to implement, enforce, and follow. (For indirect fire-control measures associated with surface danger zones and risk estimate distances, see AR 385-63 and DA PAM 385-63). Examples of risk management controls include--

- Thoroughly brief all aspects of the mission, including related hazards and controls.
- Conduct thorough PCCs and PCIs.
- Allow adequate time for rehearsals at all levels.
- Drink plenty of water, eat well, and get as much sleep as possible (at least four hours in any 24-hour period).
- Use buddy teams.
- Enforce speed limits, use of seat belts, and driver safety.
- Establish recognizable visual signals and markers to distinguish maneuvering units.
- Enforce the use of ground guides in AAs and on dangerous terrain.
- Establish marked and protected sleeping areas in AAs.
- Limit single-vehicle movement.
- Establish SOPs for the integration of new personnel.

STEP 5, SUPERVISE AND EVALUATE

A-11. During mission execution, leaders must ensure that risk management controls are properly understood and executed. Leaders must continuously evaluate the unit's effectiveness in managing risks to gain insight into areas that need improvement.

Supervise

A-12. Leadership and unit discipline are the keys to ensuring that effective risk management controls are implemented.

- All leaders are responsible for supervising mission rehearsals and execution to ensure standards and controls are enforced. In particular, NCOs must enforce established safety policies as well as controls developed for a specific operation or task. Techniques include spot checks, inspections, SITREPs, confirmation briefs, buddy checks, and close supervision.
- During mission execution, leaders must continuously monitor risk management controls to determine whether they are effective and to modify them as necessary. Leaders also must anticipate, identify, and assess new hazards. They ensure that imminent danger issues are addressed on the spot and that ongoing planning and execution reflect changes in hazard conditions.

Evaluate

A-13. Whenever possible, the risk management process should also include an after-action review (AAR) to assess unit performance in identifying risks and preventing hazardous situations. During an AAR, leaders should assess whether the implemented controls were effective. Following the AAR, leaders should incorporate lessons learned into unit SOPs and plans for future missions.

IMPLEMENTATION

A-14. Though company commanders and platoon leaders perform most formal risk management planning, all unit members share implementation responsibilities. They help ensure that hazards and associated risks are identified and controlled during planning, preparation, and execution of operations. They must look at both tactical risks and accident risks. The same risk management process is used to manage both types. In specific situations not addressed by the higher command, the senior available leader determines how and where he is willing to accept tactical risks. He is also responsible for developing and implementing mitigation procedures.

CHALLENGES

A-15. Despite the need to advise higher headquarters of a risk taken or about to be taken, the risk management process may break down. Such a failure can be the result of several factors; it can usually be attributed to the following.

- Ignoring or denying risk.
- A Soldier does not want to bother his leaders about risks that he feels are inherent to his job.
- Outright failure to recognize a hazard or the level of risk involved.
- Overconfidence on the part of an individual or the unit in being able to avoid or recover from a hazardous incident.
- Subordinates who do not fully understand the higher commander's guidance regarding risk decisions.

COMMAND CLIMATE

A-16. The commander gives direction, sets priorities, and establishes the values, attitudes, and beliefs that make up the command climate. Successful preservation of combat power requires him to embed risk management into individual behavior. To fulfill this commitment, the commander exercises creative leadership, innovative planning, and careful management. Most importantly, he must demonstrate support for the risk management process. The commander and his subordinate leaders establish a command climate favorable to risk management integration by--

- Demonstrating consistent and sustained risk management behavior through leading by example.
- Emphasizing active participation throughout the risk management process.
- Providing adequate resources for risk management. Every leader is responsible for obtaining the assets necessary to mitigate risk and for providing them to subordinate leaders.
- Understanding his own and his Soldiers' capabilities and limitations.
- Allowing subordinates to make mistakes and learn from them.
- Preventing a "*zero defects*" mindset from creeping into the unit's culture.
- Demonstrating full confidence in subordinates' mastery of their trade and their ability to execute a chosen COA.
- Keeping subordinates informed.
- Listening to subordinates.

A-17. For the commander, his subordinate leaders, and individual Soldiers, responsibilities in managing risk include--

- Making informed risk decisions. Establishing and then clearly communicating risk decision criteria and guidance.
- Establishing clear, feasible risk management policies and goals.
- Training the risk management process. Ensuring that subordinates understand the who, what, when, where, and why of managing risk. Ensuring that they know how these factors apply to their situations and responsibilities.
- Accurately evaluating program effectiveness as well as subordinates' execution of risk controls during the mission.
- Informing higher headquarters when risk levels exceed established limits.

Section II. FRATRICIDE AVOIDANCE

Fratricide is the employment of friendly weapons that results in the unintentional death or injury of friendly personnel or damage to friendly equipment. Fratricide prevention is a command responsibility, but again, all leaders across all warfighting function elements assist in planning, practicing, and enforcing fratricide avoidance measures. This section focuses on actions the commander along with all leaders take in using available resources to reduce the risk of fratricide. In any tactical situation, every Soldier must know where he is and where other friendly elements are. With this knowledge, he must anticipate dangerous conditions and take steps either to avoid or to mitigate them. The commander must always be vigilant of changes and developments in the situation that may place his platoons in danger. He must also ensure that all element positions are continuously reported for higher headquarters battle tracking. When the commander perceives a potential fratricide situation outside his company, he reports over the higher net to effect direct coordination.

EFFECTS

A-18. Fratricide results in unacceptable losses and increases the risk of mission failure. It usually affects the unit's ability to survive and function. Units experiencing fratricide suffer these consequences.

- Loss of confidence in the unit's leadership.
- Self-doubt among leaders.
- Hesitancy in the employment of supporting combat systems.
- Over-supervision of units.
- Hesitancy in the conduct of night operations.
- Loss of aggressiveness in maneuver.
- Loss of initiative.

CAUSES

A-19. The following paragraphs discuss the primary causes of fratricide. Leaders must identify any of the factors that may affect their units and then strive to eliminate or correct them.

FAILURES IN DIRECT FIRE CONTROL PLAN

A-20. These occur when units fail to develop effective fire control plans, particularly in the offense. Units might fail to designate EAs or to adhere to the direct fire plan, or they might position their weapons incorrectly. Under such conditions, fire discipline often breaks down upon contact. An area of particular concern is the additional planning that must go into operations requiring close coordination between mounted elements and dismounted teams (AR 385-63 and DAPAM 385-63).

FAILURES IN LAND NAVIGATION

A-21. Units often stray from assigned sectors, report wrong locations, and become disoriented. Much less frequently, they employ fire support weapons in the wrong location. In either type of situation, units that unexpectedly encounter another unit may fire their weapons at the friendly force.

FAILURES IN COMBAT IDENTIFICATION

A-22. Vehicle commanders and machine gun crews cannot accurately identify the enemy near the maximum range of their systems. In limited visibility, friendly units within that range may mistake one another as the enemy.

INADEQUATE CONTROL MEASURES

A-23. Units may fail to disseminate the minimum necessary maneuver control measures and direct fire control measures. They also may fail to tie control measures to recognizable terrain or events. As the battle develops, the plan cannot address branches and sequels as they occur. When this happens, synchronization fails.

FAILURES IN REPORTING AND COMMUNICATIONS

A-24. Units at all levels may fail to generate timely, accurate, and complete reports as locations and tactical situations change. This distorts the operating picture at battalion and brigade level and can lead to erroneous clearance of fires.

WEAPONS ERRORS

A-25. Lapses in individual discipline can result in fratricide. These incidents include charge errors, accidental discharges, mistakes with explosives and hand grenades, and use of incorrect gun data.

BATTLEFIELD HAZARDS

A-26. A variety of explosive devices and materiel (unexploded ordnance; booby traps; and unmarked or unrecorded minefields, including scatterable mines) may create danger on the battlefield. Failure to mark, record, remove, or otherwise anticipate these threats leads to casualties.

RELIANCE ON INSTRUMENTS

A-27. A unit that relies too heavily on systems such as GPS devices or Force XXI Battle Command Brigade and Below Systems (FBCB2) and Land Warrior will find its capabilities severely degraded if these systems fail. The unit will be unable to maintain SU because it will not have a common operations picture. To prevent potential dangers when system failure occurs, the leader must ensure that he and his elements balance technology with traditional basic Soldier skills in observation, navigation, and other critical activities.

PREVENTION

A-28. The measures outlined below provide the company with a guide to actions that reduce or prevent fratricide risk. These guidelines are not intended to restrict initiative. Leaders must learn to apply them, as appropriate, based on the specific situation and the METT-TC factors. Preventing and reducing fratricide relies on following these five key principles.

IDENTIFY AND ASSESS POTENTIAL FRATRICIDE RISKS DURING TLP

A-29. Incorporate risk reduction control measures in WARNOs, the OPORD, and applicable FRAGOs.

MAINTAIN SITUATIONAL UNDERSTANDING

A-30. Focus on areas such as current intelligence, unit locations and dispositions, obstacles, CBRN contamination, SITREPs, and the METT-TC factors.

ENSURE POSITIVE TARGET IDENTIFICATION

A-31. Review vehicle and weapons ID cards. Become familiar with the characteristics of potential friendly and enemy vehicles, including their silhouettes and thermal signatures. This knowledge should include the conditions, including distance (range) and weather, under which positive identification of various vehicles and weapons is possible. Enforce the use of challenge and password, especially during dismounted operations.

MAINTAIN EFFECTIVE FIRE CONTROL

A-32. Assure fire commands are effective and clearly understood. Rules of engagement play an important part in this process. All Soldiers must understand the circumstances that both allow and restrict engagements. Rehearsals are useful in practicing such plans and procedures. Continuously seek information on friendly forces that might be operating in and around the AO.

ESTABLISH A COMMAND CLIMATE THAT EMPHASIZES FRATRICIDE PREVENTION

A-33. Enforce fratricide prevention measures, placing special emphasis on the use of doctrinally sound techniques and procedures. Ensure constant supervision in the execution of orders and in the performance of all tasks and missions to standard.

GUIDELINES AND CONSIDERATIONS

A-34. Additional guidelines and considerations for fratricide reduction and prevention include--

- Recognizing the signs of battlefield stress. Maintaining unit cohesion by acting quickly, effectively to alleviate stress.
- Conducting individual, leader, and collective (unit) training. Covering fratricide awareness, target identification and recognition, and fire discipline.
- Developing a simple, executable plan.
- Giving complete and concise OPORDs. Including all appropriate recognition signals in paragraph 5 of the OPORD.
- To simplify OPORDs, using SOPs that are consistent with doctrine. Periodically reviewing and updating SOPs as needed.

- Striving to provide maximum planning time for leaders and subordinates.
- Using common language (vocabulary) and doctrinally correct standard terminology and control measures.
- Ensuring thorough coordination is conducted at all levels.
- Planning for and establishing effective communications.
- Planning for collocation of command posts whenever appropriate to the mission such as during a passage of lines.
- Ensuring that the ROE are clear.
- Rehearsing when the situation and time allow.
- Ensuring you are in the right place at the right time. Using position location-navigation devices, such as GPS or position navigation (POSNAV) devices, to determine your location and those of adjacent units (left, right, leading, and follow-on). Synchronizing tactical movement. If the unit or any element becomes lost, contacting higher headquarters immediately for instructions and assistance.
- Establishing, executing, and enforcing strict sleep and rest plans.

Section III. EFFECTS OF CONTINUOUS OPERATIONS

The Infantry company often operates for extended periods in continuous operations; such operations may continue at a high intensity level for extended periods. During continuous operations, leaders and Soldiers must think, decide, and act more quickly than the enemy acts. Leaders must know and implement the commander's intent. They must be able to act spontaneously and synchronously, even though the situation has changed and communications are disrupted. This continuous cycle of day and night operations and the associated stress of combat cause degradation in performance. Reducing this impact on performance is a significant challenge for the C2 system. Continuous operations force leaders and Soldiers to perform under adverse conditions that cause degradation in performance and might lead to combat stress. Table A-3 shows some combat stress behaviors.

Appendix A

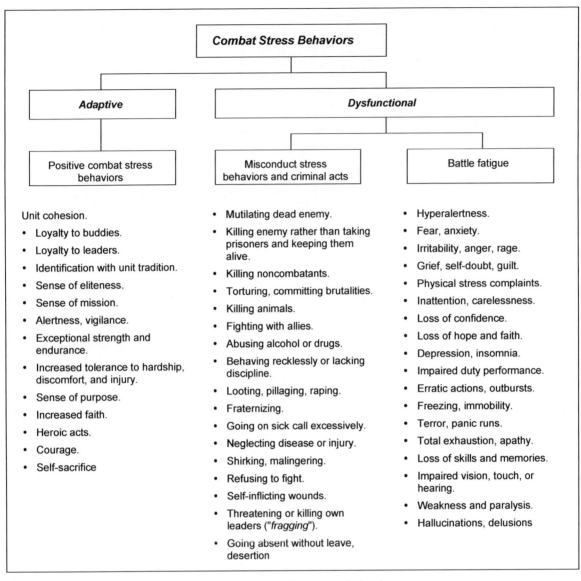

Table A-3. Combat stress behaviors.

COMBAT STRESS CONTROL

A-35. Controlling combat stress is often the deciding factor--the difference between victory and defeat--in all forms of human conflict. Stressors are a fact of combat that Soldiers must face. Controlled combat stress (when properly focused by training, unit cohesion, and leadership) alertness, strength, and endurance to accomplish the mission. It can call forth stress reactions of loyalty, selflessness, and heroism. Conversely, uncontrolled combat stress causes erratic or harmful behavior that disrupts or interferes with accomplishment of the unit mission and brings disgrace, disaster, and defeat.

A-36. The art of war aims to impose so much stress on the enemy Soldiers that they lose their will to fight. Both sides recognize this reality and each may accept severe stress in order to inflict greater stress on the other. To win, combat stress must be controlled. Table A-4 shows the measures that may reduce the negative impacts of continuous operations.

Sleep Scheduling		Countermeasures
Adequate	4 hours of continuous sleep is required every 24 hours to maintain adequate performance over several weeks.	• Give simple, precise orders. • Increase use of written orders. • Crosscheck. • Plan more time for completion of all tasks. • Enforce adequate food and water intake. • Develop and enforce sleep plans. • Enforce good physical fitness. • Increase use of confirmation briefs.
Sleep Wakefulness	A small amount of sleep relative to that lost is beneficial.	
Recovery	10 hours of uninterrupted sleep is required for full recovery after 48 to 72 hours without any.	
Catnaps (10 To 30 Minutes)	Catnaps help, but the only effective remedy is sleep.	
Timing	Consistent sleep and wakeup times help the Soldier successfully adjust to an arduous regimen.	
Note: After 48 to 72 hours without sleep, Soldiers become militarily ineffective. After 5 to 7 days of partial sleep deprivation, alertness and performance decline to the same low level as seen following 2 to 3 days without sleep.		

Table A-4. Reduction of impact of continuous operations.

RESPONSIBILITIES

A-37. Responsibility for combat stress control requires a continuous interaction that begins with every Soldier and his buddies. Combat stress control also includes unit combat lifesavers and medics. The interaction continues through the small unit leaders and extends up through the organizational leaders, both officers and NCOs, at all echelons.

UNIT COHESIVENESS DEVELOPMENT

A-38. Rigorous, realistic training for war helps assure unit readiness. Emphasis must be placed on establishing and maintaining cohesive units. Unit training and activities must emphasize development of Soldier skills. This development should focus on building trust and establishing effective communication throughout the unit.

SENIOR (ORGANIZATIONAL) LEADERS' RESPONSIBILITIES

A-39. The chain of command must ensure that the standards for military leadership are met. Senior leaders must provide the necessary information and resources to the junior leaders to control combat stress. The following are some suggestions for senior leadership considerations for combat stress control.

- Be competent, committed, courageous, candid, and caring.
- Plan to accomplish the mission with as few losses as possible.
- Set the policy and command climate for stress control, especially to build teams with high cohesion.
- Serve as an ethical role model.
- Make the system work for the Soldiers, not vice versa.
- Assure resources take care of the Soldiers.
- Plan for and conduct tough, realistic training including live fire.
- Provide as much information as possible to the Soldiers.
- Assure that medical and mental health and combat stress control personnel are assigned and trained with their supported units.
- Plan for combat stress control in all operations.
- Provide junior leaders and NCOs with necessary guidance.

- Ensure risk assessments are conducted prior to all training and combat operations.
- Supervise the junior leaders and NCOs and reward their success.
- Be visible.
- Lead all stress control by good example.
- Maintain, through positive leadership and, when necessary, with disciplinary action, the high standards of the international law of land warfare.

JUNIOR (DIRECT) LEADERS' RESPONSIBILITIES

A-40. Junior leaders, that is, squad and platoon leaders, especially NCOs, must apply the principles of stress control continuously. These crucial responsibilities overlap those of senior leaders, but include parts that are fundamentally "*sergeants' business*," supported by junior officers.

- Be competent, committed, courageous, candid, and caring.
- Build cohesive teams; integrate new personnel quickly.
- Cross-train Soldiers wherever and whenever possible.
- Plan and conduct tough, realistic training that replicates combat conditions.
- Take care of Soldiers, including leaders.
- Assure physical fitness, nutrition, hydration, adequate clothing and shelter, and preventive medicine measures.
- Make and enforce sleep plans.
- Keep accurate information flow down to the lowest level and back up again. Dispel rumors.
- Encourage sharing of resources and feelings.
- Conduct after-action debriefings routinely.
- Maintain, through positive leadership and, when necessary, with disciplinary action, the high standards of the international law of land warfare.
- Recommend exemplary Soldiers for awards and decorations.
- Recognize excess stress early and give immediate support.
- Keep those stressed Soldiers who can still perform their duties in the unit and provide extra support and encourage them back to full effectiveness.
- Send those stressed Soldiers who cannot get needed rest in their small unit back to a supporting element for brief sleep, food, hygiene, and limited duty, to return in one to two days.
- Refer temporarily unmanageable stress cases through channels for MEDEVAC and treatment.
- Welcome recovered battle fatigue casualties back and give them meaningful work and responsibilities.

SLEEP DEPRIVATION

A-41. A significant factor contributing to performance degradation is lack of sleep. Table A-5 shows the effects of sleep loss. Other contributing factors include low light levels, limited visibility, disrupted sleep routines, physical fatigue, and stress.

After 24 Hours	A deterioration in performance of tasks that are inadequately or newly learned, that are monotonous, or that require vigilance.
After 36 Hours	A marked deterioration in ability to register and understand information.
After 72 Hours	Performance on most tasks will be about 50 percent of normal.
3 To 4 Days	The limit for intensive work, including mental and physical elements. Visual *illu*sions are likely at this stage or earlier, especially in CBRN.
Between 0300 and 0600 Hours	Performance is at its lowest ebb.

Table A-6. Effects of sleep loss.

SIGNS

A-42. To minimize the effects of sleep loss, all commanders must be able to recognize the signs of sleep loss and fatigue (Table A-6).

Physical Changes	• Body swaying when standing. • Vacant stares. • Pale skin. • Slurred speech. • Bloodshot eyes.
Mood Changes	• Less energetic, alert, and cheerful. • Loss of interest in surroundings. • Possible depressed mood or apathetic and more irritable.
Early Morning Problems	• Requires more effort to do a task in the morning than in the afternoon, especially between 0300 and 0600.
Communication Problems	• Unable to carry on a conversation. • Forgetfulness. • Difficulty in speaking clearly.
Difficulty In Processing Information	• Slow comprehension and perception. • Difficulty in accessing simple situations. • Requires more time to understand information.
Impaired Attention Span	• Decreased vigilance. • Failure to complete routines. • Reduced attention span. • Short-term memory loss. • Inability to concentrate.

Table A-6. Indicators of sleep deprivation and fatigue.

LEADERS

A-43. Commanders and leaders often regard themselves as being the least vulnerable to fatigue and the effects of sleep loss. Tasks requiring quick reaction, complex reasoning, and detailed planning make leaders the most vulnerable to the effects of sleep loss. Leaders must sleep. Self-controlled leaders recognize that depriving themselves of sleep is counterproductive and take measures to avoid it.

Appendix B
TOW and Javelin Employment

This appendix addresses the organization and employment of the TOW and Javelin close-combat missile systems. The weapons company assault platoon has TOWs, and the antitank section in each rifle company line platoon has Javelins.

Close combat missile (CCM) systems include the tube-launched, optically tracked, wire-guided (TOW) missile weapon system, as well as the Javelin weapon system. Shoulder launched munitions such as the M136 AT4, the M72A7 light antitank weapon (LAW), and the M141 bunker-defeat munition (BDM) supplement and reinforce CCM fires at close ranges. The TOW-equipped Infantry battalion weapons company is often task-organized to provide rifle companies with assault platoons to provide support during combat operations.

This edition combines the discussion of TOW and Javelin employment to better address tactical employment of close combat missile systems.

Section I. OVERVIEW

For each Infantry battalion, the organic weapons company has the antitank and other hard-target destruction combat power (Figure B-1, page B-2). The weapons company commander advises the battalion commander on the tactical employment of the company and its assault platoons. This includes both the unit's CCM capability and its area-fire antipersonnel role.

INFANTRY BATTALION WEAPONS COMPANY

B-1.　The weapons company operates on the battlefield along with Infantry, armor, aviation, and other elements of the combined-arms team. Close combat missile fires by a base-of-fire force or a fixing force are important to destroy the integrity of the enemy's combined-arms team. They allow units to fix the enemy force effectively, while maintaining sufficient combat power for decisive maneuver, and sufficient depth to reduce risk and exploit success. Mass and depth are the keys to employing CCM assets. When terrain and fields of fire allow, a weapons company commander (or assault platoon leader) normally controls the CCM fires in support of the higher commander's scheme of maneuver, but commanders can also task-organize and attach the assault platoons based on the METT-TC factors.

ORGANIZATION AND EQUIPMENT

B-2.　Each weapons company has four assault platoons (Figure B-2, page B-2). Each assault platoon leader is responsible for the tactical employment of the platoon and its two sections. In addition to the TOW missile, assault platoons of the weapons company also have the capability to mount either the M2 .50 caliber machine gun or the MK 19 40-mm grenade launcher on each vehicle. Normally all three weapons systems are available within the unit. By virtue of having organic vehicles the assault platoons of the weapons company have an enormous speed and mobility advantage over other Infantry forces. The weapons company has four assault platoons. Each assault platoon has five vehicles normally mounted with two TOW systems and a combination of MK 19 40-mm grenade launchers and M2 .50 caliber machine guns.

Appendix B

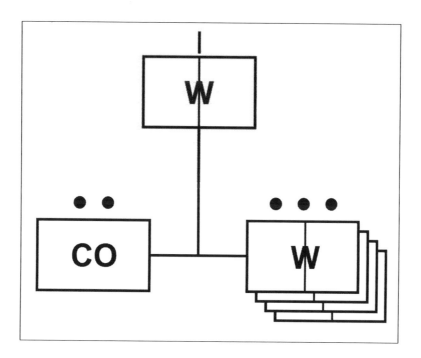

Figure B-1. Infantry battalion weapons company.

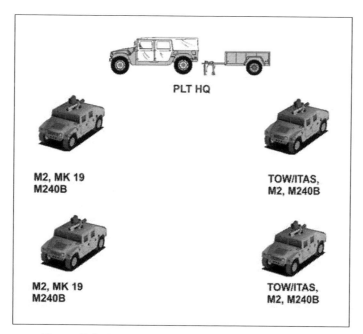

Figure B-2. Assault platoon of weapons company.

PRINCIPLES

B-3.   Following the principles for CCM employment increases the probability of destroying targets and enhances the survivability of the CCM elements. The fundamentals follow--

MUTUAL SUPPORT

B-4.   To accomplish their assigned tasks, CCM units must mutually support each other. Mutual support generates combat power by maximizing the units' capabilities and minimizing its limitations. To establish mutual support, CCMs and other heavy weapons are employed in sections with overlapping primary and secondary sectors of fire (Figure B-3). If one squad is suppressed or forced to displace, the other squad continues covering the assigned sector. To achieve this effect, the CCM squads are positioned so that fires directed at one squad can suppress only that squad.

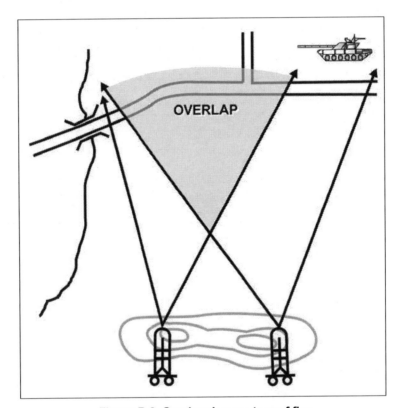

Figure B-3. Overlapping sectors of fire.

SECURITY

B-5.   When a CCM unit is attached to a rifle platoon or a company, it must be positioned where it can gain security from adjacent Infantry units for protection against possible attack by dismounted enemy infantry. Though the Infantry units are not required to collocate with the CCM unit, they must be able to cover dismounted avenues of approach to the CCM positions. CCM units moving as a unit provide their own local security. During halts, the driver or loader dismounts to secure the flank and rear sectors. Overall flank and rear security must be controlled at the platoon level.

Appendix B

FLANK SHOT ENGAGEMENTS

B-6. The CCM unit should be positioned to engage tanks or armored vehicles from the flank. Frontal engagements at enemy armor are less desirable for the following reasons.

- An armored vehicle's protection is weaker to the flank.
- An armored vehicle's firepower and crew are not normally oriented to the flank.
- A flank engagement decreases the chance of detection and suppression by enemy vehicles.
- An armored vehicle provides a larger target from the flank than it does from the front.

STANDOFF

B-7. Standoff is the difference between a friendly weapon's maximum effective range and an enemy weapon's maximum effective range (Figure B-4). For example, the TOW missile's maximum range of 3,750 meters provides it with a standoff advantage over modern, western-built tanks with maximum effective ranges of 2,800 meters and older, non-modernized tanks with maximum effective ranges of 2,000 meters. Despite this advantage, engaging enemy armored vehicles within the standoff range (2,000 to 3,750 meters) may not always be tactically feasible. The additional tracking time required to fire a TOW missile beyond 2,000 meters increases the likelihood of gunner error. This possibility gives a frontal target more time to maneuver against the friendly position and provides a flanking target more time to reach cover. Also, the terrain may not provide the fields of fire to support standoff distance engagements.

> *Note:* The T-55 (modernized), T-64B, T-72S, T-80, T-80U, and T-90 main battle tanks and the BMP-3 can fire ATGMs through their main gun tubes up to a range of 4,000 meters, which means the TOW weapon system loses the standoff advantage against them. Some of the tank-launched ATGMs can be fired while the vehicle is on the move. Also, threat armored vehicles can fire HE fragmentation rounds to suppress TOW gunners up to a range of 9,750 meters.

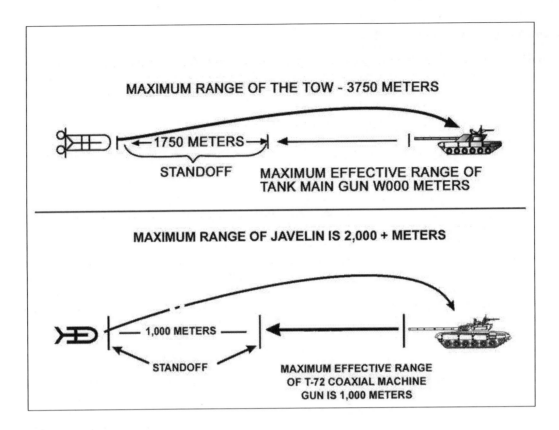

Figure B-4. Standoff ranges, TOW (top) and Javelin (bottom).

COVER AND CONCEALMENT

B-8.   Cover and concealment are critical to the survival of CCM weapon systems and must be analyzed along with the other METT-TC factors. The CCM unit must take full advantage of all natural and man-made cover and concealment in order to be effective, to survive, and to overcome the following inherent weaknesses.

- TOW, M2, and MK 19 expose gunner during firing.
- TOW exposes gunner and crew during reloading.
- TOW has long flight time, distinctive firing signature, and slow rate of fire.
- TOW takes time to track.
- MK 19 has long flight time and distinctive firing signature.

Cover

B-9.   Cover is protection from the effects of direct or indirect enemy fires. It might be natural or manmade. Natural cover includes reverse slopes, ravines, and hollows whereas manmade cover includes fighting positions, walls, rubble, and craters.

Concealment

B-10.   Concealment is the protection from enemy observation. NVDs and other detection devices penetrate darkness and prevent it from providing sufficient concealment. Leaders must choose inconspicuous positions and avoid silhouetting the vehicles or weapon systems against the skyline. The crew should avoid unnecessary movement, use all available concealment such as vegetation, terrain and buildings), stay low to observe, expose nothing that shines, keep from altering familiar outlines, and stay as quiet as possible.

Appendix B

EMPLOYMENT IN DEPTH

B-11. CCM units should be employed in depth whenever possible. This complicates the enemy's task of identifying and destroying the weapons and crews. With careful positioning and disciplined fire control, a CCM unit can mass its fires on the enemy continuously while attacking or withdrawing.

EMPLOYMENT AS PART OF A COMBINED ARMS TEAM

B-12. Skillful integration of CCM systems with Infantry squads and crew-served weapons, armored vehicles, combat engineers, indirect fire and attack helicopters improves the combat effectiveness and the survivability of the entire unit. Infantry rifle squads provide local security and engage enemy Infantry moving along covered and concealed routes. They maneuver quickly under the heavy suppressive and destructive fires of the CMM units. CCM units support the effective maneuver of tanks and Infantry fighting vehicles. By destroying the enemy's lightly armored vehicles and crew-served weapons at long range, they allow tanks and fighting vehicles to engage enemy forces with the freedom to move where their fires will be most effective. Combat engineers help shape the battlefield, creating and reinforcing obstacles that hold the enemy in position longer or force him to take more exposed routes. This means the CCM units have more time to engage and destroy him. Indirect fires from artillery and mortars suppress enemy counterfire and separate enemy Infantry from accompanying armored vehicles, which then become more vulnerable to destruction by concealed CCM units.

CONSIDERATIONS

B-13. When employing CCM weapon systems, leaders should avoid conspicuous terrain, disperse weapons laterally and in depth so that no single enemy weapon can suppress two squads, and disperse assault platoon squads to reduce casualties and equipment damage that could result from enemy mortar and artillery fires (Figure B-5). The considerations for CCM weapon system employment also apply during route selection and movement.

Offensive Considerations

B-14. Determine the routes where cover and concealment are good. Identify areas along the approaches to the objective where cover and concealment are poor. Consider using smoke or conducting missions in limited visibility to provide concealment.

Defensive Considerations

B-15. Focus on locations with good fields of fire. Determine how the enemy can use the available cover and concealment and look at it from his point of view, both in daylight and at night.

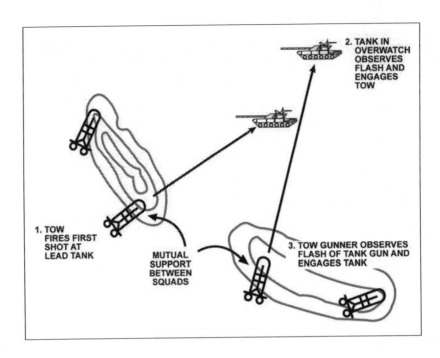

Figure B-5. Dispersion between squads.

CAPABILITIES AND LIMITATIONS

B-16. The TOW and Javelin provide direct fire against armored or other hard targets to support maneuver of Infantry.

OFFENSIVE CAPABILITIES

B-17. A CCM unit initially provides the base of fire in an attack in order to suppress, fix, or destroy the enemy in position. The CCM unit also can be employed in the offense to engage enemy in planned EAs, to isolate objectives, destroy enemy counterattacks or withdrawing enemy forces, protect flanks, or to fix enemy in place for destruction by Infantry or Armor companies. The CCM unit can be used in a reconnaissance role in the offense. The unit can perform reconnaissance itself, but is usually more effective when combined with the scout platoon. This will provide the scout platoon assets greater mobility, additional night vision capabilities and greater firepower. The CCM unit can use its powerful optical and target-acquisition systems to augment observation by the reconnaissance assets.

DEFENSIVE CAPABILITIES

B-18. A CCM unit can be positioned forward of the defensive sector to participate in security operations or to overwatch reconnaissance units or obstacles. These assets can be employed in conjunction with the reconnaissance assets in a manner similar to that of the offense. As the enemy closes, the CCM unit displaces to positions that provide good observation and fires into an EA. CCM units often are positioned throughout the depth of the AO to cover likely mounted avenues of approach. During counterattacks, the CCM unit provides overwatching fires for the maneuvering element.

ADDITIONAL CAPABILITIES

B-19. The Infantry weapons company can use the HMMWV Interchangeable Mount System (HIMS). This system enables the unit to adapt to changes in the tactical situation quickly. With the HIMS, the unit can quickly mount the MK 19 or the M2 on the HMMWV and use them to destroy light armored vehicles, field fortifications, and troops.

Appendix B

LIMITATIONS

B-20. CCM units have limitations that apply to both offensive and defensive situations. CCM units equipped with HMMWVs must consider that these vehicles lack protection against direct and indirect fires. An assault platoon squad (four Soldiers) cannot adequately defend itself when confronted with a dismounted threat for an extended period; normally the squad is able to dismount only one Soldier for any extended amount of time for security when halted. TOW missiles are accurate, but missile flight time is long. The slow rate of fire and the visible launch signature of the TOW missile increase the assault platoon squad's vulnerability, especially if a HMMWV-mounted TOW engages within an enemy's effective direct-fire range (no standoff). CCM elements can reduce this vulnerability by displacing often and by integrating their fires with those of other weapon systems (M2 and MK 19) within the CCM unit, with other CCM and shoulder launched munitions within the battalion, with obstacles, and with indirect fires. Integrated direct and indirect fires with obstacles complicate the enemy's target-acquisition process. Sustainment is limited for units conducting security missions. Additional support should be coordinated with a higher headquarters when an antiarmor unit participates in a security mission.

Section II. JAVELIN CLOSE COMBAT MISSILE SYSTEM

The Javelin provides accurate, medium-range close combat missile capability for the infantry rifle platoon and company. The weapons squad of each rifle platoon includes a two-Soldier antitank section with the Javelin as its primary weapon. The Javelin is used in offensive operations to provide precision, direct fires that suppress or destroy enemy armored vehicles, and to destroy bunkers, buildings, and other fortifications during urban combat. In defensive operations, the Javelin might be used to overwatch obstacles, destroy armored vehicles, and force the enemy to dismount prematurely, exposing his infantry to small arms and indirect fires. The Javelin can destroy targets from medium ranges (65 to 2,000 meters), including helicopters and fortified positions. The Javelin's infrared sight capability can be used to conduct surveillance in all types of weather. The Javelin (Table B-1, page B-8) is a dual-mode (top attack or direct attack), man-portable, close-combat missile that can engage and defeat tanks and other armored vehicles. The missile is contained in a disposable launch tube/container that has a reusable tracker. It is a fire-and-forget weapon system. The Javelin's soft launch reduces the visual and acoustical signature of the missile, making it difficult to identify and locate, at even moderate distances.

Type System:	Fire and Forget
Carry Weight (Total).	49.2 lb (day and night).
Command Launch Unit.	14.1 lb (day and night).
Missile (with launch tube):	35.2 lb
Crew:	Man portable
Ready to Fire:	Less than 30 sec.
Reload Time:	Less than 20 sec.
Method of Attack:	Top attack or direct attack (top attack is normal)
Range:	Top-attack mode: 150 to 2,000 meters.
	Direct-attack mode: 65 to 2,000 meters
Fighting Position Restrictions:	1 to 2 meters and ventilation is recommended
Guidance System:	Imaging Infrared
Sights:	Integrated day and nightsights
Time of Flight:	1,000 meters = about 4.6 sec.
	2,000 meters = about 14.5 sec
Sight Magnification:	4X day, 4X wide field of view and 9X narrow field of view

Table B-1. Javelin technical characteristics.

TOW and Javelin Employment

COMMAND LAUNCH UNIT

B-21.   The nondisposable section of the Javelin is the CLU (Figure B-6). The nightsight and daysight of the Javelin are integrated into one unit. The infrared sight has a 2,000-meter range, under most conditions, which greatly increases target acquisition by the infantryman. The sight can operate for over four hours on a single battery and requires no coolant bottles. It has a built-in test capability, which alerts the gunner if the system is not functioning properly during operation.

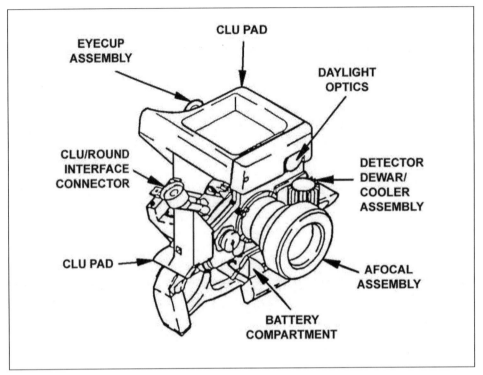

Figure B-6. Command launch unit.

MISSILE

B-22.   The Javelin missile is contained in a disposable launch tube. It has a passive imaging infrared system, which locks on to the target before launch and is self-guiding. It uses a tandem shaped charge warhead and a two-stage solid propellant with a low signature, soft-launch motor and a minimum-smoke flight motor. The launch tube assembly and missile is shown in Figure B-7.

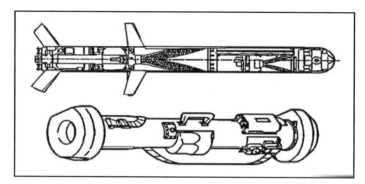

Figure B-7. Launch tube assembly and missile.

LETHALITY

B-23. The Javelin's 2,000-meter range and its tandem warhead defeats all known enemy armored vehicles. It is powerful enough to produce extensive damage to field fortifications and medium-size buildings.

B-24. In the top-attack mode, the missile strikes the thinner armor on the top of an armored vehicle rather than the thicker frontal and side armor plates. Top attack also prevents an enemy target from protecting itself by moving behind frontal cover. When used in urban areas or where obstacles might interfere with the top-attack flight path of the missile, the Javelin can also be fired in the direct attack mode.

B-25. The fire-and-forget capability of the Javelin increases the probability of a hit. Because the gunner is not exposed to enemy suppressive fires while tracking the target until impact, he can use the missile's flight time to reload, in a covered and concealed position, and begin engaging another target.

SURVIVABILITY

B-26. The Javelin's low launch signature decreases the enemy's ability to acquire gunners as they fire the missile. All gunner engagement tasks are accomplished before launching the missile, making time of flight irrelevant. The 2,000-meter range places the Javelin gunner outside the armored vehicle's effective coaxial machine gun range. However, he is still within the range of the main gun.

B-27. The Javelin uses a passive infrared system for target acquisition and lock-on. This means that it emits no infrared or radar beam (which enemy vehicles or smart munitions can detect) further increasing the survivability of the Javelin gunner.

B-28. The Javelin sight offers the infantry leader an excellent observation capability. Under ideal conditions, the Javelin sight can detect targets far in excess of 2,000 meters.

B-29. Because of the Javelin's slight back blast, it can be fired from smaller, harder to locate, better protected positions that give the gunner a greater chance of remaining undetected or, if detected, surviving any suppressive fires.

AGILITY AND FLEXIBILITY

B-30. The Javelin is man-portable and relatively lightweight for a CCM system. This allows the system to be moved about the battlefield with relative ease. The Javelin's soft launch capability allows it to be fired from inside buildings, bunkers, and other restricted spaces with less disruption to the gunner and less signature to be observed by the enemy. Although flank shots are still the preferred method of engagement, the Javelin's low signature launch and top-attack mode make frontal and oblique engagements more effective than in the past, giving the infantry leader additional options in his CCM fires planning and positioning. The capabilities of the Javelin give the leader more flexibility in the use and emplacement of his CCM systems. This new degree of flexibility challenges the leader to make a careful METT-TC analysis to ensure that he is taking full advantage of the Javelin's capabilities. The Javelin gives the leader a system that complements other CCM and shoulder fired munitions fires available, allowing him to achieve mutual support and greater overlapping fires between the systems.

LIMITATIONS

B-31. Sometimes, the Javelin system cannot engage targets. These occur either when a target is not exposed long enough for the missile seeker to achieve proper lock on or when atmospheric conditions interfere with the seeker.

LIMITED VISIBILITY

B-32. Heavy rain, smoke, fog, snow, sleet, haze, and dust are referred to as limited visibility conditions. The presence of these conditions can affect the gunner's ability to acquire and engage targets with the Javelin, especially when using the daysight of the CLU. The gunner should use the nightsight capability of the CLU to acquire targets in limited visibility conditions because it provides the best target image.

INFRARED CROSSOVER

B-33. Infrared crossover occurs at least twice in each 24-hour period when the temperatures of soil, water, concrete, and vegetation are approximately the same and the objects all emit the same amount of infrared energy. If there is little difference in the amount of infrared energy between a target and its background, then neither the Javelin CLU nor the missile seeker can detect the target well; this greatly degrades the performance of the Javelin. This situation may last as long as an hour, until either the background or the target changes temperature enough to become detectable again.

TIME-SPACE FACTOR

B-34. Just because a target appears in the open and within range does not always mean a Javelin gunner can acquire, lock on, fire, and hit the target in the time it is exposed. A vehicle must be exposed long enough for the gunner to identify it as a target and then to lock on the target with the Javelin missile seeker. This process does not occur instantly; the time needed to do this varies depending on the skill of the gunner.

EMPLOYMENT CONSIDERATIONS

B-35. The Javelin's primary role is to destroy enemy armored vehicles, but it can be used against point targets such as bunkers and crew-served weapons positions. In addition, the Javelin's CLU can be used alone as an aided vision device for reconnaissance, security operations, and surveillance. The principles of CCM employment apply to the Javelin.

- Position for mutual support.
- Consider security requirements.
- Seek flank engagements.
- Seek to exploit standoff.
- Use cover and concealment.
- Employ in depth.
- Employ as part of a combined arms team.

Appendix B

SOLDIER'S LOAD

B-36. When employing the Javelin the Soldier's load becomes important. With a total system weight of just under 50 pounds, the Javelin is admittedly heavy. Although a man-portable weapon, one Soldier cannot easily carry the Javelin cross-country for extended periods. Leaders should be aware of this problem and address it as they would any other Soldier's load difficulty. (FM 21-18 discusses Soldier's load and cross-leveling of equipment during movement to reduce the burden on Soldiers.) Leaders should develop unit SOPs that identify and describe the details of unit equipment cross-leveling.

URBAN COMBAT

B-37. Javelins provide overwatching CCM fires during the attack of a built-up area and an extended range capability for the engagement of armor during the defense. Within built-up areas, they are best employed along major thoroughfares and from the upper stories of buildings to attain long-range fields of fire. The missile's minimum arming range and flight profile could limit firing opportunities in the confines of densely built-up areas.

Restrictions

B-38. Ground obstacles and water do not restrict the Javelin with its fire-and-forget capability. However, with its unique flight characteristics, overhead obstacles can limit its use in urban terrain. In the top-attack mode, the Javelin missile requires up to 160-plus meters of overhead clearance (Figure B-8, page B-12). In the direct-attack mode, the Javelin requires up to 60-plus meters of overhead clearance (Figure B-9, page B-12). Gunners must ensure that sufficient overhead clearance is available along the missile flight path before engaging targets in an urban environment.

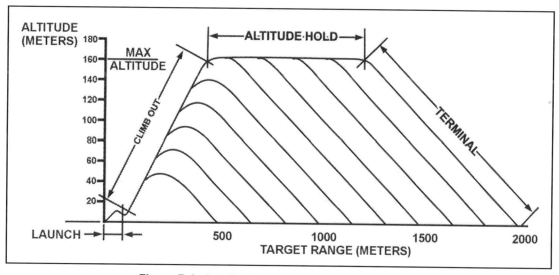

Figure B-8. Javelin flight profile in top-attack mode.

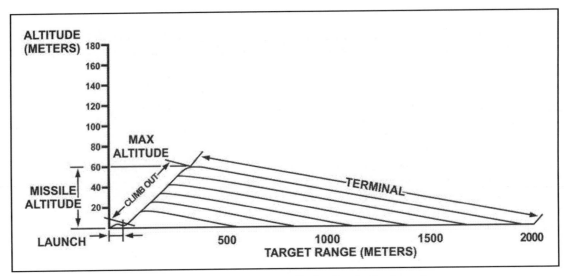

Figure B-9. Javelin flight profile in direct-attack mode.

Dead Space

B-39. The aspects of dead space that affect Javelin fires the most are arming distance and target and background temperature differences. The Javelin missile has a minimum arming window of 65 to 75 meters. Few areas in most urban environments permit fires much beyond the minimum arming distance. Ground-level long-range fires down streets or rail lines and across parks or plazas are possible. The Javelin might be used effectively from the upper stories or roofs of buildings to fire into other buildings. The Javelin gunner must take into consideration the targeting dead space that is sometimes caused by the background of the target and its heat signature. When firing from the upper stories of a building towards the ground, the missile seeker sometimes cannot discriminate between the target and surrounding rubble, buildings, or paving if that background material has the same temperature as the target.

Backblast

B-40. The Javelin's soft launch capability enables the gunner to fire from within an enclosed area with a reduced danger from back blast overpressure or flying debris (Figure B-10). Personnel within the enclosure should still wear a helmet, protective vest, ballistic eye protection, and hearing protection. To fire a Javelin from inside a room, take these safety precautions.

- Ensure ceiling height is at least 7 feet.
- The floor size of the room should be at least 15 feet by 12 feet.
- Window opening must be at least 5 square feet.
- Door opening must be at least 20 square feet.
- When launching a missile from an enclosure, allow sufficient room for the missile container to extend beyond the outermost edge of the enclosure.
- All personnel in the room must be forward of the rear of the weapon.

Appendix B

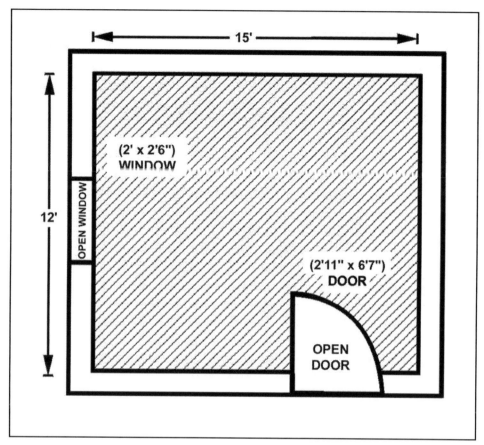

Figure B-10. Minimum room enclosure for Javelin firing.

WEAPON PENETRATION

B-41. The warhead of the Javelin can achieve significant penetration against typical urban targets. Penetration, however, does not mean a concurrent destruction of the structural integrity of a position. When engaging a position in a building, gunners should normally use the direct-attack mode to hit the target. When engaging a position or bunker in the open, use either the top- or direct-attack mode.

JAVELIN FIRING POSITIONS

B-42. Each Javelin should have a primary firing position and at least one alternate position. Depending on the factors of METT-TC, a supplementary position may also be assigned. A Javelin firing position must allow for target engagement and provide protection for the Soldiers and the weapon system. When selecting firing positions, leaders should consider the following. Avoid positions in swampy areas and very steep hillsides. Also avoid positions on or near prominent terrain features such as--

- Cover to the front, flank, and overhead.
- Concealment from ground and aerial observation.
- Good observation and fields of fire.
- Covered and concealed routes to and between positions.
- Mutual support between squad positions and with other elements.
- Position below ridgelines and crests, preferably on the sides of hills.

DETECTION, RECOGNITION, AND CLASSIFICATION OF TARGETS

B-43. Gunners must engage targets quickly and efficiently. Speed of target engagement depends on each Javelin gunner's proficiency in detecting potential targets, recognizing them as friend or foe, and determining whether they can be engaged effectively. Dust and smoke make locating and identifying the enemy difficult. As the battle progresses and friendly and enemy units become intermingled, identifying targets become crucial tasks. Gunners in the company should be trained to detect enemy targets that are camouflaged or partially concealed by terrain, vegetation, or smoke. They should also be trained to identify targets as friend or foe based on vehicle identification aspects as well as situational understanding.

PRIMARY ANALYSIS

B-44. Because the Javelin's primary targets are armored vehicles, specifically tanks, gunners should look for terrain where these targets are most likely to appear. Understanding armor tactics and the characteristics of armor vehicles can help Javelin gunners recognize the terrain where these vehicles are most likely to be employed.

ENEMY ANALYSIS

B-45. The tactics of many potential adversaries stress using speed and massive firepower to overwhelm and destroy an opposing force. This dictates a very high average daily rate of advance. To move consistently at a high rate, armored forces require firm ground to move rapidly and enough space to deploy, maneuver, and fire. High-speed avenues of approach, such as road networks, broad ridges, and flat or rolling terrain, should be observed constantly.

TERRAIN AND WEATHER ANALYSIS

B-46. A detailed analysis of the terrain and weather is useful in pinpointing armored or mechanized avenues of approach and to evaluate them from the enemy's viewpoint. Some questions that the leader should ask are "*How can the enemy use this terrain?*" and "*Where is he most likely to appear first?*" Because weather significantly affects the trafficability of terrain, a ground reconnaissance is needed to obtain current, detailed information about roads, trails, manmade objects, density of trees and brush, and the seasonal conditions of streams and rivers. If a ground reconnaissance is not possible, an aerial reconnaissance should be conducted or recent aerial photographs should be used.

ARMORED AND MECHANIZED VEHICLES' MOBILITY CHARACTERISTICS

B-47. Javelin gunners can more easily determine where to look for enemy armored vehicles if they know the vehicles' mobility characteristics. If possible, tank and motorized rifle units will avoid terrain or obstacles that can stop or impede their movement. Terrain factors that restrict armored or mechanized vehicle mobility include--

- Slopes steeper than 30 degrees.
- Sturdy walls or embankments 3 or more feet high.
- Ditches or gullies 9 or more feet wide and 3 or more feet deep.
- Hardwood trees 10 inches or larger in diameter and 10 feet or less apart.
- Water obstacles at least 5 feet deep.
- Very swampy or very rough, rocky terrain.
- Built-up areas where vehicles are restricted to moving on confined roads, through park areas, or across sports fields.

RANGE ESTIMATION

B-48. Javelin gunners do not need to know the exact range to a target before engaging; they only need to know if it is within range. To speed this determination, gunners use an MEL. A Javelin MEL is an imaginary line linked to an identifiable terrain feature. It is drawn across a sector's maximum allowable

Appendix B

range from a Javelin firing position. To determine the location of this line on the ground, the company leadership identifies terrain features at or near maximum range. Therefore, any target that crosses or appears short of this line should be within range. Establishing a maximum engagement line greatly reduces target engagement times, especially for targets that seem to be near maximum range. Several range-determination techniques can be used to find the maximum range line or the range to targets.

LASER RANGE-FINDING METHOD

B-49. Most units and all FIST teams should have laser range-finders. The range from the Javelin position to an easily identifiable terrain feature can be easily determined with the laser range-finder. Once the maximum engagement line is determined, the gunner makes a note of a terrain feature at that location on his range card. Any vehicle nearing that feature will be in range.

OBJECT RECOGNITION METHOD

B-50. Range determination by object recognition is simple and can be accurate with training. The Soldier looks at the target with his naked eye, sights through 7X binoculars, or uses a Javelin optical sight. Targets listed in Table B-2 are recognizable out to the ranges indicated. For example, if a target can be recognized with the naked eye as an armored or wheeled vehicle, it is probably within 2,000 meters. When using this method, the gunner must consider terrain, visibility conditions, and target size.

Targets	Range (Meters)	
	Naked Eye	7x Scope
Tank crewmembers	500	2,000
Soldiers, machine gun, mortar	500	2,000
Antitank gun, antitank missile launchers	500	2,000
Tank, APC, truck (by model)	1,000	4,000
Tank, Howitzer, APC, truck	1,500	5,000
Armored vehicle, wheeled vehicle	2,000	6,000

Table B-2. Range determination recognition method.

MAP AND TERRAIN ASSOCIATION METHOD

B-51. The maximum engagement line can be determined from a map. Do this for each firing position as follows.

- Draw an arc on the map across the assigned sector of fire at 2,000 meters.
- Examine the map to identify the distinctive natural or man-made terrain features that the line touches.
- Study the terrain in the sector of fire using binoculars or the Javelin CLU until all the selected terrain features are located and positively identified.
- Connect these features by an imaginary line from the maximum engagement line.

SELF-DEFENSE AGAINST HELICOPTERS

B-52. Because Javelin positions are selected to cover enemy armor avenues of approach, the medium-range fields of fire afforded by these positions may also enable Javelin gunners to engage attacking aircraft that approach along the same general avenues.

Appendix C
Heavy and Stryker Employment

The Infantry company might be employed with Heavy or Stryker forces. An estimate of the tactical situation determines the mixture and command relationship (attached versus OPCON). Tactics and techniques explained in this appendix are for Infantry working with such armored vehicles as the M1, M2, M3, or Stryker family of Infantry carrier vehicles. These tactics and techniques also apply when an Infantry rifle company is attached to a Heavy or Stryker battalion.

In this appendix, "*Heavy*" means units with BFVs and M1 Abrams tanks; "*Stryker*" means units with Stryker Infantry carrier vehicles (ICVs) and mobile gun systems (MGSs); and "mounted" means units with all of these.

"*Infantry*" means *all* Infantry units, including air assault and airborne trained Infantry and Ranger units.

This edition adds a discussion of the employment of the Infantry Company with Stryker Brigade Combat Team (SBCT) and with Heavy Brigade Combat Team (HBCT) elements to fit modular force organization concepts and to capture considerations including planning, logistics, communications, and key tactical tasks.

VEHICLES

C-1. To employ any unit effectively, the leader must understand the specific capabilities and limitations of the unit and its equipment. The most important considerations for the Infantry commander are an appreciation for the differences in the perspective (with which mounted forces view the battlefield) and how the Infantry force communicates effectively with the mounted force. Most heavy and Stryker vehicles are equipped with the same types of FM radios employed by Infantry forces but only leader's vehicles have more than one radio. Careful consideration must be given to how Infantry forces communicate with mounted forces.

TANKS

C-2. M1 tanks provide rapid mobility combined with excellent protection and highly lethal, accurate fires. They are generally more effective in open terrain with extended fields of fire. However, they can be very effective in restricted terrain, such as in urban areas, when combined with Infantry forces. The following paragraphs address the capabilities and limitations of the M1 tank.

CAPABILITIES

C-3. The tank's mobility comes from its capability to move at high speed both on and off road. The ability to cross ditches, ford streams, and shallow rivers and to push through small trees, vegetation, and limited obstructions allows effective movement in various types of terrain.

LIMITATIONS

C-4. The tank requires Infantry forces for close in and rear security in restrictive and urban terrain. Tanks consume large quantities of fuel. The tank requires dedicated Class III sustainment, most often from the tank's parent battalion, if operations will exceed eight hours. Tanks are noisy and must be started

Appendix C

periodically in cold weather or when using the thermal nightsight and radios to ensure the batteries stay charged. The noise, smoke, and dust generated by tanks make it difficult for the Infantry in their vicinity to capitalize on stealth to achieve surprise. Tanks cannot cross bodies of water deeper than 4 feet without deep water fording kits or bridging equipment. When fighting with the hatches closed the tank crews visibility in the immediate vicinity of the vehicle is severely limited. This presents an accident and fratricide risk to Infantry operating in the immediate vicinity of the tank. Infantry Soldiers must take precautions when operating in close proximity to tanks.

FIREPOWER

C-5. The tank's main gun is extremely accurate and lethal at ranges out to 2,500 meters. Tanks with stabilized main guns can fire effectively even when moving at high speeds cross country. The tank is the best antitank weapon on the battlefield. The various machine guns (tank commander's caliber .50 and 7.62-mm coax, and the loader's 7.62-mm MG) provide a high volume of supporting fires for the Infantry. The target-acquisition capabilities of the tank exceed the capability of all systems in the Infantry battalion. The thermal sight provides a significant capability for observation and reconnaissance. It can also be used during the day to identify heat sources (personnel and vehicles) even through light vegetation. The laser range finder provides an increased capability for the Infantry force to establish fire control measures, such as trigger lines and TRPS, and determine exact locations. The normal, basic load for the tank's main gun is armor-piercing discarding sabot (APDS) antitank, high-explosive antitank (HEAT), and multipurpose antitank (MPAT) rounds. Two other rounds available are an obstacle-reducing round and a canister antipersonnel round. The APDS round presents a safety problem when fired over the heads of exposed Infantrymen due to the discarded SABOT petals that fall to the ground. HEAT and MPAT ammunition provides better destructive effects on these targets, except for enemy personnel, against which the tank's machine guns work best. The resupply of all tank ammunition is generally beyond the capability of the Infantry battalion and normally requires logistic support from the heavy battalion.

PROTECTION

C-6. Generally, the tank armor provides excellent protection to the crew. Across the frontal 60-degree arc, the tank is impervious to all weapons except heavy antitank missiles or guns, and the main gun on enemy tanks. When fighting with the hatches closed, the crew is impervious to all small arms fire, artillery rounds (except a direct hit), and AP mines. The tank's smoke grenade launcher and on-board smoke generator provide rapid concealment from all but thermal observation. However, the tank is also vulnerable to lighter antitank weapons from the flanks, top, and rear. The top is especially vulnerable to precision-guided munitions (artillery or air delivered). AT mines can also destroy/disable the vehicle. When fighting with hatches closed, the tank crew's ability to see, acquire, and engage targets (especially close-in Infantry) is greatly reduced.

INFANTRY FIGHTING VEHICLE

C-7. The M2 and M3 provide good protection and mobility combined with excellent firepower. They operate best on the same terrain as the tank; however, their reduced protection (when compared to the tank) is a major employment consideration. The following paragraphs address the capabilities and limitations of the Infantry fighting vehicle.

CAPABILITIES

C-8. The mobility of the M2 and M3 is comparable to the tank. In addition to the three-Soldier crew, the vehicle is designed to carry six to seven (depending on equipment load) additional Infantrymen.

LIMITATIONS

C-9. The M2 and M3 use a lot of fuel. Also, they are louder than the M1, and must be started periodically in cold weather or when using the thermal nightsight and radios to ensure the batteries stay charged. The noise, smoke, and dust generated by mechanized forces make it difficult for the Infantry to capitalize on their ability to move with stealth and avoid detection when moving on the same approach.

FIREPOWER

C-10. The primary weapon on both the M2 and M3 is the 25-mm chain gun that fires two versions of Armor Piercing Fin Stabilized Discarding Sabot with Tracer (APFSDS-T) and High Explosive Incendiary with Tracer (HEI-T). This weapon is extremely accurate and lethal against lightly armored vehicles, bunkers, trench lines, and personnel at ranges out to 3,000 meters. The stabilized gun allows effective fires even when moving cross-country. The TOW provides an effective weapon for destroying enemy tanks or other point targets at ranges to 3,750 meters. The 7.62-mm coax provides a high volume of suppressive fires for self-defense and supporting fires for the Infantry at ranges to 900 meters (in some cases the coax can engage targets past 900 meters). The combination of the stabilized turret, thermal sight, high volume of fire, and mix of weapons and ammunition (TOW, 25-mm, and 7.62-mm), makes the M2 and M3 excellent suppression assets supporting Infantry assaults. The target-acquisition capabilities of the M2 and M3 exceed that of the other systems in the Infantry battalion. The thermal sight provides a significant capability for observation and reconnaissance. It can also be used during the day to identify heat sources (personnel and vehicles) even through light vegetation. However, the resupply of 25-mm ammunition is generally beyond the capability of the Infantry battalion and normally requires logistic support from the heavy battalion.

PROTECTION

C-11. Overall, the M2/M3 provides good protection. When fighting with the hatches closed, the crew is well protected from small-arms fire, fragmentation munitions, and AP mines. The M2/M3 smoke-grenade launcher provides rapid concealment from all but thermal observation. Though the M2 and M3 retain the ability to produce on-board smoke, this feature is seldom used, because the JP-8 fuel does not produce billowing smoke like diesel fuel did. Second, smoke reveals the vehicle's exact location to the enemy. The vehicle is also vulnerable from all directions to any AT weapons and especially enemy tanks. Antitank mines can destroy or at least disable the vehicle. When the crew is operating the vehicle with the hatches open, they are vulnerable to small-arms fire.

STRYKER INFANTRY CARRIER VEHICLE

C-12. A large part of the US Army's transformation includes fielding up to nine kinds of Strykers for the SBCT. Together, they meet the requirements of key battlefield operations and functions. They include vehicles designed for command, reconnaissance, fire support, mortars, antitank guided missiles or mobile guns, engineers, medical support, and CBRN reconnaissance. However, this paragraph focuses only on the overall design of the Stryker and on selected weapons systems.

CAPABILITIES

C-13. The Stryker family of vehicles includes the Infantry carrier vehicle (ICV), a fully mobile system that provides protected transport for an Infantry squad and direct fire support during dismounted assault. The ICV carries a nine man rifle squad plus a two-Soldier crew (vehicle commander and driver) that operates the vehicle; or a seven-Soldier weapons squad. General design characteristics and features include-- air-transportable, four to eight wheel drive, remote weapon station, FBCB2, driver vision enhancer (DVE), commander periscopes, and thermal imager display with video camera, and reduced acoustic signature. Under optimal conditions the top speed is over 60 MPH and range over 300 miles.

LIMITATIONS

C-14. The Stryker lacks shoot-on-the-move capability. Also, due to its air transport weight and space restrictions, it might not roll off fully fueled, with ammunition, or with add-on, retrofitted antiarmor protection. Off-road, Strykers must operate at greatly reduced speeds. In urban terrain, their large turning radius can affect their agility.

FIREPOWER

C-15. The ICV local defensive armament consists of a remote weapon station that mounts either an M2 50 caliber machine gun with 2,000 rounds stowed ammunition, or the MK 19 40-mm grenade launcher with 430 rounds stowed ammunition, and four smoke grenade launchers. Anti-tank armament includes the mobile gun system M68A1 105MM cannon and the antitank guided missile system employing the elevated TOW 2B. The mortar carrier gives the Stryker company very responsive, high-angle, indirect fire using either the 120- or 60-mm mortars. However, the Stryker rifle platoon ICV has limited onboard armament that can only defeat enemy thin-skinned vehicles and dismounted troops. Also, it can neither suppress nor kill armored vehicles or tanks. Nor does it have a laser range finder or gun stabilizer, of which would enhance the Stryker's lethality.

PROTECTION

C-16. The ICV standard armor provides all-round protection against up to 14.5 MM direct fire weapons and 152 MM artillery airburst fragmentation. Add-on reactive armor and the steel cage RPG-7 retrofit provide enhanced protection. Survivability is also increased by using the CBRN detection package and individual crew respirators. However, although it bridges the gap between the heavy and light forces, Stryker armor cannot withstand engagements by heavy tanks or antiarmor weapons.

SAFETY

C-17. Infantry battalion personnel may not be familiar with the hazards that may arise during operations with tanks, BFVs, and other armored vehicles. The most obvious of these include the dangers associated with main-gun fire and the inability of armored vehicle crews to see people and objects near their vehicles. Leaders of mounted and dismounted units alike must ensure that their troops understand the following points of operational safety.

DISCARDING SABOT

C-18. Tank and BFV SABOT rounds discard stabilizing petals when fired creating a downrange hazard for unprotected personnel. The aluminum petals of the tank rounds are discarded in an area extending 70 meters to the left and right of the gun-target line, out to a range of 1 kilometer (Figure C-1). The danger zone for plastic debris from BFV rounds extends 60 degrees to the left and right of the gun-target line and out to 100 meters from the vehicle. The danger zone for the aluminum base from BFV rounds extends 7 degrees to the left and right of the gun target line and out to 400 meters (Figure C-2, page C-6). Infantrymen should not be in or near the direct line of fire for the tank main gun or BFV cannon unless they are under adequate overhead cover.

Appendix C

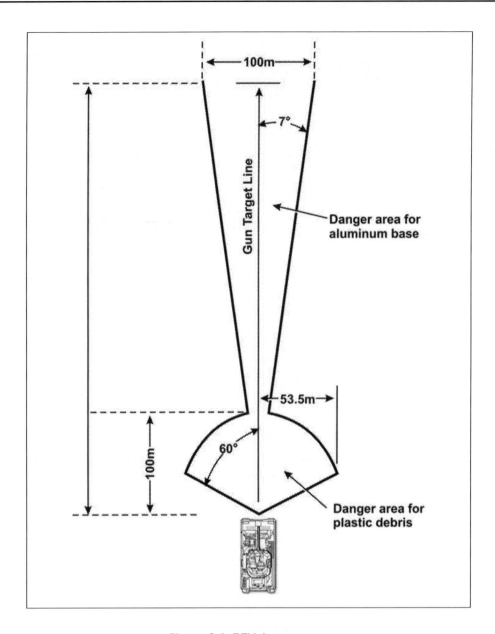

Figure C-2. BFV danger zone.

NOISE

C-19. Tank main guns create noise in excess of 140 decibels. Repeated exposure to this level of noise can cause severe hearing loss and even deafness. In addition, dangerous noise levels may extend more than 600 meters from the tank. Single-layer hearing protection, such as earplugs, allows Infantrymen to work within 25 meters of the side or rear of the tank without significant hazard.

GROUND MOVEMENT HAZARDS

C-20. Crewmembers on Strykers, tanks, and BFVs have limited ability to see anyone on the ground, on either the side or rear of the vehicle. As a result, vehicle crews and dismounted Infantrymen share responsibility for avoiding the hazards this may create. Infantrymen must maintain a safe distance from armored vehicles at all times. In addition, when they work close to an armored vehicle, dismounted Soldiers must ensure that the vehicle commander knows their location at all times.

Note: Because the Stryker and the M1-series tanks are quiet, Infantry Soldiers might have a hard time hearing them approach. Again, vehicle crews and dismounted Infantry Soldiers share the responsibility for eliminating potential dangers (managing risk) in this situation.

M1 EXHAUST PLUME HAZARD

C-21. A hot exhaust plume streams out the rear of the M1-series tank and then angles downward. This exhaust plume is hot enough to burn skin and clothing.

TOW MISSILE SYSTEM

C-22. The TOW missile system has a danger zone extending 50 meters to the rear of the vehicle in a 90-degree cone. A 25-meter deep caution zone expands the cone farther (Figure C-3).

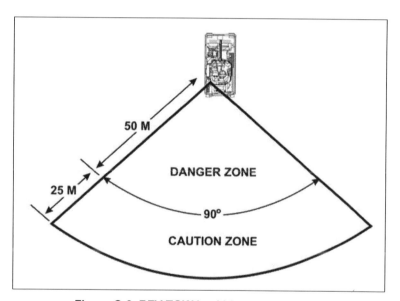

Figure C-3. BFV TOW backblast danger zone.

PLANNING CONSIDERATIONS

C-23. The Infantry company fights as part of a combined arms team, which often includes mounted units. Armored vehicles provide unique capabilities to supported Infantry units; these should be considered during the planning process. The commander must know how to employ and support these units. Generally, armored vehicles can contribute to the dismounted battle by--

- Providing suppressive fires and a mobile base of fire for dismounted Infantry. The vehicles' machine guns can suppress enemy positions, kill personnel, and destroy lightly armored targets. Vehicle main gun and antitank missile systems can destroy heavily armored targets. They might also work well against well-protected enemy forces, or against enemy forces in an urban environment.
- Using their speed and shock effect to assist the Infantry in executing an assault. Stabilized gun systems provide accurate direct fires even while the vehicle is moving at high speeds.
- Providing effective antitank fires. Main armaments can destroy tanks, armored vehicles, and fortifications such as bunkers.

Appendix C

- Providing limited mobility to the dismounted force. Armored vehicles can move cross-country over trenches, trees, and small obstacles.
- Using their technical assets, such as thermal viewers and range finders, to aid in target acquisition and ranging at long distances, day or night.
- Providing additional communication assets. The vehicle radios and the crew's arm-and-hand signals allow orders to be communicated between crews and dismounted troops.

LIMITATIONS

C-24. Armored vehicles have the following limitations and vulnerabilities that affect their employment in support of Infantry forces.

- They are vulnerable to antitank guided missiles, guns, mines, tanks, and aircraft.
- They require daily resupply of POL products in large quantities.
- They require extensive maintenance, skilled operators, and mechanics.

OBSTACLES

C-25. Existing or reinforcing obstacles can restrict or stop armored vehicle movement. Since armored vehicles often work with dismounted Infantry in dense woods, urban areas, or other restricted terrain, Infantry leaders must understand the mobility characteristics of the vehicles that are supporting the unit.

- When forced to fight buttoned-up (hatches closed), the crew's visibility is downgraded to only what they can see through their vision blocks.
- In close terrain, trees, buildings, or other obstacles can restrict turret traverse.
- In jungles or swampy areas, soft ground easily traversed by Infantry might have to be bypassed by armored vehicles.
- Depending on the situation, the ammunition basic load might also be a limitation. Bradley vehicles use a mix of 25-mm APFSDS-T (kinetic energy) rounds and HEI-T ammunition. The M1 tank's basic loads usually contain only APDS, HEAT, or MPAT rounds.

COMBINED OPERATIONS WITH ARMORED VEHICLES

C-26. Leaders must know what mounted and Infantry forces can do for each other. They must know how to communicate by radio, phone, and visual signals.

- Infantrymen help mounted forces by finding and breaching or marking antitank obstacles. They detect and destroy or suppress enemy antitank weapons. Infantrymen may designate targets for mounted forces and protect them in close terrain.
- Mounted forces help Infantry by leading Infantrymen in open terrain and providing them a protected, fast-moving assault weapons system. (This depends on the enemy's antitank capability.) They suppress and destroy enemy weapons, bunkers, and tanks by fire and maneuver. They may provide transport when the enemy situation permits.

MOVEMENT TO CONTACT

C-27. Infantry companies use the approach-march technique and the search-and-attack technique to conduct a movement.

APPROACH-MARCH TECHNIQUE

C-28. The company team uses normal movement techniques (traveling, traveling overwatch, and bounding overwatch).

- Mounted forces may follow and provide overwatch for the rifle platoons in traveling or traveling overwatch at a distance determined by the terrain and visibility. This allows the rifle platoons to move by stealth while being overwatched by the mounted forces.
- The mounted forces may lead in traveling or traveling overwatch when speed is required and when in open terrain. When mounted forces lead, they normally use (platoon) bounding overwatch. Some Infantrymen may ride with the overwatching mounted vehicle section; these men provide security for the vehicle at halts, and they dismount to clear danger areas.
- In bounding overwatch, the mounted forces are normally part of the overwatch element. In open terrain, the vehicles might be the bounding element.

SEARCH-AND-ATTACK TECHNIQUE

C-29. The mounted forces are normally employed under the battalion's scheme of maneuver. They may work with the company to concentrate combat power, isolate enemy positions, or attack enemy base camps. They may also escort convoys through terrain occupied by enemy forces.

ATTACKS

C-30. All attacks involving mounted forces and Infantry must be well-planned, thoroughly coordinated, and fully rehearsed. The communications procedures require special considerations to ensure mutual support and flexibility.

ATTACK ON CONVERGING ROUTES

C-31. In this method, the mounted force and Infantry move on separate routes that meet on the objective. Mounted forces may first support the Infantry by fire, then close on the objective in time to assault it with the Infantry (Figure C-4, page C-10). This may require the Infantry to breach obstacles and destroy certain antiarmor systems to help the mounted force reach the objective. Tanks are the only mounted forces that should assault on to the objective unless the enemy has no antiarmor capability.

Appendix C

Figure C-4. Attacks along converging routes.

ATTACK ON SAME ROUTE

C-32. When mounted force and Infantry attack on the same route (Figure C-5). The two elements may move at the same speed or at different speeds.

Figure C-5. Attacking on same route.

- They use the same speed when there are no good overwatch positions or when there is a need for close mutual support. For example, mutual support might be required when the enemy has antitank weapons and tanks, but when their locations are unknown. When attacking at the same speed, the Infantry might be slightly ahead, but not directly in front of even with, or just to the rear of, the mounted force.
- They move using different speeds when there are obstacles that their Infantry must clear for the mounted force, or when the route offers good cover and concealment for the Infantry but not for the mounted force. In these cases, the mounted force (first) support by fire while the Infantry moves to its assault position. The mounted force then move forward to assault with the Infantry. However, the mounted force might lead the Infantry against an enemy that is being suppressed, lacks well-prepared positions with overhead cover, or presents no great antiarmor threat.

MOUNTED FORCE SUPPORT-BY-FIRE

C-33. This method is used when obstacles prevent the mounted force from closing on the objective. The mounted forces occupy positions where they can support the attacking Infantry (Figure C-6). As soon as the obstacles are breached or a suitable bypass is found, the mounted force rejoins the Infantry.

Figure C-6. Mounted forces support by fire.

CONSOLIDATION AND REORGANIZATION

C-34. When a company team has seized an objective, the team consolidates. Either the company commander directs the mounted forces leader to position his vehicle in overwatch positions behind the Infantry so they are ready to move forwarded when needed, or he directs them to hull-down positions with the Infantry to block armor counterattack approaches. If the withdrawing enemy can be seen and is still in range, the mounted forces continue to fire. Throughout the attack, the team reorganizes and replaces any lost leaders.

DEFENSE

C-35. Mounted forces add strength, depth, and mobility to the defense. The company commander may initially position them forward to engage the enemy at long ranges and then move them back to cover armor approaches. However, the commander must move the vehicles where needed to concentrate fire against an enemy attack. He should also use them to add strength to the counterattack force.

C-36. The commander may temporarily position his mounted forces (with Infantry for security) forward of the company's defensive positions. When so deployed, they can force the enemy to deploy early. This forward deployment may deceive the enemy as to the location of the company's defensive positions. As soon as the enemy is close enough to threaten them, the mounted force must withdraw to their defensive positions. Smoke might be used to screen their withdrawal.

C-37. There are two basic ways for the defending Infantry company commander to employ mounted forces. In both, the commander selects their general positions and sectors of fire. The mounted forces leader advises the commander and selects the exact positions and controls fire and movement.

- The first way is to integrate the mounted force throughout the company defense, both laterally and in depth, to cover armor AAs (Figure C-7). This might be done when there are only a few good firing positions or when the terrain restricts fast vehicle movement. Each vehicle should have mutual support with at least one other vehicle. The mounted force remains under control of the mounted force leader.

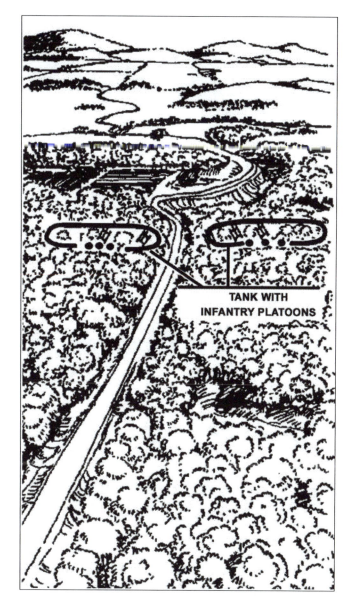

Figure C-7. Mounted forces integrated throughout position.

- The second way to employ the mounted force is to hold them in a position behind the forward Infantry platoons (Figure C-8). This might be done when there are several Armor AAs into the company sector. However, there must, be sufficient vehicle firing positions and routes to them. When the enemy appears, the mounted force moves to forward or flank firing positions. This allows quick concentration of the vehicles at a critical point to repel an attack. The commander should determine his decision points and criteria for initiating the mounted unit's move. The leader of the mounted unit must know when to move in case communication is not possible.

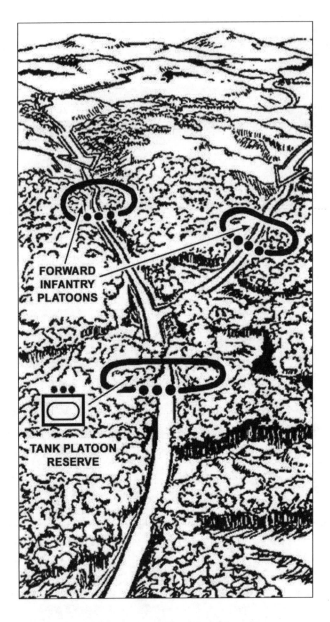

Figure C-8. Mounted force held in reserve.

- With either employment method, the mounted force leader selects covered (hull-down), primary, alternate, subsequent, and supplementary firing positions for each vehicle. If covered firing positions are not available, he may assign them hide positions.

RETROGRADE OPERATIONS

C-38. In retrograde operations, a mounted force might be used to support the Infantry when the terrain or the enemy force makes it primarily an Infantry fight. In other situations, the Infantry may protect the mounted unit or the two might be employed separately on different AAs. When fighting together on the same AA, the Infantry may first disengage to a nearby covered position. The mounted force can then disengage and move to overwatch positions where they continue to cover the Infantry's withdrawal. If the retrograde operation is conducted when visibility is poor, some Infantrymen may stay with the mounted unit to provide security if they do not have their own Infantrymen.

LOGISTICAL SUPPORT

C-39. A mounted force, OPCON to an Infantry company, receives fuel, maintenance, and recovery support as well as ammunition from its parent unit. The mounted force normally comes to the Infantry company with its own fuel tanker and ammunition vehicle. If the parent unit cannot provide recovery assets to the mounted unit, the unit must perform self-recovery. The mounted force leader can communicate with his commander for support; however, he must coordinate with the Infantry company commander for a place and time to conduct his logistical activities.

INFANTRY ON TANKS

C-40. There might be times when tanks and Infantry must move quickly from one place to another to accomplish their mission. In such cases (and providing that there is little likelihood of making enemy contact) Soldiers may ride on the turret and decks of the tank.

C-41. Riding on the outside of the tanks is hazardous. Infantry should only ride on tanks when the need for speed is great. By riding on the tank, the Infantry yields its best protection--the ability to move with stealth and to avoid detection. Soldiers on tanks are vulnerable to all types of fire. Also, Soldiers must watch out for obstacles, which could cause the tanks to turn suddenly; for tree limbs, which could knock the limbs off; and for the traverse of the turret gun, which could also knock them off.

C-42. The only advantage the Infantry gains is speed of movement and increased haul capability. In this case, the following apply.

- Avoid riding on the lead tank of a section or platoon. These vehicles are most likely to make contact, and they can react quicker without Soldiers on top.
- Position the Infantry leaders with the tank unit leaders. Discuss and prepare contingency plans for chance contact or danger areas. Infantry should dismount and clear choke points or other danger areas.
- Assign air guards and sectors of responsibility for observation. Ensure all personnel remain alert and stay prepared to dismount immediately. In the event of contact, the tank will immediately react as required for its own protection. The Infantry on top are responsible for their own safety. Rehearse a rapid dismount of the vehicle.
- Consider putting rucksacks, ammunition, and other equipment on the vehicles and having the Infantry move on a separate AA. This will increase the mobility of the Infantry, and they can move through terrain that is more suitable.

C-43. Riding on tanks reduces tank maneuverability and may restrict firepower. Infantrymen might be injured if the tank must slew its turret to return fire on a target. Consequently, Soldiers must dismount to clear danger areas (FM3-21.8) or as soon as enemy contact is made.

COMMUNICATION WITH TANKS

C-44. Before an operation, Infantry and mounted unit leaders coordinate communications means. This includes the use of radios, phones, and visual signals such as arm-and-hand signals, panels, lights, flags, and pyrotechnics.

- On the M1, the Infantryman can run communication wire to the TC through the turret. This wire can be hooked in to the tanks communication system to provide a means of communication for nearby Soldiers.
- On some versions of the BFV, there are WD-1 wire hook-ups for external communications.
- On all variants of the Stryker vehicles, there are WD-1 wire hook-ups for external communications.

Appendix D
Aviation Support

The company may conduct air assault operations either as part of the battalion or as a separate unit when conducting raids, counterguerrilla operations, or other special missions. These operations (covered in detail in FM 90-4) are planned primarily by the battalion staff. This appendix covers the information the company commander needs to know to fulfill his responsibilities during such operations.

EMPLOYMENT

D-1.   The company commander may use helicopters when inserting or extracting patrols, positioning weapons and crews, conducting resupply, and evacuating casualties. The company should have an SOP for working with helicopters. The SOP should cover--

- LZ and PZ selection.
- LZ and PZ security.
- LZ and PZ operation and activities.
- LZ and PZ marking procedures.
- Downed aircraft procedures.
- Load plan preparation.
- Loading procedures.
- Organization for an air assault operation.

D-2.   Air assaults involve assault forces (combat, CS, and sustainment) using the firepower, mobility, and total integration of helicopter assets and maneuver on the battlefield to engage and destroy enemy forces or to seize and retain key terrain. Air movement operations involve the use of Army airlift assets for other than air assaults.

HELICOPTER TYPES

D-3.   Several types of helicopters can be used in air assault operations: observation, utility, cargo, and attack. All of these types of helicopters are organic to the modular aviation brigades.

OBSERVATION

D-4.   Observation helicopters (OHs) provide--

- Command and control.
- Aerial observation and reconnaissance.
- Aerial target acquisition.

UTILITY

D-5.   The utility helicopters (UHs) are the most versatile of all helicopters and perform a variety of tasks. UHs are used to conduct combat assaults and to provide transportation, command and control, and resupply. When rigged with special equipment, they might be used to--

- Provide aero MEDEVAC s.
- Conduct radiological surveys.
- Dispense scatterable mines.

Appendix D

CARGO

D-6. These aircraft normally provide transportation, resupply, and recovery of downed aircraft.

ATTACK

D-7. Attack helicopter battalions can be task-organized to meet mission needs. Although they are seldom employed *in support* lower than Infantry battalion level operations, they might--

- Provide overwatch.
- Destroy point targets.
- Provide security.
- Suppress air defense weapons.

GROUND TACTICAL PLAN

D-8. The foundation of a successful air assault operation is the commander's ground tactical plan, around which subsequent planning is based. It specifies actions in the objective area and addresses subsequent operations. The ground tactical plan for an air assault operation is essentially the same as for any other infantry operation. It differs in that it capitalizes on speed and mobility of helicopters to achieve surprise. Army aviation assets are integrated into the plan, coordinated, and controlled by the battalion staff under the battalion commander's guidance. One additional requirement is that aircrews must know this ground tactical plan and the ground commander's intent.

LANDING PLAN

D-9. The landing plan must support the ground tactical plan. This plan sequences elements into the area of operations. It makes sure units arrive at designated places and times, and that they are prepared to execute the ground tactical plan.

FACTORS

D-10. Consider the following factors while developing the landing plan.

- The availability, location, and size of potential LZs are key factors.
- The company is most vulnerable during landing.
- Multiple insertions require multiple LZs. Do not use the same LZ twice.
- Elements must land with tactical integrity.
- Soldiers are easily disoriented if they are not informed when the briefed landing direction changes.
- There might be no other friendly units in the area initially. The company must land prepared to fight in any direction.

D-11. The landing plan should offer flexibility, so several options are available in developing a scheme of maneuver.

- Supporting fires (artillery, naval gunfire, CAS, and attack helicopters) must be planned in and around each LZ.
- Although the objective might be beyond the range of supporting artillery fire, artillery or mortars might be brought into the LZ(s) early to provide fire support on the objective for subsequent lifts.

D-12. The plan should include provisions for resupply and MEDEVAC by air.

LANDING ZONE SELECTION CRITERIA

D-13. LZs are selected by the battalion commander (or his S-3) with technical advice from the air mission commander (AMC) or his liaison officer. They do so using the following significant factors.

Location

D-14. Locate the LZ on, near, or away from the objective, depending on the situation.

Capacity

D-15. Determine how much combat power can be landed at one time by the size of the LZ. This also determines the need for additional LZs or separation between aircraft.

Alternates

D-16. Plan at least one alternate LZ for each primary LZ selected to ensure flexibility.

Enemy Disposition and Capabilities

D-17. Consider enemy troop concentrations, their air defenses, and their capability to react to a company landing nearby when selecting an LZ.

Cover and Concealment

D-18. Select LZs that deny enemy observation and acquisition of friendly ground and air elements while they are en route to or from (and in) the LZ.

Obstacles

D-19. If possible, land the company on the enemy side of obstacles when attacking, and use obstacles to protect LZs from the enemy at other times. Keep landing zones free of obstacles. Organize and attach engineers for contingency breaching of obstacles.

Identification from Air

D-20. Make landing zones readily identifiable from the air. Mark them with chemical lights (preferably infrared-type) if the assault is conducted with personnel wearing night vision goggles.

Approach and Departure Routes

D-21. Avoid continuous flank exposure of aircraft to the enemy on approach and departure routes.

Weather

D-22. Consider the weather. Reduced visibility or strong winds may preclude or limit the use of marginal LZs. Consider the impact of limited visibility and inclement weather restrictions on flying.

OPTIONS

D-23. If any options exist in selecting LZs, choose the ones that best aid mission accomplishment. This choice involves whether to land on or near the objective, to land away from it and maneuver forces on the ground to the objective, or to use single or multiple LZs. Significant factors to be considered follow--

Appendix D

- Aids control of the operation.
- Concentrates supporting fires in and around the LZ. Firepower is diffused if more than one LZ preparation is required.
- Provides better ground security for subsequent lifts.
- Requires fewer attack helicopters for security.
- Reduces the number of flight routes in the objective area, making it more difficult for enemy intelligence sources to detect the air assault operation.
- Centralizes any required resupply operations.
- Concentrates efforts of limited LZ control personnel and engineers on one LZ.

MULTIPLE LANDING ZONES

D-24. Using multiple LZs avoids grouping assets in one location, which would create a lucrative target for enemy mortars, artillery, and CAS. Multiple LZs also--

- Allow rapid dispersal of ground elements to accomplish tasks in separate areas.
- Reduce the enemy's abilities to detect and react to initial and subsequent lifts.
- Force the enemy to fight in more than one direction.
- Eliminate aircraft congestion.
- Make it difficult for the enemy to determine the size of the air assault force, the exact location of supporting weapons, or the objective of the air assault.

Note: If a number designates the objective, the LZ should be designated by a letter or code word to avoid confusion and mix-ups. This avoids having an objective and LZ with the same designator, for example, LZ 1 and OBJ 1.

LANDING ZONE OPERATIONS

D-25. Just as there is a priority of work for defensive operations, there is a priority of actions upon landing in an LZ.

Unloading

D-26. Unload the aircraft only after the crew chief or pilot directs you to do so (Figure D-1).

Aviation Support

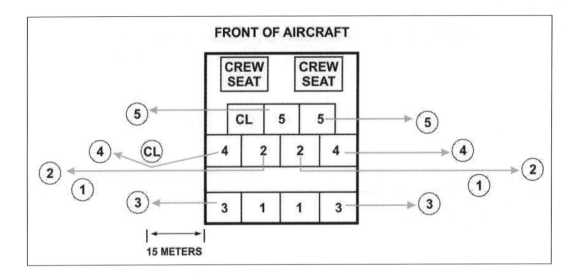

Figure D-1. UH-60 unloading diagram.

- Before leaving the aircraft, the chalk leader checks the landing direction and grid coordinates with the pilot if they were not determined during the approach. This aids in orientation to the LZ, particularly at night.
- Once the aircraft lands, the Soldiers unbuckle their seat belts and get off (with all equipment) as fast as possible.
- They move 15 to 20 meters out from the side of the aircraft and assume the prone position, facing away from the aircraft with weapons at the ready until the aircraft has left the LZ.

Immediate Action on Hot LZ

D-27. If the decision is made to use a hot LZ, or contact is made upon landing, Soldiers quickly dismount and move 15 to 20 meters away from the aircraft and immediately return fire to protect the aircraft departure.

- If the situation allows, Soldiers fire and move off the LZ to the closest cover and concealment. If this is not feasible and the enemy positions are near, they assault immediately.
- The ground or air element first detecting the enemy initiates the preplanned supporting fires.
- Once disengaged from the enemy force, the chalk leader moves the unit to a covered and concealed position, accounts for personnel and equipment, assesses the situation, and tries to link up with other elements of his lift. If unable to link up or if in a single chalk LZ, the senior man present issues a FRAGO to continue the mission or abort it.

Chalk Assembly on Cold LZ

D-28. When unloading on a cold LZ, the chalk leader moves the chalk to its preset locations using traveling overwatch movement techniques. All Soldiers move at a fast pace to the nearest concealed position. Once at the concealed assembly point, the chalk leader counts personnel and equipment and then proceeds with the mission.

Appendix D

AIR MOVEMENT PLAN

D-29. The air movement plan is based on the ground tactical and landing plans. It specifies the schedule and details for air movement of Soldiers, equipment, and supplies from PZs and LZs. It also coordinates instructions regarding air routes and air control points, and aircraft speeds, altitudes, and formations.

LOADING PLAN

D-30. This paragraph serves as a small-unit (company and below) leader's guide for the safe, efficient, and tactically sound conduct of operations in and around pickup zones.

SELECTION AND MARKING OF PZS AND LZS

D-31. Small-unit leaders should be proficient in the selection and marking of PZs and LZs, and in the control of aircraft. Tactical and technical aspects must be considered when selecting an LZ/PZ. Methods available for marking PZs and LZs include--

Day

D-32. A ground guide marks the PZ or LZ for the lead aircraft by holding his rifle over his head, by displaying a folded VS-17 panel chest high, or by some other identifiable means. Ground guides must wear eye and ear protection.

Night

D-33. Use the code letter "Y" (inverted) to mark the landing point of the lead aircraft at night (Figure D-2). Use chemical light sticks or beanbag lights to maintain light discipline. When more than one aircraft is landing in the same PZ or LZ, use an additional light for each aircraft. For observation, utility, and attack aircraft, mark each additional aircraft landing point with a single light emplaced at the exact point that each aircraft is to land. For cargo aircraft (CH-47, CH-53, and CH-54), mark each additional landing point with two lights. Place the two lights 10 meters apart and align them in the aircraft direction of flight.

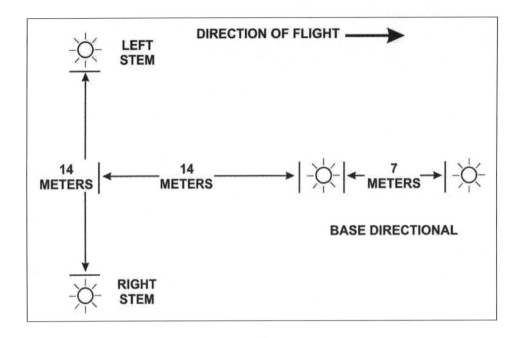

Figure D-2. Inverted "Y" marker.

D-34. Obstacles include any obstruction that might interfere with aircraft operation on the ground that cannot be reduced such as trees, stumps, and rocks. During good light, the aircrew is responsible for avoiding obstacles on the PZ or LZ. For limited visibility operations, mark all obstacles with red lights. The following criteria will be used in marking obstacles.

- If the obstacle is on the aircraft approach route, mark both the near and far sides of the obstacle.
- If the obstacle is on the aircraft departure route, mark the near side of the obstacle.
- If the obstacle protrudes into the PZ or LZ, but is not on the flight route of the aircraft, mark the near side of the obstacle.
- Mark large obstacles on the approach route by circling the obstacles with red lights.

CONTROL OF AIRCRAFT

D-35. Control approaching aircraft by the use of arm-and-hand signals to transmit terminal guidance for landing. Position the signalman to the right front of the aircraft where the pilot can see him. Give signals at night with a lighted baton or flashlight in each hand. When using flashlights, take care to avoid blinding the pilot. Keep batons and flashlights lighted at all times when signaling. The speed of arm movement indicates the desired speed of aircraft compliance with the signal.

ASSEMBLY AREAS

D-36. Before the aircraft arrives, secure the PZ, position the PZ control party, and position the Soldiers and equipment in a unit assembly area.

Occupation of Unit Assembly Area

D-37. While in a unit assembly area, unit leaders--

- Maintain all-round security of the assembly area.
- Maintain communications.
- Organize Soldiers and equipment into chalks and loads IAW the unit air-movement plan.

- Conduct safety briefings and equipment checks.
- Establish priority of loading for each man and identify bump personnel.
- Identify the locations of the straggler control points.

Organization of Units into Chalks

D-38. Make sure the chalk organization supports the ground tactical plan. Adhere to the following principles for loading the aircraft.

- Maintain tactical integrity by keeping fire teams and squads intact.
- Maintain self-sufficiency by loading a weapon and its ammunition on the same aircraft.
- Ensure key Soldiers, weapons, and equipment are cross-loaded among aircraft. This helps prevent the loss of control or all of a particular asset if an aircraft is lost.

Occupation of Chalk Assembly Areas

D-39. Linkup guides from the PZ control party meet the designated units in the unit assembly area and coordinate movement of chalks to a release point. As chalks arrive at the release point, chalk guides move each chalk to its assigned chalk assembly area. (To reduce the number of personnel required, use the same guide to move the unit from the unit assembly area to the chalk assembly area.) If part of a larger air assault, locate no more than three chalks in the chalk assembly area at one time. Maintain noise and light discipline throughout the entire movement in order to maintain the security of the PZ. Do not allow personnel on the PZ unless they are loading aircraft, rigging vehicles for slingload, or being directed by the PZ control. While remaining in chalk order, assign each Soldier a security (firing) position in the prone position weapon at the ready, and facing out (away from the PZ) to provide immediate close-in security. Figure D-3 shows an example of a large, one-sided PZ.

Aviation Support

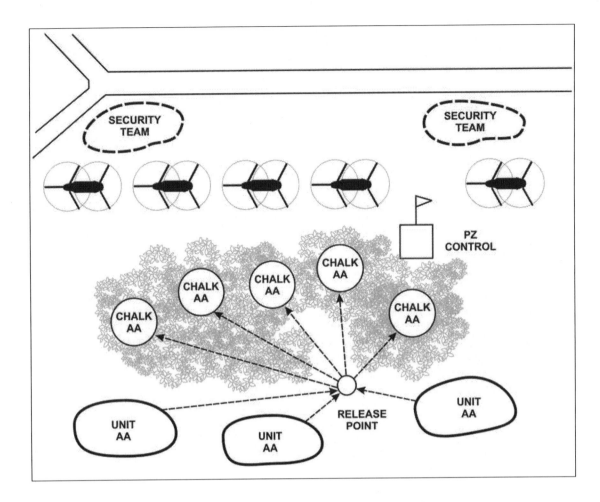

Figure D-3. Large, one-sided PZ.

Note: Artillery and mortar fire support is planned all-round the PZ, with priority to the far side of the large, open area.

D-40. Figure D-4 shows an example of a small, two-sided PZ with unit and chalk assembly areas.

Appendix D

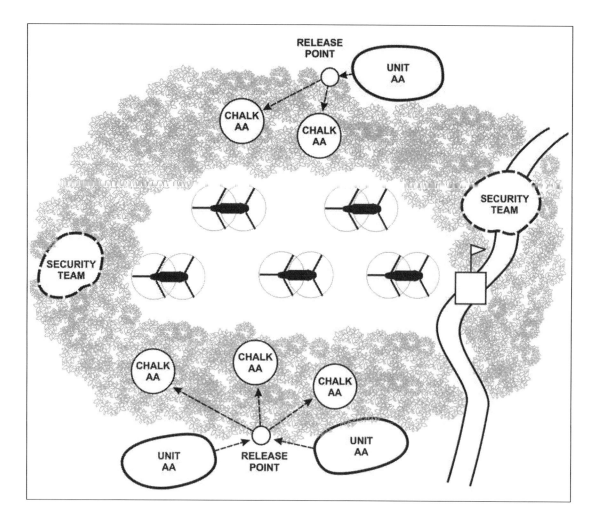

Figure D-4. Small, two-sided PZ.

D-41. Final preparations are made in the chalk assembly area. The chalk leader ensures all gear is tied down and checked, and short antenna are placed in radios, folded down, and secured before loading. He makes sure all squad and team leaders check the equipment of their men to ensure it is complete and operational. He also makes sure radios are on and a communications check is performed (unless directed otherwise). Then he assigns specific aircraft seat to each man.

Bump Plan

D-42. The least important chalk in each lift is designated for bump in case too few aircraft arrive at the PZ. These personnel report to a bump/straggler control point to be rescheduled for movement to the LZ.

Pickup Zone Closure

D-43. The CO designates a single man to be responsible for PZ closure. This might be the PZ control officer, the PZ control NCOIC, or another designated Soldier. He ensures all company men and equipment are loaded and that security is maintained.

Single Lift

D-44. The designated man positions himself at the last aircraft and collects bumped men, if required. He is the last man to board the aircraft. Once on the aircraft, he notifies the crew chief/AMC (using the CO's radio handset) that all personnel and equipment are loaded.

Multiple Lift

D-45. The duties of the PZ closure Soldier are the same as for a single lift. However, during a multiple lift, the security teams maintain security of the PZ and depart last with the PZ closure man.

UH-60 Loading Sequence

D-46. Figure D-5 shows the loading procedure for a UH-60. Up to 24 Soldiers might be loaded in a UH-60 for combat operations. (See FM 3-21.20 (FM 7-20) for a detailed discussion of "*seats-out*" operations.)

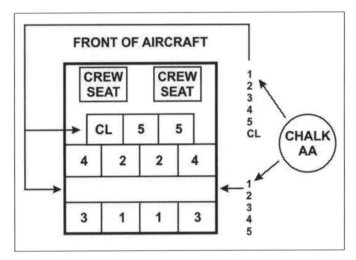

Figure D-5. UH-60 loading diagram.

- The chalk leader initiates loading once the aircraft have landed. The CO and platoon leaders normally occupy positions 5.
- The far-side and near-side groups move to the aircraft in file, with the number one Soldier leading the load to the appropriate side.

Note: The far side group always moves around to the front of the aircraft.

- The chalk leader stops at the near side of the aircraft to ensure the near-side group loads properly. Then, he moves around front of the aircraft to the far side and checks the other half of the chalk.
- All personnel buckle up as soon as they are seated in the correct seat.
- The chalk leader hands the chalk card to the pilot and answers any questions the pilot may have. They use the aircraft's intercommunication (troop commander's) handset.

Appendix D

STAGING PLAN

D-47. This plan synchronizes the arrival of Soldiers, aircraft, equipment, and logistic support at the PZs. The staging plan is based upon the loading plan. At company level, the staging plan is primarily concerned with the movement of the company to the PZ unit assembly area. It also addresses the linkup of company PZ control personnel with the battalion PZ control party (in larger operations) before the main body arrives. The staging plan should allow the company to be ready to start loading operations 15 minutes before the aircraft arrival time.

DUTIES OF KEY PERSONNEL

D-48. Key personnel are designated to perform specific duties to ensure the air assault is executed in an effective and efficient manner.

AIR ASSAULT OPERATIONS

D-49. In a company air assault, the following duties and responsibilities are assigned.

Company Commander

D-50. The CO has overall responsibility for the operation. He plans the operation, briefs subordinate leaders, issues the OPORD, and conducts rehearsals. He rides in the AMC's aircraft to ensure better command, control, and communications.

Executive Officer and First Sergeant

D-51. One of these two will--

- Set up the PZ. He supervises the marking of the PZ and the clearing of obstacles from the PZ.
- Brief all chalk leaders.
- Supervise all activity on the PZ such as PZ security, movement of troops and equipment, and placement of chalks and slingloads.
- Devise and disseminate the bump plan and control the bumped Soldiers.
- Ride in the last aircraft for control purposes and ensure that the PZ is cleared.

Chalk Leader

D-52. He briefs his personnel on their respective tasks and positions inside the aircraft. He also--

- Ensures that the lights or panels (if required) for his aircraft are properly emplaced.
- Assigns respective areas of security to his personnel. Ensures that each Soldier goes to his proper area.
- Supervises the loading of his chalk and attachments into the aircraft to ensure that all personnel assume assigned positions and buckle their lap belts.
- Keeps current on location by using his map and communicating with the aircraft crew during air movement.
- Ensures, upon landing that, all personnel exit the aircraft quickly, rush to a safe distance (15 to 20 meters) from the aircraft, assume the prone position, and prepare to return enemy fire.

PICKUP ZONE CONTROL PARTY

D-53. The PZ control party is responsible for the organization, control, and all coordinated operations in the PZ. (See FM 90-4 for more details on C2 techniques and responsibilities of key leaders.) Keeping in mind the CO's duties and responsibilities previously stated, a PZ control party for a company air assault operation could be organized as follows.

Pickup Zone Control Officer

D-54. He might be the XO, 1SG or a platoon leader.

Pickup Zone Control NCOIC

D-55. He is the 1SG, a platoon sergeant, a section sergeant, or a squad leader.

Radio Operator with Two Radios

D-56. One radio monitors the combat aviation net for communication with the aircraft. The second operates in the company command net or a PZ control net.

Chalk-Linkup Guides

D-57. There is one guide per chalk. Their primary duties are to assist in linkup and movement of chalks from the unit assembly area to the chalk assembly area. For company air assault operations, these guides should come from the same chalk they are assigned to.

Lead Aircraft Signalman

D-58. He is responsible for visual landing guidance for the lead aircraft. This signalman should come from the chalk loading on the lead aircraft.

Slingload Teams

D-59. A team includes a signalman and two hookup men.

AIR MISSION BRIEFING

D-60. The air mission briefing is the last coordination meeting of all key participants for an air assault mission. It ensures that all personnel are briefed and all details are finalized. It is coordinated by the battalion S-3 Air and normally conducted at the battalion TOC. If the company commander is the ground tactical commander, he must attend. If the battalion commander is the ground tactical commander, the company may not have a representative. The format in Table D-1 is a guide; it will help ensure that essential information is included in air assault mission briefings.

```
1. Situation.
    a. Enemy forces (especially troop concentrations and
locations and types of ADA assets).
    b. Friendly forces.
    c. Weather (ceiling, visibility, wind, temperature,
pressure and density altitude, sunrise and sunset, moonrise
and moonset, percent of moon illumination, EENT, BMNT, PZ and
LZ altitudes, and weather outlook).

2. Mission. Clear, concise statement of the task that is to be
accomplished (who, what, and when, and, as appropriate, why
and where).
```

Table D-1. Air mission briefing format.

3. Execution.
   a. Ground tactical plan.
   b. Fire support plan to include suppression of enemy air defenses.
   c. Air defense artillery plans.
   d. Engineer support plan.
   e. Tactical air support.
   f. Aviation unit tasks.
   g. Staging plan (both primary and alternate PZs).
      (1) PZ location.
      (2) PZ time.
      (3) PZ security.
      (4) Flight route to PZ.
      (5) PZ marking and control.
      (6) Landing formation and direction.
      (7) Attack and air reconnaissance helicopter linkup with lift elements.
      (8) Troop and equipment load.
   h. Air movement plan.
      (1) Primary and alternate flight routes (SPs, Army Civilian Personnel Systems [ACPS], and RPs).
      (2) Penetration points.
      (3) Flight formations and airspeeds.
      (4) Deception measures.
      (5) Air reconnaissance and attack helicopter missions.
      (6) Abort criteria.
      (7) Air movement table.
   i. Landing plan (both primary and alternate LZs).
      (1) LZ location.
      (2) LZ time.
      (3) Landing formation and direction.
      (4) LZ marking and control.
      (5) Air reconnaissance and attack helicopter missions.
      (6) Abort criteria.
   j. Laager plan (both primary and alternate laager sizes).
      (1) Laager location.
      (2) Laager type (air or ground, shut down or running).
      (3) Laager time.
      (4) Laager security plan.
      (5) Call forward procedure.
   k. Extraction plan (both primary and alternate PZs).
      (1) Pickup location.
      (2) Pickup time.
      (3) Air reconnaissance and attack helicopter missions.
      (4) Supporting plans.
   l. Return air movement plan.
      (1) Primary and alternate flight routes (SPs, ACPS, and RPs).
      (2) Penetration points.
      (3) Flight formations and airspeed.
      (4) Air reconnaissance and attack helicopter missions.
      (5) LZ locations.
      (6) 1 7 landing formation and direction.
      (7) 1 7 marking and control.
      (8) 1 7 marking and control.

Table D-1. Air mission briefing format (continued).

```
        m.  Coordinating instructions.
            (1)  Mission abort.
            (2)  Downed aircraft procedures.
            (3)  Vertical helicopter instrument flight recovery
procedures.
            (4)  Weather decision by D-hour increments and weather abort
time.
            (5)  Passenger briefing.

4.  Service Support.
        a.  Forward area arming and refueling point (FAARP) locations
(primary and alternate).
        b.  Ammunition and fuel requirements.
        c.  Backup aircraft.
        d.  Aircraft special equipment requirements such as cargo hooks
and command consoles with headsets.
        e.  Health service support.

5.  Command Signal.
        a.  Command.
            (1)  Location of commander.
            (2)  Point where air reconnaissance and attack helicopters
come under OPCON as aviation maneuver elements.
        b.  Signal.
            (1)  Radio nets, frequencies, and call signs.
            (2)  Signal operation instructions in effect and time of
change.
            (3)  Challenge and password.
            (4)  Authentication table in effect.
            (5)  Visual signals.
            (6)  Navigational aids (frequencies, locations, and
operational times).
            (7)  Identification friend or foe (radar) codes.
            (8)  Code words for PZ secure, hot, and clean; abort
missions; go to alternate PZ and LZ; fire preparation; request
extraction; and use alternate route.

6.  Time Hack.  All watches are synchronized.
```

Table D-1. Air mission briefing format (continued).

ATTACK AVIATION CONSIDERATIONS

D-61. Operations must be integrated so that air and ground forces can simultaneously work in the battlespace to achieve a common objective. Integration maximizes combat power through synergy of both forces. The synchronization of aviation operations into the ground commander's scheme of maneuver may require the integration of other services or coalition partners. It may also require integration of observation, attack, assault, and cargo helicopters.

PLANNING CONSIDERATIONS

D-62. Figure D-6 shows the minimum information required by the Army aviation team to ensure accurate and timely support. Digital transmission of information, such as coordinates, is faster and more accurate, if that method is available. Voice communications are necessary to verify information and to clarify needs and intentions.

Appendix D

- Situation including friendly forces' location, enemy situation highlighting known ADA threat in the AO, mission request, and tentative EA coordinates.
- Brigade- and battalion-level graphics update via MCS or AMPS or via radio communications, updating critical items--such as LOA, fire-control measures, and maneuver graphics--to better integrate into the friendly scheme of maneuver.
- Fire support coordination information: location of DS artillery and organic mortars, and call signs and frequencies.
- Ingress/egress routes in the AO; this includes passage points into sector or zone and air route to the HA or LZ.
- Call signs and frequencies of the battalion in contact down to the company in contact; air-ground coordination must be done on command frequencies to provide SA for all elements involved.
- GPS and SINCGARS time coordination; care must be taken to ensure that all units are operating on the same time.

Figure D-6. Minimum planning requirements.

POSITIVE LOCATION / TARGET IDENTIFICATION

D-63. The following techniques can be effective in ensuring that aircraft have positive identification of the locations of friendlies and targets (Figure D-7 through Figure D-10).

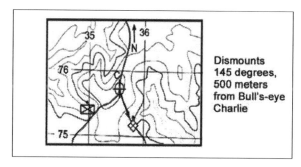

Figure D-7. Bull's-eye technique: uses a known point or an easily recognizable terrain feature.

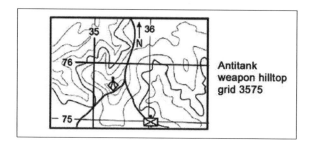

Figure D-8. Grid technique: uses grid coordinates define point.

Aviation Support

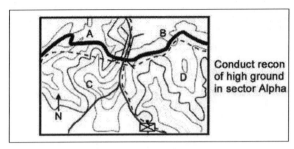

Figure D-9. Sector and terrain technique: uses terrain and graphics, which are both available to air and ground units.

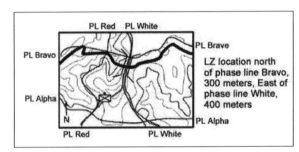

Figure D-10. Phase line technique: uses graphics, which are available to both air and ground.

MARKING

D-64. There are various ways to mark a location or target. (Table D-2, page D-18, lists various marking methods.) The effectiveness of vision systems on helicopters compares to those found on ground vehicles. During the day, the vision systems of the AH-64 and the OH-58D allow accurate identification of targets. During periods of reduced visibility, resolution is greatly degraded, requiring additional methods of verification. This situation requires extra efforts from both the ground unit and aviation element. Some US weapons can kill targets beyond the ranges that thermal, optical, and radar-acquisition devices allow positive identification. Both aviation and ground forces might become overloaded with tasks in the heat of battle. Simple, positive identification procedures must be established and known to all.

Marking US Troops

D-65. A method of target identification is direction and distance from friendly forces. Friendly forces can mark their own positions with IR strobes, IR tape, NVG lights, smoke, signal panels, body position, MRE heaters, chemical lights, and mirrors. Marking friendly positions is the least desirable method of target location information. It should be used with extreme caution. Marking friendly positions can take more time than directly marking a target, but it can reveal friendly positions to the enemy.

Marking Enemy Positions

D-66. Target marking aids aircrews in locating the target that the unit in contact desires them to attack. Ground commanders should provide the target mark whenever possible. To be effective, the mark must be timely, accurate, and easily identifiable. Target marks might be confused with other fires on the battlefield, suppression rounds, detonations, and marks on other targets. Although a mark is not mandatory, it improves aircrew accuracy, enhances SA, and reduces the risk of fratricide.

Appendix D

METHOD	DAY	NIGHT	NVG	NVS	FRIENDLY MARKS	TARGET MARKS	REMARKS
Smoke	Go	No Go	Marginal	No Go	Good	Good	Easy ID. May compromise friendly position, obscure target, or warn of FS employment. Placement may be difficult because of terrain, trees, or structures.
Smoke (IR)	Go	Go	Go	No Go	Good	Good	Easy ID. May compromise friendly position, obscure target, or warn of FS employment. Placement may be difficult because of terrain, trees, or structures. Night marking is greatly enhanced by the use of IR reflective smoke.
Illumination, Ground Burst	Go	Go	Go	No Go	NA	Good	Easy ID. May wash out NVDs.
Signal Mirror	Go	No Go	No Go	No Go	Good	NA	Avoids compromise of friendly location. Depends on weather and available light. May be lost in reflections from other reflective surfaces such as windshields, windows, or water.
Spot Light	No Go	Go	Go	No Go	Good	Marginal	Highly visible to all. Compromises friendly position and warns of FS employment. Effectiveness depends on the degree of ambient lighting.
IR Spot Light	No Go	No Go	Go	No Go	Good	Marginal	Visible to all NVGs. Effectiveness depends on the degree of ambient lighting.
IR Laser Pointer (below .4 watts)	No Go	No Go	Go	No Go	Good	Marginal	Effectiveness depends on the degree of ambient lighting.
IR Laser Pointer (above .4 watts)	No Go	No Go	Go	No Go	Good	Good	Less affected by ambient light and weather conditions. Highly effective under all but the most highly lit or worst weather conditions. IZLID-2 is the current example.
Visual Laser	No Go	Go	Go	No Go	Good	Marginal	Highly visible to all. High risk of compromise. Effective, depending upon degree of ambient light.
Laser Designator	Go	Go	No Go	Go	NA	Good	Highly effective with precision-guided munitions. Very restrictive laser-acquisition cone and requires LOS to target. May require precoordination of laser codes. Requires PGM or LST equipped.

Table D-2. Techniques for marking of target or location.

Aviation Support

METHOD	DAY	NIGHT	NVG	NVS	FRIENDLY MARKS	TARGET MARKS	REMARKS
Tracers	Go	Go	Go	No Go	No Go	Marginal	May compromise position. May be difficult to distinguish mark from other gunfire. During daytime use, may be more effective to kick up dust surrounding target.
VS-17 Panel	Go	No Go	No Go	No Go	Good	NA	Easy to see when visibility is good. Must be shielded from the enemy.
IR Paper	No Go	No Go	No Go	Go	Good	NA	Must be shielded from the enemy. Affected by ambient temperature.
AN/PAQ-4C IR Aiming Light	No Go	No Go	Go	No Go	NA	Good	Effective to about 600 meters.
AN/PEQ-2A IR Aiming Light, Pointer, Illuminator	No Go	No Go	Go	No Go	NA	Good	Effective to about 1,300 meters. Can illuminate the target.
Chem Light	No Go	Go	Go	No Go	Good	NA	Must be shielded from enemy observation. Affected by ambient light. Spin to give unique signature.
IR Chem Light	No Go	No Go	Go	No Go	Good	NA	Must be shielded from enemy observation. Affected by ambient light. Spin to give unique signature.
Strobe	No Go	Go	Go	No Go	Excellent	NA	Visible to all. Affected by ambient light.
IR Strobe	No Go	No Go	Go	No Go	Excellent	NA	Effectiveness depends on ambient light. Coded strobes aid acquisition. Visible to all with NVGs.
Flare	Go	Go	Go	Marginal	Excellent	NA	Visible to all. Easily seen by aircrew.
IR Flare	No Go	No Go	Go	No Go	Excellent	NA	Easily seen by aircrews with NVGs.
Glint/IR Panel	No Go	No Go	No Go	Go	Good	NA	Not readily detected by enemy. Effective except in high ambient light.
Combat ID Panel	Go	No Go	No Go	No Go	Good	NA	Provides temperature contrast on vehicles or building.
Chemical Heat Sources, MRE Heater	No Go	No Go	No Go	Go	Poor	NA	Can be lost in thermal clutter. Difficult to acquire. Best to contrast a cold background.
Briefing Pointer	No Go	Go	Go	No Go	Fair	Poor	Short range.
Electronic Beacon	NA	NA	NA	NA	Excellent	Good	Ideal friendly marking for AC-130 and some USAF CAS. Not compatible with Navy/Marines. Can be used as a TRP. Coordination with aircrew essential.
Hydra 70 Illumination	Go	Go	Go	Go	NA	Good	Assists with direct fire and adjustment of indirect fire.

Table D-2. Techniques for marking of target or location (continued).

Appendix D

CLOSE COMBAT ATTACK BRIEFING

D-67. The CCA briefing follows the joint standard nine-line format with minor modifications for Army helicopters (Figure D-11). The briefing provides clear and concise information in a logical sequence that enables aircrews to employ their weapons systems. It also provides appropriate control to reduce the risk of fratricide. Figure D-12 shows an example of a briefing.

```
CLOSE COMBAT ATTACK BRIEFING

(Omit data not required. Do not transmit line numbers. Units of measure are standard
unless otherwise specified. *Denotes minimum essential in limited communications
environment. BOLD denotes readback items when requested.)

Terminal controller: _____  This is _____
                     (Aircraft call sign)                  (Terminal controller)

*1. IP/BP/ABF or friendly location: _____
                                    (Grid, known point or terrain feature)

*2. Heading to target: _____ (magnetic)
                      (Specify from IP/BP/ABF or friendly location)

*3. Distance to target: _____ (meters)
                       (Specify from IP/BP/ABF or friendly location)

 4. Target elevation: _____ (feet mean sea level)

*5. Target description: _____
                        _____

*6. Target location: _____
                     (Grid, known point or terrain feature)

 7. Type of target mark: _____  Code: _____ (day/night)
                         (WP, laser, IR, beacon)    (Actual code)

    Laser to Target Line: _____ degrees

*8. Location of friendlies: _____
                            (Omit if previously given--grid, known point, or terrain feature)

    Position Marked By _____

 9. Egress direction: _____
                      (Cardinal direction not over threats)

    Remarks (as appropriate): _____
                              _____
          (Threats, restrictions, danger close, attack clearance, SEAD, abort codes, hazards)

    Time on target (TOT): _____

    or time to target (TTT):  Standby _____ plus _____ hack.

Note: When identifying position coordinates for joint operations, include the map datum
data. DESERT STORM operations have shown that simple conversion to latitude/longitude
is not sufficient. The location may be referenced on several different databases;
for example, land-based versus sea-based data.
```

Figure D-11. Example CCA nine-line briefing.

Aviation Support

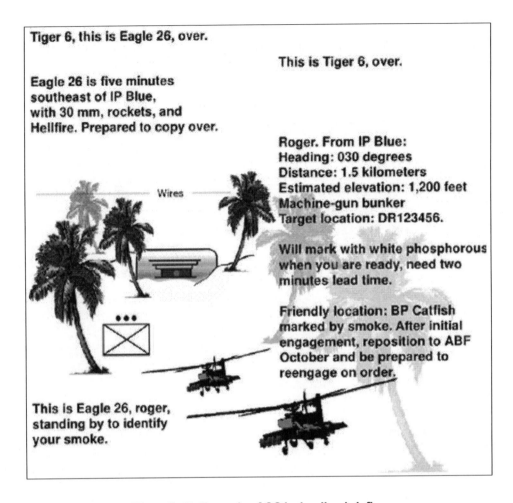

Figure D-12. Example of CCA nine-line briefing.

SAFETY

D-68. The CO and his subordinate leaders must enforce strict safety measures when working with helicopters. Primary safety measures include--

- Keeping the body low when approaching and departing a helicopter, especially on slopes.
- Keeping safety belts fastened when airborne (for training).
- Keeping weapons unloaded (no round in chamber) and on SAFE. Keeping the muzzle down on UH-60, OH-58, and CH-47, but up on the UH-1.
- Keeping radio antennas down and secured.
- Keeping hand grenades secured.
- Not jumping from a hovering helicopter until told to do so by a crewmember.
- Not approaching from, or departing to, the rear of a helicopter.

Appendix E
Sniper Employment

This edition introduces concepts of sniper employment during tactical operations.

Snipers and observers play a critical role in Infantry company operations. Since snipers are seldom employed below battalion level, each Infantry squad has one designated marksman. Unit snipers are assigned through an authorized position in the unit TOE. Well-trained snipers provide the commander accurate, discriminating, long-range small-arms fire, and direct observation of key terrain and avenues of approach. The two best uses of sniper fire or long-range precision fire are against key targets beyond the range of organic rifles and automatic weapons, or against any targets that other weapon systems cannot destroy due to range, size, location, visibility, or security and stealth requirements. Sniper TTP enable them to directly gather and relay critical, detailed enemy information. Sniper effectiveness is measured by more than casualties or destroyed targets. Commanders know snipers also affect enemy activities, morale, and decisions. Knowing snipers are present hinders the enemy's movement, and creates confusion and continuous personal fear. It also disrupts enemy operations and preparations, and compels the enemy to divert forces to deal with the snipers (FM 3-22.10 (FM 23-10)).

SNIPER TEAM

E-1.   Snipers avoid sustained battles. They typically operate in three-Soldier teams, each with at least one sniper and one observer, normally cross-trained. The observer carries an M4 rifle; the sniper carries the sniper weapon system; and each member carries a side arm. Team members help each other with range estimation, round adjustment, and security. Sniper activity should be planned and controlled by the sniper employment officer. The SEO is designated by the commander and is usually the HHC commander or a member of the S-3 staff.

SQUAD DESIGNATED MARKSMAN

E-2.   The squad designated marksman can seldom perform as well as well-trained snipers, so commanders and platoon leaders should avoid employing them as such. However, the marksman is a valuable asset who can contribute in many ways. Leaders should remember the value of the marksman versus the sniper, and use each to the best of their respective abilities in any situation (FM 3-21.8 (FM 7-8)).

EMPLOYMENT

E-3.   The commander, S-3, SEO, or sniper squad leader controls sniper teams from a central location. Once deployed, sniper teams generally operate independently. To accomplish the assigned unit mission, they must understand the commander's intent, concept of the operation, and purpose for their assigned tasks. Snipers are effective only in areas with good fields of fire and observation. They must have the freedom of action to choose their own positions once on the ground. The number of sniper teams participating in an operation depends on their availability, on the expected duration of the mission, and on the enemy's strength and disposition.

Appendix E

SECURITY ELEMENT

E-4. Sniper teams should move with a security element (squad or platoon) whenever possible. Initially, sniper teams can also move with a mounted element, which allows them to enter an area more quickly and more safely than if they operated alone. The security element also protects the snipers during the operation. When moving with a security element, snipers follow these guidelines.

- The leader of the security element leads the sniper team.
- Snipers must appear to be an integral part of the security element. Whenever possible, based on METT-TC, snipers conceal their sniper-unique equipment, such as optics, radios, and ghillie suits, from view.
- Snipers must wear the same uniforms as the members of the security element. Snipers and element members maintain proper intervals and positions in the element formation.

Mission

E-5. The sniper's primary mission is to support combat operations by delivering precise rifle fire from concealed positions. The mission assigned to a sniper team for a particular operation consists of the task(s) the commander wants the sniper team to accomplish and the reason (purpose) for it. The commander must decide how he wants his sniper team to affect the battlefield. Then he must assign missions to achieve this effect.

- The commander assigns target priorities so snipers can avoid involvement in sustained engagements. Sniper teams are free to change targets to support the commander's intent.
- The commander describes the effect or result he expects and allows the sniper team to select key targets. Since either the M24 or M107 weapon system is available to the sniper team, they can select the best one to use to achieve the desired effect.
- The commander may also designate the sniper to act as an observer of a target or an area rather than task conventional forces to do so. The sniper's ability to remain undetected for long periods may make this a more practical mission than dedicating other forces to do so.
- The commander may assign specific types of targets to achieve an effect. He may task snipers to kill bulldozer operators and other engineer equipment operators to disrupt enemy defensive preparations. Or, he may task snipers to disable enemy command or supply vehicles, or to engage enemy soldiers digging defensive positions.
- The commander may assign specific point targets such as bunkers, CPs, or crew-served weapons positions. These can include enemy leaders, command and control operators, antitank guided missile gunners, armored vehicle commanders, weapons crews, or selected individuals. Snipers may also be assigned countersniper roles.

Enemy

E-6. The commander must consider the following characteristics, capabilities, strengths, weaknesses, and disposition of the enemy:

- Is the enemy force heavy or light, rested or tired, disciplined or not?
- Is it motorized infantry or towed artillery?
- Is it well or poorly supplied?
- Is it patrolling aggressively or is it lax in security?
- Is it positioned in assembly areas or dug in?

E-7. The answers to questions like these help the commander determine the enemy's susceptibility and reaction to effective sniper operations. Obviously, a well-rested, well-led, well-supplied, and aggressive enemy with armored protection poses a greater threat to snipers than one that is poorly led, poorly supplied, lax, and unprotected. The commander needs to know if enemy snipers are present and effective, since they can pose a significant danger to his operations and his snipers. The commander must also consider the enemy's directed-energy weapons capability. Snipers' optical devices make them particularly vulnerable to this kind of weapon.

Terrain

E-8. The commander must evaluate and consider the terrain to and within the sniper's AO, the time and effort snipers will need to get into position, and the effects of weather on the sniper and his visibility. Snipers prefer positions at least 300 meters from their target area. Operating at this distance allows them to avoid effective fire from enemy rifles, while retaining much of the 800- to 1,000-meter effective range of the sniper rifle. Snipers need areas of operations with good observation, fields of fire, and firing positions.

Troops

E-9. The commander must decide how many sniper teams to use depending on their availability, the duration of the operation, expected opposition, and the number and difficulty of tasks and targets assigned. Commanders must consider the snipers' level of training and physical conditioning, and must remember the effects of these human factors on sniper operations.

Time Available

E-10. The commander must consider how much time the snipers have to achieve the result he expects. He must allocate time for snipers to plan, coordinate, prepare, rehearse, move, and establish positions. He must understand how the snipers' risk increases with inadequate time to plan or to perform other tasks such as moving to the AO. The length of time a sniper team can remain in a position without loss of effectiveness due to eye fatigue, muscle strain, or cramps depends mostly on the type of position the team occupies. Generally, snipers can remain in an expedient position for six hours before they must be relieved. They can remain in belly positions or semi-permanent hides for up to 48 hours before they must be relieved. The average mission takes about 24 hours. Movement factors for snipers moving with a security element are the same as for any Infantry force. When snipers move alone in the AO, they move slowly; their movement can be measured in feet and inches. The sniper team is the best resource in determining how much time is required for their movement.

OFFENSIVE EMPLOYMENT

E-11. Offensive operations carry the fight to the enemy to destroy his capability and will to fight. The sniper plays a major role in offensive operations by killing enemy targets that threaten the success of the attack, or by providing timely intelligence in support of offensive operations.

OFFENSIVE MISSIONS

E-12. During offensive operations, snipers--

- Conduct countersniper operations.
- Conduct offensively oriented reconnaissance operations.
- Overwatch movement of friendly forces and suppress enemy targets that threaten the moving forces.
- Place precision fire on enemy crew-served weapons teams and into exposed apertures of bunkers.
- Place precision fire on enemy leaders, armored-vehicle drivers or commanders, FOs, and other designated personnel.
- Place precision fire on small, isolated, bypassed forces.
- Place precision fire on targets threatening a counterattack or fleeing.
- Assist in screening a flank using supplemental fires.

MOVEMENT TO CONTACT

E-13. During a movement to contact, snipers move with the lead element, or they can be employed 24 to 48 hours before the unit's movement--

- To choose positions.
- To gather information about the enemy.
- To deny enemy access to key terrain through controlled precision fires, preventing enemy surprise attacks.

ASSAULT

E-14. Snipers can provide effective support during an assault.

- Snipers placed with lead elements move to positions that allow them to overwatch the movement of the element and to provide long-range small arms fire. Sniper teams are most effective where supporting vehicles cannot provide overwatching fires.
- Snipers may also be placed in a position to suppress, fix, or isolate the enemy on the objective. The sniper rifle's precision fire and lack of blast effect allow the sniper to provide close supporting fires for assaulting Soldiers.
- If time permits, snipers might be deployed early in the operation. Because the snipers' weapons have better optics and longer ranges than other types of small arms, they can provide additional long-range observation and precision fire on any enemy targets that may appear.
- Snipers may move with the assault force toward the objective; occupy a close-in, support-by-fire position where they can help suppress or destroy targets threatening the assault force or move onto the objective to provide close-in, precision fire against enemy fortified positions, bunkers, and trench lines. Selection of the sniper support-by-fire position depends on METT-TC. The closer snipers are to the objective area, the greater the chance they will be discovered and lose their effectiveness.
- If elements appear on the battlefield at the same time snipers arrive, the snipers' security and potential for surprise are degraded. To increase security and surprise, snipers may move covertly into position in an objective area well before the main attack arrives. Ideally, a sniper team moves with infiltrating dismounted Infantry. This is faster and more secure than moving alone. After the snipers are in position, Infantrymen may remain nearby as additional security, but they are more likely to have other supporting tasks to perform.
- After their fires are identified, snipers reposition as soon as possible. The commander must carefully evaluate where snipers will be most useful. If he wants to use snipers in several different places, or if he wants them to contribute throughout the attack, he provides transportation to enable them to move quickly, stealthily, and safely on the battlefield.
- Upon consolidation, snipers may displace forward to new positions that are not necessarily on the objective. From these positions, the snipers provide precision fire against bypassed enemy positions, enemy counterattack forces, or other enemy positions that could degrade the unit's ability to exploit the success of the attack.

ACTIONS AGAINST FORTIFIED AREAS

E-15. Assaulting forces usually encounter some type of fortified positions prepared by the defending force. These can range from hasty, field-expedient positions produced with locally available materials to elaborate steel and concrete emplacements complete with turrets, underground tunnels, and crew quarters. Most are field expedient. More elaborate positions are likely only when the enemy has had enough time to prepare his defense. He might have fortified weapons emplacements or bunkers; protected shelters; or reinforced natural or constructed caves, entrenchments, and other obstacles.

ENEMY DEFENSIVE POSITIONS

E-16. The enemy tries to locate these positions so they are mutually supporting and arrayed in depth across the width of his sector. He tries to increase his advantages by covering and concealing positions and by preparing fire plans and counterattack contingencies. Fortified areas should be bypassed and contained by a small force.

SNIPER SUPPORT

E-17. The sniper's precision fire and observation capabilities are invaluable in the assault of a fortified area. Precision rifles can easily detect and destroy pinpoint targets that are invisible to the naked eye. The snipers' role during the assault of a fortified position is to deliver precision fire against OPs; exposed personnel; and the embrasures, air vents, and doorways of key enemy positions. The commander plans the order in which snipers should destroy targets. Their destruction should systematically reduce the enemy's defense by destroying the ability of enemy positions to support each other. Once these positions are isolated, they can be reduced more easily. The commander must decide where he will try to penetrate the enemy's fortified positions, and then employ his snipers against those locations. When operating from positions near the breach point on the flanks, snipers can provide continuous fire support for both assaulting units and other nearby units. Sniper fires add to the effectiveness of the entire unit. The commander can employ snipers where he cannot use other resources, for various reasons.

SNIPER PLAN

E-18. The sniper team bases its plan on information available. The enemy information it needs includes--

- Extent and exact locations of individual and underground fortifications.
- Fields of fire, directions of fire, locations and numbers of embrasures, and types of weapon systems in the fortifications.
- Locations of entrances, exits, and air vents in each emplacement.
- Locations and types of existing and reinforcing obstacles.
- Locations of weak spots in the enemy's defense.

DEFENSIVE EMPLOYMENT

E-19. Snipers may effectively enhance or augment any unit's defensive fire plan. After analyzing the terrain, the sniper team should recommend options to the commander.

DEFENSIVE TASKS

E-20. The sniper team can perform the following tasks during defensive operations.

Cover obstacles, minefields, roadblocks, and demolitions.
Perform counterreconnaissance, that is, identify or destroy enemy reconnaissance elements.
- Engage enemy OPs, armored vehicle commanders exposed in turrets, and ATGM teams.
- Damage enemy vehicles' optics to degrade their movement.
- Suppress enemy crew-served weapons.
- Disrupt follow-on units with long-range small-arms fire.

PRIMARY POSITIONS

E-21. Snipers generally position themselves where they can observe or control one or more avenues of approach into the defensive position. Sniper employment can increase all-round security and allow the commander to concentrate his combat power against the most likely enemy avenue of approach. Snipers may support the battalion by providing extra optics for target acquisition and precise long-range fires to complement the fires of other weapon systems. This arrangement takes advantage of the effectiveness of all of the unit's weapon systems. Snipers in an economy-of-force role may cover dismounted enemy avenues of approach into task force positions.

ALTERNATE AND SUPPLEMENTARY POSITIONS

E-22. Snipers establish alternate and supplementary positions for all-round security. Positions near the FEBA are vulnerable to concentrated attacks, enemy artillery, and obscurants. Multiple sniper teams, if used, can be positioned for surveillance and mutual fire support. If possible, they should establish positions in depth for continuous support during the fight. The sniper's rate of fire neither increases nor decreases as the enemy approaches. Instead, sniper teams systematically and deliberately shoot specific targets, never sacrificing accuracy for speed.

KEY TERRAIN

E-23. The commander can position snipers to overwatch key obstacles or terrain such as river-crossing sites, bridges, minefields, or anything that canalizes the enemy directly into engagement areas. Snipers are mainly used where weapon systems are less effective due to security requirements or terrain. Even though he commander has weapon systems with greater ranges and optical capability than the snipers' weapons, the commander might be unable to use them for any of several reasons. They might present too large a firing signature, be difficult to conceal well, create too much noise, or be needed more in other areas. Sniper team members provide the commander with better observation and greater killing ranges than do other Soldiers.

FORCE SECURITY

E-24. Snipers can be used as an integral part of the security effort. They can help acquire and destroy targets, augment the security element by occupying concealed positions for long periods, observe and direct indirect fires (to maintain their security), and engage targets. Selective long-range sniper fires are difficult for the enemy to detect. A few well-placed shots can disrupt enemy reconnaissance efforts, force him to deploy into combat formations, and deceive him as to the location of the MBA. The sniper's stealth skills counter the skills of enemy reconnaissance elements. Snipers can be used where scout or rifle platoon mobility is unnecessary, freeing the scouts and riflemen to cover other sectors. Snipers can also be used to direct ground maneuver elements toward detected targets. This also helps maintain security so ground maneuver elements can be used against successive echelons of attacking enemy.

STRONGPOINT EMPLOYMENT

E-25. Snipers should be tasked to support any unit defending a strongpoint. The sniper team's characteristics enable it to perform independent harassing and observation tasks in support of the force in the strongpoint, from either inside or outside the strongpoint.

REVERSE SLOPE DEFENSE

E-26. If the unit is occupying a reverse slope defense, snipers can provide effective long-range fires from positions forward of the topographical crest or on the front slope.

RETROGRADE EMPLOYMENT

E-27. The sniper team must know the concept, intent, scheme of maneuver, withdrawal times or conditions and priorities, routes, support positions, rally points, and locations of obstacles. Both engagement and disengagement criteria must be planned and coordinated to ensure snipers achieve the desired effect without compromising their positions.

FORCE ENEMY DEPLOYMENT

E-28.   Snipers can help the delaying force cause the enemy to deploy prematurely during retrograde operations. They help by inflicting casualties with accurate, long-range, small-arms fire. When the enemy receives effective small-arms fire from unknown positions, he is likely to assume he is near an enemy position--most likely one with ATGMs--and to begin maneuvering to a position of advantage against the perceived threat. Using a sniper team, the commander can achieve the same effect he could with another Infantry unit. The snipers' stealth also gives them a better chance of infiltrating out of positions close to the enemy.

SELECTION OF NEW POSITIONS

E-29.   Delaying forces risk being bypassed or overtaken by attacking enemy forces. Commanders may provide transportation to move snipers to successive positions. Vehicles must remain in defilade positions to the rear of the sniper position, or they must occupy different positions away from the sniper's AO. This keeps the vehicles from compromising the sniper's position. In either case, a linkup point, egress routes, and conditions for executing the linkup must be fully coordinated. Commanders may also provide communications assets to the sniper team to facilitate control and movement.

INFILTRATION

E-30.   Snipers might be required to infiltrate back to friendly positions. Their infiltration plans must be fully coordinated to avoid fratricide during reentry of a friendly position. When planning successive positions, the commander must realize the sniper team might be unavailable for use if it is destroyed, or if it is having trouble disengaging from an enemy force. The commander must carefully consider how and where he wants snipers to contribute to the operation. Planning too many positions for the sniper team in a fast-paced retrograde may result in failure.

TASKS

E-31.   Snipers might be assigned any of the following specific tasks.

- Delay the enemy by inflicting casualties.
- Observe avenues of approach.
- Cover key obstacles with precision fire.
- Direct artillery fire against large enemy formations.

URBAN OPERATIONS

E-32.   The sniper's value to a unit conducting UOs depends on several factors, including the type of operation, level of conflict, and ROE. Where ROE allow destruction, the snipers may not be needed since other weapon systems have greater destructive effect. But where ROE prohibit collateral damage, snipers might be the commander's most valuable tool. During stability operations in urban terrain, the sniper or marksman can provide greatly enhanced observation of an area or population, and can apply precise firepower within the limits of the ROE more easily than can the Infantry force.

URBAN TERRAIN

E-33.   Sniper effectiveness depends partly on the terrain. The characteristics of an urban area degrade control. To provide timely and effective support, the sniper must understand the scheme of maneuver and commander's intent.

- Observation and fields of fire are clearly defined by roadways. However, rooftops, windows, and doorways limit surveillance, because each requires constant observation. The effects of smoke from military obscurants and burning buildings can degrade what otherwise appears to be an excellent vantage point. All-round defense becomes more important, because the enemy can fire from many directions. His infiltration attempts must be countered.

- Cover and concealment are excellent for both the attacker and defender. The defender normally has an advantage, because the attacker normally exposes himself when moving through the area.
- Avenues of approach inside buildings are best, because movement in a building is less easily detected than movement through the streets. The sniper must be conscious of all avenues of approach, and must be prepared to engage targets that appear on any of them.

SELECTION OF POSITIONS

E-34. Snipers should be positioned in buildings of mass- or heavy-clad frame construction that offer long-range fields of fire and all-round observation. The sniper has an advantage because he need not move with, or be positioned with, lead elements. He may occupy a higher position to the rear or flanks and some distance away from the element that he is supporting. By operating far from the other elements, a sniper avoids decisive engagement, but remains close enough to kill distant targets threatening the unit. Snipers should not be placed in obvious positions, such as church steeples and rooftops, since the enemy often observes these and targets them for destruction. Indirect fires can generally penetrate rooftops and cause casualties in top floors of buildings. Snipers should not be positioned where there is heavy traffic, because these areas invite enemy observation as well.

MULTIPLE POSITIONS

E-35. Snipers should operate throughout the AO, moving with and supporting the company teams as necessary. Some teams may operate independently from other forces. They search for targets of opportunity, especially for enemy snipers. Since a single position may not afford adequate observation for the entire team without increasing the risk of detection by the enemy, the team may occupy multiple positions. Separate positions must maintain mutual support. Each team should also establish alternate and supplementary positions.

TASKS

E-36. The commander may assign the following tasks to snipers.

- Conduct countersniper operations.
- Kill targets of opportunity. The sniper team assigns priorities to these targets based on their understanding of the commander's intent, which might include, for example, to engage enemy snipers, leaders, vehicle commanders, radio men, sappers, and machine gun crews, in that order.
- Deny enemy access to certain areas or avenues of approach. In other words, control key terrain.
- Provide fire support for barricades and other obstacles.
- Maintain surveillance of flank and rear avenues of approach (screen).
- Support local counterattacks with precision fire.

STABILITY AND RECONSTRUCTION OPERATIONS

E-37. In stability and reconstruction operations, the sniper can dominate an AO by delivering selective precision fire against specific targets IAW the ROE. Since the ROE normally limit collateral damage and civilian casualties, snipers selectively kill or wound key individuals who pose a threat to friendly forces. Targets often hide in the civilian populace, which makes them nearly invulnerable to US forces, who cannot destroy these targets without causing innocent casualties. The sniper may also be employed to gather information in an S&R operation.

TASKS

E-38. Some of the specialized tasks that commanders may assign to snipers follow.

- If and as authorized by local orders or instructions, snipers engage dissidents involved in such activities as hijacking, kidnapping, and hostage taking.
- Snipers engage dissident snipers as opportunity targets or as part of a deliberate clearance operation.
- Snipers covertly occupy concealed positions to observe selected areas.
- Snipers record and report all suspicious activity in the area of observation.
- Snipers help coordinate the activities of other elements from their hidden observation positions.
- Snipers protect other elements of the controlling forces, including key civilian noncombatants such as judges, politicians, fire fighters, and repair crews.

ANONYMITY

E-39. Commanders must carefully protect the anonymity of unit snipers, even from other Soldiers in the unit. This is especially true of successful snipers, because dissidents will target them. Ideally, snipers are held in a central reserve and employed only after shooting starts. If needed, snipers may deploy in hidden observation posts.

SPECIAL CONSIDERATIONS

E-40. Ideally, a sniper should deploy where he can receive the order to fire from the appropriate local commander. This is often difficult. Due to the typical remoteness of the sniper's position, direct communication with the commander is often impossible. Therefore, all orders, to include targets and Rules of Engagement, must be clear to the sniper team before it deploys. Before that, the team must rehearse when to open fire in all possible scenarios. They learn how to determine when their fire constitutes reasonable force, regardless of circumstances:

- If the sniper is away from the local commander, then he must positively identify and engage his targets based on his written orders.
- If he is physically near or in radio contact with the local commander, he identifies and engages based on the verbal orders of the local commander.
- For accuracy in actual operations, snipers zero their weapons daily, before being placed on standby. They zero at a minimum range of 100 meters just before their standby shift. They should zero again just before deploying to a covert OP.

PEACE OPERATIONS

E-41. The tasks of the sniper team during peace operations generally consist of gathering intelligence, overwatching, and reporting, but may also include countersniping. For peace operations, snipers are employed in various types of observation posts.

COVERT RURAL

E-42. This is just like a conventional OP except that, depending on the nature and duration of the task, the team should have--

- Weapons and other equipment to suit the task, based on METT-TC.
- Smoke and pen flares as an alternate means of communication.
- Close support during the insertion. This might mean acting as a radio relay or providing any other means of support needed in case the team has to extract.

Appendix E

- A method of insertion appropriate to the task. Insertion is usually coordinated through the battalion S-3. Common methods of insertion include foot, vehicle, or helicopter.
- The team will need at least 24 hours to prepare for a long-term OP.

COVERT URBAN

E-43. A covert urban OP requires more preparation time than does an overt OP. Reconnaissance for suitable OP locations can take two to four days. Some of that time is used to determine the local habits in the area such as patterns of foot traffic and children at play. Children present the greatest compromise threat. The team must also learn what local security is in place, and where unexploded ordnance (UXO) is located. Finally, they must allow time to infiltrate, set up security, and exfiltrate.

OVERT URBAN

E-44. Commanders use snipers overtly in urban operations as deterrents. Overt urban OPs should cover the target area and have both flank and rear security. Higher vantage points reduce sniper team exposure. The commander should only place snipers in overt OPs if the enemy sniper threat is low and *if no other assets can achieve the desired results*. Just as they do when snipers operate anywhere, commanders should aggressively protect the identities of the sniper team.

RIVER CROSSINGS

E-45. Sniper teams, by virtue of their observation and precision-fire capabilities, are uniquely adaptable to the initial stages of a river crossing. They are normally employed in general support of the TF both before and during the crossing.

SELECTION OF POSITIONS

E-46. Snipers assume positions across the total width of the crossing area (if possible) before the crossing. Their main task is to observe. They report all sightings of enemy positions and activity immediately, and they provide a stealthy observation capability not otherwise available to the commander. Their stealth prevents the enemy from learning key facts like what type of unit is trying to cross. The snipers supplement normal reconnaissance assets.

CROSSING SUPPORT

E-47. Snipers provide support during the crossing by continuing to observe and suppress enemy OPs and other key targets that heavier supporting elements might overlook. The snipers' ability to continue to provide close-in suppressive fire makes continuous fire support possible. They can continue providing suppressive fire until the elements reach the far side and start moving to establish the bridgehead line. Snipers should be positioned as early as possible, preferably as part of the reconnaissance force. Their movement across the river must also be planned. The means of crossing, and their subsequent positions, must be coordinated. Generally, the snipers displace once friendly elements reach the far side.

INSERTED FORCE SUPPORT

E-48. Snipers allow the inserted force to engage threatening targets at long ranges. Once on the far side, snipers may screen the flank or rear of the crossing force, infiltrate to destroy key targets, such as a demolition guard or fortified emplacement, or operate OPs well to the front of the crossing force. This placement increases both early warning time and the crossing force's ability to disrupt enemy counterattack forces.

PATROLS

E-49. With any size or type of patrol, only the terrain and the patrol leader's ingenuity limit how he can effectively employ sniper teams. Snipers must know and be able to apply all aspects of patrolling.

RECONNAISSANCE PATROLS

E-50. Snipers normally remain with the security element during reconnaissance patrols. If terrain permits, snipers can provide long-range support to enable the reconnaissance element to patrol farther from the security element. To avoid compromising the reconnaissance element's position, snipers fire only in self-defense or when ordered by the patrol leader. Normally, the only appropriate time to fire at a target of opportunity is when extraction or departure from the position is imminent and firing will not endanger the success of the patrol.

RAID PATROLS

E-51. How snipers are employed on a raid depends on the time of day and the size of the patrol. When the patrol needs maximum firepower, yet its size is limited, snipers are not employed. However, they might be employed with raid patrols as follows.

Security Element

E-52. If the patrol needs long-range precision fire, and the patrol size permits, sniper teams might be attached to the raid patrol's security element. When attached to the security element, the sniper team may provide observation, or may help prevent the enemy from escaping the objective area.

Support Element

E-53. If appropriate, the sniper team might be attached to the raid patrol's support element to help provide long-range supporting fires.

Stay-Behind Element

E-54. It can also help cover the withdrawal of the assault force (raid patrol) to the rally point. When the element withdraws from the rally point, the sniper team may stay behind to delay and harass enemy counteraction or pursuit.

AMBUSH PATROLS

E-55. During ambushes, snipers are positioned in areas with observation and fields of fire on terrain features the enemy might use for cover after the ambush starts. The snipers' long-range capability allows them to position themselves away from the main body. Sniper fires are coordinated into the fire plan. Once the signal to initiate fires is given, snipers add their fires to those of the rest of the patrols. Snipers shoot leaders, radio operators, and crew-served weapons teams. If the enemy is mounted, the snipers try to kill the drivers of the lead and trail vehicles in order to block the road, prevent escape, and create confusion. Again, snipers may stay behind to cover the withdrawal of the ambush patrol.

Appendix F
Operations with Army Special Operations Forces

At the Infantry company level, operations with and near SOF are unique.

This edition discusses the integration of Infantry Company operations with Army SOF. It addresses types, organization, capabilities and limitations, and considerations for planning and execution.

Examples of SOF and Infantry integration and cooperation might include SOF precision attacks against an enemy target inside an Infantry unit's area of operation. It also might include an Infantry unit's capture of an SOF host nation informant. Both demand constant coordination and communication between Infantry and SOF to achieve the shared goal of defeating the enemy.

This appendix discusses the organization, capabilities, and limitations of the various SOF and, most importantly, considerations the Infantry company commander must take into account whenever SOF are present.

UNITED STATES SPECIAL OPERATIONS COMMAND

F-1. The United States Special Operations Command (USSOCOM) operators have specialized skills, equipment, and tactics. They are organized with regional focus to take advantage of language skills, political skills, and cultural sensitivity training. The totality of their full-spectrum, multi-mission force critical specialties include civil affairs, psychological operations (PSYOP), combat controllers, combat weathermen, pararescue Soldiers, Rangers, SEALS (sea, air, land), Special Forces (SF), and special operations aviation (both Air Force and Army).

UNITED STATES ARMY SPECIAL OPERATIONS COMMAND

F-2. The United States Army Special Operations Command (USASOC) is comprised of five types of Army special operations (ARSOF) units including Special Forces (SF), Rangers, special operations aviation (SOA), psychological operations (PSYOPS), and Civil Affairs (CA) units. To facilitate ARSOF/conventional coordination, integration, synchronization, or interoperability the controlling SOF headquarters will dispatch C2 or liaison teams. Such teams may also be used when Army conventional forces must operate with sister service SOF. Figure F-1, page F-2, lists ARSOF missions and collateral activities.

F-3. As part of the brigade combat team, the infantry battalion may conduct operations with or in support of SOF in the OE. Detailed planning and coordination is required at the brigade level. On today's noncontiguous battlefield, the battalion may find SOF forces operating in close proximity to its AO. To maximize their combined combat power, these forces must share an appreciation and understanding of each others mission, purpose, capabilities, and limitations.

Appendix F

Missions	Collateral Activities
• Unconventional Warfare (UW) • Foreign Internal Defense (FID) • Psychological Operations (PSYOP) • Civil Affairs (CA) • Information Operations (IO) • Direct Action (DA) • Special Reconnaissance (SR) • Combatting Terrorism (CBT) • Counterproliferation (CP) of Weapons of Mass Destruction (WMD)	• Coalition Support • Combat Search and REscue (CSAR) • Counterdrug (CD) Activities • Countermine (CM) Activities • Humanitarian Assistance (HA) • Security Assistance (SA) • Special Activities

Figure F-1. ARSOF missions and collateral activities.

SPECIAL FORCES

F-4. Special Forces (SF) operations are inherently joint and often controlled by higher echelons, with little involvement by the intermediate HQ. The basic building block of SF is the 12-Soldier SF operational detachment-A (SFODA), known as an ODA or A-team (Figure F-2).

CAPABILITIES

- Infiltrate and exfiltrate specified operational areas by air, land, or sea.
- Have foreign language ability and cultural training.
- Operate at all levels across the entire spectrum of military operations.
- Conduct operations in remote areas and non-permissive environments for extended periods with little external direction and support.
- Develop, organize, equip, train, advise, and direct indigenous military and paramilitary units or personnel.
- Train, advise, and assist allied and indigenous forces.
- Conduct reconnaissance, surveillance, and target acquisition.
- Conduct direct-action operations that include raids, ambushes, sniper, special munitions, and guidance for precision weapons.
- Conduct rescue and recovery operations.

LIMITATIONS

- Depend on the resources of the theater army to support and sustain operations.
- Cannot conduct conventional combined armed operations on a unilateral basis. Their abilities are limited to advising or directing indigenous military forces conducting this type of operation.
- Do not have organic combined arms capability. They habitually require the support or attachment of other combat, CS, and sustainment assets.
- Cannot provide security for operational bases without severely degrading operational and support capabilities.

Figure F-2. Special forces operational detachment A.

75TH RANGER REGIMENT

F-5. The Ranger Regiment is structured roughly along the lines of a conventional infantry brigade and as such has similar limitations. The Ranger force has a robust regiment headquarters that includes organic reconnaissance, signal, and military intelligence detachments, and three Ranger battalions. The force is organized, equipped, and trained to fight at the Ranger platoon, company, battalion, or regiment level but possesses the flexibility to provide tailored elements to joint special operations task forces (or other headquarters) and to employ other conventional or special operations forces placed under Ranger command and control. Table F-1, page F-4, shows Ranger capabilities.

Appendix F

- Plan and conduct joint special operations in conjunction with Army, Air Force, and Navy special operations forces.
- Conduct, or support, a forcible entry in conjunction with other joint special operations assets to establish lodgment for inserting follow-on forces deep in enemy or denied territory.
- Maintain a Ranger Force in an alert posture prepared for immediate deployment.
- When properly augmented, form an Army Special Operations Task Force (ARSOTF) headquarters or serve as the ground component of a joint task force (JTF).
- Provide liaison teams for up to three higher controlling headquarters. Each team is equipped and staffed to communicate with each command's deployed Ranger unit and integrate the Ranger units into the warfighting functions of the supported command.
- Employ cross-functional teams (CFT) to serve the Rangers as intermediate headquarters between company-size elements and battalion headquarters. The CFT is task organized according to METT-TC to perform numerous functions including: fuse intelligence and operations, perform liaison, conduct C2, synchronization with conventional forces, and the full targeting cycle.
- Employ sniper teams in support of tactical operations to increase force protection and minimize collateral damage with precision fires in limited visibility.
- Conduct special reconnaissance (SR) in support of Ranger operations.
- Conduct urban combat. Rangers are highly trained in urban combat and operate primarily at night, maximizing the advantages of state-of-the-art technology for night vision and target acquisition. They operate under very restrictive rules of engagement to minimize collateral damage and noncombatant casualties.
- Compress military decision-making process (MDMP) and ttroop-leading procedures (TLP). Rangers are capable of compressing the MDMP and TLP.
- Create an environment in which other special operations forces have freedom to operate.
- Conduct operations to safeguard and evacuate Americans, or protect property abroad.
- Provide a Ranger deployable planning team (RDPT) on short notice to any warfighting commanders, JTF, or JSOTF headquarters to plan potential Ranger operations in support of an emerging or ongoing contingency operation.
- Conduct terminal guidance operations against high-value targets, either in support of operations by a larger Ranger Force or as the primary Ranger mission in support of direct action conducted by other forces.
- Move small Ranger elements, small numbers of evacuees, resupply, or casualties through urban terrain in armor-protected vehicles.
- Operate exclusively in a digital environment maximizing situational awareness and understanding of every Ranger.
- Employ man-portable air defense weapons for force protection in forward support bases (FSB) and on targets.
- Operate in a chemical, biological, radiological, or nuclear (CBRN) contaminated environment in conjunction with other JSOTF forces.

Table F-1. Ranger force capabilities.

SPECIAL OPERATIONS AVIATION

F-6. The SOA rotary-wing aircraft include the AH/MH-6 Cayuse; the MH-60 Blackhawk; the MH-60 variant, known as the direct action penetrator (DAP); and the MH-47 Chinook. ARSOA units are designed to plan, conduct, and support SO missions unilaterally or jointly in all theaters and all levels of conflict. To accomplish this mission, ARSOA units are task organized according to the unit they will support, the theater of operations, and expected missions.

F-7. The AH/MH-6 Little Bird's immediate ancestors are the OH-6A light observation helicopters used during the Vietnam War. The AH-6 is an attack version, used in close-air ground support and direct action.

The MH-6 is a utility aircraft, used to insert or extract small combat teams. The MH-60 variants of the Black Hawk are utility aircraft typically equipped with aerial refueling capability, infrared suppressive exhausts, and other special operations-specific technology. The MH-47E Chinook is the 160th SOAR's long-distance, heavy-lift helicopter, which is equipped with aerial refueling capability, a fast-rope rappelling system and other upgrades or operations-specific equipment.

CAPABILITIES

- Plan and conduct air operations in all operational environments across the spectrum of conflict.
- Conduct SO as part of an Army special operations task force (ARSOTF) or joint special operations task force (JSOTF).
- Provides the commander a means to infiltrate, resupply, and exfiltrate Army special operations forces (ARSOF) engaged in all core missions and collateral activities.
- Prefer to operate at night. They use night vision goggles (NVG) or night vision systems (NVS) and low-level flight profiles.
- Can operate in all operational environments and terrain: desert, mountain, jungle, urban, and over water. Inherent in their training is the ability to operate from maritime platforms. Training emphasizes precise navigation over long-range and under adverse weather conditions.
- Aircraft are modified to add the capability for aerial refueling; they are modified to enhance precise navigation, secure communications, long-range flight performance, and increased weapons lethality.
- Are specifically trained to provide close air support (CAS) and terminal guidance for precision munitions and support of SOF.

LIMITATIONS

- Not equipped or manned to provide its own food service or water storage; requires food service 24 hours a day due to varied aircrew schedules.
- Cannot secure its aircraft or operating base; operates only from a secure base and airfield.
- Is not equipped or manned to effect its own integration into the airspace control system; requires support or augmentation for airspace deconfliction and tactical air support coordination.
- Cannot accept supply point distribution or to conduct moves; lacks the ground support assets necessary to accept supply point distribution or to conduct moves; to conduct unit moves, requires the unit distribution method of resupply and ground transportation support.
- Is not equipped to provide sufficient billeting for its personnel; requires climate-controlled facilities that must be compartmented and lighted to accommodate varied aircrew schedules.
- Is not equipped, manned, or apportioned to the theater in sufficient quantities to provide even its own aerial resupply or to conduct its own unit movement; requires GP aviation aerial resupply and aerial movement support.
- Requires stove-pipe requisition and distribution systems for resupply of ARSOA-peculiar Class II, V, and IX items; resupply of these items cannot be met through normal requisition and distribution systems.

CIVIL AFFAIRS

F-8. CA units establish, maintain, influence, or exploit relations between military forces and civil authorities (both government and non-government) and the civil populace in a friendly, neutral, or hostile AO to facilitate military operations and consolidate operational objectives. CA units are designed for employment independently, attached, OPCON, or tactical control (TACON) to other forces. The most commonly encountered element from a CA organization is Civil Affairs Team Alpha (CAT-A). The CAT-A is structured to meet the immediate needs of the host nation populace by executing civil-military operations in support of the overall plan. A civil affairs assessment team (CAAT) can also be sent down

Appendix F

from the Joint Special Operations Task Force (JSOTF) or the ARFOR (Army force) command element to make a determination of the needs within the brigade AO prior to, or in conjunction with, a CAT-A. At the platoon level, the typical relationship is one of providing security support in high threat areas to the CATs.

CAPABILITIES

F-9. The Civil Affairs (CA) company of the CA battalion (USAR and Active Army) can—.

- Provide the CATs (five of them) with tactical-level civil reconnaissance (CR). Be able to plan, execute, and transition CAO in all environments. Have communications capability that links directly into the supported unit's communications architecture.
- Provide the civil military operations center (CMOC) cell with tactical level planning, management, coordination, and synchronization of key civil-military operations (CMO) functions and activities within the supported commander's environment. Operate (over the horizon) away from the supported unit as required, serving as a "*standing capability*" by providing CMOC support to the BCT level HQ.
- A mechanism for civil-military coordination, collaboration, and communication.
- An initial entry and rapid deployment capability (Active Army).
- Limited functional specialty capability for initial assessment of the civil component of the operational environment, assess the mission planning requirements, and develop and coordinate the resources to meet immediate requirements to mitigate civil threats to the supported commander's mission (only RC USAR has functional specialists organic in the CA company CMOC).
- The CAT conducts Civil affairs operations (CAO) and provides CMO planning and assessment support to tactical maneuver commanders.

FUNCTIONS

F-10. The functions of the CAT are—.

- To conduct civil reconnaissance.
- To engage key leaders by constantly vetting contacts to identify elites within the CAT's AO.
- To plan, coordinate, and enable CAO and project management.
- To provide civil information to the supported unit and CMOC for inclusion of civil inputs to the supported commander's COP.

EMPLOYMENT

F-11. The CAT is deployed--

- To infiltrate rapidly by a variety of means, including static-line parachute (Active Army).
- To provide CMO staff augmentation and CA planning and assessment support to tactical maneuver commanders.
- To maintain direct data and voice communications with conventional, SOF, IPI, international organizations, NGOs, and interagency elements with classified and unclassified connectivity.
- To provide cross-cultural communications and limited linguistic support to supported commanders.
- To plan and support CMO conducted by military forces.
- To conduct liaison with civilian authorities and key leader engagement.
- To minimize interference between civil and military operations and synchronize CMO to enhance mission effectiveness.
- To conduct area studies and area assessments.

F-12. Civil affairs forces coordinate with military and civilian agencies. CA forces have extensive capabilities in all forms of communications; thus, requiring very little, if any, augmentation from the supported command. By table of organization and equipment, CA units are authorized the latest in conventional and Special Operations (SO) communications equipment and computers. This allows them to send secured and unsecured Internet communication, over-the-horizon (OTH) radios, satellite-capable radios, and laptop computers with Internet access. Also, CA units must be equipped with the current and most common civilian communications equipment to allow them to interface with international organizations, NGOs, and IPI in the AO. Specific requirements beyond these capabilities are determined during mission analysis and forwarded to the supported command as a statement of requirements (SORs).

LIMITATIONS

F-13. The small size of the CA usually requires security by infantry. Also, CA typically needs transportation due to their lack of internal transportation assets.

PSYCHOLOGICAL OPERATIONS

F-14. Tactical PSYOP teams (TPTs) normally provide PSYOP support at battalion level and below. They are the most common PSYOP elements that Infantry squads and platoons are likely to meet. When attached to a maneuver battalion, the TPT chief, most commonly an E-6, acts as the PSYOP staff advisor and planner to the battalion commander. The TPT is a nonlethal, fire support combat multiplier, best employed by the S-3. The TPT chief coordinates with the S-3 to employ the team to best support the Commanders overall objectives effectively. He also coordinates with the tactical PSYOP detachment for developing and producing PSYOP products to meet the battalion commander's requirements. At the discretion of the battalion commander, TPTs might be attached to platoons. In these instances, platoon leaders must have a clear understanding of the commander's intent to ensure the TPT is properly employed. It is the goal of PSYOP to influence foreign populations by expressing information subjectively to influence their attitudes and behavior. The TPT will advise the supported commander through the targeting process regarding psychological actions (PSYACTs), PSYOP-enabling actions, and targeting restrictions to be executed by the military force. The TPT will also provide public information to foreign populations to support humanitarian activities; serve as the supported military commander's voice to foreign populations to convey intent; and counter enemy propaganda, misinformation, and opposing information.

CAPABILITIES

F-15. PSYOP capabilities include--

- Providing PSYOP staff support from the battalion level to the GCC.
- Deploying globally with conventional and SOF.
- Disseminating products, conducting face-to-face communications with the targeted population, conducting loudspeaker broadcasts, such as surrender appeals, introduction of forces, harassment and deception, and provide linguistic and cultural expertise.
- Coordinating PSYOP support requirements with the supported commander or staff.

LIMITATIONS

F-16. PSYOP must rely on the supported unit to provide security for teams. Preapproved but unavailable PSYOP products might take 24 to 72 hours to obtain. For the TPT to provide the commander with communications with the foreign population, they will need an interpreter.

PLANNING CONSIDERATIONS

F-17. Combining the various forms of infantry with special operations elements is a combat multiplier. Such operations take advantage of the infantry unit's ability to operate in restricted and severely restricted terrain such as urban areas, forests, and mountains. Special operations forces (SOF) provide the units with force multipliers, especially in information operations, effects, and intelligence.

Appendix F

COORDINATION

F-18. When operating with or near SOF, the Infantry company commander should coordinate, at a minimum, the following with the SOF unit leader.

- C2 relationship.
- Communication information (frequencies, call signs, challenge and passwords, emergency signals and codes).
- Safehouse locations.
- Number and types of vehicles.
- Control measures being used.
- Battle handover criteria.
- Liaisons.
- Sustainment plans.
- Contingency plans for mutual support.

F-19. SOF may operate with the infantry or within the infantry AO as well as with infantry units conducting operations inside a JSOA. Physical contact between infantry units and SOF may range from short-term direct action operations to sustained combat operations. It is essential to conduct adequate coordination and integration to accomplish the specific mission. SOF have several elements to aid in coordination at the battalion level and above.

SPECIAL OPERATIONS COMMAND AND CONTROL ELEMENT

F-20. A special operations command and control element (SOCCE) might be used as an intermediate command element between the ODAs and the theatre SOC. The SOCCE typically will be embedded in conventional forces at brigade or higher levels and serves as a conduit to ensure special operations activities meet the needs of the SOC as well as the conventional force's campaign plan. A SOCCE consists of a Special Forces company headquarters with possible augmentation from ODAs to meet the SOCCE's support requirements.

CIVIL AFFAIRS PLANNING TEAM A

F-21. CA planning team A (CAPT-A) is a CA communications (CACOM) asset designed to provide responsive civil-military operations (CMO) staff augmentation of functional commands and corps-level or JTF-level commands. They are capable of conducting initial area assessments for CMO and providing recommendations for CA force structure to support the maneuver commander's CMO objectives.

RANGER DEPLOYABLE PLANNING TEAMS AND CROSS-FUNCTIONAL TEAMS

F-22. The 75th Ranger Regiment forms RDPT and CFT to function as described in Table F-1.

REQUEST FOR SUPPORT

F-23. Commanders can request direct support of SOF from the unified command's SOC. The SOC forms joint special operations task forces as required IAW the unified commander's guidance and operational needs. Based on operational needs and complexity of operations, conventional and SOF units may exchange liaison cells or, depending on proximity of headquarters and habitual relationships, commanders might be comfortable with daily coordination meetings.

Appendix G
Improvised Explosive Devices, Suicide Bombers, Unexploded Ordnance, and Mines

Improvised explosive devices (IEDs), mines, car bombs, unexploded ordnance (UXO), and suicide bombers pose deadly and pervasive threats to Soldiers and civilians in operational areas all over the world. Infantrymen at all levels must know about these hazards, and they must know how to identify, avoid, and react to them properly. Newly assigned leaders and Soldiers should read everything they can find on current local threats, and they should learn the unit's policies such as those found in the unit's standing operating procedures (SOP) and in locally produced Soldier handbooks and leader guidebooks.

This edition introduces discussions of Improvised Explosive Devices (IEDs), homicide bombers, Unexploded Ordnance (UXO) and mines. It incorporates tactical-level countermeasures learned from recent combat operations.

Section I. IMPROVISED EXPLOSIVE DEVICES

IEDs are nonstandard explosive devices used to target US Soldiers, civilians, NGOs, and government agencies. IEDs range from crude homemade explosives to extremely intricate remote-controlled devices. The devices are used to instill fear in US Soldiers, coalition forces, and the local civilian population, and to diminish US national resolve with mounting casualties. The sophistication and range of IEDs continue to increase as technology continues to improve and as terrorists gain experience.

TYPES

G-1. Some of the many types of IEDs follow.

TIMED EXPLOSIVE DEVICES

G-2. These can be detonated by remote control such as by the ring of a cell phone; by other electronic means; or by the combination of wire and either a power source or timed fuze (Figure G-1, page G-2).

IMPACT DETONATED DEVICES

G-3. These detonate after being dropped, thrown, or impacted in some manner.

VEHICLE BOMBS

G-4. These may include explosive-laden vehicles detonated with electronic command wire or wireless remote control, or with timed devices. They might be employed with or without drivers.

Appendix G

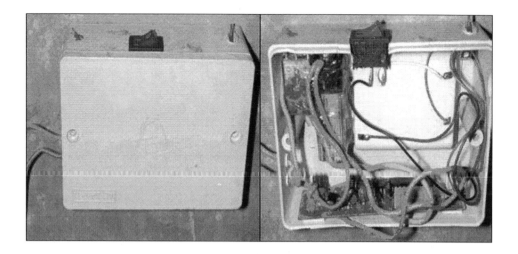

Figure G-1. Example of IED detonation device with explosive.

CHARACTERISTICS

G-5. Key identification features and indicators of suspected IEDs or the presence of IEDs include--

- Exposed wire, cord, or fuze protruding from an object that usually has no such attachment.
- An unusual smell, sound, or substance emanating from an object.
- An item that is oddly light or heavy for its size.
- An object that seems out of place in its surrounding.
- An object or area that the locals are obviously avoiding.
- An object used with written or verbal threats, or an object that is thrown at personnel or facilities, or both.

INGREDIENTS

G-6. Anything that can explode will be used to make IEDs, for example--

- Artillery rounds containing high explosives or white phosphorous.
- Any type of mine (antitank or antipersonnel).
- Plastic explosives such as C4 or newer.
- A powerful powdered explosive.
- Ammonium nitrate (fertilizer) combined with diesel fuel in a container. The truck bomb that destroyed the Oklahoma City Federal Building used ammonium nitrate and diesel fuel.

CAMOUFLAGE

G-7. An IED can vary from the size of a ballpoint pen to the size of a water heater. They are often contained in innocent-looking objects to camouflage their true purpose. The type of container used is limited only by the imagination of the terrorist. However, containers usually have a heavy metal casing to increase fragmentation. Figure G-2 shows some of the types of camouflage that have been used to hide IEDs in Iraq. Some of the more commonly used containers are--

- Lead, metal, and PVC pipes with end caps (most common type).
- Fire extinguishers.
- Propane tanks.

- Mail packaging.
- Wood and metal boxes.
- Papier-mâché or molded foam or plastic "*rocks*," which are containers that look like rocks, usually employed along desert roads and trails).
- Military ordnance, or rather modified military ordnance, which uses an improvised fuzing and firing system.

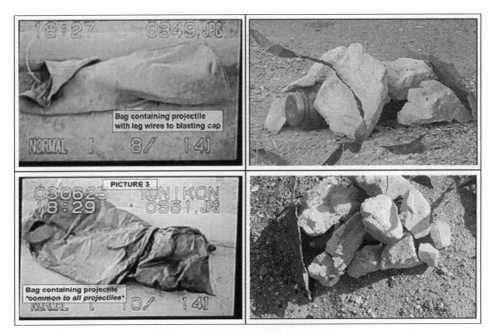

Figure G-2. Camouflaged UXO.

VEHICLE-BORNE DEVICES (CAR BOMBS)

G-8. Car bombs obviously use a vehicle to contain the device. The size of the device varies by the type of vehicle used—anywhere from a small sedan to a large cargo truck (Figure G-3, page G-4). Larger vehicles can carry more explosives, so they can cause more damage than smaller vehicles. Device functions, like package types, vary. Some of the signs of a possible car bomb include--

- A vehicle riding low, especially in the rear, and especially if the vehicle seems empty. Explosive charges can also be concealed in the panels of the vehicle to distribute the weight of the explosives better.
- Suspiciously large boxes, satchels, bags, or any other type of container in plain view such as on, under, or near the front seat in the driver's area of the vehicle. One sign is wires or rope-like material coming from the front of the vehicle and leading to the rear passenger or trunk area.
- A timer or switch in the front of a vehicle. The main charge is usually out of sight, and as previously stated, often in the rear of the vehicle.
- Unusual or very strong fuel-like odors.
- An absent or suspicious-behaving driver.

ATF	Vehicle Description	Maximum Explosives Capacity	Lethal Air Blast Range	Minimum Evacuation Distance	Falling Glass Hazard
	Compact Sedan	500 pounds 227 Kilos (In Trunk)	100 Feet 30 Meters	1,500 Feet 457 Meters	1,250 Feet 381 Meters
	Full Size Sedan	1,000 Pounds 455 Kilos (In Trunk)	125 Feet 38 Meters	1,750 Feet 534 Meters	1,750 Feet 534 Meters
	Passenger Van or Cargo Van	4,000 Pounds 1,818 Kilos	200 Feet 61 Meters	2,750 Feet 838 Meters	2,750 Feet 838 Meters
	Small Box Van (14 Ft. box)	10,000 Pounds 4,545 Kilos	300 Feet 91 Meters	3,750 Feet 1,143 Meters	3,750 Feet 1,143 Meters
	Box Van or Water/Fuel Truck	30,000 Pounds 13,636	450 Feet 137 Meters	6,500 feet 1,982 Meters	6,500 Feet 1,982 Meters
	Semi-Trailer	60,000 Pounds 27,273 Kilos	600 feet 183 Meters	7,000 Feet 2,134 Meters	7,000 Feet 2,134 Meters

Figure G-3. Vehicle IED capacities and danger zones.

EMPLOYMENT

G-9. IEDs have been used against the US military throughout its history. Operation Enduring Freedom (Afghanistan) and Iraqi Freedom (OIF) has seen the use of IED attacks on a significant scale targeting not only US, coalition, and Iraqi Security forces, but also civilian gatherings and concentrations as well. Some threat TTPs might include--

- An IED dropped into a vehicle from a bridge overpass. An enemy observer spots a vehicle and signals a partner on the overpass when to drop the IED. Uncovered soft-top vehicles are the main targets. These IEDs are triggered either by timers or by impact (Figure G-4).
- An IED used in the top-attack mode and attached to the bottom of a bridge or overpass. This IED is command-detonated as a vehicle passes under it. This method gets around the side and undercarriage armor used on US vehicles.
- An IED used with an ambush. Small arms, RPGs, and other direct-fire weapons supplement the IED, which initiates the ambush (Figure G-5, page G-6; and Figure G-6 and Figure G-7, page G-7). Terrorists sometimes use deception measures, such as dummy IEDs, to stop or slow vehicles in the real kill zone.
- The driver of a suicide or homicide vehicle, such as a taxicab, feigns a breakdown and detonates the vehicle when Soldiers approach to help. The vehicle with IEDs might also run a checkpoint and blow up next to it.
- Suicide bombers sometimes approach US forces or other targets and then self-detonate. Children might approach coalition forces wearing explosive vests.

Appendix G

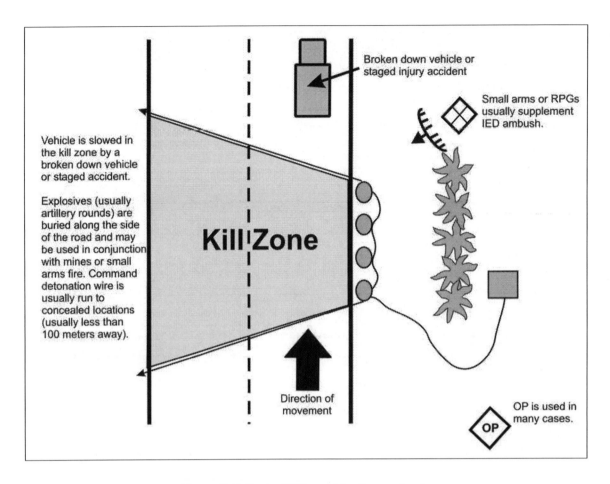

Figure G-5. Typical IED combination ambush.

Figure G-6. IED combination ambush in Iraq.

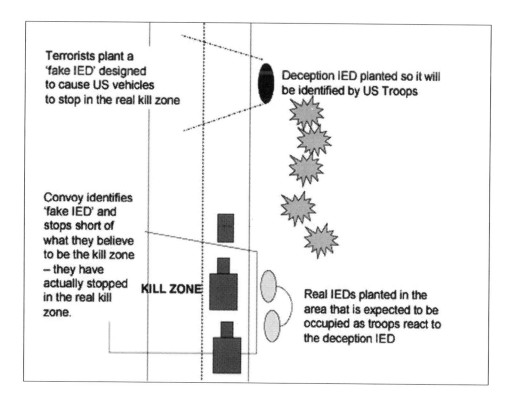

Figure G-7. Deception or fake IED used to stop convoy in kill zone.

COUNTERMEASURES

G-10. The enemy continues to adapt as friendly countermeasures evolve. The following are some measures used to counter an IED threat.

AVIATION SUPPORT

G-11. Operate with army aviation support when possible. Terrorists employing command-detonated IEDs generally rely on a quick escape after detonating an IED or executing an ambush. Recent trends have shown that OH-58D support deters attacks because terrorists are unable to break contact easily.

ALL-ROUND SECURITY

G-12. Remain alert. Maintain all-round security at all times. Scan rooftops and bridge overpasses for enemy activity.

CONVOY SECURITY

G-13. When possible, travel in large convoys. Vary road speed to disrupt the timing of command-detonated devices. However, terrorists often target convoys (or specific vehicles within convoys) with poor security postures. All occupants of convoy vehicles should have and keep their weapons pointed in an alert and defensive posture. Maintain a strong rear security element or a follow-on "*shadow*" trail security element. This force can more quickly be brought to bear on an enemy attacking the rear of a convoy. Use armed vehicles to speed ahead of a convoy to overwatch overpasses as the convoy passes. The lead vehicle in a convoy should have binoculars to scan the route ahead. All convoys should have with extra tow bars or towing straps to recover broken-down vehicles quickly.

ADAPTATION

G-14. Be aware of evolving enemy tactics and procedures and design countermeasures (Figure G-8, page G-11). To the maximum possible extent, avoid becoming predictable; vary routes, formations, speeds, and techniques.

TURNS

G-15. Avoid moving toward or stopping for an item in the roadway. Give wide clearance to items in the road. Turn to the outside of corners because terrorists will often plant IEDs on the insides of turns to close the distance to the target. Turning to the outside also allows a longer field of view past the turn.

AUDIBLE SIGNALS

G-16. At night, be aware of flares, gunfire lights going off, or horn honking, which can be used to signal the approach of a convoy.

ENEMY OBSERVERS

G-17. Be alert for people who seem overly interested in your convoy, especially those using cell phones while watching your convoy.

UNUSUAL SILENCE

G-18. Be aware of unusually quiet areas. Often, local civilians have been warned of an enemy attack on coalition forces.

USE OF HEADLIGHTS

G-19. Do not use service drive headlights during the day. Having lights on during daylight makes the military vehicles stand out and easier to identify at a greater distance.

VEHICLE PROTECTION

G-20. Harden all vehicles.

OTHER TRAVELING PRECAUTIONS

G-21. Do not stop for broken down civilian vehicles, vehicle accidents, or wounded civilians along a convoy route.

CIVILIAN VEHICLE THREATS

G-22. Be alert to civilian vehicles cutting in and out or ramming vehicles in a convoy as if attempting to disrupt, impede, or isolate the convoy. Current ROE might permit you to fire warning shots or to engage threatening vehicles.

FIVES C'S TECHNIQUE

G-23. Using the five C's (confirm, clear, call, cordon, control) technique helps to simplify both awareness and reaction to a suspected IED.

CONFIRM

G-24. The first step when encountering a suspected IED is to confirm that it is an IED. If Soldiers suspect an IED while performing 5- and 25-meter searches of their positions, they should act as if it could detonate at any moment, even if it turns out to be a false alarm. Using as few people as possible, troops should begin looking for telltale signs such as wires, protruding ordnance, or fleeing personnel.

CLEAR

G-25. If an IED is confirmed, the next step is to clear the area. The safe distance is determined by several factors: the tactical situation, avoidance of predictability, and movement several hundred meters away. Everyone within the danger zone should be evacuated. If more room is needed, such as when the IED is vehicle-born, Soldiers should clear a wider area and continuously direct people away. Only explosive ordnance disposal (EOD) personnel or their counterparts may approach the IED. While clearing, avoid following a pattern and look out for other IEDs. If you find any more, reposition to safety and notify a ranking member on the scene.

CALL

G-26. While the area around the IED is being cleared, a nine-line IED/UXO report should be called in. The report is much like the nine-line MEDEVAC report. It includes the necessary information for the unit's TOC to assess the situation and prepare an appropriate response.

CORDON

G-27. After the area has been cleared and the IED has been called in, Soldiers should establish fighting positions around the area to prevent vehicle and foot traffic from approaching the IED. They assure the area is safe by checking for secondary IEDs. They use *all* available cover. The entire perimeter of the effected area should be secured and dominated by all available personnel. Available obstacles should be used to block vehicle approach routes. Scan near and far for enemy observers who might try to detonate the IED. Insurgents often try to hide where they can watch their target area and detonate at the best moment. To deter attacks, randomly check the people leaving the area.

Appendix G

CONTROL

G-28. Since the distance of all personnel from the IED directly affects their safety, Soldiers should control the site to prevent people from straying too close until the IED is cleared. No one may leave the area until the EOD gives the "*all clear*." While controlling the site, assure all Soldiers know the contingency plans in case they come under attack by any means, including direct-fire small arms or RPGs, or indirect fires.

Section II. SUICIDE BOMBERS

These are different from all other terrorist threats, and require specific guidance on actions, particularly the interpretation of the ROE.

DEFINITION

G-29. A suicide attack is so called because it is an attack that means certain death for the attacker. The terrorist knows that success depends on his willingness to die. He conducts this kind of attack by detonating a worn, carried, or driven portable explosive charge. In essence, the attacker is himself a precision weapon. Suicide bombers aim to cause the maximum number of casualties, or to assassinate a particular target. Stopping an ongoing suicide attack is difficult. Even if security forces stop him before he reaches his intended target, he can still activate the charge and kill or injure those around him at the time. An additional benefit is the simplicity of such an attack. Neither escape nor extraction is an issue. Nor is intelligence, for no one will be left to interrogate. The only way to prepare for a suicide attack is to train Soldiers to react immediately to it with competence and confidence. They also train to avoid overreacting with unnecessary or inappropriate lethal force. The following are potential high-value targets for suicide bombers.

- High-signature forces such as uniformed military and security elements; military vehicles; civilian vehicles used for military purposes; military bases; checkpoints; patrols; liaison personnel; or supportive host nation personnel.
- Members and facilities of the international community such as ambassadors and other diplomats; embassy, UN, and NGO buildings; and diplomatic vehicles and staffs.
- National and provincial leaders and government officials.
- Civilians in public places such as markets, shops, and cafes. Although civilians in these locations are seldom primary targets, some groups do attack them.

DELIVERY METHODS

G-30. The two main methods of employing devices are by person or by vehicle.

- A person-borne suicide bomb usually has a high-explosive and fragmentary effect and uses a command-detonated firing system such as a switch or button the wearer activates by hand. A vest, belt, or other specially modified clothing can conceal explosives with fragmentation (Figure G-8, page G-11).
- A vehicle-borne suicide bomb uses the same methods and characteristics of other package or vehicle bombs, and is usually command detonated.

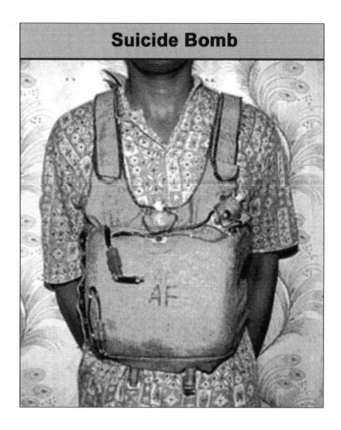

Figure G-8. Suicide bomber vest.

INDICATORS

G-31. Suicide bombers can be either gender and any age. For example, recent Palestinian bombers were female teenagers. However, you might be looking at a suicide bomber if you see someone who--

- Tries to blend in with the (target) environment.
- Wears ordinary, nondistinctive clothing, military or religious garb, or an oversized, bulky, or unseasonably heavy coat or jacket.
- Demonstrates fanatical religious beliefs by behaviors such as praying fervently, possibly loudly, in public.
- Has a shaved head (Muslim males); or wears their hair short and their face clean shaven; or wears fragrance, which is unusual for an Arab man.
- Behaves nervously, that is, sweats, or glances about anxiously.
- Has religious verses from the Quran written or drawn onto their body, hands, or arms.
- (Islamic males) dresses as and pretends to be a woman.
- Carries a bag tightly, clutched close to the body, and in some cases squeezes or strokes it.

SPECIAL CONSIDERATIONS

G-32. Consider the following when dealing with potential suicide bombers:

- Most will try to detonate the device if they believe they have been discovered.
- Suicide bombers are of any nationality, not necessarily of direct Middle Eastern descent. They may simply sympathize with the terrorist group's cause(s).

Appendix G

- If you determine that a suspect is a suicide bomber, then you will probably have to use deadly force. Prepare for and expect a detonation. Shoot from a protected position from as far away as possible.
- Many suicide bombers use pressure-release-type detonation devices that they hold in their hands. They apply the pressure before they begin their final approach to the target. The explosive payload will detonate as soon as the bomber relaxes his grip, so it will go off even if you kill him.
- Some bombers also have a command-detonated system attached to their bomb, and a second person observes and tracks him to the target. This also allows the terrorists to control and detonate the bomb, even if the bomber dies or his trigger is destroyed or disabled.
- The suicide bomber may also use a timed detonation system, and again this works whether or not you kill him before he reaches his target.

COMPLICATIONS

G-33. Dealing with a suicide bomber is one of the toughest situations a Soldier can face. In just a few seconds, he must identify the bomber, assess the situation, consider how to comply with the ROE, and act decisively. There is seldom time to think beyond that, or to wait for orders. The only possible way to stop the bomber short of his target is to immediately incapacitate him with lethal force. Challenging him would probably cause him to trigger his device at once. The suicide bomber is trained and prepared to carry out his mission. Some experts believe that a suicide bomber considers himself already dead when setting out on an attack. The Soldier and leader must continually be aware that--

- A pressure release switch can detonate the device as soon as the bomber is shot.
- A device could be operated by remote control or timer even after the bomber is incapacitated.
- Another person observe and command-detonate the bomb.
- A second suicide bomber might be operating as a backup or to attack the crowd and assistance forces that normally gather after a detonation.

Section III. UNEXPLODED ORDNANCE

UXO are made up of both enemy and friendly force ordnance that have failed to detonate. UXO sometimes pose no immediate threat, but they can cause injuries, loss of life, and damage to equipment if appropriate actions are not taken. UXO can be found on the battlefield, in urban areas, caves, and almost anywhere in an AO. UXO can be a result of a recent battle or war, or left over from past conflicts. During Operation Enduring Freedom (OEF), US Soldiers, coalition forces, and the local population were in danger of encountering an estimated 10,000,000 pieces of UXO and mines left over from 23 years of war in Afghanistan. Soldiers in Bosnia and Soldiers fighting in Operation Iraqi Freedom have been exposed to an estimated 8 million antipersonnel mines and 2 million antitank mines, as well as UXO. Soldiers can expect to encounter UXO in any future conflict.

RECOGNITION

G-34. Soldiers' knowledge of UXO is essential to help prevent the risk of injury. Soldiers are generally familiar with the appearance of ammunition and munitions used in their own weapons. They seldom recognize what the actual projectile looks like once it has been fired, especially if it is discolored or deformed by impact. Also, Soldiers might not be able to easily recognize UXO from USAF-delivered weapons or from non-US weapons. In general, leaders should caution their Soldiers against disturbing any unknown object on the battlefield.

G-35. FM 21-16 provides detailed *illus*trations and identifying characteristics of the four categories of UXO, including projected, thrown, placed, and dropped.

PROJECTED ORDNANCE

G-36. This includes--

- Projectiles such as HE, chemical, illumination, and submunitions.
- Mortar rounds such as HE, chemical, WP, and illumination.
- Rockets such as self-propelled projectiles, no standard shape.
- Guided missiles such as missiles with guidance systems.
- Rifle grenades similar to mortars but fired from rifles.

THROWN ORDNANCE

G-37. Thrown ordnance including fragmentation, smoke, illumination, chemical, and incendiary hand grenades.

PLACED ORDNANCE

G-38. This category includes--

- AP mines, generally small, of various shapes and sizes, and made of plastic, metal, or wood. Might have trip wires attached.
- AT mines, large, of various shapes and sizes, and made of plastic, metal, or wood. Might have antihandling devices.

DROPPED ORDNANCE

G-39. Dropped ordnance include--

- Bombs, small to very large, with metal casings, tail fins, lugs, and fuzes. May contain HE, chemicals, or other hazardous materials.
- Dispensers that look similar to bombs but may have holes or ports in them. Do not approach as sub-munitions might be scattered around.
- Very sensitive submunitions such as small bombs, grenades, or mines.

DANGER

DO NOT TRY TO TOUCH OR MOVE UXO. ORDNANCE FAILS FOR MANY REASONS, BUT ONCE FIRED OR THROWN, THE FUZING SYSTEM WILL LIKELY ACTIVATE. THIS MAKES THE ORDNANCE TOO UNSTABLE TO HANDLE. IF THE ROUND FAILED TO FUNCTION INITIALLY, ANY SUBSEQUENT STIMULUS OR MOVEMENT MIGHT SET IT OFF.

IMMEDIATE ACTION

G-40. Many areas, especially previous battlefields, might be littered with a wide variety of sensitive and deadly UXO. Soldiers need to follow these precautions on discovering a suspected UXO:

- Do not move toward the UXO. Some types of ordnance have magnetic or motion-sensitive fuzing.
- Never approach or pick up UXO even if identification is impossible from a distance. Observe the UXO with binoculars if available.

- Send a UXO report to higher HQ (Figure G-9). Use radios at least 100 meters away from the ordnance. Some UXO fuzes might be set off by radio transmissions.
- Mark the area with mine tape or other obvious material at a distance from the UXO to warn others of the danger. Proper markings will also help EOD personnel find the hazard in response to the UXO report.
- Evacuate the area while carefully scanning for other hazards.
- Take protective measures to reduce the hazard to personnel and equipment. Notify local people in the area.

1. *DTG:* Date and time UXO was discovered.
2. *Reporting Unit or Activity, and UXO Location:* Grid coordinates.
3. *Contact Method:* How EOD team can contact the reporting unit.
4. *Discovering Unit POC:* MSE, or DSN phone number and unit frequency or call sign.
5. *Type of UXO:* Dropped, projected, thrown, or placed, and number of items discovered.
6. *Hazards Caused by UXO:* Report the nature of perceived threats such as a possible chemical threat or a limitation of travel over key routes.
7. *Resources Threatened:* Report any equipment, facilities, or other assets threatened by the UXO.
8. *Impact on Mission:* Your current situation and how the UXO affects your status.
9. *Protective Measures:* Describe what you have done to protect personnel and equipment such as marking the area and informing local civilians.

Figure G-9. Nine-line UXO incident report.

BOOBY TRAPS

G-41. Booby traps typically are hidden or disguised explosive devices rigged on common items to go off unexpectedly (Figure G-10, page G-15). They may also be employed as antihandling devices on UXO, emplaced mines, or as improvised explosive devices (IED). Identify, mark, and report using the nine-line UXO incident report (Figure G-9). Field-expedient booby traps have also been employed with some success during most conflicts.

Improvised Explosive Devices, Suicide Bombers, Unexploded Ordnance, and Mines

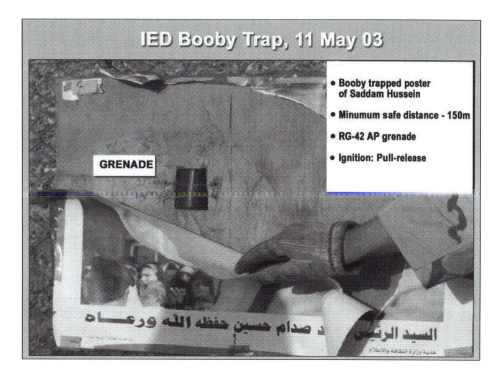

Figure G-10. Example booby trap.

Section IV. MINEFIELDS

Minefields are used by most adversaries and can be generally categorized into known or charted minefields and uncharted minefields. All minefields pose a danger to Soldiers.

TYPES

G-42. The two types of mines follow.

- Antipersonnel (AP) mines are generally small and of various shapes and sizes. They can be made of plastic, metal, or wood and may have trip wires attached.
- Antitank (AT) mines are larger than AP mines; of various shapes and sizes; and made of plastic, wood, or metal. They may have antihandling devices.

Note: Current US policy limits the use of non-self-destructing antipersonnel mines to defending the US and its allies from aggression across the Korean demilitarized zone.

STANDARD MINEFIELDS

G-43. Some of the many types of standard minefields used by our enemies and potential enemies follow.

- Protective minefields such as those used to impede and slow an attack, to provide for a counterattack, to channel an attacking force, and to provide early warning.
- Defensive minefields such as those typically used to prevent a penetration of defensive positions against armor or to reinforce the defense positions.

- Barrier, screening, restrictive, or obstructive minefields such as those emplaced to repel an attack from the flanks and to channel the attacker toward a selected kill zone.
- Harassment, disruptive, and nuisance minefields such as those used to delay, disrupt, or confuse an attacker.
- False or dummy minefields such as those used to deceive opposing troops into believing that a given area is mined.
- Scatterable minefields such as those normally delivered by rockets or other means. Scatterable minefields follow no pattern, and are usually unmarked.
- Air assault and airborne landing denial minefields such as those placed in potential air assault landing zones or in airborne drop zones to prevent seizure of the terrain.

MINEFIELD PATTERNS AND MARKINGS

G-44. Some enemy armies will emplace their standard minefields according to a standard pattern and mark them according to a standard method.

UNCHARTED MINEFIELDS

G-45. Many uncharted, randomly seeded minefields exist in the current AOs of units, and they will probably also be present in future US combat and stability operations. Soldiers may encounter antipersonnel and antitank mines of every type and origin. For example, in the northern regions of Iraq alone, 25 different types of mines have been documented, mostly in uncharted minefields.

MINE INDICATORS

G-46. When mines are properly laid and camouflaged, visual location is nearly impossible. However, three groups of possible visual indicators of minefields follow.

Man-Made Indicators

G-47. These indicate more typical use of conventional mine-marking systems or materials. The friendly sides of minefields have standard markings such as metal stakes 10 to 15 meters apart with two strands of barbed wire at a distance of four feet off the ground with rectangular metal signs. The enemy side of a minefield has a single strand of barbed wire attached to short metal stakes, not more than one foot off the ground.

Improvised Markings

G-48. Improvised markings are deliberate markings of mines using naturally available materials such as tree limbs or painted or unpainted rocks.

Natural Indicators

G-49. Over time, armed and buried mines become difficult to keep camouflaged. Some of the indicators include--

- Dead animals with missing or damaged limbs. Note that the animal may have walked several miles before dying.
- Mines surface laid or exposed by weather and soil erosion.
- Depressions in the ground (regular or odd spacing).
- Raised patches of earth (regular or odd spacing).
- Wilting or dead patches of vegetation.

- Available trees and bushes not collected for firewood.
- Overgrown fields and footpaths.

REPORTS

G-50. All mines and minefields should be marked, recorded, and reported to higher headquarters (Figure G-11). Avoid mines and minefields during unit operations. If you cannot, then take appropriate measures to reduce hazards to personnel and equipment.

Date: _____ From: _____ Thru: _____ To: _____ Reference SIR #. _____		
A. Incident DTG	A1. DD/time/zone/MM/YY	
B. Incident location	B1. Map sheet/UTM/grid reference (8 digit) B2. Location (road, field, building) B3. Emplacement (buried, surface laid, off route)	Include a site sketch as an attachment.
C. Effects (to complement information already in SIR)	C1. Casualties (name, rank, brief description of injuries) C2. Vehicle damage (number, type, extent of damage)	
D. Device suspected	D1. Type of mine (AT, AP, make, model) D2. Type of booby trap (pull, release, pressure) D3. Type of UXO (dropped, thrown, projected) D4. Unknown (detail, color shape, size)	
E. Circumstances	E1. Activity at the time of the incident E2. Degree of previous use of the route, area, location E3. Date of previous clearance and proofing by engineers E4. Where the route, area, or location is monitored	
F. Recommendations	F1. Recommendations to prevent reoccurrence	
G. Miscellaneous	G1. Any other pertinent data	

Figure G-11. Example format for a mine incident report.

EXTRACTION

G-51. Soldiers use different methods to extract from minefields, depending on the situation.

EXTRACTION WHEN NOT IN CONTACT

G-52. When you or your patrol is not in contact but you mistakenly enter a minefield, immediately stop and radio for assistance. Use mine detectors to clear a safe path out of the mined area. Mark and report the area. When help is unavailable and you must rely on your own resources, use the following procedures.

Note: In an emergency, you can use radios in and around a minefield. Some mines can be fuzed to detonate on a specific radio frequency, but these mines are rare and you are unlikely to encounter them. The potentially life-saving advantage of using a radio to call for help far outweighs the unlikely threat from radio frequency-sensitive mines.

Appendix G

EXTRACTION WHEN FOOTPRINTS CAN BE SEEN

G-53. Where you have detected a mine or tripwire, no casualties have occurred, and you do see footprints--

- Stop. Warn the rest of the unit.
- Call for help. If help is available, do not move.
- If no help is available and you can see footprints, then follow them out of the minefield.
- Take care to step exactly in the footprints already on the ground.
- Once out of the mined area, mark and report it.

EXTRACTION WHEN FOOTPRINTS CANNOT BE SEEN

G-54. Where you have detected a mine or tripwire, but no casualties have occurred, and footprints cannot be easily identified--

- Stop. Warn the rest of the unit.
- Call for help. If help is available, do not move.
- If no help is available and you see no footprints, then start from a standing position. Begin clearing an area immediately around you using the "*look, feel, probe*" procedures.

Look

G-55. Without moving your feet, look all around you, including looking forward and to the sides for tripwires. Then look closely around your feet for signs of fuzes, mine parts, disturbed ground, or slack trip wires.

Feel

G-56. Find or make a tripwire feeler from a 2- to 3-foot long straight wooden rod, stick, light gauge wire, or anything else that will allow you to feel a trip wire without engaging it. Use the feeler to check for tripwires to your front left, front right, and front center, from ground level up to head height. Move your fingers in a slow, sideways sweep around your feet to feel for exposed mine fuze prongs or other mine parts.

Probe

G-57. Using any rigid, sharp, long, thin instrument, probe the ground in a regular pattern at a 30-degree angle. Push in the probe at least 3 inches, with no more than 2 inches between probes, side to side. When you finish probing an 18-inch wide row, start a new row 2 inches farther along the direction of travel. Probe as gently as gentle as possible to achieve the desired depth. If the probe hits a solid object, investigate the obstacle to identify if it is a mine or other explosive device. Using a bayonet or other tool, dig down to the object's depth, and then slowly toward the side of the object. Dig in a side-to-side sweep rather than downward. If you find a mine, do not dig around it. Leave it. Tell the others, and then mark your cleared, 18-inch footpath to guide the Soldiers who follow. Once out of the mined area, mark and report it.

EXTRACTION OF A SINGLE CASUALTY

G-58. Resist the urge to race into a minefield to help a casualty. Doing so is extremely hazardous. However, to extract the casualty--

- Stop and warn the rest of the unit.
- Call for help.

Appendix H

Operations in a Chemical, Biological, Radiological, or Nuclear Environment

Chemical, biological, radiological, and nuclear (CBRN) weapons cause casualties, destroy or disable equipment, restrict the use of terrain, and disrupt operations. They are used separately or in combination to supplement conventional weapons. The Infantry company must be prepared to operate on a CBRN-contaminated battlefield without degradation of the platoon's overall effectiveness. This appendix prescribes active and passive protection measures to avoid or reduce the effects of CBRN weapons.

This edition updates discussions of Chemical, Biological, Radiological, and Nuclear (CBRN) defense operations. It introduces current concepts, terms, procedures, and equipment.

DEFENSE

H-1. Protection of the force requires adherence to four rules of CBRN defense.

- Contamination avoidance.
- Reconnaissance.
- Protection.
- Decontamination.

CONTAMINATION AVOIDANCE

H-2. Avoiding CBRN attacks and hazards is the first rule of CBRN defense. Avoidance allows commanders to shield Soldiers and units, thus shaping the battlefield. It involves both active and passive measures. Passive measures include training, camouflage, concealment, hardening of positions, and dispersion. Active measures include detection, reconnaissance, alarms and signals, warnings and reports, markings, and contamination control.

RECONNAISSANCE

H-3. CBRN reconnaissance is detecting, identifying, reporting, and marking CBRN hazards and consists of search, survey, surveillance, and sampling operations. Due to the limited availability of the M93 Fox reconnaissance vehicle, commanders should consider, as a minimum, the following actions when planning and preparing for this type reconnaissance.

- Use the IPB process to orient on CBRN threat NAIs.
- Pre-position reconnaissance assets to support requirements.
- Establish command and support relationships.
- Assess the time and distance factors for the conduct of CBRN reconnaissance.
- Report all information rapidly and accurately.
- Plan resupply activities to sustain CBRN reconnaissance operations.
- Determine possible locations for post-mission decontamination.
- Plan fire support.
- Enact fratricide prevention measures.

Appendix H

- Establish MEDEVAC procedures.
- Identify CBRNWRS procedures and frequencies.

PROTECTION

H-4. CBRN protection is an integral part of operations. Techniques that work for avoidance also work for protection, for example, shielding Soldiers and units and shaping the battlefield. Other protection activities involve sealing or hardening positions, protecting Soldiers, assuming MOPP (Table H-1), reacting to attack, and using collective protection. Individual protective items include the protective mask, battle dress overgarments (BDOs), green vinyl overboots, and gloves. The corps or higher level commander establishes the minimum level of protection. Subordinate units may increase this level as necessary, but they may not decrease it.

Equip	MOPP Ready	MOPP0	MOPP1	MOPP2	MOPP3	MOPP4	Mask Only
Mask	Carried	Carried	Carried	Carried	Worn	Worn	Worn***
BDO	Ready*	Avail**	Worn	Worn	Worn	Worn	
Overboots	Ready*	Avail**	Avail**	Worn	Worn	Worn	
Gloves	Ready*	Avail**	Avail**	Avail**	Avail**	Worn	
Helmet Cover	Ready*	Avail**	Avail**	Worn	Worn	Worn	

* Items available to Soldier within two hours with replacement available within six hours.

** Items must be positioned within arm's reach of the Soldier.

*** Never "*mask only*" if a nerve or blister agent has been used in the AO.

Table H-1. MOPP levels.

DECONTAMINATION

H-5. Using CBRN weapons creates unique residual hazards that may require decontamination. In addition to such weapons, collateral damage, natural disasters, and industrial emitters may require decontamination. Contamination forces units into protective equipment that degrades performance of individual and collective tasks. Decontamination restores combat power and reduces casualties that may result from exposure, thus allowing commanders to sustain combat operations.

Principles

H-6. Four principles of decontamination are used in planning decontamination operations.

- Decontaminate as soon as possible.
- Decontaminate only what is necessary.
- Decontaminate as far forward as possible (METT-TC dependent).
- Decontaminate by priority.

Levels

H-7. The three levels of decontamination are immediate, operational, and thorough (Table H-2). (See Appendix C, FM 3-11.4 for information on BDO risk assessment.)

Immediate

H-8. Immediate decontamination requires the least planning. It is a basic Soldier survival skill that Soldiers perform IAW STP 21-1-SMCT. A personal wipe-down removes contamination from individual equipment using the M291.

Operational

H-9. Operational decontamination involves MOPP gear exchange and vehicle spray-down. When a thorough decontamination cannot be performed, MOPP gear exchange should be performed within six hours of contamination.

Thorough

H-10. Thorough decontamination involves detailed troop decontamination (DTD) and detailed equipment decontamination (DED). Thorough decontamination is normally conducted by company-size elements as part of restoration or during breaks in combat operations. These operations require support from a chemical decontamination platoon and a water source or supply.

Level	Technique	Best Start Time	Responsibility	Advantages
Immediate	Skin decontamination	Within 1 minute of contamination	Individual	Prevents agents from penetrating*
	Personal wipe-down	Within 15 minutes	Individual or crew	
	Operator spray-down	Within 30 minutes		
Operational	MOPP gear exchange**	Best done within 6 hours, but must be done within 24 hours	Contaminated unit	Provides temporary relief from MOPP4. Limits agent spread
	Vehicle wash-down***			
Thorough	DED	When mission allows/ reconstitution	Platoon leader or senior unit leader	Reduces MOPP long-term with minimal risk
	DTD			
*	The techniques become less effective the longer they are delayed.			
**	Performance degradation and risk must be considered when exceeding 6 hours.			
***	Vehicle washdown is most effective if started within one hour.			

Table H-2. Comparison data for decontamination levels.

Planning Considerations

H-11. Leaders should include the following when planning for decontamination.

- Plan decontamination sites throughout the width and depth of the sector (identify water sources or supplies throughout the sector as well).
- Tie decontamination sites to the scheme of maneuver and templated CBRN strikes.
- Apply the principles of decontamination.
- Plan for contaminated routes.
- Plan for logistics and resupply of MOPP, mask parts, water, and decontamination supplies.
- Plan for medical concerns to include treatment and evacuation of contaminated casualties.
- Maintain site security.

Appendix H

CHEMICAL AGENTS

H-12. Chemical agents can cover large areas and might be delivered as a liquid, vapor, or aerosol. They can be disseminated by artillery, mortars, rockets, missiles, aircraft spray, bombs, land mines, and covert means. Table H-3 shows the symptoms and treatment of exposure to chemical agents.

Agent	Nerve	Blister	Blood	Choking
Protection	Mask and BDO	Mask and BDO	Mask	Mask
Detection	M8A1, M256A1, chemical agent monitor (CAM), M8 and M9 paper	M256A1, CAM, M8 and M9 paper	M256A1	Odor (freshly mowed hay)
Symptoms	Difficult breathing, drooling, nausea, vomiting, convulsions, and blurred vision	Burning eyes, stinging skin, irritated nose	Convulsions and coma	Coughing, nausea, choking, headache, and tight chest
Effects	Incapacitates	Blisters skin, damages respiratory tract	Incapacitates	Floods and damages lungs
First aid	Mark 1 nerve agent antidote kit (NAAK)	As for 2d and 3d degree burns	None	Keep warm and avoid movement
Decontamination	M291 and flush eyes with water	M291 and flush eyes with water	None	None

Table H-3. Characteristics of chemical agents.

H-13. Toxic industrial chemicals (TIC) and toxic industrial materials (TIM). Asymmetric warfare is expected to pose increasing risks and hazards to the Infantry force from TIC and TIM. The chemical corps continues to develop procedures that deal with these emerging threats in both foreign and domestic situations. TIM is a generic term for toxic radioactive compounds in solid, liquid, aerosol, or gas form. These may be used, or stored for use, for industrial, commercial, medical, military, or domestic purposes. Toxic industrial materials may be chemical, biological, or radioactive and described as toxic industrial chemical (TIC), toxic industrial biological (TIB), or toxic industrial radiological (TIR). Examples of TIC and TIM include--

- Fuels.
- Oils.
- Pesticides.
- Radiation sources.
- Fertilizers.
- Arsenic.
- Cyanide.
- Metals such as mercury and thallium.
- Phosgene.

TREATMENT OF CHEMICAL CASUALTIES

H-14. Survival in a chemically contaminated area requires Soldiers to perform protection and life-saving tasks. They must check for casualties, give first aid, identify the agent, send an NBC-1 or NBC-4 report, request permission to move, schedule decontamination operations, and mark the area to warn friendly Soldiers. A discussion of first aid under such conditions should include principles for the use of nerve agent antidotes. The nerve agent antidote kits, Mark I and CANA (convulsant antidote for nerve agent, Figure H-1, page H-7) are used by the Army and Air Force to treat nerve agent poisoning.

MARK I

H-15. The following principles apply to the administration of the Mark I (Figure H-1, page H-7).

Self-Aid

H-16. If you experience most or all of the *mild* symptoms of nerve agent poisoning, you should *immediately* hold your breath (do not inhale) and put on your protective mask. Then, administer one set of Mark I injections into your lateral thigh muscle (or buttocks) as in Figure H-2, page H-7; and Figure H-3, page H-8. (Procedures are listed on the autoinjector.)

H-17. Wait 10 to 15 minutes after the first set of injections to allow the antidote to take effect. If you can get around (ambulate) are able to ambulate, and know who and where you are, then you will *not* need a second set of Mark I injections.

> **WARNING**
> Giving yourself a second set of injections may create a nerve agent antidote overdose, which could result in incapacitation.

H-18. If symptoms of nerve agent poisoning are not relieved after administering one set of Mark I injections, seek someone else to check your symptoms. A buddy must administer the second and third sets of injections, if needed.

Buddy Aid

H-19. If you encounter a service member suffering from *severe* signs of nerve agent poisoning, render the following aid.

Step 1

H-20. Mask the casualty, if necessary. Do not fasten the hood.

Step 2

H-21. Administer, in rapid succession, three sets of the Mark I. Follow administration procedures outlined in the kit.

> **CAUTION**
> Use the casualty's own antidote autoinjectors when providing aid. Do not use your injectors on a casualty. If you do, you might not have any antidote available when needed for self-aid.

Appendix H

Combat Lifesaver

H-22. The CLS must check to verify if the individual has received three sets of the Mark I. If not, the CLS performs first aid as described for buddy aid. If the person has received the first three sets of Mark I, then the CLS may administer additional atropine injections at approximately 15 minute intervals until atropinization is achieved (that is, a heart rate above 90 beats per minute; reduced bronchial secretions; and reduced salivation). Administer additional atropine at intervals of 30 minutes to 4 hours to maintain atropinization or until the casualty is placed under the care of medical personnel. Check the heart rate by lifting the casualty's mask hood and feeling for a pulse at the carotid artery. Request medical assistance as soon as the tactical situation permits.

Combat Medic or Corpsman

H-23. A casualty has received three sets of Mark I; however, atropinization has not been achieved. Administer additional atropine at approximately 15 minute intervals until atropinization is achieved (that is a heart rate above 90 beats per minute; reduced bronchial secretions and reduced salivations). Administer additional atropine at intervals of 30 minutes to 4 hours to maintain atropinization or until the casualty is evacuated to an MTF. Check the heart rate by lifting the casualty's mask hood and feeling for a pulse at the carotid artery. Provide assisted ventilation for severely poisoned casualties, if equipment is available. Monitor the patient for development of heat stress.

CONVULSANT ANTIDOTE FOR NERVE AGENTS

H-24. The following principles apply to the administration of a convulsant antidote for nerve agent (CANA) (Figure H-1).

Self-Aid

H-25. The CANA is *not* for self-aid. If you know who and where you are and what you are doing, you do *not* need CANA. If symptoms fail to subside after self-administering one Mark I, seek assistance from a buddy.

Buddy Aid

H-26. In addition to administering the Mark I antidotes for nerve agents as buddy aid, also administer the CANA. Mask the casualty, if necessary. Do not fasten the hood. Administer the CANA with the third Mark I to prevent convulsions. *Do not* administer more than one CANA.

CAUTION

Use the casualty's CANA when providing aid. Do not use your own CANA. If you do, you might not have any antidote available when needed for self-aid.

Combat Lifesaver and Medic or Corpsman

H-27. The CLS, medic, or corpsman should administer additional CANA to casualties suffering convulsions.

- Administer a second and, if needed, a third CANA at 5 to 10 minute intervals up to three injections, which is a total of 30 mg of diazepam.
- Follow the steps and procedures described in buddy aid for administering the CANA. Do not give more than two additional injections for a total of three (one buddy aid plus two by CLS, medic, or corpsman).

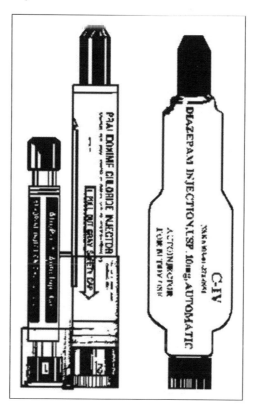

Figure H-1. Nerve agent antidote Mark I and CANA.

Appendix H

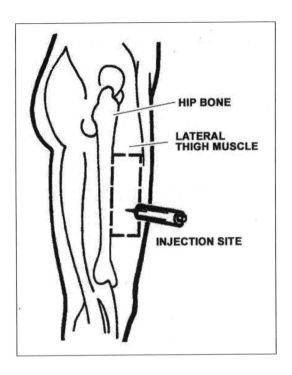

Figure H-2. Thigh injection site.

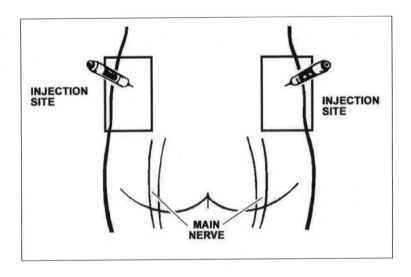

Figure H-3. Buttocks injection site.

BIOLOGICAL AGENTS

H-28. A biological agent is a microorganism that causes disease in personnel, plants, or animals or that causes the deterioration of material.

TYPES

H-29. Biological agents can be classified as toxins, pathogens, bioregulators, or prions.

Toxins

H-30. Toxins are poisonous substances produced from living organisms and--

- Can be synthesized (artificially produced).
- Mirrors the symptoms of nerve agents.
- Presents 8 to 12 hours of tactical concern before the sun destroys them.
- Can be fast acting (neurotoxins) or slower acting (cytotoxins).

Pathogens

H-31. These are infectious agents, such as bacteria, viruses, and rickets, which cause disease in man and animals. Pathogens have the following characteristics.

- Delayed reaction (incubation 1 to 21 days).
- Capability of multiplying and overcoming natural defenses.
- Vectors (disease-infected insects) that can get around, into, or through protective clothing and prolong hazards.

Bioregulators

H-32. Bioregulators include biochemical compounds that regulate cell processes and physiologically active compounds such as catalysts and enzymes. Although they can be found in the human body in small quantities, introduction of large quantities can cause severe adverse effects or death.

Prions

H-33. Prions are proteins that cause neurodegenerative diseases in humans and animals. Proteins have a unique, genetically defined amino acid sequence that determines their specific shapes and functions. When prions enter brain cells, they apparently convert normal proteins into prions. Ultimately, the infected brain cells die and release prions into the tissue, causing continuing cell destruction. Prions entered the public's consciousness during the mad cow epidemic that hit England in 1996.

PROTECTIVE MEASURES

H-34. Protective measures against biological attack include--

- Up-to-date immunizations.
- Good hygiene.
- Area sanitation.
- Physical conditioning.
- Water purification.

RADIOLOGICAL WEAPONS

H-35. A radiological dispersal device (RDD) is a conventional bomb, not a yield-producing nuclear device. An RDD disperses radioactive material to destroy, contaminate, and injure. An RDD can be almost any size.

TYPES

H-36. The types of RDDs follow.

Passive

H-37. A passive RDD is unshielded radioactive material that is dispersed or placed manually at the target.

Explosive ("*Dirty Bomb*")

H-38. An explosive RDD--often called "*dirty bomb*" is any system that uses the explosive force of detonation to disperse radioactive material. A simple explosive RDD consisting of a lead-shielded container--commonly called a "*pig*"--with a kilogram of explosive attached could easily fit into a backpack.

Atmospheric

H-39. An atmospheric RDD is any device that converts radioactive materials into a form that is easily transported by air currents.

EMPLOYMENT

H-40. Use of an RDD by terrorists could result in health, environmental, political, social, and economic effects. It would cause fear, injury and, possibly, levels of contamination requiring costly and time-consuming cleanup efforts. Hospitals, universities, factories, construction companies, and laboratories are possible sources for the materials needed to produce an RDD.

H-41. Friendly force avoidance and defensive measures are much the same as for other terrorist attacks, but prevention relies heavily on successful intelligence efforts.

H-42. Decontamination and casualty treatment are much the same as for exposure to the radiation in nuclear radiation.

NUCLEAR WEAPONS

H-43. Soldiers should know the effects of nuclear explosions and how to protect themselves from them.

EFFECTS

H-44. The following are effects of nuclear detonations.

Blast

H-45. High-pressure shock wave crushes structures and causes missiling damage.

Thermal Radiation

H-46. Intense heat and extremely bright light cause burns, temporary blindness, and dazzle.

Nuclear Radiation

H-47. Energy released from nuclear detonation produces fallout in the form of initial and residual radiation, both of which can cause casualties.

Electromagnetic Pulse

H-48. A surge of electrical power occurs within seconds of a nuclear detonation, damaging electrical components in equipment (radios, radar, computers, and vehicles) and weapons systems (TOW, Javelin).

Appendix I
Media Considerations

This appendix introduces the concepts of media considerations. Today, the media is present in most military operations and can immediately transmit what they see and hear. The powerful images and words they project can affect national policy. With our form of government, the media have the right to cover operations, and the public has a right to know what the media have to say. Unfortunately, many in the media lack a full understanding of the military, yet they are the key conduits of information about the Army to the public. Many good things about the Army are unknown to the public. Commanders and public affairs personnel have a responsibility to tell the Army story.

Freedom of the press does not negate the requirement for OPSEC and the accomplishment of the military mission. This appendix discusses how to deal appropriately with the media.

OBJECTIVE

I-1. In dealing with the media, the objective of all Infantry leaders is to ensure that operations are presented to the American public and audiences around the world in the proper context. Commanders and leaders can best achieve this goal by educating Soldiers and subordinate leaders about the positive aspects of a well-informed public.

REALITY

I-2. Today, keeping secret the fact of a large-scale military movement or operation is almost impossible. Our society as a whole understands and accepts freedom of the press and understands that different or competing viewpoints on issues are normal, indicating a healthy, open society.

I-3. News coverage for deployments is immediate and worldwide. The Army cannot and should not control--or try to manipulate--media messages or stories.

I-4. The media will go everywhere it can to uncover unique angles and stories. They try to gather their information firsthand. Some members of the media have not served in the military. Many do not understand military nuances. However, during Operation Iraqi Freedom, members of the media were embedded with combat units. This strategy succeeded: the reporters understood and reported combat better at the small unit level. They were able to see and experience first-hand what the Soldiers did. Embedding reporters will most likely be used again in future combat.

OBJECTIVES AND INTERESTS

I-5. The media wants access to Soldiers and units, which embedded reporters provide. The media accepts media pools if required. They seek fresh stories every day and expect daily authoritative briefings from operators and leaders. The press also wants the Soldiers' perspective. They want to discuss the ROE and issues related to them. Also, large on the press list are any US casualty figures, both actual and projected. These events should neither be emphasized or deemphasized. They are part of continuing military operations. Simply tell the media the truth, but focus only on events at your level and on known facts.

Appendix I

CAPABILITIES

I-6. With available technology, the media have the capability to collect and transmit images and sounds worldwide and live from any location. They can cover events quickly, and can influence the public either positively or negatively. With interest high in worldwide deployments, the media can send large numbers of reporters to cover operations in detail.

COMMAND CONSIDERATIONS

I-7. Commanders must anticipate, prepare, and respond within 12 to 24 hours to breaking events. Otherwise, it is difficult or impossible to explain or counter what has already appeared on the television or in print. When the released report is inaccurate, the commander should aggressively counter the false report with timely and accurate information backed up by subject matter experts. It is also important to coordinate statements among agencies. Bad news does not get better with time and ignoring the media does not make them go away but usually creates or fosters an adversarial relationship. If the commander refuses to talk to the media, he only guarantees the military's perspective will not be seen or heard. The commander must balance his time with the news media to avoid being overexposed or ignored.

GUIDELINES

I-8. The following are general guidelines for dealing with the media. These guidelines must be tempered with the public's right to know and the requirements of OPSEC.

SECURITY CONSIDERATIONS

I-9. It is important that all Soldiers understand what is classified and not discuss it with the press. Embedded media agreed prior to Operation Enduring Freedom not to report classified or current operations, and except for a very few exceptions, the media did not report OPSEC material. Soldiers should also understand that they are not required to talk to the media if it is against their wishes. Precautions should be taken to protect classified information from the news media that are not embedded. If someone accidentally reveals classified information, the reporter should be informed, the situation should be explained, and he or she should be asked not to report the information. All such incidents must be reported to the chain of command. All Soldiers represent the military; they should not guess or speculate on things they do not know. Anything said could be in the hands of the enemy in minutes. Grumbling or thoughtless complaining provide the enemy with propaganda to use against the military. The media must be prevented from televising nearby recognizable landmarks, sensitive equipment, or operational or classified information contained in the CP. The reasons for interfering with the telecast should be explained to the press.

MEDIA CONTROLS

I-10. Media in the AO should be checked to ensure that they are certified, and a military escort should escort them at all times for their safety. An interview should not be scheduled when it could interfere with the mission. Media material or equipment should not be confiscated, even in an effort to prevent the disclosure of classified information.

TOPICS AND STATEMENTS TO AVOID

I-11. Neither the commander nor any member of his command should discuss political or foreign policy matters. These are outside the direct purview of the military and would be purely speculative. No Soldier should discuss matters about which he does not have direct knowledge. Operational capabilities, including exact numbers or troop strengths, numbers or types of casualties, types of weapons systems, and plans, should not be discussed with the press.

INTERVIEWS

I-12. Any Soldier might be requested to grant an interview. This should be considered an opportunity to ensure that the needs of the media are met by providing accurate, timely, and useful information.

MAINTAIN A PROFESSIONAL ATTITUDE

I-13. Know exactly what your message is before you begin the interview. You should have a series of "*talking points*" that you are comfortable discussing. Keep your comments in line with these points. Remain in control even when the media are aggressive or ask silly questions. Be polite but firm. Be brief and concise; use simple language. Control the tempo of the questions if more than one reporter is present and asking questions. Do not use jargon or acronyms, the public does not know what they mean. Do not speculate and do not talk about issues or subjects that are not applicable to your position or level, for example, a rifleman should not talk about battalion or brigade issues. Instead, he should focus on his team and squad.

MAKE A GOOD IMPRESSION

I-14. Relax and be yourself. Ignore the cameras and talk directly to the reporter. Remove your sunglasses so the audience can see your eyes. Do not smoke or dip during the interview. Do not use profanity. Use appropriate posture and gestures.

THINK FIRST

I-15. Always stop and think before answering; questions need not be answered instantly. Answer only one question at a time. Do not be badgered or harassed. Correct answers are more important than deadlines. Do not get angry.

KNOW QUESTION

I-16. If you do not understand the question, ask the reporter to rephrase it. Know the question you are answering. Do not answer "*what if*" questions or render opinions. Reporters may ask the same question in different ways so stay consistent.

REMEMBER THAT EVERYTHING COUNTS

I-17. Everything is on the record. You may be friendly, but stick to business. The interviewer chooses the questions; but *you choose the answers*.

ASK YOURSELF--CAN MY ANSWER STAND ALONE?

I-18. Videotape and print media will not include the question, just your answer. Your answer should stand alone. If the interviewer uses a catch phrase, such as "*assassination squad*," carefully resist using it in your answer. For example, someone asks you, "*What are you doing about the assassination squads?*" A good answer might be, "*We are committed to investigating this matter. We will take the necessary and appropriate action.*" You might also include the original question in your answer to help keep your answers in context.

SPEAK ABOUT WHAT YOU KNOW

I-19. If you do not know the answer, simply say so, "*I do not know.*" That answer rarely appears in print. Avoid speculation or answering a question more appropriate for the Secretary of Defense. Talk about your area of expertise.

TELL TRUTH

I-20. Tell the truth. Never cover embarrassing events with a security classification. Never lie to the media.

TRAINING FOR MEDIA AWARENESS

I-21. Units should train for media awareness in two parts, first in a classroom, then in the field.

CLASSROOM PHASE

I-22. OPSEC should be covered thoroughly. Many of the things outlined in this appendix should be discussed with Soldiers and leaders. If a media card is available in the command, it should be explained in detail. Soldiers should be instructed on how to give an interview and their right to refuse to do so. Leaders should understand their responsibility to tell the Army's story truthfully, and in a way that the public will understand it.

FIELD PHASE

I-23. Soldiers should be given an opportunity to participate in an interview using Soldiers who role-play as reporters. If possible, the role-playing Soldiers should be qualified in public affairs training. This training should be included in regular field training exercises. If a video camera is used during the interview, the tape can be replayed during an AAR. Due to the possible far reaching effects of interviews, this training should receive considerable command emphasis.

MEDIA CARDS

I-24. If higher headquarters has not developed a media card, the commander should ask the PAO to do so. If they do not or cannot, the commander should consider making one for the company. A media card should state--

- Who Soldiers should contact and how to contact them if reporters show up in the area unannounced.
- Responsibilities of a media escort.
- What information you may or may not discuss.
- When to allow a media interview.
- How to treat reporters.
- How to conduct an interview.
- The best techniques to use in telling the Army's story.

Appendix J
Pattern Analysis and Situational Understanding

This chapter introduces concepts of pattern analysis and situational understanding to assist the commander in gaining rapid tactical insight and enable planning in a diminished time frame.

Developing valid and relevant situational understanding (SU) of the operational environment (OE) requires the personal attention and effort of each individual commander.

Pattern analysis and information-gathering operations enhance and contribute to the commander's force-protection program, SU, and battlefield vision. They portray relevant environmental and threat information that could affect the commander's operational and tactical environment.

For the commander to develop a relevant and timely understanding of the OE, he must logically and diligently collect, organize, and study all the information he can, from all sources, about all facets of the OE. This includes, among other things, information about friendly, enemy, and noncombatant forces; about cultural and ethnic considerations; about short and long term goals; and about IO campaigns.

GATHERING OF INFORMATION

J-1. The first steps to developing SU of the operational environment are information gathering and pattern analysis. For the commander to assume an offensive orientation rather than a reactionary one, he must have developed enough SU to perform accurate pattern and predictive analysis. This is likely to be a relatively long term and potentially time consuming process, but it will yield actionable information that will allow the commander to take the fight to the enemy and substantially increase the commander's ability to protect his force. Information gathering occurs continuously during the conduct of day-to-day operations. This information, which might be police, criminal, or combat, is provided as input to the intelligence collection effort and turned into action or reports. The information collected during the conduct of operations is analyzed at the commander's level and reported up through the proper channels so that it can be analyzed by higher headquarters. Every Soldier in the unit should know what information is required and how to report it. Every member of every patrol or mission should be specifically tasked to collect information and each member should be debriefed after every mission accordingly. Information is gathered while conducting combat operations and other operations such as--

- Patrolling. Every Soldier in every patrol should be tasked to collect information.
- Checkpoints and roadblocks.
- Traffic control points.
- Field interviews.
- Reconnaissance (zone, area, and route).
- Cordon and search.
- Soldier and leader observations of the daily operational environment.
- Local and international media sources such as newspapers, flyers, radio, TV, gossip, and graffiti.

Appendix J

SOURCES

J-2. Collection can result primarily in combat information, but may lend to police or criminal information. Units collect information throughout the entire AO. Soldiers and leaders gather information from contacts that are often very valuable in substantiating or verifying other sources of information. These sources include--

- Daily contact with the local populace.
- Combined patrols with HN police, military police, and civilian police agencies.
- Close liaison with local, HN, and multinational police agencies.
- Field interviews.
- Nongovernmental organizations (NGOs).
- Private volunteer organizations (PVOs).

ASSESSMENT

J-3. Pattern analysis and the information assessment process (IAP) are tools used to contribute to SU. Information gained through the IAP may contribute independently or simultaneously to the all source analysis product (ASAP) and the IPB process. Pattern analysis, the IAP, the ASAP, and the IPB enhance and support the commander's force protection program, SU, and battlefield vision. The IAP independently or collectively--

- Provides the commander with information necessary to improve measures to protect the forces.
- Provides information that clarifies the threat and operational situation.
- Reduces opportunities for threat forces to disrupt military operations and inflict friendly casualties.

OPERATIONAL VARIABLES

J-4. Commanders continually monitor their operational environment at the tactical level consistent with METT-TC. They apply the military aspects of terrain (OAKOC) as a means of protecting the force. Commanders will also find it useful to use the operational environment variables as a method to analyze information. Information is used to clarify the evolving operational, tactical, and criminal threat picture for commanders through pattern analysis and the IAP. This helps planners predict threat COAs against our forces or protected populations. Operational environment variables include--

- The physical environment (status of or changes in).
- Nature and stability of the enemy state, sponsor, and population.
- Military capabilities of the enemy state or sponsor.
- Technology available to US, enemy forces, and noncombatant population.
- Information warfare/PSYOPs considerations.
- External organizations (government and nongovernment).
- Sociological demographics and population considerations.
- Regional and global political and strategic relationships.
- US and enemy national will.
- Time.
- Economics.

COLLECTION

J-5. Units gather information actively or passively. Active collecting efforts result from a direct tasking, while passive collecting efforts result from normal, daily operations.

Pattern Analysis and Situational Understanding

ACTIVE MODE

J-6. Units gather information in the active mode when the commander directs or when directed by the higher headquarters. In this mode, the unit conducts specific missions with the intent to actively collect information. Specific activities, such as setting up a checkpoint or roadblock, are performed to fulfill a specific requirement. Examples of such requirements include looking for people who have weapons or military property, or who know others who do. Valuable information is gathered during these operations. The leader determines the scope of the IAP by--

- Reviewing pattern analysis and trends developed to that point.
- Conducting a detailed mission analysis.
- Reviewing the mission of the higher headquarters and the commander's intent.
- Reviewing the CCIR (PIR and FFIR).
- Reviewing the EEFI.
- Reviewing mission priorities.
- Determining the AO and AI.
- Reviewing the IPB estimates of the higher headquarters.
- Determining the required information products.

J-7. Assemble the working aids and--

- Post the applicable maps with all applicable operational overlays.
- Post population and other pertinent demographic maps or overlays.
- Acquire crime statistics and other related data.
- Obtain language aids such as cultural references, and arrange for interpreters.
- Acquire the necessary automation equipment.

J-8. Recommend and supervise police, criminal, and combat collection efforts by--

- Coordinating with the S-2, the G-2, the SJA, the CID, adjacent or supporting Joint or SOF unit, and other applicable agencies before initiating a collection effort.
- Determining the criteria to satisfy information requirements.
- Providing collectors with reporting instructions such as the reporting frequency and the report format.
- Monitoring information collection efforts to prevent duplication of efforts.
- Determining the best methods (memory, photography, digital) to physically record the information.

J-9. Process police, criminal, and combat raw data by--

- Assembling and assessing the reliability of the data.
- Integrating information from the collectors (other organizations collecting information in the field).
- Conducting pattern analysis (developing trends and indicators).
- Always conducting immediate AARs of the collection effort.

J-10. Report and disseminate information assessment by--

- Recommending actions to improve and focus future collection efforts.
- Reporting information assessments to the S-2 or the G-2.
- Disseminating information to the unit.
- Modifying or improving the unit's force protection status and actions based on the information.
- Reviewing and ensuring that the release of information assessments do not violate established guidelines and constraints.

J-11. Standardized checklists enhance the collection effort and aid in the analysis of the information collected. They also give Soldiers a means by which to organize their thoughts and personal collection efforts and aid in debriefings. The checklist indicates a pattern in the behavior of the local population or enemy forces. It shows, among other things, what the local nationals are transporting, and where they are transporting the items. The checklist may include--

- The number and types of vehicles stopped. Identifying marks, license plate numbers, and any signs displayed on the vehicles are recorded and reported.
- The number of passengers in the vehicle. The nationality, age, and gender mix of the passengers are recorded and reported.
- The type and quantity of cargo.
- The vehicle's point of origin and destination.
- The stated reason for travel by the passengers.
- The description of arms, ammunition, explosives, communications equipment, and sensitive items found, observed, or confiscated from the vehicle.
- The possible or actual sightings of weapons, explosives, or threat forces by the passengers.
- The condition of the passengers.
- Apparent or perceived nationality or ethnicity/clan of passengers.
- Reports of anything unusual by the passengers.
- Location and time of incident, attack, or ambush.
- A change or trend noticed in apparent enemy tactics, weapons, or composure.

J-12. Digitally photographing or recording the collection effort and information acquired will allow the commander and higher headquarters to more efficiently and effectively study and disseminate the information. Digitally capturing information also allows the rapid sharing of information via electronic means. Commanders should also consider using recording or filming methods to monitor relatively stationary activities remotely. For example, they could set up cameras in a parking garage. Information recorded this way can prove useful in studying or recreating events.

PASSIVE MODE

J-13. Every Soldier gathers information in the passive mode during his normal day-to-day operations. In the passive mode, information gathering is not a stand-alone function and as such, it cannot be separated from other operations. If while conducting operations, a Soldier receives, observes, or encounters police, criminal, or combat information, he immediately submits a SALUTE, SPOTREP, or other appropriate report to relay information up the chain of command. Soldiers should report any suspicious activity without trying to distinguish whether the information is police, criminal, or combat in nature. The information is then integrated into the ongoing IAP and pattern analysis and forwarded to the higher echelon S-2 and G-2 for IPB applications.

RECORDS

J-14. The commander should construct a time pattern analysis worksheet to record the date and time of each relevant event, collection, or serious incident in his AO or AI. In these examples (Figure J-1 and Figure J-2), the rings show the days of the month, the segments show the hours of the day. (*See also* Chapter 3, FM 34-130.) Similar tools help distinguish patterns in activity that are tied to particular days, dates, or times. The commander can adopt this format or modify as he sees fit. It will likely be useful to make the pattern analysis wheel as large as possible and post it in an area where subordinate leaders and Soldiers can study it while maintaining OPSEC. The commander also maintains maps of the AO showing the location of events linked to pattern analysis.

J-15. As the unit begins to collect information and the commander receives additional information from higher and adjacent units, he begins to post or display this information on the pattern analysis wheel and on maps of the AO/AI. Using this displayed information; intuition and experience of his subordinates, local authorities, or HN personnel; and historical examples, the commander can begin to identify trends, patterns, and specific AIs or danger within the assigned AO/AI.

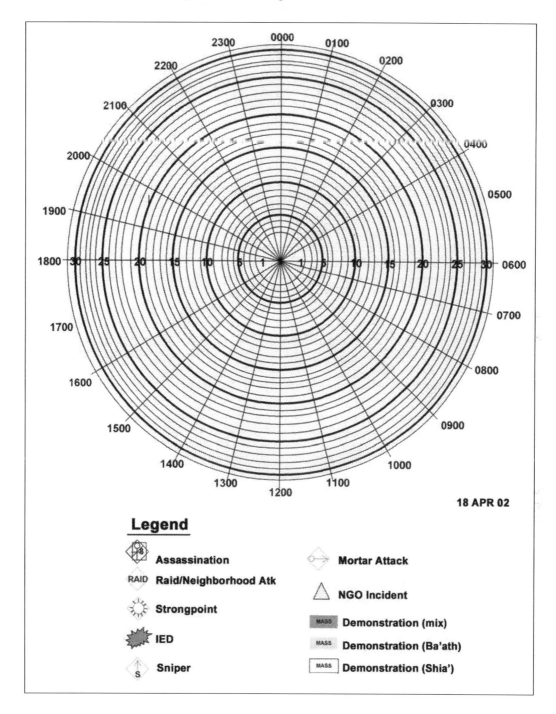

Figure J-1. Example 1, pattern analysis.

Appendix J

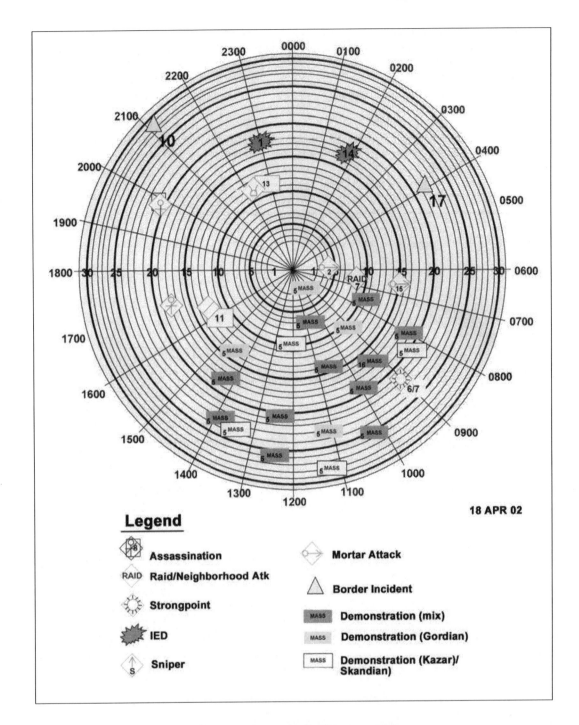

Figure J-2. Example 2, pattern analysis.

POPULATION CONSIDERATIONS

J-16. The commander will probably determine that the local population is one of the most important characteristics of the operational environment/battlefield, and will likely have a major influence on both friendly and enemy COAs. Accordingly, he should prepare one or more population status overlays showing their political, ethnic, tribal, or racial sympathies and affiliations. Figure J-3 (page J-7) and Figure J-4 (page J-8) show examples; the commander should adopt whatever format best suits his needs.

Traditionally, the larger the scale of the overlay, the greater its usefulness, and the greater the detail. He will use the overlay later to analyze patterns, to determine enemy trends and likely COAs, and to determine areas inside his AO or AI that warrant increased attention. The nature of the operational environment and urban operations requires the commander to include factors other than OAKOC into the definition of the battlefield environment. The politics and ethnic considerations of the urban area and its populace will have a considerable impact on the ease with which US forces can conduct operations. The interactions between rival forces, tribes, clans, or religious groups and their interaction with US forces, coalition forces, HN forces, and other GOs and NGOs are also critical aspects of the definition of the operational environment. Also, the commander must consider the impact of higher headquarters or competing parties' IO and PSYOPs campaigns or the missions and activities of any friendly SOF that operates within the AO/AI.

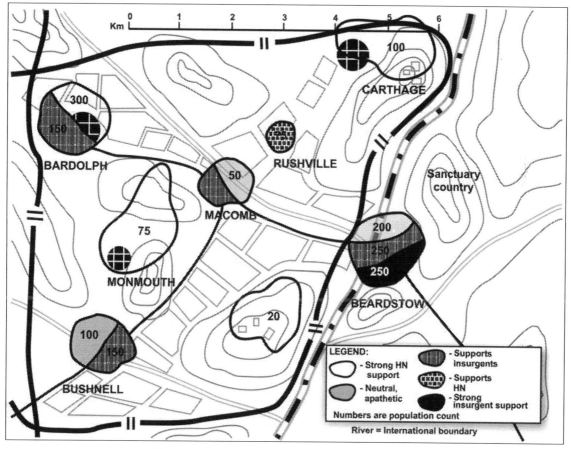

Figure J-3. Example 1, population status overlay.

Appendix J

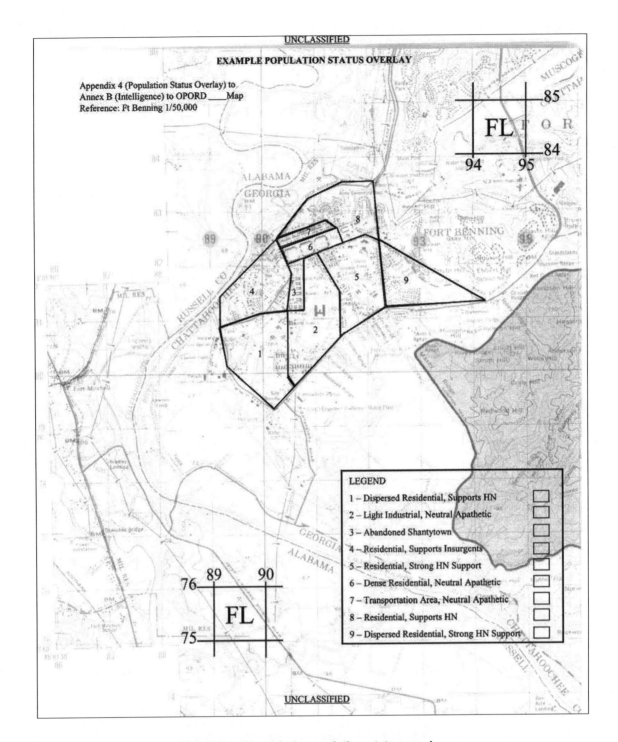

Figure J-4. Example 2, population status overlay.

RESPONSIBILITIES

J-17.   The commander has the overall responsibility of supervising the collecting, analysis, and reporting of information to the higher headquarters. The commander should report both the raw information and his own analysis conclusions to higher headquarters. Training scenarios can be developed to enhance the information collecting process of both the active and passive mode. Also, the scenarios

Pattern Analysis and Situational Understanding

stress collective training such as processing police information. They also stress individual training such as improving interpersonal communication and interview skills.

PLATOON LEADER

J-18. When the platoon leader is tasked to conduct collecting and reporting, he initiates TLP for the mission. The platoon leader also--

- Participates in the actual pattern analysis conducted by the commander and draws on his own observations/experiences to help identify patterns or trends.
- Coordinates with the higher headquarters for the CCIR, police and criminal information requirements, and threat estimates.
- Establishes liaison with civil and MP forces and law enforcement agencies in the AO.
- Coordinates with the local and HN police to determine the existence of organized crime in the AO and the identification of current and emerging criminal leaders and associates.
- Coordinates with the local and HN police and the populace to identify the types of criminal activity, such as smuggling, counterfeiting, trafficking in narcotics, or engaging in extortion in the AO.
- Reports information of potential intelligence value by participating in the debriefing process of his Soldiers, gathering collected information from squads and teams, and consolidating original reports and sketches of potential intelligence value and forwarding them through the company headquarters for analysis.

PLATOON MEMBERS

J-19. Soldiers may observe more relevant information than the combined technical intelligence can collect. This realization lead to the Army formalizing the *Every Soldier as a Sensor* (ES2) concept. Soldiers are exposed to information that would be of significant value if collected, processed, and integrated into a common operating picture. Therefore, all platoon members from rifleman to platoon leader collect and report information that supports the commander's collection plan. They accomplish this by--

- Continually reviewing the commander's information needs and any checklists or information distributed by the company or higher headquarters.
- Noting terrain information pertaining to streets, roads, canals, subterranean systems, built-up areas, cities and villages, and the impacts of weather on the terrain.
- Collecting information on pro-government and anti-government individuals and groups who might cause disruptions during protests, strikes, riots, and other spontaneous or organized efforts.
- Identifying private establishments that might be a target or whose presence or operations contribute to the disruption. Examples of these establishments include gun shops, pawnshops, religious sites, culturally sensitive areas, liquor stores, and pornography stores.
- Identifying critical infrastructures such as power stations, water works, radio and television stations, telephone and communication facilities, public transportation, and other establishments that might be critical to the sustainment of the community.

J-20. Platoon members identify EPWs, stragglers, and others who may have information of potential intelligence value and reporting it to the chain of command.

J-21. Platoon members use a SPOTREP, SITREP, or SALUTE report or a format directed by the chain of command to report information. SOPs and checklists may also be developed and used.

Appendix K
Motorized Operations

Throughout history, success in battle has gone to combat leaders who can build and effectively fight combat organizations with the right blend of mobility, firepower, and protection. The Infantry company commander might have to conduct motorized operations in various environments to increase his company's tactical or operational mobility, or to increase his firepower or protection.

This edition introduces discussions on motorized operations to address planning considerations, and urban motorized patrolling.

This appendix introduces and provides basic information for the Infantry company commander about motorized operations. This includes considerations in employing various wheeled vehicles.

This appendix also provides a selection of TTPs from current operations and emerging threat trends. Depending on his METT-TC analysis, the commander may adapt these TTP to suit the tactical requirements of the company.

Section I. WHEELED VEHICLE PLANNING CONSIDERATIONS

Two major considerations for employing light wheeled vehicles during tactical operations are, first, the risks associated with the reduced protection and firepower of more heavily armored and "*up-gunned*" systems; and, second, the benefits of vehicle speed and maneuverability, crew observation and agility, and weapon engagement reaction times. In balancing these risks and benefits, the commander thoroughly evaluates the factors of METT-TC. Company leadership must know the latest developments in threat tactics against wheeled patrols and convoys. Other sources to consider for TTP include the unit TSOP and combined arms lessons learned (CALL) publications. This section addresses several considerations for balancing offensive and defensive vehicle characteristics. These considerations are based on historical and current observations from both US and foreign military engagements, and on both OEF and OIF.

- All-round security and observation--coupled with an aggressive crew and passenger security and weapons posture-is normally the biggest enemy deterrent.
- With regard to weight and thickness, protection against 5.56-mm and 7.62-mm rounds is easier to achieve than protection against RPGs. Even the M2 BFV and Stryker are not always fully RPG proof.
- Vehicle suspension must be enhanced to avoid deterioration and failure under the weight of additional armor. Vehicle passenger or material loads are greatly reduced. For each vehicle, commanders should identify the types and capabilities of the suspensions for each vehicle.
- Trends indicate that vehicles without all-round security are individually targeted for attack. The enemy can even target individual vehicles with poor security from within larger convoys.
- Most vehicle passenger seating configurations place passengers in a seated position with their backs to the enemy--modifying vehicle seats, both cargo area and rear passenger, so that they mount in the center of the vehicle and allow Soldiers to face outward.

MAINTENANCE

K-1. Vehicle maintenance is critical to mission accomplishment. Poorly maintained vehicles will fail. As they do so, the OPTEMPO for functional vehicles increases dramatically. Potential adverse effects can result:

- Reduced cargo- or personnel-carrying capacity, which reduces the combat power ratio in the field.
- Crews who perform proper maintenance suffer greater exposure to enemy fire than do crews who do not.
- Reciprocal maintenance problems for overused vehicles.

VEHICLE WEIGHT, SURVIVABILITY, AND ARMOR

K-2. Force protection is irrevocably linked to mission success. Consequently, it must always be an important consideration in the planning and execution of missions that employ a soft motorized force.

K-3. Again, the balance between the protection of vehicles and crews, observation, and the employment of weapons is critical. Normally, heavily armored vehicles, especially wheeled vehicles with extra armor such as the up-armored HMMWV, severely limit crew and passenger observation in restrictive and urban terrain. They can also limit weapons employment at close ranges. Both RPGs and IEDs can defeat many armored vehicles, and will likely defeat any wheeled vehicle at the point of detonation, with or without an armor package. At times, insurgent and terroristic enemy forces target vehicles with poor security, because they seem easier to destroy and less likely to respond effectively. Commanders must analyze enemy trends and events in their AOs before deciding on the appropriate levels of armor versus offensive capabilities, mission demands, and crew survivability. Other considerations might include--

- Can the vehicle suspension support additional armor and still carry the payload?
- Can the vehicle crew and passengers provide all-round security for themselves?
- Can the vehicle crew secure itself if the passengers dismount?
- Will additional armor affect vehicle mobility over rough terrain or in restrictive urban areas?
- Can the vehicle crew and passengers quickly and safely mount or dismount? Can they do so under fire?
- If the vehicle has a turret-mounted weapon system, does the gunner have enough protection? Consider his legs and lower body, which will be exposed through the middle of the vehicle as well, and which are not covered by body armor. (Figure K-1A and Figure K-1B show examples of vehicle gunner armor.)
- What is the primary enemy threat? Is it 7.62-mm guns, RPGs, IEDs, or mines, for example, how well can the vehicle's armor protect the crew from each threat? What threats must be protected by offensive capabilities?
- Determine whether the enemy is employing mines. Consider whether the armor and weight associated with mine strike protection is practical versus armor, which protects against other threats.
- Consider armoring critical pieces of the vehicle itself such as the fuel system, communication system, and cooling system.
- Use Kevlar blankets from disabled HMMWVs or the M2 Bradley spall liner to 'armor' HMMWV seats and to add light weight armor to other areas of the vehicle.

Motorized Operations

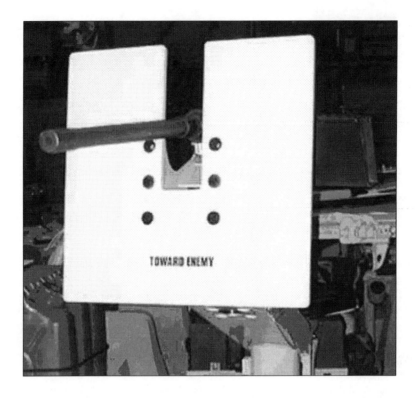

Figure K-1A. Commercially produced and available gunner armor protection.

Figure K-1B. Army-issue, roof-mounted, gun-ring armor.

Appendix K

> *Note:* While the gunner's upper body is protected, his lower torso is exposed through the middle of the vehicle. Also, note the 'screen' window on the rear doors. These prevent items from being thrown into the rear of the vehicle.

K-4. One technique units may use in balancing armor and observation is to install armored half doors (Figure K-2) rather than a full door or no door. Benefits include protection against either 5.56-mm or 7.62-mm rounds, depending on armor thickness; protection for crewmembers' legs, groins, and sides (areas not covered by conventional body armor); increased observation, mounting and dismounting capabilities and speed; the ability to effectively employ personal or crew-served weapons; and reduced vehicle weight. Limitations include less upper body protection for the vehicle crew than if fully armored doors and windows were installed. Similar products are available commercially, or they might be fabricated using on-hand or captured materials.

Figure K-2. Commercially produced HMMWV armored half-door and double-articulating, swing-arm mount for crew-served weapons.

K-5. Note the door-mounted and running board-mounted ammunition storage areas. Units can also sandbag all or portions of vehicles, or they can construct effective vehicle armor using on-hand or captured materials. This can include armor from destroyed enemy vehicles.

FIREPOWER AND OBSERVATION CONSIDERATIONS

K-6. In combination with previous improvements, the following modifications can enhance engagement capability and protection of the crew and on-board personnel.

- Consider removing vehicle windshields and glass windows. Windshields greatly restrict visibility when dirty or cracked; restrict mobility and weapons employment; and create secondary missile hazards.
- Install armor plates for crew-served weapons on roof-mounted gun rings. Consider adding armor beneath the gun ring to protect the gunner's abdomen and femoral artery.
- If operating vehicles without roof-mounted gun rings, consider constructing field-expedient roof mounts.
- Consider removing vehicle roofs or cutting away sections of the vehicle roof not required to support roof-mounted gun rings. Roofs greatly restrict observation, and often limit crew mobility, while providing little or no armored protection.
- If transporting troops in the cargo area of vehicles, remove all tarpaulins and bows to allow full visibility and security for the passengers.
- Replace all conventional passenger and cargo area seating with outboard facing, center-mounted seats.
- Install seats from destroyed or disabled vehicles in the cargo areas of HMMWVs (M2 Bradley seats or HMMWV seats).

COMMUNICATION CONSIDERATIONS

K-7. Vehicle crew communication is paramount to smooth vehicle operation. Commanders must consider how dismounts will communicate with the mounted or dismounted crew. Drivers and TCs can normally communicate by voice in most wheeled vehicles, but might not be able to do so if in contact. Passengers and gunners have a hard time communicating with the driver or TC under normal operating conditions, and most likely cannot do so during contact. Once passengers dismount, voice communications is nearly impossible. Commanders should consider the following.

- Equip the driver, TC, and gunner (if applicable) with headset radios for internal communication.
- All passengers and dismounted elements must have some form of effective internal communication with the crew while mounted and dismounted.

CREW AND PASSENGER DESIGNATED VEHICLE POSITIONS

K-8. The TSOP typically dictates positions for the crew and passengers. The commander adjusts them as needed based on METT-TC. When moving to a tactical objective area, he employs principles similar to those used in load planning for air assaults. That is, based on the ground tactical plan, he assigns positions in addition to the crewmembers based on unit integrity, the bump plan, and the cross-leveling of personnel and equipment.

K-9. Commanders must balance vehicle and crew survivability; vehicle weight and payload,; the offensive capabilities of the crew and passengers; and their ability to quickly and efficiently mount and dismount the vehicle. The positions will likely change once the passengers dismount. Vehicle crews normally remain with the vehicles while passengers dismount. This allows them to operate the vehicles and any mounted weapon systems. Figure K-3A and Figure K-3B, page K-6, show two possible vehicle-manning configurations for an Infantry platoon employing the M1025 variant with the hard shell rear cover removed (Figure K-4). The K-3A example maintains squad integrity among vehicle sections, but not if Soldiers dismount. The seating plan in Figure K-3B dedicates one organic squad to man the vehicle, so that two organic squads can dismount. The actual vehicle and crew manning configuration employed by the Infantry platoon or company should be driven by careful METT-TC analysis. Additional manning and seating considerations include--

Appendix K

- What types of vehicles are available? Will the unit use HMMWVs, LMTVs, FMTVs, or some mix of these? What version(s) of HMMWV is available?
- Will the vehicles have dedicated operators and crews, or will the Infantry platoon or company provide drivers and crews?
- Will the vehicles mount M2 or MK19s, or will the Infantry platoon or company use its organic, vehicle-mounted, automatic or crew-served weapons? Does the commander anticipate dismounting the crew-served or automatic weapons from the vehicles at the objective or if contact is made?
- Do the vehicles have roof-mounted gun rings designed to support crew-served weapons?

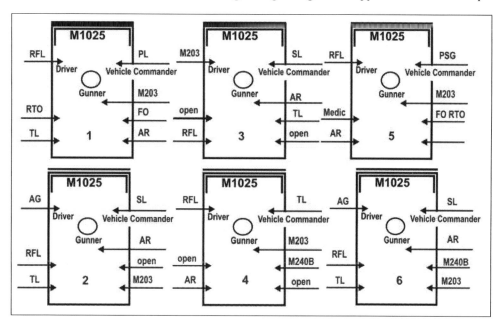

Figure K-3A. Possible platoon-seating technique.

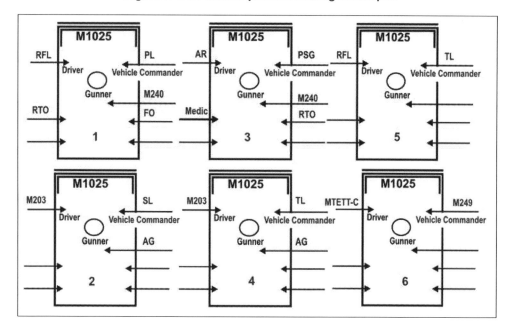

Figure K-3B. Variation of platoon-seating technique.

Motorized Operations

Figure K-4. Modified M1025 (turtle-shell HMMWV).

K-10. Will the unit maintain team and squad integrity within vehicles and vehicle sections? Or, will the unit dedicate an element to vehicle crews and maintain squad and team integrity only when dismounted? Will assistant gunners travel with their gunners in the same vehicle or in the same section?

- Is the mission primarily focused on the mobility aspect of the vehicles? Or, are the vehicles merely a means of transportation to the objective, with the focus on dismounted operations at or around the objective?
- Does the commander anticipate using the vehicles' crew-served weapons as a base of fire or as a support-by-fire element?
- Passengers in forward facing rear seats in most HMMWVs will have a difficult time effectively employing their weapons or quickly dismounting. Commanders should consider modifying rear passenger seats. They can either cut away under-seat storage and move the seat and seat back to the center console so passengers can face outward, or they can simply move the seat pad and seat back to the center. (Figure K-5, page K-8, shows an example of rear seat modifications that allow the passenger to face outwards.)

Appendix K

Figure K-5. Commercially produced version of an outboard-facing rear passenger seat and double-articulated swing arm for crew-served weapons.

K-11.   This technique maintains squad integrity across each vehicle section, but does not do so if forces dismount. It dedicates one squad and the M240B section to vehicle crews, and it allows two organic squads to dismount. An up-armored HMMWV (M1109 or M1114 model) is not designed for quick-dismount as are the Stryker or BFV. Commanders should consider this when planning their use in the urban environment. All passengers in the cargo bed or rear of any vehicle must face outwards. Seats must be modified to allow for this. If the HMMWV has a hard shell, that is, if it is a model M1025 or M1043 and has a turtle shell or turtle back, then consider removing the hard shell and installing seats for additional crewmembers and passengers, and to ensure all-round observation. Each passenger sits facing outboard, which improves his observation and his ability to employ a personal or crew-served weapon. Units can achieve this same benefit in one of several ways.

- Remove the rear passenger seat and cut away the under-seat storage; move the seat to the center console.
- Remove the vehicle roof, and stand on the passenger seat or center console.
- Identify the modifications required, and contract the work prior to deployment.

VEHICLE EQUIPMENT LOAD PLANS

K-12.   Vehicle commanders must ensure that any externally stowed items are secured from theft and do not constitute a fire hazard if the vehicle is attacked by IED, RPG, or other flammable device. External stowage should be minimized or modified to lessen the threat of vehicle fire and not restrict the view or movement of gunners or passengers providing security. All loose items stored inside the vehicle must be secured to prevent theft or becoming secondary missiles in the event of a mine or IED strike or a roll over.

Commanders should consider stowing flammable items that are mission essential inside the vehicle and behind armored portions of the vehicle, and securing non-mission-essential and nonflammable items outside the vehicles. Other considerations follow.

- Remove windshields, which when dirty often restrict mobility and visibility. Windshields also shatter when struck, creating secondary missiles or fragments (shrapnel).
- Consider lashing pioneer tools and crew-served weapon tripods to the front hoods of vehicles. Extend them fully before lashing them down; this eases employment.
- Use on-board ammunition storage containers such as 60-mm mortar ammunition cans. These hold several types of ammunition. This saves the crew a lot of time when they have to switch between ammunition for the 7.62-mm.50 caliber, and MK19 40-mm crew-served weapons.

Carry complete spare wheel and tire assemblies rather than just spare tires. This reduces the time needed to change a flat, and will often allow a crew to repair a vehicle after a mine strike.

- Consider equipping every vehicle or every other vehicle with wheeled vehicle tow bars, so that vehicles can recover or tow each other. Tow bars are better than cables, since no driver is needed in the towed vehicle.
- Consider emplacing civilian or military fire extinguishers in fixed positions inside the vehicle. Normally, locate them to protect the crew rather than the vehicle. This helps ensure crew survivability. Carry additional loose fire extinguishers to fight vehicle fires. Never use a halon fire extinguisher inside an armored vehicle.

K-13. Commanders should establish load plan SOPs for sensitive items. They should account for ammunition and additional special equipment such as breach kits, demolitions, and first aid equipment. They should also account for any additional weapons, such as rifles, in case no automatic or crew-served weapons are required after dismounting.

Section II. PATROLLING CONSIDERATIONS IN URBAN OPERATIONS

Infantry companies and platoons may conduct urban operations or patrols using wheeled vehicles as the sole means of mobility, or along with dismounted patrols.

URBAN PATROLLING CONSIDERATIONS

K-14. The basic planning considerations for both mounted and dismounted patrolling are almost identical. Although wheeled vehicles typically have the advantage of mobility and firepower over dismounted elements, as previously noted, communications and protection pose difficulties. The more heavily armored vehicles might be at greater risk in the urban environment unless closely supported by dismounted forces. Therefore, this environment calls for integrating dismounted with mounted patrol. elements. The combination of the two capabilities enhances conditions for successful urban operations.

Note: Sometimes, a psychological dimension exists in the employment of armored fighting vehicles (the M2 Bradley and the M1 tank) in urban operations. To avoid antagonizing noncombatants, even when they think using armored vehicles would help, leaders should carefully consider whether to do so.

TACTICAL VEHICLE EMPLOYMENT AND URBAN PATROLS

K-15. Vehicles operating as part of a patrol in stability operations should always operate in sections of at least two vehicles. For all stability operations, the commander should consider employing vehicles to augment dismounted patrolling. The enhanced mobility allows greater and faster AO saturation and expansion of control. Integrating vehicles with foot patrols also allows for increased sustainment loads. Lighter, faster Infantry forces have a greater chance of capturing or killing lightly armed insurgents;

mounted firepower provides support or moves to block escaping or flanking enemy. Mounted elements increase patrol flexibility and versatility. Other considerations for urban operations include--

- To avoid contact with overhead power lines, tie down vehicle antennas.
- Avoid streets and alleys that are too narrow for particular vehicles. Enemy forces can quickly identify this fact and plan their own routes accordingly.
- Consider civilian vehicle and pedestrian traffic flow when planning patrol routes. For each operation, consider the effects of military vehicles on civilian traffic flow and patterns.
- Tracked vehicles can damage or even destroy civilian roads in the AO. Carefully plan the use of tracked vehicles with this in mind.
- Avoid using night vision devices or blackout driving around civilian traffic using white lights. Doing so endangers you and the civilians, and they can see you anyway!.

BRADLEY AND STRYKER CONSIDERATIONS

K-16. When halting Strykers and BFVs, the commander should leave at least three vehicle lengths to the front and sides of each vehicle. This allows for emergency movement. In addition to patrolling, vehicles can enhance or assist in the following common stability operations.

- Hasty checkpoints.
- Cordon or security during searches.
- Deployment or redeployment of foot patrols.
- Baseline for riot or crowd control incidents.
- Support or overwatch obstacle reduction.
- Mounted reserve or quick-reaction force.

MOUNTED HASTY CHECKPOINT OPERATIONS

K-17. This is a generic technique for example only. Leaders must always consider the factors of METT-TC (Figure K-6).

Motorized Operations

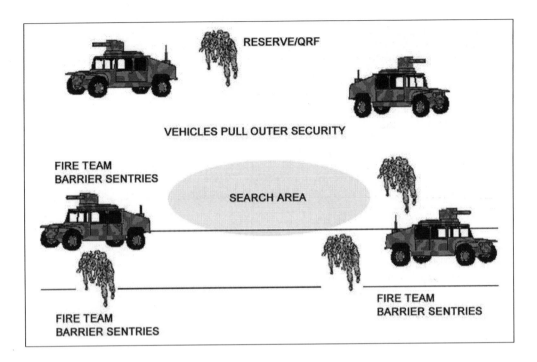

Figure K-6. Hasty checkpoints.

ACTIONS ON CONTACT

K-18. The actions on contact for mounted and dismounted forces are normally the same--seek cover and return fire, then develop the situation. Detrucking or dismounting troops should occur in a covered or concealed position when possible, but the speed of dismounting and the need to get Infantry on the ground quickly once in contact must also be considered. Once in contact, based on the enemy situation, the vehicles may or may not maneuver in order to dismount troops or form a base of fire. The commander must consider the following.

- Rehearse unwieldy dismounting and remounting procedures before mission execution.
- Avoid dismounting troops directly into the line of fire.
- Take measures to prevent troops from becoming casualties while mounted.

K-19. Vehicles can maneuver quickly with troops mounted or dismounted) either to occupy covered and concealed positions or to form a base of fire. The vehicles might be required to form an immediate cordon while the Infantry clears or searches the area.

K-20. Vehicles might be required to maneuver immediately, with or without support, to take advantage of speed and shock action, or to move to a position to gain a tactical advantage.

K-21. When troops dismount, or when the vehicle maneuvers away from immediate troops, the vehicle crew must remain vigilant and protect the vehicle from further enemy actions or isolation.

Section III. LONG-RANGE OPERATIONS CONSIDERATIONS

The Infantry company might have to conduct long-range movements or patrols using with wheeled vehicles. Special operations forces are likely to conduct these operations. However, this section contains considerations for the Infantry company commander in case he is tasked to do so. (FM 31-23 discusses long-range vehicular operations in detail.)

Appendix K

LONG-RANGE OPERATIONS

K-22. If the commander must conduct long-range operations, then he must be reasonably sure of the range and expected time of the operation. This allows him to estimate logistics support accurately. The logistical resources required to support long-range vehicular operations will drive many aspects of the mission planning (Classes III and IX, specifically). Additional considerations for long-range operations include the following.

- Mission range (mileage), duration, and expected vehicle fuel consumption.
- Availability of resupply or support during operations.
- Planned vehicle loads (personnel and equipment).
- Expected enemy situation.
- Terrain considerations such as---
  - Road conditions.
  - Off-road travel conditions.
  - River crossings.
  - Expected rainfall or snowfall.
- Presence of civilian vehicles in area--are vehicles normally encountered in the AO?
- Communications ranges.
- Navigational considerations.
- Movement times.
  - Day or night?
  - White light or NVDs?
  - Road or cross-country?
- Recovery.
  - Can the unit self-recover?
  - How will the unit deal with nonrecoverable or disabled vehicles?
- Load plan for casualties or additional personnel in case vehicles are destroyed or disabled.
- Fuel cans, 5 or 10 gallon, with long fuel lines to attach directly to engine intakes, to allow the vehicle to operate even if the fuel tank is punctured.

ROUGH TERRAIN DRIVING

K-23. Using good off-road driving techniques is the best way to limit broken vehicle parts and getting stuck. All drivers must be well trained in judging terrain and negotiating various ground conditions. Many operations and movements are at night, so driver's training should focus on using night vision devices. In addition, drivers should develop the following skills.

- Selecting proper gear ratio and shifting.
- Using momentum and understanding the effects of vehicle speed.
- Knowing the vehicle's capabilities and the impact of the on board payload.
- Estimating and using proper speeds for the appropriate terrain conditions.
- Avoiding sudden forward and braking thrusts.
- Applying traction theory.

K-24. Drivers must become familiar with the various terrain conditions in the AO and considerations for crossing the conditions encountered.

VEHICLE RECOVERY

K-25. Many recovery operations consist of self-recovery methods, either when the vehicle becomes stuck, or when it has a mechanical or enemy caused [what?]. Vehicle recovery is easiest when the tires still have traction and when crewmembers can help the vehicle move through the original tire tracks. When the vehicle is stuck in snow, sand, or mud, the crew can lower tire pressure to increase traction. Using a second vehicle to help winch or pull a stuck vehicle is normally the quickest recovery method. However, winches are used only to assist in recovery., They are never used as the sole source of power for vehicle recovery. The company should always carry tow straps or chains. Braided rope (three 12feet long by 5/16-inch diameter pieces) or a heavy 20-foot chain work well. They should have hooks or clevises attached to both ends for anchoring to the vehicles. If possible, the company carries at least one tow bar for each vehicle section to assist in long-range recovery or to tow a vehicle at high speed. When a vehicle is stuck in mud or sand, the unit uses pioneer tools to emplace dry or solid matter under the tires for traction. Sandbags or other materials can be dug into and under the wheels to assist traction. Vehicles normally carry empty sandbags for this purpose. When conducting recovery, one section provides security while the other vehicle makes the recovery. The recovery section decides before making the recovery where the vehicle can go after it breaks loose. When they use a military winch, they should remember these do's and don'ts.

DO--
- Use the stuck vehicle's wheel power to help the winch.
- Carefully prepare the winching operation.
- Position personnel where they will not be injured should the cable snap or unhook.
- Ensure the anchor points are solid.
- Use artificial surfaces for traction when the vehicle is stuck in water or soft sand.

DO NOT--
- Overtake the cable.
- Exceed the maximum angle of pull.

K-26. The commander should make contingency plans for vehicles that cannot be repaired or recovered. The company will make every attempt to recover the vehicle and return it where it can be repaired, if needed. However, if the company cannot recover the vehicle, then they may destroy it in place to prevent the enemy from capturing it.

Section IV. OEF AND OIF VEHICLE MODIFICATIONS

This section contains examples of vehicles used by various units in different stages of both OEF and OIF. These pictures show some of the ways that units, leaders, and Soldiers have modified wheeled vehicles using a variety of line item, commercial, and on-hand materials to meet mission requirements.

Note: This section neither endorses nor approves the use of any of the equipment or techniques discussed.

UNIT-INSTALLED WIRE CUTTERS AND WIRE GUARDS

K-27. Wire cutters and wire guards (Figure K-7, page K-14) are used in areas with low-lying power wires or wires laid across the roads to allow insurgents to ambush troops. Armor plating on the sides of the cargo bed help protect the Soldiers' lower bodies. The armor height protects the areas not covered by body armor, yet provides excellent observation, and does not affect the use of the crew-served weapon.

Appendix K

Figure K-7. Unit-installed wire cutters and wire guards.

URBAN PATROL

K-28. The Soldiers in Figure K-8 are conducting a vehicular patrol in an urban area. To increase observation, the Soldiers in the rear seat are standing, rather than sitting. Possible modifications include adding running boards to the vehicle as foot rests for seated passengers facing outboard; removing the rear seat assemblies and storage so Soldiers can stand on the vehicle floor; adding an armored half-door; or attaching sections of armored or steel plating to the doors to protect Soldier's legs.

Figure K-8. Urban patrol.

UNIT COMMAND VEHICLE

K-29. Figure K-9 shows the addition of four BFV seats (two facing forward, two rearward) and a large wooden footlocker for secure storage. Also, communications gear has been moved to the rear of the vehicle.

Figure K-9. Unit command vehicle.

UNIT-PRODUCED TRIPOD AND MODIFIED ARMOR

K-30. Figure K-10 shows a unit-designed and -produced elevated tripod. The unit has bolted the tripod into the rear bed of a cargo HMMWV. This allows the weapon to be employed over the rear and flanks of the vehicle, but does not look high enough to employ the weapon over the front. A standard-issue foot locker has been added for storage and seating. BFV seats have been added for passengers, but do not face outward. The unit has removed sections of the interior spall lining (Figure K-11) from a disabled M2 BFV and installed it as exterior armor to protect passengers.

Figure K-10. Unit-produced tripod and modified armor.

Figure K-11. Closeup of sections of Bradley fighting vehicle spall liner armor applied to exterior of cargo bed.

UNIT-INSTALLED STEEL ARMOR

K-31. Figure K-12 shows unit-fabricated and -installed plate steel armor and a tripod mount, which is visible through the front door. The tripod looks high enough to allow a mounted weapon system to fire in an all-round fan. Also note that the unit has not added any additional armor beneath the seating area--this leaves Soldiers' legs relatively unprotected (Figure K-13).

Figure K-12. Unit-installed steel armor.

Figure K-13. Closeup, reverse-angle view of unit-fabricated and -installed steel plate armor.

Glossary

Acronyms and Abbreviations

1SG	first sergeant		
	A		**C**
AA	avenue(s) of approach	C2	command and control
AAR	after-action review	CACOM	CA communications
ACPS	Army Civilian Personnel System	CANA	convulsant antidote for nerve agent
ACU	Army combat uniform	CAPT-A	CA planning team A
ADA	air defense artillery	CAS	close air support
ADAM	area-denial artillery munition	CASEVAC	casualty evacuation
ADW	air defense warning	CAT	civil affairs team
AFATDS	Advanced Field Artillery Tactical Data System	CBRN	chemical, biological, radiological, or nuclear (replaces NBC except when referring to existing reports and reporting systems)
AHD	antihandling devices	CBRNE-CM	chemical, biological, radiological, nuclear, and high-yield explosive consequence management
AMC	air mission commander		
AO	area of operations	CCA	close combat attack
AP	antipersonnel	CCIR	commander's critical information requirements
APC	armored personnel carrier		
APDS	armor-piercing discarding sabot	CCP	casualty collection point
APOD	aerial port of debarkation	chem	chemical
ARFOR	Army forces	CIC	combat in cities
ARNG	Army National Guard	CLS	combat lifesaver
ARTEP	Army Training and Evaluation Program	CLU	command launch unit
		CMO	civil-military operations
aslt psn	assault position	CMOC	civil-military operations center
ASOC	air support operations center	CO	commanding officer
AT	antitank	COA	course of action
ATGM	antitank guided missile	COE	contemporary operational environment
atk psn	attack position		
	B	COMSEC	communications security
BDAR	battle damage assessment and repair	CONOP	concept of operations
		CONUS	continental United States
BDO	battle dress overgarments	COP	common operational picture
BFV	Bradley fighting vehicle	CP	command post
BHL	battle handover line	CS	combat support
BMNT	begin morning nautical twilight		**D**
BP	battle position	DA	Department of the Army
BSA	brigade support area	DED	detailed equipment decontamination
BUA	built-up area		

Glossary

DEUCE	deployable universal combat earthmover		FSB	forward support battalion
DLIC	detachment left in contact		FSCM	fire support coordination measure
DOD	Department of Defense		FSE	fire support element
DPICM	dual-purpose improved convention munition		FSEM	fire support execution matrix
DRAW-D	defend, reinforce, attack, withdraw, delay		FSO	Fire Support Officer

G

G/VLLD	ground/vehicle laser locator designator
GPS	Global Positioning System
GS	general support
GSR	ground surveillance radar
GS-R	general support-reinforcing
GT	gun target
GTAO	graphic terrain analysis overlay

DTD	detailed troop decontamination
DVE	driver's vision enhancer
DZ	drop zone

E

EA	engagement area
ECOA	enemy course of action
EEFI	essential elements of friendly information
EENT	end evening nautical twilight
EFST	essential fire support task
EMT	emergency medical treatment
EOD	explosive ordnance disposal
EPW	enemy prisoner of war

H

HAZMAT	hazardous materials
HCA	humanitarian and civic assistance
HCP	health and comfort pack
HDC	headquarters distribution company
HE	high explosive
HEDP	high explosive, dual purpose
HHC	headquarters and headquarters company
HN	host nation
HPT	high-payoff target
HSS	health service support
HUMINT	human intelligence

F

FA	field artillery
FASCAM	family of scatterable mines
FBCB2	Force XXI battle command brigade and below
FCL	final coordination line
FDC	fire direction center
FEBA	forward edge of battle area
FFIR	friendly force information requirements
FHA	foreign humanitarian assistance
FIBUA	fighting in built-up areas
FIST	fire support team
FLOT	forward line of own troops
FM	frequency modulated
FO	forward observer
FPF	final protective fire
FPL	final protective line
FRAGO	fragmentary order
FS	fire support

I

I2R	imaging infrared
IAW	in accordance with
IBCT	Infantry brigade combat team
ICV	Infantry carrier vehicle
IED	improvised explosive device
illum	illumination
IM	information management
INFOSYS	information systems
IO	information operations
IPB	intelligence preparation of the battlefield
IR	infrared

ISU	integrated sight unit		MOPMS	modular pack mine system
IV	intervisibility		MOPP	mission-oriented protective posture
			MP	military police
J			MPAT	multipurpose antitank
JFC	joint force commander		MR	moonrise
JWARN	joint warning and reporting network		MRB	motorized rifle battalion
			MRE	meal, ready to eat
			MRP	motorized rifle platoon
K			MS	moonset
KIA	killed in action		MSD	minimum safe distance
			MSL	minimum safe line
L			MSR	main supply route
LCE	load-carrying equipment		MTC	movement to contact
LD	line of departure		MTF	medical treatment facility
L-MOPP	laser mission-oriented protective posture			
LOA	limit of advance		**N**	
LOC	length of column		NAAK	nerve agent antidote kit
LOGPAC	logistics package		NAI	named area(s) of interest
LOS	line of sight		NBC	nuclear, biological, and chemical (replaced by CBRN except when referring to reports or to existing reporting system)
LRP	logistics release point			
LZ	landing zone			
			NBCWRS	Nuclear, Biological, and Chemical Warning and Reporting System
M				
MANPADS	Man-Portable Air Defense System		NCA	National Command Authority
MBA	main battle area		NCO	noncommissioned officer
MBC	mortar ballistics computer		NCS	net control station
MCOO	modified combined obstacle overlay		NEO	noncombatant evacuation operations
MDI	modernized demolition initiator		NFA	no-fire area
MDMP	military decision-making process		NGO	nongovernmental organization
MEDEVAC	medical evacuation		NLT	not later than
MEL	maximum engagement line		NVD	night vision device
METT-TC	mission, enemy, terrain (and weather), troops (and support) available, time available, and civil considerations			
			O	
			OAKOC	observation and fields of fire, avenues of approach, key terrain, observation, and cover and concealment
MEV	medical evacuation vehicle			
MGS	mobile gun system		OBSTINTEL	obstacle intelligence
MIB	motorized infantry battalion (used to train threat model)		OH	observation helicopter
			OIC	officer in charge
MIP	motorized infantry platoon (used to train threat model)		OP	observation post
			OPCON	operational control
MOE	measure of effectiveness			

OPORD	operation order		RPG	rocket-propelled grenade
OPSEC	operations security		S&R	stability and reconstruction
ORP	objective rally point			
OTM	on-the-move			

P

PA	physician's assistant
PAO	public affairs officer
PC	personnel carrier
PCC	precombat check
PCI	precombat inspection
PEWS	platoon early warning system
PIR	priority intelligence requirements
PL	phase line
PLD	probable line of deployment
PMCS	preventive maintenance checks and services
PME	peacetime military engagement
POC	point of contact
POF	priority of fire
POL	petroleum, oils, and lubricants
pos	position
POSNAV	position navigation
PSG	platoon sergeant
PSYOP	psychological operations
PUC	personnel under control
PVO	private voluntary organization
PZ	pickup zone

R

R&S	reconnaissance and surveillance
RAAM	remote antiarmor mine
RCPA	relation combat power analysis
RCU	remote control unit
RD	round
RDD	radiological dispersal device
RED	risk estimate distance
REDCON	readiness condition
RFL	restrictive fire line
ROE	rules of engagement
ROI	rules of interaction
RP	release point

S

S-1	personnel staff officer (adjutant)
S-2	intelligence staff officer
S-3	operations staff (and training) officer
S-4	supply officer
SBCT	Stryker brigade combat team
SCATMINE	scatterable mine
SEAD	suppression of enemy air defenses
SEE	small emplacement excavator
SF	special forces
SFODA	special forces operational detachment-A (*also known as* A-Team)
SITEMP	situation template
SITREP	situation report
SOC	special operations command
SOF	special operations forces
SOI	signal operating instructions
SOP	standing operating procedure
SOSRA	suppress, obscure, secure, reduce, and assault
SP	start point
SPOD	sea port of debarkation
SPOTREP	spot report
sqd	squad
SR	sunrise
SS	sunset
SU	situational understanding

T

TACP	tactical air control party
TC	tank commander
TCP	traffic control posts
TDA	table of distribution and allowances
TF	task force
TLP	troop-leading procedures
TOC	tactical operations center

TOE	table of organization and equipment		**V**	
TPME	task, purpose, method, and effects	**VPK**	vehicles per kilometer
TRP	target reference point	**VPM**	vehicles per mile
TSM	target synchronization matrix		**W**	
TSOP	tactical standing operation procedures	**WARNO**	warning order
		WCS	weapons control status
TTP	tactics, techniques, and procedures	**WFF**	warfighting function
	U	**WIA**	wounded in action
UAS	unmanned aerial system	**WMD**	weapons of mass destruction
UH	utility helicopter	**WP**	white phosphorus
UN	United Nations		**X**	
UO	urban operations	**XO**	executive officer
US	United States			
USACE	United States Army Corps of Engineers			

References

SOURCES USED

These are the sources quoted or paraphrased in this publication.

DEPARTMENT OF DEFENSE PRODUCTS

DOD Directive 5100.77, *DOD Law of War Program.*

DOD Directive 5240.1-R, *Procedures Governing the Activities of DOD Intelligence Components that Affect US Persons.*

DD Form 2745, *Enemy Prisoner of War (EPW) Capture Tag.*

DOCUMENTS NEEDED

These documents must be available to the intended users of this publication.

ARMY TRAINING EVALUATION PROGRAMS (MTPS AND DRILL BOOKS)

ARTEP 7-8-Drill. *Battle Drills for the Infantry Rifle Platoon and Squad.* 25 June 2002.

ARTEP 7-10-MTP. *Mission Training Plan for the Infantry Rifle Company.* 1 June 2002.

DEPARTMENT OF THE ARMY FORMS

DA Form 1156, *Casualty Feeder Card.*

DA Form 2028, *Recommended Changes to Publications and Blank Forms.*

DA Form 2404, *Equipment Inspection or Maintenance Worksheet.*

DA Form 2823, *Sworn Statement.*

DA Form 4137, *Evidence/Property Custody Document.*

DA Form 5988-E, *Maintenance Request Register (EGA).*

FIELD MANUALS

FM 1-02. *Operational Terms and Graphics.* 21 September 2004.

FM 3-0. *Operations.* 14 June 2001.

FM 3-05.70. *Survival.* 17 May 2002.

FM 3-06.11. *Combined Arms Operations in Urban Terrain.* 28 February 2002.

FM 3-07. *Stability Operations and Support Operations.* 20 February 2003, with Change 1, 30 April 2003.

FM 3-07.31. *Peace Operations Multi-Service Tactics, Techniques, and Procedures for Conducting Peace Operations.* 26 October 2003.

FM 3-09.32. *(J-FIRE) Multiservice Procedures for the Joint Application of Firepower.* 29 October 2004.

FM 3-7. *NBC Field Handbook.* 29 September 1994.

FM 3-11. *Multiservice Tactics, Techniques, and Procedures for Nuclear, Biological, and Chemical Defense Operations.* 10 March 2003.

References

FM 3-11.4. *Multiservice Tactics, Techniques, and Procedures for Nuclear, Biological, and Chemical (NBC) Protection.* 2 June 2003.

FM 3-19.40. *Military Policy Internment/Resettlement Operations.* 1 August 2001.

FM 3-21.8. *Infantry Rifle Platoon and Squad.* TBP.

FM 3-21.20. *The Infantry Battalion.* TBP.

FM 3-21.91. *Tactical Employment of Antiarmor Platoons and Companies.* 26 November 2002.

FM 3-22.10. *Sniper Training and Operations.* TBP.

FM 3-22.27. *MK 19, 40-Mm Grenade Machine Gun, Mod 3.* 28 November 2003.

FM 3-22.40. *Multiserivce TTP (MTTP) for Tactical Employment Of Nonlethal Weapons.* 15 January 2003.

FM 3-34.2. *Combined-Arms Breaching Operations.* 31 August 2000, with Changes 1 Through 3, 15 November 2000 through 11 October 2002.

FM 3-90. *Tactics.* 4 July 2001.

FM 3-90.2. *The Tank and Mechanized Infantry Battalion Task Force.* 11 June 2003.

FM 5-0. *Army Planning and Orders Production.* 20 January 2005.

FM 5-170. *Engineer Reconnaissance.* 5 May 1998, with Change 1, 13 July 1998.

FM 5-250. *Explosives and Demolitions.* 30 July 1998, with Change 1, 30 June 1999.

FM 7-85. *Ranger Unit Operations.* 9 June 1987.

FM 7-98. *Operations in a Low-Intensity Conflict.* 19 October 1992.

FM 17-95. *Cavalry Operations.* 24 December 1996.

FM 21-18. *Foot Marches.* 1 June 1990.

FM 27-10. *The Law of Land Warfare.* 18 July 1956.

FM 31-23. *Special Forces Mounted Operations Tactics, Techniques, and Procedures.* 5 May 1999.

FM 34-130. *Intelligence Preparation of the Battlefield.* 8 July 1994.

FM 90-4. *Air Assault Operations.* 16 March 1987.

FM 90-7. *Combined Arms Obstacle Integration.* September 29 1994, with Change 1, October 2003.

INTERNET WEB SITES

Some of the documents listed in these References may be downloaded from Army websites:

Army Knowledge Online	https://akocomm.us.army.mil/usapa/doctrine/index.html
Army Publishing Directorate (forms)	http://www.apd.army.mil/usapa_PUB_formrange_f.asp
Reimer Digital Library	http://www.train.army.mil
Air Force Publishing	http://afpubs.hq.af.mil/
NATO Online Library (ISAs)	http://www.nato.int/docu/standard.htm

JOINT PUBLICATIONS

JP 3-07.2. *Joint Tactics, Techniques, and Procedures for Antiterrorism.* 17 Mar 1998.

JP 3-07.3. *Joint Tactics, Techniques, and Procedures for Peace Operations.* 12 Feb 1999.

TRAINING CIRCULARS

TC 7-98-1. *Stability and Support Operations Training Support Package.* 5 June 1997.

Index

A

120-mm mortar, 10-29
155-mm howitzer, 10-24
60-mm mortar, 10-26
81-mm mortar, 10-29
AAR. *See* after-action review
actions
   on contact, 4-12
   against fortified areas, E-4
   on the objective, 4-18
ADA. *See* air defense artillery
ADAM. *See* area-denial artillery munition
adaptive phase, 1-3
adjacent unit coordination, 5-41
adjutant (S-1), 11-1, 11-21
advantages and disadvantages
   conventional indirect fire, 10-33
   direct alignment, 10-33
   direct lay, 10-32
   hip shoot, 10-34
ADW. *See* air defense warning
aerial port of debarkation (APOD), 1-5, 3-1
after-action review (AAR), A-5, A-13
air defense artillery (ADA), 2-25, 2-45, 10-45, 12-65, D-12
air defense warning (ADW), 10-46
air mission briefing format, D-12
air mission commander (AMC), 10-38, D-3, D-10
air movement plan, D-6
air support operations center, 10-34
all-round security, 3-15
ambush, 8-44
   IED, G-6
ammunition, 12-9
   blank, 1-7, 2-9, 2-16
   mortar, 10-28
analysis
   "IV" line, 2-23
   course of action, 2-45
   key terrain, 2-22

analysis (*continued*)
   mission, using METT-TC, 2-10
   mobility corridors and avenues of approach, 2-20
   obstacles and restricted terrain, 2-18
analysis of mission, 2-10 (*illus*)
antihandling device (AHD), 8-27, G-13
antipersonnel (AP), 5-8, 8-1, 8-27, 10-4, 10-22, 10-41, 12-15, 12-53, B-1, C-2, G-2, G-12, G-15
antitank guided missile, 9-2, 9-9, 9-18, 12-7, 12-29, 12-56, B-4, C-3, C-8, E-2, E-5
AO. *See* area of operations
approach march technique, 4-43
area-denial artillery munition (ADAM), 10-43
area of interest, 2-14 (*illus*)
area
   numbering system, 12-37
   reconnaissance patrol, 8-37
area of operations (AO), 1-19, 2-14 (*illus*), 4-40, 6-15, 6-29, 6-30, 8-1, 8-8, 8-21 8-47, 9-3, 11-17, 11-24, 12-5, D-2, F-1
   battlefield organization, 2-5
   command and control, 1-19
   commander's critical information requirements (CCIR), 1-7
   enemy, 1-2
   shaping operations, 2-5
armored personnel carrier (APC), 10-28, 10-41, 12-60, B-16
armored vehicles, 12-15
   combined operations, C-8
   considerations, 12-28
   employment, 12-16
   positions, 12-22
armorer, 1-16
armor-piercing discarding sabot (APDS), C-2

arms control, 6-8
Army aviation, 10-48
Army Civilian Personnel System (ACPS), D-13
Army forces, 2-31, F-6
Army planning process, 2-2
Army Special Operations Command, F-1
Army Special Operations Forces, F-2
artillery
   direct-fire role, 12-35
   response times, 10-9
assault, 4-10
   of a building, 12-45, 12-46
   platoon, weapons company, B-2
   position, 2-31, 4-8, 4-18, 4-23, 4-30, 4-43, 11-7
assessment, J-2
   risk, 2-38
attack
   helicopters, 10-48
   position, 4-8, 4-21, 4-25, 8-15, 12-38
attacks, 4-15, 4-23, C-9
   aviation considerations, D-14
   buildings, 12-48
   characteristics, 4-16
   converging routes, C-10
   same route, C-11
   special purpose, 4-19
   types, 4-16
attrition approach, 12-13
authentication statements, 10-35
avenue-in-depth technique, 2-47, 2-48 (*illus*)
avenues of approach, 2-19, 2-20 (*illus*), 12-53
   airspace, 12-3
   analysis, 2-19 (*illus*), 2-20 (*illus*)
   dismounted, 5-9
   enemy, 5-14, 5-19, 5-22, 5-29, 9-8, E-6
   secondary, 5-3
   towed howitzer, 12-35
aviation support, D-1
   employment, D-1

Index

aviation support (*continued*)
key personnel, D-11
safety, D-21

B

backblast, C-7
backbrief, 2-52
bangalore torpedoes, 10-42
battalion fire support execution matrix, 10-10 (*illus*)
battle command, 2-1
battle damage assessment and repair (BDAR), 11-19
battle dress overgarments (BDO), H-2
battle dress uniform, 11-8
battlefield, 2-13
battle handover line (BHL), 5-45, 8-14
battle position (BP), 2-32, 2-49, 5-3, 5-12 (*illus*), 5-13 (*illus*), 5-14, 5-17 (*illus*), 12-55 (*illus*), 12-63
BCT. *See* brigade combat team
begin morning nautical twilight (BMNT), 2-25, 6-26, D-12
belt technique, 2-46, 2-47 (*illus*)
BFV. *See* Bradley fighting vehicles
biological agents, H-8
booby traps, 8-1, 12-8, 12-31, 12-46, A-8, G-14, G-15 (*illus*)
boundaries, 12-38
bounding overwatch technique, 3-5
box technique, 2-46 (*illus*)
Bradley fighting vehicles, 12-17, 12-29, C-1, K-1, K-8, K-10, K-15
danger zone, C-6, C-7
planning considerations, K-10
task organization, 12-29
breaches, 8-26
breaching organization, 8-29
breaching and securing of a foothold, 4-27
brigade combat team (BCT), 1-9, 10-38, 10-48, 11-24, F-6
brigade support area (BSA), 1-16, 11-7, 11-15, 11-19

building
analysis, 12-6
assault, 12-45, 12-46
attack, 12-48
hide position, 12-25
built-up area (BUA), 4-6, 4-53, 6-26, 8-2, 12-2, 12-30, 12-40, 12-50, 12-52 (*illus*), B-11, B-15, J-9
bull's eye technique, D-16
buttocks injection site, H-8 (*illus*)
bypass, 4-53

C

C2. *See* command and control
calculation, 11-7
camouflaged UXO, G-3
CANA. *See* convulsive nerve agent antidote
capabilities and limitations, 1-9, B-7, I-2, 1-9, B-7
casualties, 12-9
casualty collection point, 1-14, 5-11, 5-26, 11-2, 11-20, 11-21
casualty evacuation (CASEVAC), 1-13, 4-31, 4-46, 5-11, 5-26, 8-14, 11-22, 12-1, 12-28, 12-30, 12-43
CCIR. *See* commander's critical information requirements
checkpoint (CP), 2-45, 3-14, 3-19, 4-18, 4-44, 4-47, 4-51, 6-3, 6-17 (*illus*), 6-22 (*illus*), K-11 (*illus*)
chemical, biological, radiological, or nuclear (CBRN), 1-16, H-1, H-4
CIC. *See* combat in cities
civil
affairs, F-5, F-8
considerations, 2-36
reconnaissance, F-6
support operations, 7-1
civil affairs team (CAT), 6-4
civil-military operations (CMO), 6-2, F-6
classes of supply, 11-11
clearance
buildings, 12-43
objective, 4-53

close air support, 1-14, 2-27, 4-40, 6-26, 10-23 (*illus*), 10-24 (*illus*), 10-34, 10-35 (*illus*), 10-37 (*illus*), 10-38 (*illus*), F-5
close combat, 12-12, 12-13
close combat attack (CCA) nine-line briefing, D-20 through D-21 (*illus*)
COE. *See* contemporary operational environment
collateral damage, 12-10, 12-13
column formation, 3-7
combat
patrol, 8-37
stress, A-11
system, 1-4
combat in cities (CIC), 11-2, 12-2
combat lifesaver, 1-17, 6-26, 11-4, 11-20, 12-9, A-12, H-5
combat patrol organization, 8-35
combat power, 1-17
effects, 1-5
relative, 2-41
combat support (CS), 1-9, 1-12, 2-43, 3-14, 4-47, 5-12, 5-46, 6-25, 8-7, 8-15, 8-22, 10-1, 10-38, 11-26, 12-4, 12-10, 12-30, D-1, F-2
combat trains command post (CTCP), 11-2
combination technique, 4-48
combined arms, 12-10, 12-14
combined operations with armored vehicles, C-8
command
climate, A-6
considerations, I-2
launch unit, 2-50, B-9, B-11, B-16
relationships, 10-1
command and control (C2), 1-19, 2-43, 2-50, 3-19, 4-11, 5-5, 6-4, 7-6, 12-36, 12-60, D-11, F-8
enemy, 2-31, 4-2

Index

command post (CP), 1-12, 2-40, 3-10, 3-20, 4-16, 4-42, 5-22, 5-37, 6-4, 6-16, 8-9, 8-19, 10-16, 10-34, 12-60, A-9, I-2
   enemy, 2-25, 4-49, 5-4, E-2
commander's critical information requirements (CCIR), 1-4, 2-8, 2-48, 8-3, 11-24, J-3
commander's intent, 2-40
commands, 9-20
commercially produced
   gunner armor, K-3
   HMMWV mount, K-4
common activities, 4-48
common fire-control measures, 9-9
common operational picture (COP), 2-30, F-6
communications
   command post, 3-13
   considerations, 10-17, 12-9, K-5
   fire support team, 10-18
   security (COMSEC), 1-15, 11-8, 11-23
   tanks, C-16
company
   delay from alternating positions, 5-45 (*illus*)
   dismounted delay from subsequent positions, 5-44 (*illus*)
   formations, 3-7 through 3-11 (*illus*)
   lodgment area using existing facilities, 6-10 (*illus*)
   timeline, 2-36 (*illus*)
company commander, 1-12, 5-38, 8-4, 8-35, D-11
company trains, 3-20, 4-11, 4-30, 5-11, 5-36, 8-19, 11-9, 11-19, 12-46, 12-60
compliance with an agreement, 6-13
concept of operations (CONOP), 2-4, 2-12, 2-42, 5-48, 10-12
conditions, 12-2
   assault, 4-18
   asymmetric environment, 1-1

battlefield, 2-32
changes, 12-8
conditions (*continued*)
   end state, 2-40
   forward planning, 2-7
   limited visibility, 9-2
   operational environment, 1-6
   shaping operations, 2-5
   synchronization, 2-45
   time available, 2-34
confirmation brief, 2-52
consolidation and reorganization, 4-10, 5-4, 12-49
contact, forms of, 4-12, 4-45
contamination avoidance, H-1
contemporary operational environment (COE), 1-1, 2-28, 6-2
Continental United States (CONUS), 6-1
contingency operation, 4-41, F-4
continuous operations, A-10
control, 9-9
   crowds, 6-29
   essential, 12-13
   measures, 2-6, 5-12
   techniques, 3-14
   transition, 12-14
convoy
   briefing checklist, 6-26 (*illus*)
   escorts, 6-25
convulsive nerve agent antidote (CANA), H-4, H-6
coordination, F-8
   adjacent unit, 5-41
counterattack, 5-4
countermeasures, G-8
course of action, 2-40
   sketch, 2-44 (*illus*)
cover and concealment, 2-24 (*illus*), 3-4, 3-21, 4-9, 4-49, 5-6, 5-28, 8-2, 8-27, 8-47, 9-3, 9-30, 10-47, 11-22, 12-3, 12-54, B-5
crew-served weapons, 2-32 (*illus*), 4-23, 5-11, 5-33, 6-15, 6-24, 8-9, 9-10, 10-41, 11-19, 12-61, B-6, E-2, K-4 (*illus*), K-8 (*illus*)

critical points, 1-13, 2-46, 3-17, 5-2, 8-39, 9-1, 12-10, 12-34, C-14
cross-leveling of personnel, 4-30, 5-37, 11-17, 11-23, B-11, K-5

D

danger areas, 3-12, 3-18, 4-51, 8-4, 10-9, 12-27 (*illus*), 12-40, A-2, C-9, C-15
danger zones
   Bradley fighting vehicle rounds, C-4, C-6, C-7
   M1 tank, C-5
   surface, 3-39, A-5
   vehicle IED, G-4
decisive action, 4-14
decisive operation, 2-5, 2-11, 2-39, 2-42, 4-8, 4-21, 4-40, 5-26, 8-29, 10-6, 12-10
decisive terrain, 2-21, 4-36, 5-2, 5-29, 12-34, 12-62
decontamination, 1-9, 1-14, H-2
   levels, H-3 (*illus*)
   planning, 7-6
   support, 5-40, 10-48
defend, reinforce, attack, withdraw, delay (DRAW-D), 2-27
defense, 12-50, C-12, H-1
   battle position, 5-14
   block or group of buildings, 12-63
   key terrain, 12-64
   reverse slope, 5-27
   strongpoint, 5-18
   urban strongpoint, 12-65
   village, 12-62
defensive
   employment, E-5
   operations, 5-1
   sector sketch, 5-39
   techniques, 5-11
delay, 12-67
   from alternating positions, 5-45
deliberate
   attack, 12-41
   checkpoint layout, 6-17
   observation post, 6-14 (*illus*)
delivery methods, G-10
demolitions, 10-42

Index

deployable universal combat earthmover (DEUCE), 10-44
deployment, 4-9
depression and elevation, 12-11
designated marksman, E-1
detachment left in contact (DLIC), 1-13, 5-46, 8-8, 10-31
detailed equipment decontamination (DED), H-3
detailed troop decontamination (DTD), H-2
detainees, 11-24
direct fire
 control, 9-1
 integration with indirect fire, 5-36
direction-of-attack technique, 12-45
dismounted delay
 reserve, 4-54
 responsibilities, 11-2
 strongpoint, 5-19
 subsequent positions, 5-44
 tasks, 6-8
 team, task organization of Bradleys with, 12-29
 timeline, 2-36
 vee formation, 3-9
 wedge formation, 3-8
dispersion between squads, B-7
driver vision enhancer (DVE), C-3
drop zone (DZ), 5-21, 11-17, 12-51, G-16
dual-purpose improved convention munition (DPICM), 10-4, 10-21
duties and responsibilities of key personnel, 1-12

E

early warning procedures, 10-46
echelonment of fires, 10-20
echelon right formation, 3-11
EEFI. *See* essential elements of friendly information
effects of combat power, 1-5
elements of combat power, 1-18

emergency medical treatment (EMT), 1-16, 11-3
emplacement
 authority, 10-43
 of weapons systems, 5-33
employment, 10-46
 checkpoints, OPs, and patrols, 6-22
 considerations, B-11, G-4
 Heavy and Stryker, C-1, C-4, C-7
 TOW and Javelin, B-1
 zone of separation, 6-22
end evening nautical twilight (EENT), 2-25, 6-26, D-12
end state, 2-6
enemy
 analysis, 2-27
 assault, 5-4
 capabilities, 2-31
 composition, 2-29 (*illus*)
 disposition, 2-30
 locations, probable operations in the OE, 1-2
 preparatory fires, 5-3
 probable locations, 9-4
 scheme of maneuver, 9-4
 situation template, 2-32, 2-34
 strength, 2-30
enemy course of action (ECOA), 2-32, 5-14
enemy prisoner of war (EPW), 4-34, 4-53, 5-4, 5-39, 6-25, 8-38, 11-2, 11-24, 12-4, 12-39, J-9
 detainee tag, 11-27 (*illus*)
engagement
 area development, 5-29
 ranges, 12-11
 times, 12-11
engagement area (EA), 2-1, 2-22, 2-42, 5-3, 5-15, 5-29, 8-23, 9-10, 10-7, 10-28, E-6
 multiple, 5-18
engineers, 10-38
 missions, 10-41
envelopment, 4-4
essential elements of friendly information (EEFI), 1-5, 2-8, 2-49, J-3
essential fire support task (EFST), 5-35, 10-11 (*illus*). *See also* fire support
essential task, 2-12

ethnic dynamics, 2-37
example formats
 air mission briefing, D-12
 close air support check-in, 10-35
 mine incident report, G-17
 nine-line close air suppport briefing, 10-37
 situation update, 10-36
 warning order, 2-8
excessive load, 11-5
execution, 10-17, 10-23
executive officer (XO), 1-12
exploitation of the penetration, 4-29
explosive ordnance disposal (EOD), 1-19, G-9, G-14
extraction, G-17

F

factors in analyzing
 key terrain, 2-21
 mobility corridors and avenues of approach, 2-19
 observation and fields of fire, 2-24
 obstacles and restricted terrain, 2-17
family of scatterable mines (FASCAM), 10-4, 12-59
fatigue, A-14, G-11
field artillery (FA), 1-14, 5-24, 6-3, 10-3, 10-9, 10-15, 10-27, 12-31
fighting in built-up areas (FIBUA), 12-2
file formation, 3-10
final coordination line (FCL), 4-24
final protective fire (FPF), 10-16
final protective line (FPL), 5-9
fire commands, 9-20
fire control, 9-1, 9-3, 9-9
fire direction center (FDC), 3-20, 5-21, 5-37, 6-3, 10-10, 10-16, 10-30
fire planning process, 10-11
firepower and observation considerations, K-5
fire support (FS), 1-14, 1-18, 3-13, 4-11, 4-44, 5-10, 7-5, 8-17, 8-29, 10-3
 essential tasks, 5-35

Index

plan, 6-27, 10-5, 10-9, 10-11
fire support (*continued*)
   plans and coordination, 10-5
   preparation, 10-7
   team, 10-4
fire support coordination (FSC), 10-4
fire support coordination measure (FSCM), 10-10
fire support element (FSE), 1-14, 4-8, 4-11, 4-23, 4-34, 8-38, 10-5, 10-9, 10-35, 12-36
fire support execution matrix (FSEM), 10-9, 10-10 (*illus*), 10-14, 10-15 (*illus*)
fire support officer (FSO), 1-14, 3-13, 5-40, 6-3, 8-8, 8-19, 10-3, 10-4, 10-10
fire support team (FIST), 1-16, 3-13, 10-4, 10-18
first response, 11-20
first sergeant, 1-13
five Cs technique, G-9
five Ss and T method, 11-25
flank guard, 4-44, 8-25
follow and support, 4-52
foothold, breaching and securing of, 4-27
force orientation, 9-5
Force XXI battle command brigade and below (FBCB2), 1-16, 10-18, 12-9, 12-36, A-8, C-3
foreign humanitarian assistance (FHA), 6-22, 8-34
foreign internal defense, 6-6
format. *See* example formats
formations, 3-6
forms of contact, 4-12
forms of maneuver, 4-3
forward area arming and refueling point (FAARP), 2-25, D-14
forward edge of the battle area (FEBA), 2-49, E-6
forward line of own troops (FLOT), 8-18
forward observer (FO), 3-13
forward operating base (FOB), 6-4, 6-20

fragmentary order (FRAGO), 2-1, 2-10, 2-52, 4-15, 4-25, 4-40, 5-40, 12-50, A-8, D-5
fratricide avoidance. *See* safety
friendly
   based quadrants, 9-12
   exposure, 9-2
   fire, 12-11
friendly forces information requirements (FFIR), 2-49, J-3
frontal
   attack, 4-7
   fire, 9-13

G

gathering of information, J-1
general support (GS), 10-2, 12-56, E-10
general support-reinforcing (GS-R), 10-2
global positioning system (GPS), 2-53, 3-14, 8-41
graphic control measures for Infantry and Heavy, 12-22
graphic terrain analysis overlay (GTAO), 2-13
grid technique, D-16
ground surveillance radar (GSR), 12-67
ground tactical plan, D-2
ground/vehicle laser locator designator (G/VLLD), 10-8
guard, 8-21
guidelines, I-2
gun-ring armor, K-3

H

hasty
   checkpoint, K-10
   defense, 12-61
HBCT. *See* Heavy brigade combat team
health and hygiene, 11-20
health service support (HSS), 11-20
Heavy brigade combat team (HBCT), 1-9, 10-8, C-1
Heavy employment, 1-10, C-1
helicopters
   machine gun aim points, 10-48
   self-defense against, B-16
   types, D-1
hide position, 12-24

high explosive (HE), 4-30, 10-4, G-2
high-explosive antitank, C-2
high explosive, dual purpose (HEDP), 9-17
high-payoff target (HPT), 10-5, 10-12, 10-19
high-performance aircraft, 10-48
Hornet, 10-44
host nation (HN), 6-7, 6-16, 6-22, 8-41, 12-2, 12-20, A-2, F-1, F-5, G-10
hull-down position, 12-23
human dimension, 12-14
human intelligence (HUMINT), 6-2, 12-10, 12-41
humanitarian and civic assistance (HCA), 6-7

I

IBCT. *See* Infantry brigade combat team
immediate action, G-13
implied tasks, 2-12
improvised explosive device (IED), G-1
   detonator, G-2 (*illus*)
   dropped into vehicles, G-5 (*illus*)
indirect-fire
   capabilities, 10-4
   close support, 10-19
   integration with direct fires, 5-36
Infantry battalion weapons company, B-1
Infantry brigade combat team (IBCT), 1-8, 5-11, 10-38
   Engineer company, 10-37
Infantry company
   forward passage of lines, 8-15
   linkup, 8-7
   lodgment area using existing facilities, 6-10
   rearward passage of lines, 8-16
Infantry fighting vehicle, C-2
Infantry on tanks, C-15
Infantry vehicles, 12-15
   employment, 12-16
infiltration, 4-6, 4-48
   lanes, 4-50

Index

information operations (IO), 12-12
information systems (INFOSYS), 2-19
in-position method, 11-15
integration of direct and indirect fires, 5-36
intelligence, 1-18
  requirements, 8-4
intelligence officer (S-2), 2-28, 8-35, J-3
intelligence preparation of the battlefield (IPB), 2-28, 4-10, 4-39, 6-2, A-1, H-1, J-2
intelligence, surveillance, and reconnaissance (ISR), 2-6, 12-12
interviews, I-3
intervisibility (IV), 2-15, 2-22, 2-23 (*illus*), 9-10, 9-17
inverted "Y" marker, D-6
isolation
  of urban objective, 12-43, 12-44
  of objective, 4-26

J

Javelin, B-1
  Close Combat Missile System, B-8
  firing positions, B-14
  flight profile
    direct attack mode, B-12
    top attack mode, B-12
  minimum room enclosure, B-13
  technical characteristics, B-8

K

key personnel, duties and responsibilities, 1-12
key terrain, 2-20, 2-21 (*illus*), 2-22 (*illus*), 12-53
killed in action, 11-23

L

landing plan, D-2
landing zone (LZ), 1-13, 2-53, 4-40, 4-43, 6-26 (*illus*), 5-21, 8-11, 8-25, 11-17, 12-51, D-1, D-3, D-5, D-13 (*illus*)

launch tube assembly and missile, B-9
lead company movement, 4-44
leadership, 1-17
light matrix, 2-26
limit of advance (LOA), 1-14, 4-22, 4-38, 8-18
limited visibility conditions, extreme, 9-2
line of departure (LD), 2-49, 4-8, 4-18, 4-24, 4-31, 4-37, 8-14, 10-12, 10-22, 12-38
line formation, 3-8, 4-35, 5-23
linkup, 8-5
load echelon diagram, 11-6
loading plan, D-6
local security, 8-25
location, marking of, D-18
lodgment area, 6-8
logistical support, C-15
logistics package (LOGPAC), 1-4, 2-31, 11-3, 11-18
logistics release point (LRP), 5-39, 11-3, 11-12
long-range operations, K-11

M

M1 tank danger zone, C-5
M112 charge, 10-42
M183 satchel charge, 10-42
main battle area (MBA), 5-1, 5-3, 10-17, 12-60, E-6
main supply route (MSR), 6-16, 6-29, 11-12
maintenance operations, 11-19
maneuver, 4-8, 12-31
  commander's intent, 10-10
  forms of, 4-3
  space, 12-10
  support, 10-1
  techniques, 3-2
man-made structures, 12-12
map, strip, 3-19
Mark I nerve agent antidote, H-5
marking of target or location, 12-48 (*illus*), D-18
massed effects of fire, 9-1, 9-5
maximum engagement line (MEL), 5-37, 9-9, 9-12, B-15
media, 6-5
  cards, I-4
  considerations, I-1
medic, 1-16

medical evacuation (MEDEVAC), 1-15, 7-3, 7-6, 8-19, 10-48, 11-3, 11-22, A-13, D-1
METT-TC. *See* mission, enemy, terrain, troops, time available, and civil considerations
military aspects of terrain, 2-15 (*illus*)
military decision-making process (MDMP), 2-1, 2-10, 4-3, F-4
minefields, G-15
minimum safe distance (MSD), 10-20
minimum safe line (MSL), 10-9
missile, B-9
mission, 1-8, 10-40, 12-31
  analysis, 2-9
  analysis of, 2-10 (*illus*), 12-32
  common, 12-31
  convoy briefing checklist, 6-26
  Engineer, 7-5, 10-41
  execution, 4-49
  follow on, 8-19
  future, 12-42
  higher headquarters, 2-11
  receipt, 2-7
  restated, 2-12
  subsequent, 8-11
  types, characteristics, capabilities, limitations, and organization, 1-8
mission, enemy, terrain (and weather), troops (and support), time available, and civil considerations (METT-TC), 2-7, 2-35, 2-49, 3-4, 3-13, 3-18, 4-9, 4-18, 4-27, 4-33, 4-39, 4-44, 5-2, 5-6, 5-29, 6-1, 6-4, 6-13, 6-27, 8-2, 8-17, 8-27, 9-3, 10-39, 11-2, 11-8, 11-15, 11-21, 12-1, 12-8, 12-13, A-2 (*illus*), B-1, E-2, F-4, H-2, J-2, K-1

Index

mission, enemy, terrain (and weather), troops (and support), time available, and civil considerations (METT-TC, *continued*)
   analysis of individual factors, 2-10 (*illus*), 12-31, 12-50
   terrain analysis, 2-15
   visual aids, 2-15
mission-oriented protective posture (MOPP), 1-16, 5-40, 11-10, 11-23, H-2
mobility corridors, analysis, 2-19 (*illus*), 2-20 (*illus*)
modern buildings, 12-12
modernized demolition initiator (MDI), 10-42
modified combined obstacle overlay (MCOO), 2-13
modified M1025 (turtle shell HMMWV), K-7
modified Y-shaped perimeter defense, 5-23
Modular Pack Mine System (MOPMS), 5-6, 10-44
MOPP. *See* mission-oriented protective posture
mortars, 10-27
mortar section leader, 1-16
motorized operations, K-1
mounted
   forces, G-12
   hasty checkpoint operations, K-10
movement, 3-1, 12-40
   battalion, 3-16
   to contact, 4-38, C-8
   formations, 3-6, 3-11
   to line of departure, 4-8
   and maneuver, 1-18, 4-10, 5-5, 6-3, 7-4
   to objective, 4-24
   planning considerations, 4-39
   techniques, 3-2
movement to contact (MTC), 4-3, 4-17, 4-39, 4-43, 4-46, 8-18, 8-25 (*illus*), C-8, E-4
Multiple Delivery Mine System (Volcano), 10-43
multiple engagement areas, 5-18
multipurpose antitank (MPAT), 12-28, C-2

munitions considerations, 12-28

N

named area(s) of interest (NAI), 4-39, 4-47, 8-18, 8-23, 10-13, H-1
negotiations, 6-11
nerve agent antidotes, H-7
nested concepts, 2-5
nesting of concepts, 2-6
nine-line
   briefing, D-20
   UXO incident report, G-14
no-fire area (NFA), 10-14, 12-21
noncombatant evacuation operations (NEO), 6-8, 6-22, 6-34
noncombatants, 12-9
   evacuation of, 6-8
nongovernmental organization (NGO), 6-5, J-2, 7-2, F-6
nonlinear defense, 5-25
Nuclear, Biological, and Chemical Warning and Reporting System (NBCWRS), 1-16
nuclear weapons, H-10
numbering system, 12-37 (*illus*)

O

OAKOC. *See* obstacles, avenues of approach, key terrain, observation and fields of fire, and cover and concealment
objective rally point (ORP), 3-16, 4-19, 4-51, 8-6, 8-34, 8-41, 11-7
observation and fields of fire, 2-22, 2-24 (*illus*)
observation post (OP), 2-25, 2-49, 4-15, 6-3, 6-13, 6-14 (*illus*), E-5
observer positions, 10-17
obstacle intelligence (OBSTINTEL), 8-27
obstacles, 2-15, 2-17 (*illus*), 2-18 (*illus*)
   analysis, 2-15 through 2-18 (*illus*)
   effects, 5-8

obstacles (*continued*)
   plan, 5-35
   urban, 12-53
obstacles, avenues of approach, key terrain, observation and fields of fire, and cover and concealment (OAKOC), 2-13, 2-15 (*illus*), 4-10, 12-52, J-2, J-7
occupation and preparation, 5-3
offense, 12-30
   characteristics, 4-1
   intelligence, 4-10
   offensive employment, E-3
   offensive operations, 4-1
   planning considerations, 4-10, 12-30
   sequence, 4-81
   types, 4-3
one-third/two-thirds rule, 2-5
operational control (OPCON), 8-12, 10-1, 10-2, 12-9, 12-17, 12-39, C-1, C-15, D-14, F-5
operational environment, 1-1
   definition, 1-1
   operational variables, 1-6, J-2
operation order (OPORD), 1-12, 2-1, 2-7, 2-11, 2-40, 4-8, 4-24, 5-40, 6-25, 8-6, 10-7, 10-15, 11-2, 11-23, 12-50, A-5, D-11
operations
   approach sequencing, 2-5
   with outside agencies, 6-5
operations security (OPSEC), 3-12, 4-19, 4-49, 6-5, 8-11, 12-12, I-1, J-4
operations and training officer (S-3), 5-46, 10-3, 10-34, D-3, E-1
organization, 1-9, 8-35, 10-38
   battlefield, 2-5
   elements, 8-39
   equipment, B-1
   target acquisition, 9-6
organizations of influence, 2-38
overlapping sectors of fire, B-3
overwatch, 4-52

Index

P

parallel planning, 2-4 (*illus*)
passage of lines, 8-13
passive air defense, 10-47
pathogens, H-8
patrol, 6-21, E-11, 8-32
    base, 8-49
    reconnaissance, 8-4
    types, 8-32
    urban operations, K-9
pattern analysis, J-1, J-5 (*illus*), J-6 (*illus*)
peace operations, E-9
penetration, 4-7
    exploitation of, 4-29
perimeter defense, 5-20
personnel, replacement of, 11-23
petroleum, oils, and lubricants (POL), 11-3, C-8
phase lines, 12-38, D-16
phases of the conflict, 1-2
pickup zone (PZ), 1-13, 2-53, 4-40, 4-43, 6-26 (*illus*), 5-21, 8-11, 8-25, 11-17, 12-51, D-1, D-3, D-5, D-8 through D-9 (*illus*), D-13 (*illus*)
PIR. *See* priority intelligence requirements
planning
    Bradley and Stryker, K-10
    cover and concealment, 2-25
    key concepts, 2-3
    minimum requirements, D-15
    process, 10-11
platoon battle position
    in a company sector, 12-55
    defense in sector, 5-13
    leader, 1-13, J-9
    members, J-9
    mutual support, 5-17
    seating technique, K-6
    sergeant, 1-14
platoon medical trauma specialist, 11-21
population
    considerations, J-6
    perceptions, 2-36
    positions for Infantry riding on a tank, 12-26
    status overlays, J-7 through J-8 (*illus*)

positions
    for Infantry riding on a tank, 12-26 (*illus*)
    primary and alternate, 5-15
    support-by-fire, 12-11
potential hazards, A-2 (*illus*)
precipitation, 2-26
precombat checks, checklists, and inspections, 1-14, 2-52, 2-53 (*illus*), 2-54 (*illus*), 4-8, 6-25, 11-23, 12-20, A-2
    convoys, 6-25
preparation
    and integration, 5-11
    for war, 1-6
presence patrol, 6-22
prestockage operations, 11-16
preventive maintenance checks and services (PMCS), 5-40, 11-19
primary and alternate positions, 5-15
principles of fighting, 1-5, B-3
priority of fire (POF), 5-26, 10-10, 10-14, 10-30
priority intelligence requirements (PIR), 2-34, 2-48
priority of work, 5-37
prisoners of war, 11-24
    detainee tag, 11-27
    document and special equipment tag, 11-28
probable enemy locations, 9-4
probable line of deployment, 4-18
probe, G-18
protection, 1-18, 4-11, 5-11, 6-3, 7-5
protective wire obstacles, 5-9
psychological operations (PSYOP), 1-3, 1-12, 6-2, 12-1, 12-65, F-7
public affairs officer (PAO), 6-5, I-4

Q

QuickFire channel, 10-19

R

radiological weapons, H-9
radio operator, 1-7, 1-15, 3-13, 5-40, 10-4, 11-8, 12-22, D-12, E-11
raid, 8-39

rain, 2-26
range determination
    recognition method, B-16
Ranger, 1-8, 1-11
    deployable planning teams and cross-functional teams, F-9
    Regiment, F-3
reaction procedures, 10-47
readiness condition (REDCON), 3-6, 6-29
rear guard, 4-44
recent activities, 2-32
recognition, G-12
reconnaissance, 4-8, 8-1
    chemical, biological, radiological, nuclear, H-1
    and security operations, 5-3
reconnaissance and surveillance (R&S), 2-28, 2-49, 4-39, 8-1, 12-4
records, J-4
rehearsal, 2-51, 10-17
relational combat power analysis (RCPA), 2-48
release point (RP), 3-14, 3-17, 4-36, 5-39, 8-9, 8-46, 11-3, D-7
relief in place, 8-8, 8-10, 8-8
remote antiarmor mine (RAAM), 10-43
remote control unit (RCU), 10-44
reorganization and weapons replacement, 11-23
reports, G-17
request for support, F-9
reserve operations, 6-29
restated mission, 2-12
restoration of essential services, 12-13
restricted terrain, 2-17 (*illus*), 2-18 (*illus*)
restrictive fire line (RFL), 8-6, 8-15, 9-9, 9-12
resupply
    considerations, 11-17
    resupply, 11-16
    routine, 11-12
retrograde operations, 5-41, C-15
    employment, E-6
    purpose, 5-41
    types, 5-41

Index

reverse
  planning, 2-7
  slope defense, 5-27
risk
  assessment, 2-4, 2-38
  challenges, A-6
  considerations, A-9
  guidelines and implementation, A-6
  levels, A-3
  risk estimate distance, 10-21
  risk management, A-1
  risk management worksheet, A-4 (*illus*)
  steps, A-2
  types, A-1
risk estimate distance (RED), 2-39, 10-20, 10-21 (*illus*), A-5
river crossings, E-10
rocket-propelled grenade (RPG), 12-8
rough terrain driving, K-12
routes, 6-29
  classification of, 3-20
rules of engagement (ROE), 6-4, 7-1, 9-9, 9-17, 12-2, A-9, F-4
rules of interaction (ROI), 6-5, A-2

S

safety
  DANGER (death)
    120-mm discarding sabots, 12-27
    barrel swap, 1-14, 2-2
    blank ammunition, 1-7, 2-9, 2-16
    blank firing attachment, 1-7
    bolt position, 2-13
    destruction of weapon, 1-42
    feed cover, 1-30
    hot weapon, 1-10, 1-30
    purpose, 5-2
    sequence, 5-3
    spring guide, bolt, 1-12, 1-16
    types, 5-1
    unexploded ordnance, G-13

safety (*continued*)
  fratricide avoidance, 9-2, A-7
  WARNING (injury)
    120-mm overpressure, 12-27
    batteries, G-8, G-36
    bolt assembly, 2-16
    bolt position, 2-6, 3-31
    cocking handle, 2-6, 3-15, 3-42
    drive spring assembly, 3-15
    laser, G-2, G-24
    overdose, nerve agent antidote, H-5
    risk estimate distance, 10-21
    rod assembly, 3-14
sapper squad, 10-40
SBCT. *See* Stryker brigade combat team
scatterable mine (SCATMINE), 5-7, 5-10, 5-37, 8-29, 10-4, 10-41, 10-42, 10-43 (*illus*), 12-59, A-8, D-1, G-16
scheme of maneuver, 5-31, 9-3
  enemy, 9-4
scoop loader, 10-45
sea port of debarkation, 1-5
search, 6-19
  and attack, 4-39
sector
  defense, 5-12
  of fire, 9-10
  terrain technique, D-16
security operations, 8-16
  assistance, 6-7
  during movement, 3-15
  movement into position, 8-43
  planning considerations, 8-17
  screen, 8-20
  security, 11-9
  types, 8-16
selectable lightweight attack munition (SLAM), 10-41
selection of control measures, 5-12
self-defense against helicopters, B-16
senior radio operator, 1-15
senior trauma specialist, 11-21

separation of combatants and noncombatants, 12-13
service station resupply method, 11-14
shaping operation, 2-5
shifting of fires, 9-6
shock effect, 8-44
show of force, 6-8
situation template (SITEMP), 2-8, 2-27, 2-32, 2-33 (*illus*), 2-34 (*illus*), 2-35, 9-3, 10-7
situational understanding (SU), 1-6, 2-28, 2-37, 4-52, 6-2, 6-24, 9-7, 12-3, A-8, B-14, J-1
situation report (SITREP), 1-15, A-5
sleep deprivation, A-14, G-11
small emplacement excavator (SEE), 10-40, 10-44
small-unit battles, 12-8
small, unmanned aerial system (SUAS), 4-1, 4-8, 4-32, 4-41, 6-3, 8-1, 8-12
smoke support, 10-17
snipers, 12-10, E-1
Soldier's load, 11-4
spall liner, cargo bed, K-16
special forces, F-2
special forces operational detachment (SFOD), F-3
specialized teams, 8-42
special munitions, 10-16
special operations
  aviation, F-4
  command and control element, F-1, F-8
  forces, F-1
special-purpose operations, 8-5
specified tasks, 2-11
spectrum of attacks, 4-16
squad designated marksman, 6-4, 6-13, E-1
stability operations, 6-1
  fire support, 6-3
  intelligence, 6-2
  peace operations, 6-6
  planning considerations, 6-2
  sustainment, 6-4
  types of operations, 6-6
stability and reconstruction operations, E-8
staging plan, D-10
standard minefields, G-15

Index

standing operating procedure (SOP), 2-3, 2-42, 3-18, 4-11, 5-6, 5-37, 6-20, 7-6, 8-30, 9-5, 9-20, 10-47, 11-1, 11-19, 12-11, 12-48 (*illus*), 12-61, A-5, B-11, D-1, J-9, K-9. *See also* tactical standing operating procedure
standoff ranges, TOW (top) and Javelin (bottom), B-5
start point (SP), 2-40, 3-19
stationary guard, 8-24
steel plate armor, K-17
Stinger, man-portable and mounted on a HMMWV, 10-45
strategic and operational principles of fighting, 1-5
stress, combat, A-11
strip map, 3-19
strongpoint defense, 5-18, 12-66
Stryker, 1-11
  employment, C-1
  Infantry carrier vehicle, C-3
  planning considerations, C-7, K-10
  safety, C-4
Stryker brigade combat team (SBCT), 1-9, C-1
suicide bombers, G-10
supplementary position, 5-16
supply
  and transportation operations, 11-9
  sergeant, 1-15
supply officer (S-4), 1-15, 11-1, 11-7, 12-20
support
  counterdrug operations, 6-7
  insurgency, 6-7
  relationships, 10-2
support-by-fire positions, 12-11
suppression of enemy air defense (SEAD), 5-10, 10-35, D-13
suppress, obscure, secure, reduce, and assault (SOSRA), 4-27, 8-27
survivability, B-10
sustaining operations, 2-5

sustainment, 1-19, 2-5, 2-43, 4-2, 4-8, 4-11, 5-1, 5-11, 5-18, 5-22, 5-41, 6-4, 6-25, 7-6, 10-39, 12-4, 12-20, 12-60, B-8, C-2, D-1, F-2
sustainment operations, 11-1
  overview, 11-1
  planning considerations, 11-1
swing arm for crew-served weapons, K-8
symbols, 3-3
synchronization, 2-45
systems based warfare, 1-4 (*illus*)

T

table
  of distribution and allowances (TDA), 10-1
  of organization and equipment (TOE), 1-9, 8-11, 8-46, 10-1, F-7
tactical air control party (TACP), 10-34
tactical enabling operations, 8-1
  categories, 8-1
  definition, 8-1
  execution, 8-3
  planning considerations, 8-3
  types, 8-2
tactical movement and enemy contact, 3-1
tactical operations center (TOC), 1-12, 6-25, 10-35, 11-21, D-12, G-9
tactical standing operation procedures (TSOP), 1-13, 3-13, 5-6, 5-37, 6-20, 6-27, 8-5, 11-4, 12-19, 12-47, K-1
tactical vehicle employment and urban patrols, K-9
tactics, techniques, and procedures (TTP), 2-42, 6-15, 7-4, 11-1, 12-41, E-1, G-4, K-1
tailgate resupply method, 11-14
tank commander (TC), 12-19, 12-26, 12-28, C-2, C-15, K-5
tanks, 12-17, C-1
target
  acquisition, 9-5

  force orientation, 9-5
  best weapon for, 9-2
target (*continued*)
  Javelin, B-14
  marking, D-18
  overkill, 9-2
  task organization
    with Bradleys, 12-29
    with tanks, 12-17
    into three elements, 12-38
target reference point (TRP), 2-24, 2-39, 2-45, 3-14, 5-16, 6-32, 8-11, 9-5, 9-9, 9-20, 10-8, 12-60, C-2
task force, 7-4, 10-23, 10-34, 12-1, 12-10, 12-16, 12-67, E-6, F-3
task, purpose, method, and effects (TPME), 10-5, 10-12, 10-13 through 10-14
tasks, 1-9
  civil support, 7-2
  company, 6-8
  direction-of-attack, 12-45
  duties and responsibilities of key personnel, 1-12
  essential, 2-12
  fire support officer, 1-14
  implied, 2-12
  specified, 2-11
  tactical, 1-13
  warfighting, 1-17
terrain, 2-20
  analysis, 2-15, 2-25
  restricted, 2-17 (*illus*), 2-18 (*illus*)
  weather and, 2-12
terrain-based quadrants, 9-11
terrorism, 6-7
threat, destruction, 9-1
three-dimensional terrain, 12-10
time, 2-35
TOW employment, B-1
toxins, H-8
traffic control point (TCP), 7-4, J-1
training
  ambush, 8-46
  for media awareness, I-4
  program, 1-7
  risk estimate distances, 10-20
trains, 11-9
  company, 3-20, 4-11, 4-30,

Index

5-11, 5-36, 8-19, 11-9, 11-19, 12-46, 12-60
transportation, 11-18, 12-25
traveling overwatch technique, 3-4
traveling technique, 3-3
treatment of chemical casualties, H-4
troop-leading procedures, 1-12, 2-1, 2-5, 4-12, 4-15, 4-30, 4-49, 10-5, 12-61, A-2, A-8, F-4, J-9
  steps, 2-7
troop requirements, 12-30
troops and support available, 2-34
"TTFACOR" technique, 10-35
turning movement, 4-5
types and characteristics of Infantry rifle companies, 1-8

U

UH-60 helicopter
  loading diagram, D-10
  unloading diagram, D-5
unassisted withdrawal, 5-47 (*illus*)
unexploded ordnance, G-3 (*illus*)
  immediate action, G-13
unit-installed or produced equipment
  modified armor, K-15
  steel armor, K-17
  tripod, K-16
  wire cutters, K-13
  wire guards, K-14

unmanned aerial system (UAS), 1-12, 4-1, 4-32, 5-1,
unmanned aerial system (*continued*), 5-11, 6-3, 7-4, 8-17, 10-38, 10-47, 12-3, 12-12. *See also* small unmanned aerial system
urban
  battlespace, 12-3
  command vehicle, K-15
  considerations, K-9
urban (*continued*)
  obstacles, 12-53 (*illus*)
  operations, 12-1, E-7
  patrols, K-9, K-14
  record copy, 11-27
  strongpoint, 12-66
UXO. *See* unexploded ordnance

V

vehicle-borne devices (car bombs), G-3
vehicle, C-1
  equipment load plans, K-8
  gunner armor, K-3
  improvised explosive devices, capacities and danger zones, G-4
  modifications, OEF and OIF, K-13, K-14 through K-17
  positions, K-5
  recovery, K-13
  weight, survivability, and armor, K-2
vehicular traffic stop, 6-19

visibility, 2-25
  and noise, 12-11
visual aids, 2-15

W

warfighting function, 1-17, 2-1, 2-30, 2-43, 2-52, 4-1, 4-10, 5-5, 10-40
war-gaming techniques
  belt, 2-47
  box, 2-46
warning orders, 2-1, 2-8, 2-51, 5-46, A-8
weapons
  considerations, 12-28
  demolitions, 12-11
  replacement, 11-23
  safety posture levels, 9-18
weapons control status (WCS), 9-2, 9-17, 10-45, B-16
weapons of mass destruction (WMD), 7-2
weather
  five military aspects of, 2-25
wheeled vehicles, K-1
white phosphorus (WP), 10-4, 10-16, G-2, G-13
wounded in action (WIA), 5-11, 11-9, 11-23

Y

Y-shape perimeter defense, 5-22

Z

zone, 12-5, 12-38
  reconnaissance patrol, 8-37
  of separation, 6-22

Made in the USA
Lexington, KY
07 February 2019